BioHydrogen

BioHydrogen

Edited by

Oskar R. Zaborsky
University of Hawaii
Honolulu, Hawaii

Associate Editors

John R. Benemann, *Consultant, Walnut Creek, California*
Tadashi Matsunaga, *Tokyo University of Agriculture and Technology, Tokyo, Japan*
Jun Miyake, *National Institute for Advanced Interdisciplinary Research, Tsukuba, Japan*
and
Anthony San Pietro, *Indiana University, Bloomington, Indiana*

Plenum Press • New York and London

Library of Congress Cataloging in Publication Data

BioHydrogen / edited by Oskar R. Zaborsky; associate editors, John R. Benemann ... [et al.].
    p.   cm.
   "Proceedings of an International Conference on Biological Hydrogen Production, held June 23–26, 1997, in Waikoloa, Hawaii"—T.p. verso.
   Includes bibliographical references and index.
   ISBN 0-306-46057-2
   1. Hydrogen—Biotechnology—Congresses. I. Zaborsky, Oskar R. II. International Conference on Biological Hydrogen Production (1997: Waikoloa, Hawaii)
TP248.65.H9B55   1998   98-46065
665.8′1—dc21   CIP

Proceedings of an International Conference on Biological Hydrogen Production, held June 23–26, 1997, in Waikoloa, Hawaii

ISBN 0-306-46057-2

© 1998 Plenum Press, New York
A Division of Plenum Publishing Corporation
233 Spring Street, New York, N.Y. 10013

http://www.plenum.com

10 9 8 7 6 5 4 3 2 1

All rights reserved

No part of this book may be reproduced, stored in a retrieval system, or transmitted in any form or by any means, electronic, mechanical, photocopying, microfilming, recording, or otherwise, without written permission from the Publisher

Printed in the United States of America

PREFACE

The world needs clean and renewable energy and hydrogen represents an almost ideal resource. Hydrogen is the simplest and most abundant molecule in the universe, yet one that is a challenge to produce from renewable resources. Biohydrogen, or hydrogen produced from renewable resources such as water or organic wastes by biological means, is a goal worthy of increased global attention and resources.

The purpose of BioHydrogen '97 was to bring together leaders in the biological production of hydrogen from the United States, Japan, Europe, and elsewhere to exchange scientific and technical information and catalyze further cooperative programs. Participants came from at least 10 different countries representing academia, industry, and government. Especially important participants were young research scientists and engineers: the next generation of contributors.

The conference consisted of plenary presentations, topical sessions, posters, and mini-workshop discussions on key areas of biohydrogen. It was designed to maximize information exchange, personal interaction among participants, and formulate new international initiatives.

BioHydrogen '97 was an outgrowth of an international workshop convened by the Research Institute of Innovative Technology for the Earth (RITE) and was held in Tokyo, Japan, November 24–25, 1994. The RITE workshop was highly successful but largely limited to traditional biochemical and biological studies and not engineering research topics. At the Tokyo meeting, an international organizing committee called for the next meeting to be hosted and held in the United States, with the final selection being Hawaii due to its strategic location and its prominent programs in alternative energy sources.

An important feature of BioHydrogen '97 was to elevate the importance of engineering research and technology development. The time had come for a transition from research to engineering and more active involvement by industry. Essentially all previous workshops and conferences, including the 1994 Japan meeting, had focused on the biology, physiology, and biochemistry of various hydrogen-evolving systems. The entire biohydrogen community was at a critical junction and the need for more engineering research was realized.

There was also a growing need for innovative ideas and the participation of young research engineers. As such, one critical purpose of this meeting, from a U.S. biohydrogen community perspective, was to engage young engineers, those already in the area, or those who may be in allied fields of relevance, yet would like to pursue this area of endeavor. An especially germane area in the biological production of hydrogen is bioprocess engineering and in particular photobioreactor design, modeling, and evaluation. Very little work has been done on photobioreactors, yet this is one of the most important barriers to the commercialization of marine microbial resources to produce commodity chemicals like hydrogen and even specialty bioproducts.

The conference was held from June 23 to 26, 1997, in Hawaii, with the conference venue being the Royal Waikoloan Hotel located on the Big Island of Hawaii—a location adjacent to the home of Hawaii's marine biosolar industry in Kona, hailed by some as the new "Silicon Valley" of the microalgae industry, and the Natural Energy Laboratory of Hawaii Authority.

BioHydrogen '97 was planned by three international committees and its general chairman, Oskar R. Zaborsky, the HECO Williamson-Matsunaga FREE (Fellow in Renewable Energy Engineering) Scholar at the University of Hawaii. The three international committees were Program, International, and Publication.

The Program Committee consisted of leading research scientists and engineers active in biohydrogen production and included John Benemann, Elias Greenbaum, Tadashi Matsunaga, Jun Miyake, Kazuhisa Miyamoto, Roger Prince, Michael Seibert, and Koichi Takasaki.

The International Committee consisted of distinguished scientists and engineers or key administrators and included Paolo Carrera, Martin Gibbs (Brandeis University), David O. Hall, Isao Karube (University of Tokyo), Jiro Kondo (RITE), Shigetoh Miyachi, Tatsuya Mizukoshi, and Anthony San Pietro.

The Publication Committee consisted of active scientists and engineers who oversaw the early stages of the conference proceedings and the selection of the publisher. The members included Yasuo Asada, Dulal Borthakur, J. Grant Burgess, Maria Ghiradi, Haruko Takeyama, and Jonathan Woodward.

This volume is the culmination of the conference and the labor of many authors, reviewers, and Associate Editors John Benemann, Tadashi Matsunaga, Jun Miyake, and Anthony San Pietro. The purpose of BioHydrogen '97 was to produce a peer-reviewed proceedings because of the immense need at this time for a quality publication on the subject and to document current and future directions. The review process entailed an independent review of each manuscript by two experts, a review and deliberation by one or several editors on comments offered by the reviewers and the disposition of the manuscript, the incorporation of modifications suggested by reviewers or editors, and the final copy editing by Gregg Hirata. Inevitably and as was anticipated, some delays occurred in the process. However, the final product reflects BioHydrogen '97 quite well, in addition to the current state of knowledge in this field. A meeting of all editors was held in April 1998 in Hawaii, at which time many manuscripts were further scrutinized and further modifications incorporated.

BioHydrogen '97 was possible only through the assistance of many individuals and organizations that worked diligently to make the conference and the proceedings a reality. Many deserve praise but some deserve special mention for their assistance:

Mary Kamiya, conference coordinator, for her exceptional assistance on the logistics of the conference; Gregg Hirata for his outstanding editorial capabilities in the final stages of the proceedings; Eileen Kalim for her assistance with marketing and conference logistics; Pamela Walton for her assistance at the conference and with many follow-up actions dealing with the associate editors; Tracie Nagao for her graphics assistance; Brandon Yoza and Mitsufumi Matsumoto for their help in conference logistics; William Y. Kikuchi for his beautiful artistic representation of the Biohydrogen '97 theme; Darcie Mayeshiro for her assistance with university financial matters; Patrick Takahashi for his early support and encouragement; Catherine Gregoire Padro for her planning and logistical support; and

Neil Rossmeissl for his encouragement and support of the entire activity—both the conference and the proceedings.

I am also indebted to Edward E. David for his most inspiring presentation as keynote speaker, John Benemann for his many hours of assistance on scientific and international aspects, and Tony San Pietro for his unending dedication to excellence with the proceedings. Special praise also goes to Jun Miyake, Tadashi Matsunaga, and Koichi Takasaki for their outstanding and dedicated assistance on Japanese scientific and logistical matters and without whose help BioHydrogen '97 would not have been such a success.

All of us engaged with BioHydrogen '97 deeply appreciate the support of the U.S. Department of Energy's Hydrogen Program, the primary sponsor of this event. We are also grateful for the sponsorship of key organizations, including the Cyanotech Corporation (Gerald Cysewski); Food and Agriculture Organization of the United Nations (Morton Satin); International Energy Agency (Bjorn Gaudernack and Carolyn Elam); Marine Biotechnology Institute Co., Ltd. (Shigetoh Miyachi); National Renewable Energy Laboratory (Catherine Gregoire Padro); National Science Foundation (Fred Heineken); New Energy and Industrial Technology Development Organization (Tatsuya Misukoshi); and the Research Institute of Innovative Technology for the Earth (Koichi Takasaki).

The challenge was to bring together the leading practitioners at the edge of a new millennium, in which hydrogen will achieve its goal as the clean, green, and sustainable energy resource. It is my sincere hope that BioHydrogen '97 has fulfilled its objective and will contribute in making this long-standing dream a reality in the next decades.

Oskar R. Zaborsky
General Chairman and Editor-in-Chief
HECO Williamson-Matsunaga FREE Scholar

CONTENTS

I. Plenary Sessions

1. Biohydrogen 97: Keynote Address 1
   Edward E. David, Jr.

2. The Science of Biohydrogen: An Energetic View 7
   Jun Miyake

3. The Technology of Biohydrogen 19
   John R. Benemann

4. Marine Genomes 31
   Tadashi Matsunaga and Haruko Takeyama

5. Commencement Challenge 39
   Patrick Takahashi

II. Fundamentals

6. Maximizing Photosynthetic Productivity and Light Utilization in Microalgae by Minimizing the Light-Harvesting Chlorophyll Antenna Size of the Photosystems 41
   Anastasios Melis, John Neidhardt, Irene Baroli, and John R. Benemann

7. *Nostoc* PCC 73102 and H_2: Knowledge, Research, and Biotechnological Challenges 53
   Peter Lindblad, Alfred Hansel, Fredrik Oxelfelt, Paula Tamagnini, and Olga Troshina

8. Molecular Biology of Hydrogenases 65
   Claudio Tosi, Elisabetta Franchi, Francesco Rodriguez, Alessandro Selvaggi, and Paola Pedroni

9. Effect of Hydrogenase 3 Over-Expression and Disruption of Nitrate Reductase on Fermamentive Hydrogen Production in *Escherichia coli:* A Metabolic Engineering Approach 73
   Koji Sode, Mika Watanabe, Hiroshi Makimoto, and Masamitsu Tomiyama

10. Improvement of Bacterial Light-Dependent Hydrogen Production by Altering the Photosynthetic Pigment Ratio 81
    Masato Miyake, Makoto Sekine, Lyudmila G. Vasilieva, Eiju Nakada, Tatsuki Wakayama, Yasuo Asada, and Jun Miyake

11. A Toolkit for Metabolic Engineering of Bacteria: Application to Hydrogen Production 87
    J. D. Keasling, John Benemann, Jaya Pramanik, Trent A. Carrier, Kristala L. Jones, and Stephen J. Van Dien

12. Electron Transport as a Limiting Factor in Biological Hydrogen Production 99
    Patrick C. Hallenbeck, Alexander F. Yakunin, and Giuseppa Gennaro

13. Reconstitution of an Iron-Only Hydrogenase 105
    Hugh McTavish

14. Attempt at Heterologous Expression of Clostridial Hydrogenase in Cyanobacteria 111
    Yoji Koike, Katsuhiro Aoyama, Masato Miyake, Junko Yamada, Ieaki Uemura, Jun Miyake, and Yasuo Asada

III. Photosynthetic Bacteria

15. Study on the Behavior of Production and Uptake of Photobiohydrogen by Photosynthetic Bacterium *Rhodobacter sphaeroides* RV 117
    Reda M. A. El-Shishtawy, Yoji Kitajima, Seiji Otsuka, Shozo Kawasaki, and Masayoshi Morimoto

16. Characterization of a Novel Light-Harvesting Mutant of *Rhodobacter sphaeroides* with Relation to Photohydrogen Production 123
    Lyudmila Vasilyeva, Masato Miyake, Masayuki Hara, Eiju Nakada, Satoshi Nishikata, Yasuo Asada, and Jun Miyake

17. Hydrogen and 5-Aminolevulinic Acid Production by Photosynthetic Bacteria .. 133
    Ken Sasaki

18. Continuous Hydrogen Production by *Rhodobacter sphaeroides* O.U.001 143
    İnci Eroğlu, Kadir Aslan, Ufuk Gündüz, Meral Yücel, and Lemi Türker

19. Photobiological Hydrogen Production by *Rhodobacter sphaeroides* O.U.001 by Utilization of Waste Water from Milk Industry 151
    Serdar Türkarslan, Deniz Özgür Yigit, Kadir Aslan, Inci Eroglu, and Ufuk Gündüz

20. Polyhydroxybutyrate Accumulation and Hydrogen Evolution by *Rhodobacter sphaeroides* as a Function of Nitrogen Availability 157
    Emir Khatipov, Masato Miyake, Jun Miyake, and Yasuo Asada

21. Photosynthetic Bacteria of Hawaii: Potential for Hydrogen Production 163
    Mitsufumi Matsumoto, Brandon Yoza, JoAnn C. Radway, and
        Oskar R. Zaborsky

22. Conversion Efficiencies of Light Energy to Hydrogen by a Novel *Rhodovulum* sp. and Its Uptake-Hydrogenase Mutant 167
    Akiyo Yamada, Tomoyuki Hatano, and Tadashi Matsunaga

IV. Cyanobacteria

23. Hydrogenase-Mediated Hydrogen Metabolism in a Non-Nitrogen-Fixing Cyanobacterium, *Microcystis aeruginosa* 173
    Yasuo Asada, Masato Miyake, Youji Koike, Katsuhiro Aoyama,
        Ieaki Uemura, and Jun Miyake

24. Identification of an Uptake Hydrogenase Gene Cluster from *Anabaena* sp. Strain PCC7120 181
    Jyothirmai Gubili and Dulal Borthakur

25. Hydrogenase(s) in *Synechocystis*: Tools for Photohydrogen Production? 189
    Jens Appel, Saranya Phunpruch, and Rüdiger Schulz

26. Detection of Marine Nitrogen-Fixing Cyanobacteria Capable of Producing Hydrogen by Using Direct Nested PCR on Single Cells 197
    Haruko Takeyama and Tadashi Matsunaga

27. Programmed DNA Rearrangement of a Hydrogenase Gene During *Anabaena* Heterocyst Development 203
    Claudio D. Carrasco, Joleen S. Garcia, and James W. Golden

28. Construction of Transconjugable Plasmids for Use in the Insertion Mutagenesis of *Nostoc* PCC 73102 Uptake Hydrogenase 209
    Alfred Hansel, Fredrik Oxelfelt, and Peter Lindblad

29. Effect of Exogenous Substrates on Hydrogen Photoproduction by a Marine Cyanobacterium, *Synechococcus* sp. Miami BG 043511 219
    Yao-Hua Luo, Shuzo Kumazawa, and Larry E. Brand

V. Green Algae

30. Development of Selection and Screening Procedures for Rapid Identification of H_2-Producing Algal Mutants with Increased O_2 Tolerance 227
    Michael Seibert, Timothy Flynn, Dave Benson, Edwin Tracy, and
        Maria Ghirardi

31. Photosynthetic Hydrogen and Oxygen Production by Green Algae: An Overview ... 235
    Elias Greenbaum and James W. Lee

32. Light-Dependent Hydrogen Production of the Green Alga *Scenedesmus obliquus* ... 243
    Rüdiger Schulz, Jörg Schnackenberg, Kerstin Stangier, Röbbe Wünschiers, Thomas Zinn, and Horst Senger

33. Association of Electron Carriers with the Hydrogenase from *Scenedesmus obliquus* ... 253
    Jörg Schnackenberg, Wolfgang Reuter, and Horst Senger

VI. Fermentations and Mixed/Hybrid Systems

34. Algal CO_2 Fixation and H_2 Photoproduction ... 265
    Akiko Ike, Ken-ichi Yoshihara, Hiroyasu Nagase, Kazumasa Hirata, and Kazuhisa Miyamoto

35. Hydrogen Production by Facultative Anaerobe *Enterobacter aerogenes* ... 273
    Shigeharu Tanisho

36. Artificial Bacterial Algal Symbiosis (Project ArBAS): Sahara Experiments ... 281
    Ingo Rechenberg

37. The Effect of *Halobacterium halobium* on Photoelectrochemical Hydrogen Production ... 295
    Vedat Sediroglu, Meral Yucel, Ufuk Gunduz, Lemi Turker, and Inci Eroglu

38. Photosynthetic Bacterial Hydrogen Production with Fermentation Products of Cyanobacterium *Spirulina platensis* ... 305
    Katsuhiro Aoyama, Ieaki Uemura, Jun Miyake, and Yasuo Asada

39. Hydrogen Photoproduction from Starch in CO_2-Fixing Microalgal Biomass by a Halotolerant Bacterial Community ... 311
    Akiko Ike, Naohumi Toda, Tomoko Murakawa, Kazumasa Hirata, and Kazuhisa Miyamoto

VII. Photobioreactors: Photobiology

40. Hydrogen Production by Photosynthetic Microorganisms ... 319
    Yoshiaki Ikuta, Tohru Akano, Norio Shioji, and Isamu Maeda

41. Development of Efficient Large-Scale Photobioreactors: A Key Factor for Practical Production of Biohydrogen ... 329
    James C. Ogbonna, Toshihiko Soejima, and Hideo Tanaka

42. Light Penetration and Wavelength Effect on Photosynthetic Bacteria Culture for Hydrogen Production ... 345
    Eiju Nakada, Satoshi Nishikata, Yasuo Asada, and Jun Miyake

43. Cylindrical-Type Induced and Diffused Photobioreactor: A Novel Photoreactor for Large-Scale H_2 Production 353
Reda M. A. El-Shishtawy, Shozo Kawasaki, and Masayoshi Morimoto

44. Analysis of Compensation Point of Light Using Plane-Type Photosynthetic Bioreactor 359
Yoji Kitajima, Reda M. A. El-Shishtawy, Yoshiyuki Ueno, Seiji Otsuka, Jun Miyake, and Masayoshi Morimoto

45. Hydrogen Production by a Floating-Type Photobioreactor 369
Toshi Otsuki, Shigeru Uchiyama, Kiichi Fujiki, and Sakae Fukunaga

46. Photohydrogen Production Using Photosynthetic Bacterium *Rhodobacter sphaeroides* RV: Simulation of the Light Cycle of Natural Sunlight Using an Artificial Source 375
Tatsuki Wakayama, Akio Toriyama, Tadaaki Kawasugi, Takaaki Arai, Yasuo Asada, and Jun Miyake

VIII. Photobioreactors: Algae

47. Bioreactors for Hydrogen Production 383
Sergei A. Markov

48. A Tubular Integral Gas Exchange Photobioreactor for Biological Hydrogen Production: Preliminary Cost Analysis 391
Mario R. Tredici, Graziella Chini Zittelli, and John R. Benemann

49. A Tubular Recycle Photobioreactor for Macroalgal Suspension Cultures 403
Ronald K. Mullikin and Gregory L. Rorrer

50. Technoeconomic Analysis of Algal and Bacterial Hydrogen Production Systems: Methodologies and Issues 415
M. K. Mann and J. S. Ivy

51. Environmental Aspects of Large-Scale Microalgae Cultivation: Implications for Biological Hydrogen Production 425
Roger Babcock, Jr.

52. An Automated Helical Photobioreactor Incorporating Cyanobacteria for Continuous Hydrogen Production 431
Anatoly A. Tsygankov, David O. Hall, Jian-guo Liu, and K. Krishna Rao

53. Internal Gas Exchange Photobioreactor: Development and Testing in Hawaii ... 441
James P. Szyper, Brandon A. Yoza, John R. Benemann, Mario R. Tredici, and Oskar R. Zaborsky

IX. Related Topics

54. International Collaboration in Biohydrogen: "An Opportunity" 447
Neil P. Rossmeissl

55. Principles of Bioinformatics as Applied to Hydrogen-Producing Microorganisms ... 451
   Lois D. Blaine and Oskar R. Zaborsky

56. Production of Sulfolipids from Cyanobacteria in Photobioreactors 459
   Seher Dagdeviren, Karen A. McDonald, and Alan P. Jackman

57. Practical Considerations in Cyanobacterial Mass Cultivation 467
   JoAnn C. Radway, Joseph C. Weissman, John R. Benemann, and
   Edward W. Wilde

58. Secreted Metabolite Production in Perfusion Plant Cell Cultures 475
   Wei Wen Su

59. Strategies for Bioproduct Optimization in Plant Cell Tissue Cultures 483
   Susan C. Roberts and Michael L. Shuler

60. The *Renilla* Luciferase-Modified GFP Fusion Protein Is Functional in
   Transformed Cells 493
   Yubao Wang, Gefu Wang, Dennis J. O'Kane, and Aladar A. Szalay

X. Workshop Summaries

61. RITE Biological Hydrogen Program 501
   Tadaaki Kawasugi, Paola Pedroni, Masato Miyake, Akio Toriyama,
   Sakae Fukunaga, Koichi Takasaki, and Teruaki Kawamoto

62. Standards Workshop 507

63. Roundtable 511

XI. Appendices

BioHydrogen 97 Program 519

Participant Roster 529

Subject Index 543

BIOHYDROGEN 97

Keynote Address

Edward E. David, Jr.

1. INTRODUCTION

It is a pleasure to join you here today. This meeting is, of course, aimed at reviewing the status of an advancing technology, namely biohydrogen production. But in the larger picture of a sustainable world energy system for the coming millennium, hydrogen production is only beginning to be taken seriously. It is a technology that requires a vision reaching out 50 years or more. So it is extraordinary that all of you in this audience have come together to study such a long-range, speculative, yet promising prospect. It is not yet clear that hydrogen alone will play a central role in the world's energy system, but it is clear that a fundamental research approach, plus engineering ingenuity, is essential to assessing its prospects. In reviewing the literature on a sustainable energy future, I have been impressed with the serious thought that has been applied to the possibilities. It appears that decades of effort, organized to produce the knowledge and technologies, will be required.

Too few conferences in this era of here-and-now motivations have been willing to look at the pathways leading to the far horizon. Too few conferences in this era of national interests have been willing to engage the wide participation that this one has. Too few conferences in this era of contracting financial resources have been willing to propose the bold actions that this one promises to put forward. Too few conferences in this era of contention have been willing to reach consensus on the matters discussed. So this is a special occasion animated by the opportunities to initiate meaningful and well considered programs aimed at a sustainable future.

In the remainder of my time with you I will discuss first the status of research in the U.S., then some ideas about the use of hydrogen as a fuel, and finally some views of the international scene. I hope these perspectives will set the stage for your discussions and will give you a view of the situations which will determine the success of long-term, capital-intensive, initiatives of the kind this conference represents.

2. RESEARCH AND THE UNITED STATES

Research in the U.S. is in a state of flux—some would say chaos. Rapidly changing prospects for research stem from several causes. Among the most discussed is the end of the cold war. In the U.S. at least, research (and development) inspired by military requirements has played a major role over the past 40 years in animating commercial activities beyond the defense industries. Computers, advanced communication, jet aircraft, semiconductors, satellites for navigation, communication, and earth surveillance all owe their parentage to the imperatives of the cold war. There have been arguments made that these developments would have appeared regardless of the cold war, and that may be true. But the pressures at least accelerated developments, and probably changed their character as well. In any case, large amounts of money were spent by the U.S. government on technological research, much of which turned out to have dual uses, that is, both military and commercial or civilian applications. That wellspring of technology is now being phased down and there is no agreed-upon replacement.

The federal government has made at least two attempts to support commercial technology, the Technology Reinvestment Program (TRP) and the Advanced Technology Program (ATP). Neither promises a long life. Politically, the TRP and ATP have been attacked as inappropriate for federal funding. The idea of direct technology development for the commercial, private sector has been labeled as "corporate welfare" by economic conservatives and both TRP and ATP may well be phased out.

However, TRP especially, and ATP perhaps, will be reborn through a dual-use logic. Perhaps this is merely a renaming in the spirit of the rather cynical, but effective directive, "if the power structure doesn't like a program, don't abolish it, just change the name." There is a legitimate side to this tactic in this case, however. The military and associated civilian agencies have perceived that only by using off-the-shelf technologies and products from the commercial sector can those agencies afford the procurements they require to carry out their missions. Examples abound. One is the use of Intel microprocessors and commercial systems software as a principal resource in new military electronics equipment. Another is the use of commercial navigation receivers utilizing the Global Positioning Satellites (GPS) for military troops and ships. Significantly the Pentagon has set up a dual-use office to encourage the production and federal use of dual-use equipment. This strategy is being adopted by other federal departments as well. Thus, TRP and ATP by other names may well survive as favored implements for procuring affordable equipment.

The idea of specially developed equipment for federal missions is out of favor. The flow of technologies from federal programs to the private sector is likely to be reversed. That means the prime source of new science-based technology is increasingly in the hands of industry, with all of the uncertainties that brings. Among these is the role for academic research. Will industry make use of academic research capabilities and will it support financially the research performers?

Federal financing of research has been declining in inflation-adjusted dollars for some years. The latest budget agreement between the Congress and the President projects further declines (about 7% by the year 2002). So the funding of research seems likely increasingly to depend on other sources. But you should note that such forward projections of funding are not binding and could be changed, upward or downward. The uncertainty is disturbing.

The character of research is also changing. Since the end of World War II, U.S. researchers have been accustomed to initiating their research from their own ideas. But this is much less the case today. The federal government and industrial labs now regularly program their research from the demand side, that is, from national or corporate needs. Industry has long so programmed its development, but central research labs were allowed at least an element of free play in research. This element is disappearing, although it is not likely to vanish completely. The same is true of the U.S. Department of Defense, which in the past has been a major supporter of ground-breaking research. It is significant that the Japanese government has announced what appears to be a major expansion in its support of fundamental research. Japan and the U.S. appear to be going in opposite directions. Whether they settle at a similar balance will be interesting.

Another trend in U.S. federal research is increasing emphasis on the life sciences and those related to human health in particular. The National Institutes of Health is the favorite of the Congress and is by far the largest research operation. Then, too, the biotechnology industry is in the fast growth mode, and much of its activities can be described as research. Most if not all the activity is related to molecular biology. Further, the medical devices industry is growing explosively. Both diagnostic and therapeutic dimensions are contributing to this growth.

An exception to the trend away from investigator-initiated research is in the government's Small Business Innovation Research (SBIR) grants. These are funded by a 1.25–2.5% set-aside from all federal agencies' budgets. This program, which amounted to several billion dollars in 1996–1997, has created a cohort of small innovative companies, many incorporating new technologies still in the research phase. This research is a little-known component of U.S. policy, as is DOD's Independent R&D (IR&D) program. That program aids companies in preparing for their next program by allowing an overhead charge on their current program. This R&D can be as much as 2% of the current contract. These research initiatives are a major source of innovation in many industries and provide an outlet for investigator-initiated activities. Regardless, the environments for research in the U.S. is more constrained today than it has been traditionally.

3. HYDROGEN AND A SUSTAINABLE FUTURE

The future of hydrogen as a fuel has been much argued and is not agreed upon by energy experts. Interestingly, it seems that the longer the lead time under consideration, the clearer hydrogen's role becomes. That could be due to assumptions about the long-term future specifying a sustainable energy supply with minimal or zero use of hydrocarbons, or it could arise from the lack of critical thinking about a 50–100-year horizon. Regardless, responsible people do seriously propose near-term introduction of hydrogen as a fuel for mobile applications, and spade work has been done on automotive use of hydrogen. The critical question is the realistic time for introduction. As this audience well knows, there are no commercial or near-commercial systems today for biological hydrogen production. Of course, most oil refineries use hydrogen generated on-site using the carbon-steam reaction. And there are other technologies for producing hydrogen from water or methane, some using electrolysis. None of these technologies address the sustainable future since they use fossil fuels as feeds. Thus, biological production fills a void, although other renewables could become elements in hydrogen production systems. Economics will eventually determine which will become the favored solution.

There is much research and development remaining before bioproduction can become an attractive technology. In judging future possibilities, it is well to remember that the conventional wisdom is that the costs double and time line increases fourfold. To be a bit more conservative, let us just say that developments take longer and cost more then usually anticipated. The current level of effort on bioproduction is not likely to produce a viable technology any time soon. Regardless, biohydrogen may very well play a critical role in the attempt to establish a sustainable energy future.

Let me just review briefly the thinking behind this conclusion. Today, the world's energy supply system is closely tied to existing oil and gas markets. Prices are stable or declining and the global reserve base is growing. There is excess production capacity in most areas of the world. This situation and the loss of confidence in the nuclear option has impeded establishing a coherent effort to find a strategy for a sustainable energy supply, one not requiring fossil fuels or other exhaustible energy sources. Federal energy policy in recent years has been focused on energy security. This is now out-dated and needs to be refocused on sustainable energy supply.

The strategy for achieving the end involves two basic requirements. One is an inexhaustible base load resource supplying electricity using distributed generation to accommodate the high-tech renewables including solar, wind, and advanced electrical storage facilities. Biomass has a role but it is ill-defined as of now primarily because of the inefficiency of photosynthetic conversion of solar energy to electricity and to hydrogen. It does seem clear that there is an increasing electrification of the world's energy supply as the prime resource for end-uses. However, while this satisfies the needs for stationary power uses, the problem of mobile power is more questionable. Electric drives for trains and tracked vehicles of all kinds is a no-brainer but is less so for personal transportation. The PNGV program seems to be settling on the hybrid-electric drive with the prime mover being a light diesel engine. Superficially, this seems sensible given the deadline that the program adopted, but it is clearly not on a sustainable track. With biohydrogen as a fuel, it would be. While electrification is moving forward, the development of a renewable mobile fuel supply is not. The federal R&D strategy needs revision toward an integrated effort for sustainable energy. It is a good bet that such a strategy will rest upon electricity and hydrogen as end-use supplies. Thinking along this line is going on in DOE laboratories prodded by several advisory committees, but that thinking is yet to show up in budgets.

On the industrial side, the restructuring of the U.S. electrical utility industry is the major focus of attraction. Beyond deregulation and retail wheeling of electricity, major objectives are three: reduced cost to consumers, increased reliability of supply, and new services at the end use destination. This track, too, is not aimed sustainability. However, it does address the here-and-now needs of consumers and producers, and will require major upgrades of the generation and transmission elements of the national and worldwide network as it comes into being.

There are several ongoing efforts to define the required knowledge and technology development to implement the deregulated, growing electrical supply. Among these is an electricity road map for R&D being prepared by the Electric Power Research Institute (EPRI). Hopefully, it will represent, when it appears, the consensus of users, manufacturers, utilities, and governments. It will be interesting to see how the recommended efforts will be funded given the outlook that I projected earlier in this talk. Just let me say that sustainability has not yet become a guideline for this effort. However, the results so far are not inimical to that objective. Let me just recommend that you keep your eye on progress because for hydrogen to mature as a major fuel for mobile and other uses, it must be integrated with the electrical supply system.

4. THE INTERNATIONAL SCENE

The most recent fashion in collaborative research is the "partnership." Partnerships are being used to couple corporate research efforts between companies, government and industrial research, government and academic research, and, most relevant to this meeting, coupling research efforts across national boundaries. The partnership theme is new only in its emphasis on major international efforts. There have been smaller, but quite effective, joint efforts for many years. Among these have been bilateral arrangements, for example, the U.S.-Japan program and the U.S.-Korea program; there also have been effective multilateral programs, for example, the NATO Science Committee's activities that over the past decade have involved 75,000 scientists and engineers from 40 countries working together on fundamental science topics in the physical, life, and social sciences. The objectives of this program are to reinforce the research capabilities and the science bases of the NATO countries, but it has provided a widely circulated series of books and publications and informative events that have invigorated both science and technology within Europe and across the Atlantic to include Canada and the U.S. In addition, NATO has pursued a program of education in technological management aimed at Greece, Portugal, and Turkey. These countries have been involved each with 20 or more development projects using sophisticated management tools introduced by the consortium of NATO countries. The economic and social benefits have been profound. Project management on a high level has become a welcome resource in these countries.

The partnership movement is moving beyond these pioneering efforts. Among the developing multilateral efforts is the particle accelerator, the LHC at CERN, the NASA Space Station, and the multilateral efforts on global warming and/or climate change. Increasingly, the size and cost of research is pushing countries to combine forces. However, there are barriers to achieving effective efforts.

During the 1970s and 1980s, the issue of military security monopolized the thinking of governments and their allied industries. Critical technologies and the science behind them were closely held and nations were reluctant to join others in partnerships lest inadvertent leaks of critical knowledge and technique would be damaging. With the end of the cold war, many people expected this reluctance to fade, and that has to some extent. Replacing the security problem, however, is the current concern with competitiveness. Countries and their competitive industries now are protective lest they lose advantage in world markets through technological leakage. Competitiveness is coupled in peoples' minds to domestic jobs and employment. Thus, again, there is reluctance to join others in partnerships. A related concern is how to share costs between partners and how to assure that corresponding benefits flow to the participants. The matter of patent and copyright ownership is another controversy. Regardless, partnerships between nations aiming at agreed goals is essential to achieve cross-cutting objectives such as a sustainable energy supply worldwide. I don't have to tell you that no such partnership exists today. The future for hydrogen and biohydrogen especially is tied to the willingness of nations to collaborate in planning and executing a resolute program over many years. I personally believe that industry must play a significant role in any such program. Governments can accelerate commercialization and capital investment using various incentives, but cannot sustain these subsidies without industry participation. It is highly preferable that biohydrogen production become part of a market-based solution to the sustainable energy imperative.

5. CONCLUSION

Biohydrogen as a fuel or fuel element is a tantalizing possibility. There has been recent experimental and conceptual progress that seem to hold promise and justify resolute efforts to clear pathways for biophotolysis processes and systems. But the current state of such visionary efforts is less than encouraging because of the constrained funding situation for R&D in the U.S. The situation seems more promising in Japan, Germany, and some of the Asian countries.

However, the success of any effort of this long-range kind will depend heavily on three conditions. One of these is the finding of a champion for the program in the political power structure. The loss of such a person, namely, Robert Walker, then-chairman of the U.S. House of Representative's Committee on Science, Space, and Technology. He retired last year, and no similar convinced politician has emerged. Developing such a champion in the political know is essential for the public support of biohydrogen production in the U.S. I think Bob Walker could be of great assistance in finding a new person or people to occupy the champion's position. Perhaps this meeting could designate a small group to talk with Bob Walker about this possibility. The U.S. DOE may have already made such an approach.

The second condition is the establishment of an international coalition of laboratories and institutions to plan and execute a program of research aimed at overcoming the known obstacles to economic biohydrogen. Such a coalition should include both scientific and engineering capabilities, and should include industrial interests as well.

The third condition is the recognition that a sustainable energy option is a valid and a principal objective for the nations of the world. As I have already said, that recognition has not yet been adopted. There needs to be a determined effort to convince societies to move away from the current goals of finding energy supplies to satisfy the demands of an expanding economy, and to the goal of a sustainable energy system, realizing that it too must allow for an expanding economy.

These three conditions do not address the science and engineering involved with demonstrating the feasibility, technological and economic, of biohydrogen. Those subjects are, of course, at the center of this meeting and the discussions that will proceed here. You, the participants, are fully qualified to put forward the scientific and technical basis for the resolute effort required to make biohydrogen a realistic candidate for a market-based element of a sustainable energy system.

Thank you for the opportunity to attend this important meeting.

2

THE SCIENCE OF BIOHYDROGEN

An Energetic View

Jun Miyake

National Institute for Advanced Interdisciplinary Research and
National Institute of Biological Science and Human-Technology
Agency of Industrial Science andTechnology
Ministry of International Trade and Industry
Higashi 1-1, Tsukuba, Ibaraki 305, Japan

1. INTRODUCTION

The importance of biological solar energy conversion to hydrogen is discussed from the viewpoint of energy and entropy. The historical view of the nature of the present system clarifies its limitations and the necessity for a brand-new energy accumulation technology, replacing the Industrial Revolution-based one. The capability of biological energy conversion system is described.

2. BIOLOGICAL METHOD ENABLES ENTROPY ENGINEERING

Hydrogen is a powerful energy source that does not produce carbon dioxide. If we can produce hydrogen from renewable resources, such as biomass, energy use will not have a harmful impact on the environment. In this presentation, I will describe briefly the reason biological systems are of use for creating a safe and efficient energy production system.

2.1. The Nature of the Present Energy System

The Industrial Revolution began in the 18th century. Although this revolution is considered to have been spurred by advancements in mechanical engineering, it was actually an energy revolution. Before the revolution, we were totally dependent on natural energy sources derived from solar energy stored in agricultural by-products and timber and collected from nearby forests. This was a closed, recycling energy system, limited by the so-

BioHydrogen, edited by Zaborsky *et al.*
Plenum Press, New York, 1998

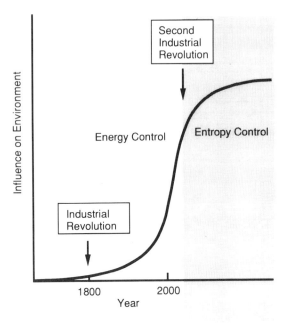

Figure 1. The need for entropy engineering. The use of fossil energy sources initiated by the Industrial Revolution affects the environment and grows exponentially. Decreases in resources and increases in environmental problems should be termed the "entropy problem." A technological revolution to reduce entropy, or at least not to increase it, is required.

lar energy stored in biomass over a period of months to hundreds of years. This solar energy-dependent society lasted for eons and ended only about 200 years ago.

The invention of the steam engine drastically changed the situation. We began tapping into seemingly limitless reserves of fossil fuels, which represented solar energy stored over a period of millions of years. Such energy was readily available and readily used, and so energy use escalated rapidly. The balance of energy flow was forever lost. This also created a vast increase in exhaust materials, carbon dioxide, and other industrial wastes. As the accumulated substances were released and expanded irreversibly, entropy increased. Figure 1 shows the concept of the irreversible growth of the effect of fossil energy that has to be stopped to ensure our survival.

2.2. Entropy Control Is Required

Natural energy, like biomass and solar energy, is considered to have low density. We have to collect it before using it. It means this energy is in a high entropy state. On the other hand, the Industrial Revolution provided people with a way to draw from single, high-density energy sources, first coal and then oil. These high-density energy sources represent low-entropy energy reserves because they are concentrated in relatively small areas. It is much easier to tap these low-entropy energy sources because we can now remove large amounts of energy from a single location. The subsequent rapid increase in energy usage means that these energy sources caused a rapid increase in entropy production. From this perspective, the Industrial Revolution was the point where we switched from using high-entropy, renewable energy sources, to using low-entropy, fossil energy sources.

The accompanying industrial expansion saw exponential growth in both the amount of energy consumed and the waste produced. The phenomenon should be called the "entropy problem" (Rifkin, 1981). Products of the entropy problem are carbon dioxide and greenhouse gases, fluorocarbons, industrial and home wastes, and so forth. We could be

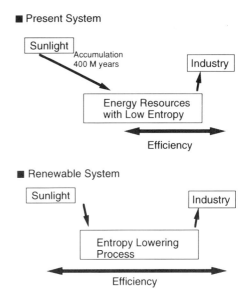

Figure 2. Energy problem is an "entropy" problem. Fossil energy resources are abundantly available at one location. The enthalpy of such energy could be used. However, renewable energy resources require as a first step their accumulation in diverse states. This entropy-reducing process requires much energy and is not competitive with low-entropy (fossil) resources.

facing a catastrophe because of their irreversible effects. To decrease the entropy problem, we must totally change the system of energy use because the problem comes from the present system. But this task is not simple and probably not possible using current technologies. Such technologies require a revolutionary leap from current ones, a leap we could call the "the second industrial revolution." The first revolution involved energy engineering, the second must involve entropy engineering (Figure 1). As Schrodinger (1945) has pointed out, a biological system has a special ability to control entropy. Application of biological reactions for this purpose should be studied.

2.3. Collection of Solar Energy: A Typical Example of Entropy Engineering

To reduce the use of fossil energy resources, many people believe we must use renewable energy sources, all of which represent solar energy stored in some form or another. But the efficiency of converting solar energy into any of a number of possible renewable energy carriers is very limited, representing inefficient use of the available solar energy. Collecting solar energy with photovoltaic cells or thermal solar collectors yields a somewhat higher conversion efficiency, but because solar radiation at the surface of the earth is relatively low, we need a huge area to collect a sufficient amount of energy.

Fossil energy has been accumulating for more than 4 million years. We are using these accumulated energy sources simply by digging a hole and removing them. When we talk of "efficiency," we usually only consider it from the point of removal to the industrial application. We are apt to forget the time and energy necessary to create the fossil energy sources. The ultimate amount of fossil energy sources is estimated to be only 3×10^{24} J, which is almost equivalent to the annual energy from sunlight (ca. 4×10^{24} J).

So-called renewable sources have a significant drawback in that we have to accumulate solar energy and then supply it to the industrial application. Biological energy conversion systems like photosynthesis might not be so efficient in the utilization of solar energy. However, it is a remarkable process from the viewpoint of entropy engineering. Plants and

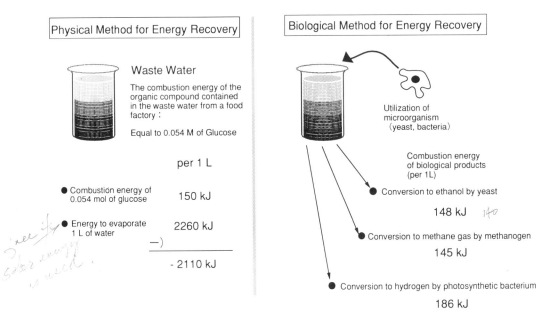

Figure 3. Recovery of energy from diluted resources; an example of organic wastewater treatment. Wastewater from a tofu factory is employed. Potential combustion energy of the organic substrates could not be used by most of the engineering methods available today. Biological conversion on the substrates enables the phase separation to overcome the entropy growth during the energy extraction process.

photosynthetic microorganisms grow by themselves and collect energy. Such spontaneous energy accumulation system is not obtainable in man-made industrial technologies.

3. BIOLOGICAL LOW-ENTROPY PRODUCTION PROCESSES

3.1. Biological Energy Conversion Characteristics

Here is an example of how biological systems produce energy from diverse, in energy recovery from organic wastewater. The wastewater from a tofu factory has various organic acids and their total combustion enthalpy is calculated to be 150 kJ/L (Zhu et al., in press). It is impossible to use this wastewater as fuel because the organic substances are dissolved and diluted in the water (in a high-entropy state). We must first extract these substances. There are methods to do this, but they require energy. For example, if we evaporate the water and recover the dried substances, the enthalpy necessary to evaporate water is about 2200 kJ/L, yet the available energy of the substance is only 150 kJ, so the net energy produced is negative. Therefore, it is very difficult to obtain energy from diluted solutions by mechanical means.

On the other hand, biological systems enable the efficient recovery of energy. Bacteria can convert such organic material in a diluted solution. Using yeast, glucose could be converted to ethanol. In this case, the ethanol produced contains 140 kJ/L. The ethanol produced is soluble in water. Again, we need energy to separate the ethanol and water. The net energy produced in this case is small or negative. Using methanogenic bacteria, we

Table 1. Equations of reactions

Hydrolysis
$$2H_2O \rightarrow 2H^+ + 1/2 O_2 \quad \triangle G = -242 \text{ kJ} \tag{1}$$

Reduction of Proton (Final Stage of H2 Production)
Nitrogenase
$$2H^+ + 2Fd_{red} + 4ATP \rightarrow H_2 + 2Fd_{ox} + 4ADP + 4Pi \tag{2}$$

Hydrogenase
$$2H^+ + 2e^- \leftrightarrow H_2 \tag{3}$$

Electron Supply
Photosynthetic bacteria
$$(1)\ Glucose + 6H_2O \rightarrow 24H^+ + 6CO_2 + 24\ e^- \rightarrow 12H_2 + 12CO_2 \quad \triangle G = -33.8 \text{ kJ} \tag{4}$$
$$(2)\ C_3H_6O_3 + 3H_2O \rightarrow 12H^+ + 3CO_2 + 12\ e^- \rightarrow 6H_2 + 3CO_2 \quad \triangle G = 51.2 \text{ kJ} \tag{5}$$

Anaerobic bacteria
$$(1)\ Glocose + 2H_2O \rightarrow 2Acetate + 2CO_2 + 4H_2 \quad \triangle G = -184.2 \text{ kJ} \tag{6}$$
$$(2)\ Glucose \rightarrow Butyrate + 2CO_2 + 2H_2 \quad \triangle G = -257.1 \text{ kJ} \tag{7}$$

---

can obtain methane with an energy content of 145 kJ/L. Because methane is a gas, it can be easily separated from the solution, releasing the stored chemical energy in solution without any additional energy input. By using photosynthetic bacteria and illuminating the solution, we can convert glucose into hydrogen that contains 185 kJ/L of chemical energy. In this way, the diluted energy can be extracted in a usable form.

From the viewpoint of entropy engineering, biological methods enable the efficient collection of diverse, low-density solar energy. Cells grow themselves to cover the large areas. Using only a simple box or plastic tube, we can free ourselves of our dependence on industrial products and fossil energy. The most important point for our future is independence from fossil energy sources. Though higher efficiency of solar energy conversion will be necessary, we should recognize the essential merits of the biological system.

There are three types of microorganisms for hydrogen production. The first is cyanobacteria. These organisms split water into hydrogen and oxygen gas by photosynthesis. Because of the direct conversion of sunlight into usable chemical energy, we do not need to accumulate solar radiation or accumulate organic substances to feed the bacteria. For future applications, hydrogen production using cyanobacteria should be significantly increased. The second is anaerobic bacteria, which use organic substances as the sole source of electrons and energy and convert them into hydrogen. The reaction is rapid and the process does not require solar radiation, making it useful for treating large quantities of wastewater. The third is photosynthetic bacteria, which are somewhat between anaerobic bacterial and cyanobacterial systems, because although photosynthetic bacteria also convert organic substances to hydrogen at a fairly high rate, they require light energy to assist or promote the reactions involved in the hydrogen production.

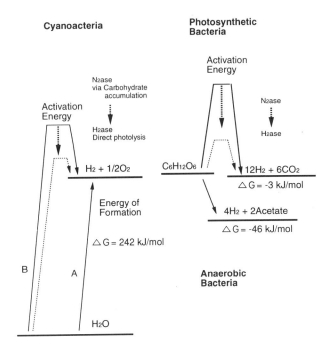

Figure 4. Energetic view of phototrophic bacterial hydrogen production. Free energy change from substrates and activation energy are illustrated for algal and photosynthetic bacterial reactions. Activation energy could be reduced by the alteration of the reaction mechanism; however, there are many factors that contribute to activation energy.

All of these processes have their respective advantages and drawbacks, and we need to match these processes to the applications for which they are best suited. Therefore, we should not focus only on a single energy production technology, but consider all of them.

3.2. Basis for Biohydrogen

3.2.1. Mechanisms of Biological Hydrogen Production. In nature, only bacteria and algae have the capability to produce hydrogen. Among these organisms, those currently selected for research are anaerobic bacteria and photosynthetic microorganisms such as photosynthetic bacteria and cyanobacteria (and green algae).

Cyanobacteria directly decompose water to hydrogen and oxygen with light energy (Equation 1), the process that generated oxygen for the earth. The reaction requires only water and sunshine and is very attractive from the viewpoint of environmental protection. However, natural-born organisms of those species examined thus far show rather low rates of hydrogen production due to the complicated reaction system needed to overcome the large free energy (+237 kJ/mol hydrogen). Another drawback is that a carrier gas is required to collect the evolved gas from the culture. Separation of oxygen and hydrogen is also a subject for future study.

Photosynthetic bacteria do not utilize water as the starting compound for hydrogen production but, rather, use organic acids. Equations 4 and 5 show photosynthetic bacterial hydrogen production from organic substrates. In the case of lactate, the free energy change is +8.5 kJ/mol hydrogen. Compared to algal hydrolysis, much less free energy is required to produce hydrogen from photosynthetic bacteria using organic substances. However, activation energy must be high to drive nitrogenase. There are various kinds of photosynthetic bacteria and many kinds of organic substrates, such as fatty acids, sugars, starches, cellulose, and so forth, that could be the starting materials for hydrogen production. Photosynthetic bacteria can de-

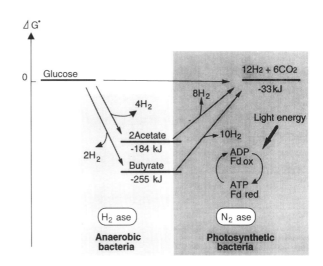

Figure 5. Energetic view of anaerobic bacterial hydrogen production. Anaerobic bacteria release hydrogen during a process in which the substrates lose free energy. Because no other energy is supplied (as in photosynthesis or oxidative phosophorization), complete decomposition of the substrates (using glucose as an example) is impossible. The combination of photosynthetic bacteria solves the problem because the light-driven system can overcome the uphill reaction. Note that the total free energy change from glucose to hydrogen + carbon dioxide is nearly zero.

compose these organic substances completely. Hydrogen production could be combined with organic waste treatment (waste fluids from food factories, pulp factories, etc.).

Anaerobic bacteria such as *Clostridium* also produce hydrogen from organic substrates (Equations 6 and 7). However, they cannot utilize light energy and so the decomposition of organic substrates is incomplete. The free energy of the above equation is -46 kJ/mol hydrogen reaching to the bottom of the system. No further decomposition of the remaining organic substance (acetic acid) is possible under anaerobic conditions (Figure 5). The advantage of anaerobic hydrogen production is the high reaction rate. Insofar as the reaction is not driven by light, a large fermenter could be used. This reaction should be suitable for the initial step of wastewater treatment and hydrogen production. The combination with photosynthetic bacteria creates synergy for efficient hydrogen production (from light and substrate).

3.2.2. Efficiency of Light Energy Conversion. Next we examine the maximum efficiency of light energy conversion to hydrogen. Calculation of the efficiency or quantum requirement (QR) is difficult. The reaction mechanism of light to hydrogen in phototrophic organisms is not precisely elucidated. There are various ways of estimation from the viewpoints of free energy, enthalpy, or engineering.

Here, we will focus on light energy conversion, for which engineering calculations are suitable. In this calculation, energy from substrates is not counted. The reasons are: (1) free energy change from the substrate to hydrogen is highly positive, meaning no energy supply (see Equation 5) and (2) the entropy of the diluted solution of substrates is too high to give any substantial energy (see discussions above). The efficiency of light to hydrogen conversion is defined as:

Efficiency = Combustion enthalpy of produced H_2/Light energy supplied to the reactor
(Combustion enthalpy of H_2 = 242 kJ/mol [g] or 268 kJ/mol [L])

The reaction mechanism of photosynthetic bacterial hydrogen production is shown in Figure 6. Light energy is converted to the potential energy of the electron, which then

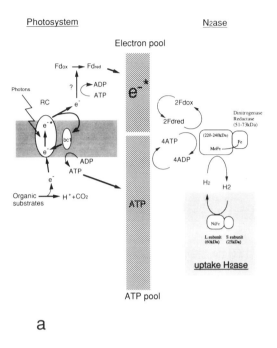

a

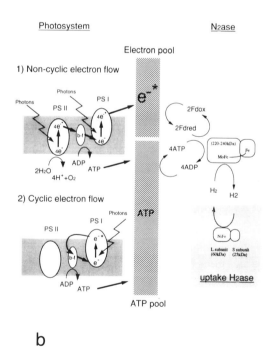

b

Figure 6. Schematic view of the mechanism of biological system for light to hydrogen: (a) photosynthetic bacterial reaction, (b) algal reaction (cyanobacteria and algae), and (c) replacement of nitrogenase by hydrogenase reduces the requirement of ATP.

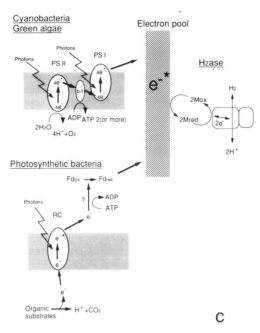

Figure 6. (*Continued*)

forms ATP. In another process, the elevated electron is needed to reduce ferredoxin (Fd), a typical electron carrier to the hydrogen-producing enzyme, nitrogenase. ATPs produced here are supplied to the enzyme together with the electron carrier. Nitrogenase requires 4 ATP and 2 Fd_{red} for hydrogen production (Equation 2).

Photons activate the photosystem in the reaction center to pump protons. Proton transport couples with the generation of ATPs. It has not been well-elucidated how many protons are required to give a mole of ATP (Oosawa, 1984). Two to three protons are used to give an ATP (Berry and Hinkel, 1983). In Gobel's (1978) experiments, 1.5 photons generate an ATP. Another unknown factor for hydrogen production is the energy required to elevate electrons for the reduction of Fd. Electrons emitted from the RC of non-sulfur purple bacteria (in quinone) have a potential from 0 to -0.2 V (Dutton, 1978) that is much lower than the Em of Fd (-0.4 V). It means that any additional energy should be supplied to elevate the electron potential. The number of photons required for the reaction is not clear.

If nitrogenase mediates the reaction (4 ATP), 1 ATP is used for the elevation of 1 electron and 1.5 photons for one ATP. Eleven photons are used to give one H_2 (QRH_2 = 11). It is equivalent to the light energy conversion efficiency of about 15% at 850 nm (monochromatic light). In the most favorable case, if no ATPs are required for Fd_{red} formation, QRH_2 could be 8. Energy conversion efficiency at 850 nm monochromatic light is about 20%.

Figure 6b shows the reaction mechanism of cyanobacteria. Photosystems (PS) I and II provide electrons at high energy levels and ATPs. Cyclic electron flow in PSI provides ATPs. The combination of these photoreactions supplies electrons and ATPs for hydrogen production by nitrogenase. The exact number of QRs for hydrogen is not obtained because of the uncertainty of the process described above.

Table 2. Efficiency of light to H_2

Light source	Energy*	Organisms	QRH$_2$	Efficiency (%)**	References
Indoor experiments					
Monochromatic	20	R. sphaeroides	17	10	Nakada et al.***
Solar simulator	75	R. sphaeroides		7.9	Miyake and Kawamura
	1000	R. sphaeroides		1.9	
W-lamp	–	R. pulstris		6.7	Vinventini et al.
Monochromatic	2-5	R. rubrum	7.5		Warthmann et al.
		Chlorella moewusii		6-24	Greenbaum
Fluorescent tube	25	Synechococcus		2.6	Mitsui
Fluorescent tube	25	Mastigocladus laminosus		2.7	Miyamoto et al.**
Outdoor experiments					
Sunlight 36 days		Anabaena cylindrica		0.2	Miyamoto et al.

*W/m^2
**%, Combustion enthalpy of H_2/Light energy
***Unpublished data

Some examples of the conversion efficiency of light to hydrogen are listed in Table 2. The width of the range of the efficiencies indicates the diversity of the measuring methods. The lowest QR of 4–5 is recorded both for photosynthetic bacteria and cyanobacteria, which is a large discrepancy compared to the estimation above. The outdoor experiments show much lower values. Cyanobacteria gave 0.2% (Miyamoto et al., 1979) and photosynthetic bacteria below 1% (data not shown). Before comparing records from various laboratories, we need studies on the mechanism of the reactions for photohydrogen production and the standardization of measuring conditions, such as illumination and culture conditions.

3.2.3. Means of Improvement. The mechanism of energy and electron flow in photosynthetic organisms is highly complicated. Figures 6a and 6b show a simplified nitrogenase reaction, although the computation of the photon to hydrogen molecule requires further elucidation of the mechanism. The substitution of nitrogenase by hydrogenase might provide a simple system (Figure 6c). In a nitrogenase system, much energy is consumed to overcome the activation energy (Figure 4). Efforts have been made to identify the method (see papers in these proceedings). However, it is impossible to transfer all the photosynthetic products to hydrogen production. We have to spare some energy for the survival of the bacteria. Devoting all energy to hydrogen production might reduce the total activity of the cells. In the case of nitrogenase, ATP levels in the cell regulate the activity so it does not lose too much energy. We should consider how to control the energy distribution in the event strong hydrogenase for hydrogen production successfully replaces nitrogenase by genetic engineering. Lowering the activation energy lessens control of the electron flow; directing all energy for hydrogen production would be fatal for the cells. An artificial regulation system is needed to balance the electron supply.

The real metabolism in the cell is not as simple as depicted in Figure 6 because it contains an accumulation of redox power as carbohydrates (Figure 7). Improvements in hydrogen production require a more detailed understanding of the metabolism. PHA accumulation and uptake hydrogenase should be controlled to provide the maximum energy and electrons for hydrogen production. Genetic engineering to suppress the activities has been studied (see these proceedings).

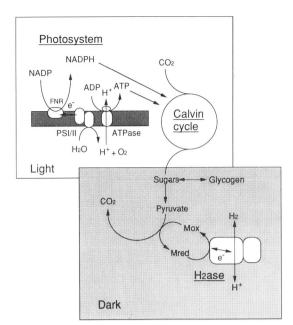

Figure 7. Factors affecting hydrogen production. Uptake hydrogenase and carbohydrate accumulation are possible factors that compete with the reaction of hydrogen production. Hydrogenase-substituting nitrogenase might affect the reaction and harm cell survival. Engineering of cells requires attention so the overall reaction can be controlled.

ACKNOWLEDGMENTS

I express my sincere appreciation to Dr. Yasuo Asada for his critique and Dr. Masato Miyake and Mr. Y. Morimoto for the illustrations.

REFERENCES

Berry, E.A., and Hinkel, P.C., 1983, Measurement of the electrochemical proton gradient in submitchondrial particles, *J. Biol. Chem.,* 258:1474–1484.

Dutton, P.L., and Prince, R.C., 1978, Reaction-center-driven cytochrome interactions in electron and proton translocation and energy coupling, *Photosynthetic Bacteria,* Clayton, R.K., and Sistrom, W.R. (eds.), Plenum Press, New York and London, pp. 525–570.

Gobel, F., 1978, Quantum efficiencies of growth, *Photosynthetic Bacteria,* Clayton, R.K., and Sistrom, W.R. (eds.), Plenum Press, New York and London, pp. 907–925.

Greenbaum, E., 1988, Energetic efficiency of hydrogen photoevolution by algal water splitting, *Biophys. J.,* 54:365–368.

Kondratieva, E.N., and Gogotov, I.N., 1983, Production of molecular hydrogen in microorganisms, *Adv. Biochem. Bioeng/Biotech.,* 28:139–191.

Kumazawa, S., and Mitsui, A., 1994, Efficient hydrogen photoproduction by synchronously grown cells of a marine cyanobacterium, *Synechococcus* Sp. Miami BG 043511, under high cell density conditions, *Biotechnol. Bioeng.,* 44:854–858.

Miyamoto, K., Hallenbeck, P.C., and Benemann, J.R., 1979, Solar energy conversion by nitrogen-limited cultures of *Anabaena cylindrica, Appl. Environ. Microbiol.* 37:454–458.

Miyamoto, K., Hallenbeck, P.C., and Benemann, J.R., 1979, Nitrogen fixation by thermophilic blue-green algae, *J. Ferment. Technol.,* 57:287.

Miyake, J., and Kawamura, S., 1987, Efficiency of light energy conversion to hydrogen by the photosynthetic bacterium *Rhodobacter sphaeroides, Int. J. Hydrogen Energy,* 12:147–149.

Nakada, E., Asada, Y., Arai, T., and Miyake, J., 1995, Light penetration into cell suspensions of photosynthetic bacteria; relation to hydrogen production, *J. Ferment. Bioeng.,* 80:53–57.

Miyake, J., Asada, Y., and Kawamura, S., 1989, Nitrogenase, *Biomass Handbook*, Hall, C.W., and Kirani, O. (eds.), Gordon and Breach Science Publishers, pp. 363–370.

Oosawa, A., and Hayashi, F., 1984, A loose coupling mechanism of synthesis of ATP by proton flux in the molecular mechanism of living cells, *J. Phys. Soc. Jpn.*, 53:1575–1579.

Rifkin, J., 1981, *Entropy*, Bantam Books, New York.

Schrodinger, E., 1945, *What Is Life? The Physical Aspect of the Living Cell*, Cambridge University Press, United Kingdom.

Vincenzini, M., Balloni, W., Mannelli, D., and Florenzano, G., 1981, A bioreactor for continuous treatment of wastewaters with immobilized cells of photosynthetic bacteria, *Experientia*, 37:710–711.

Warthmann, R., Pfennig, N., and Cypionka, H., 1993, The quantum requirement for H_2 production by anoxygenic phototrophic bacteria, *Appl. Microbiol. Biotechnol.*, 39:358–362.

Zhu, H., Suzuki, T., Nakada, E., Asada, Y., and Miyake, J., Hydrogen production from tofu wastewater by *Rhodobacter sphaeroides* immobilized in agar gel, *Int. J. Hydrogen Energy*, in press.

THE TECHNOLOGY OF BIOHYDROGEN

John R. Benemann

Consultant
343 Caravelle Drive
Walnut Creek, California 94598

1. ABSTRACT

Hydrogen produced by microalgae and bacteria is biohydrogen. There are currently no practical biohydrogen production processes. However, several concepts have promise for near- to long-term process development.

The conversion of CO to H_2, the microbial shift reaction, operates at ambient temperatures in a single-stage process, compared to the two-stage, high-temperature, chemical catalyst processes currently used. Process development is just beginning, but this concept appears promising for near- to mid-term practical applications.

H_2 yields from dark fermentations of organic wastes are typically less than 20% (on a heating value basis) compared to CH_4 fermentations. Higher yields might be possible at elevated temperatures, with nutrient limitations, and through metabolic engineering of the bacteria. Photofermentations, the conversion of organic substrates to H_2 by nitrogen-fixing photosynthetic bacteria, achieve high H_2 yields, but low solar conversion efficiencies. The inefficiency of the nitrogenase enzyme suggests that biohydrogen processes must be based on reversible hydrogenases. Even then, dark fermentations of wastes to H_2 would be preferable to light-driven processes, in part due to the high cost of photobioreactors. The production of H_2 from organic wastes is of mid-term potential but limited by resources and competition from other processes.

Larger-scale biohydrogen production requires biophotolysis processes—H_2 production from water and sunlight. Photobioreactor costs and solar conversion efficiencies are main challenges in the development of practical processes. Direct biophotolysis couples the reductant produced by photosynthesis directly to hydrogenase, producing O_2 and H_2 simultaneously, while indirect processes separate these basically incompatible reactions through intermediate CO_2 fixation. Direct, but not indirect, biophotolysis processes require hydrogenase activity in the presence of high O_2 levels, something not known to occur. Hybrid indirect processes using both algae and photosynthetic bacteria have been proposed and even tested outdoors, but are complex and inefficient. The simplest indirect

BioHydrogen, edited by Zaborsky *et al.*
Plenum Press, New York, 1998

process would use the same algal cells for both CO_2 fixation/O_2 evolution and H_2 production, at separate times or even in different reactors. The H_2 reactions would take place both in the dark and light. Light-driven H_2 evolution requires suppression of the O_2 evolution process. Biophotolysis process must achieve the highest possible solar conversion efficiencies, which will require development of algal strains with reduced light-harvesting ("antenna") pigment content.

A preliminary economic analysis of a large-scale, two-stage, indirect process suggests that biohydrogen costs close to $10/GJ, lower than for PV-electrolysis processes. This assumed use of very large, low-cost (<$10/m^2) open ponds for the first (CO_2 fixation) stage operating at a 10% solar conversion efficiency, and much smaller (approximately 10% of the total area) photobioreactors (<$150/m^2) for the second H_2 production stage. Single-stage biophotolysis systems require low-cost (<$50/m^2) photobioreactors, but may be suitable for much smaller-scale (perhaps even roof-top) applications.

The development of practical biohydrogen production systems will require focusing on the more promising alternatives and critical R&D issues and could be accelerated through international basic and applied R&D collaborations.

2. HISTORICAL INTRODUCTION

Applied R&D for the production of biohydrogen, H_2 production using bacteria and microalgae, started some 25 years ago with a meeting on "Biological Energy Conversion" (Hollaender et al., 1972). The late Prof. Lester Krampitz reported that experiments were underway in his laboratory to produce H_2 by coupling chloroplasts with methyl viologen and a bacterial hydrogenase (see Krampitz, 1977). Similar studies were also being carried out by Benemann (1973), who reported that an illuminated mixture of spinach chloroplasts, *Clostridium kluyveri* hydrogenase and clostridial ferredoxin, produced H_2 in a light-driven reaction sensitive to inhibitors of the photosynthetic water-splitting reaction (PSII) and stimulated by O_2 absorbers (Benemann et al., 1973). This suggested that the H_2 was derived from water and was inhibited by the O_2 produced during the reaction (although simultaneous O_2 and H_2 production by this system was not actually demonstrated until later work by Greenbaum[1980]). Further attention was focused on biohydrogen and, specifically, photobiological H_2 production at another workshop in late 1973 (Gibbs et al., 1973).

Hydrogen metabolism, both uptake and evolution, by microorganisms had been a subject of scientific inquiry for many decades. Hydrogen metabolism, evolution, and uptake by fermentative bacteria was reported starting in the 1930s, followed by reports of H_2 evolution in the light by green microalgae (Gaffron and Rubin, 1942) and photosynthetic bacteria (Gest and Kamen, 1949). Perhaps the earliest report of microbial H_2 evolution dates to the last century, when Jackson and Ellms (1886) collected some "pond scum," a culture of the heterocystous cyanobacterium *Anabaena cylindrica*, and found it produced gas consisting of almost pure H_2 immediately upon being placed in a container,. This same microalga has been the subject of extensive applied research for biological hydrogen production by the author and colleagues (Benemann and Weare, 1974; Weissman and Benemann, 1977) and many others.

Even prior to the 1970s, considerable basic R&D had been carried out in this field. For example, although much of the H_2 produced by green algae derives from carbohydrates, Spruit (1958) demonstrated that these algae could also produce H_2, at least transiently, by a direct biophotolysis process, that is, through an electron transport chain from water to hydrogenase, without intermediate CO_2 fixation. Other studies, for example by Healy (1970), showed that H_2 production by anaerobically adapted *Chlamydomonas moewsuii* was maximal at lower light intensities, being inhibited at high light levels due to resulting O_2 evolution.

Indirect biophotolysis processes use algal cultures that fix CO_2 into stable organic compounds, which are then converted to H_2 either by the algae themselves or by bacterial fermentations. The nitrogen-fixing heterocystous cyanobacteria mentioned above are an example of such indirect biophotolysis: the vegetative cells fix CO_2 while the heterocysts use the organic substrates to produce H_2 (in the absence of the normal substrate, N_2). Non-heterocystous nitrogen-fixing cyanobacteria also can produce H_2 by alternating periods of aerobic photosynthesis with anaerobic nitrogenase-mediated H_2 production (Weare and Benemann, 1974).

The potential for the production of H_2 by dark fermentative bacteria using organic substrates, particularly wastes, was also evaluated in the early 1970s. Thauer (1976) suggested that the free energy released during the complete dissimilation of glucose to H_2 ($C_6O_6H_{12} + 6H_2O \rightarrow 12H_2 + 6CO_2$) would not generate sufficient ATP to support cell growth. He concluded that anaerobic metabolism would be limited at most to about four H_2 and two acetates per glucose, (e.g., $C_6O_6H_{12} + 2H_2O \rightarrow 4H_2 + 2CO_2 + 2CH_3COOH$) and that, therefore, dark anaerobic H_2 fermentations of organic substrates would not be practical. This limitation of dark fermentation processes focused attention on photosynthetic bacteria, as these can essentially stoichiometrically convert organic acids and some other substrates to H_2 through a nitrogenase-mediated, light-driven process. The use of photosynthetic bacteria for photobiological H_2 production was suggested by several researchers, including the "photofermentation" of organic wastes as an alternative to methane fermentations (Benemann, 1977). H_2 production from wastes by photosynthetic bacteria was first demonstrated with sugar processing wastewater by Zurrer and Bachofen (1979).

During the 1970s, there thus was a great deal of applied research in biohydrogen production in the United States and other countries (for early reviews see Mitsui et al., 1974; Lien and San Pietro, 1975; Hallenbeck and Benemann, 1979; Weaver et al., 1980). Although several groups demonstrated H_2 production using cyanobacteria, green algae, and photosynthetic bacteria in small photobioreactors, both indoors with artificial lights and even outdoors under sunlight (e.g., Miyamoto et al., 1979), no scale-up studies were carried out and economic issues discussed only superficially (e.g., Benemann et al., 1980). Starting in the early 1980s, support for renewable energy R&D generally, including biohydrogen, declined in the United States and elsewhere.

By the early 1990s, however, environmental issues came to the forefront, with global warming, regional air pollution, and economic/ecological sustainability being the major driving forces behind the renewed interest in alternative energy sources. Biohydrogen R&D programs were again initiated in Japan, Germany, and the United States.

In Japan, an eight-year program, currently in its final year, emphasizes the development of practical processes for conversion of wastes to H_2, mainly using photosynthetic bacteria. Under this program, several industrial companies are developing alternative approaches to pretreatment/fermentations of wastes and photobioreactor designs. Supporting R&D is also ongoing at research institutes and universities, and one project at a power plant in Osaka demonstrated a combined algal-bacterial process (see these proceedings for reports on these projects).

In Germany, a broad-based, mainly basic R&D effort was undertaken from 1990 through 1994, with over 40 projects supported, mainly at universities, ranging from the biochemistry, physiology, and genetic engineering of H_2-producing microbes to biomimetic systems. However, this program was terminated after only half its initially scheduled term due to, at least in part, lack of industry interest in the relatively long-term R&D effort required to develop this technology.

In the United States, a more modest applied biohydrogen R&D effort has been supported by the U.S. Department of Energy, as part of its Hydrogen R&D Program, involving a handful of research groups at the DOE's National Renewable Energy Laboratory and Oak Ridge National Laboratory and at the universities of Hawaii and Miami (with these latter two activities recently combined after the untimely death of Prof. Akira Mitsui, a pioneer in biohydrogen research; Mitsui, 1992).

A century after the first report of hydrogen production by algae, over 50 years after the demonstration of hydrogen evolution by most major types of microbes, and after a quarter-century of applied R&D, biohydrogen has not yet resulted in any practical applications. The reported rates, stabilities, efficiencies, scales, and other attributes and metrics have been, typically, an order of magnitude or more lower than would be required for practical processes. Major technical problems remain to be solved and the economics of such processes have only begun to be addressed. And a consensus has not yet developed regarding the most promising R&D approaches in this field.

However, the basic sciences on which biohydrogen technologies and process development must be based have greatly advanced over the past two decades, with an ever more intimate understanding of the genetics, biochemistry, and physiology of microbial hydrogen metabolism. For some recent examples only, an active site of a hydrogenase has been proposed (Happe et al., 1997) and protein engineering of hydrogenases has demonstrated increased resistance to O_2 (McTavish et al., 1995).

This fundamental scientific information must now be applied to the development of production processes that could deliver biohydrogen at acceptable costs. Conversely, we can also identify those approaches that are less promising to allow focusing limited R&D resources on the most promising alternatives. These issues are the focus of this presentation (see also Benemann and Zaborsky, 1996; Benemann, 1994, 1996, for other recent reviews).

3. CONVERSION OF CARBON MONOXIDE TO HYDROGEN

Most commercial H_2 currently derives from reforming natural gas and gasification of coal, resulting in gaseous mixtures containing CO and H_2, along with CO_2, H_2O, and other gases. The CO is "shifted" by catalysts to H_2 in a multistage process. As mentioned above, some bacteria can carry out this transformation in the dark due to the presence of an inducible CO dehydrogenase enzyme complex that is coordinately expressed with a CO-insensitive hydrogenase enzyme (see Kerby et al., 1992, and references therein). The enzyme complex is characterized by a reasonable affinity for CO (Km of 32 μM) and high maximal velocity (V_{max} ca. 7 mmol CO oxidized/mg protein/min). Photosynthetic bacteria have become the organisms of choice in applied R&D of this process (Klasson et al., 1993; Weaver et al., 1995, 1997).

The CO shift reaction is thermodynamically favored at the ambient temperatures at which microbial processes operate, resulting in essentially complete conversion of CO to H_2 in a single-stage process, compared to the two-stage processes required at the high temperatures (>200 °C) at which chemical catalysts operate. However, the economics of the microbial process are presently uncertain; they depend on the size bioreactors required and the losses inherent in such ambient pressure-temperature conversion process. Microbial CO conversion may be particularly competitive in biomass gasification processes to produce H_2 for fuel cells, where small scales would reduce the cost penalties while accentuating the advantages of the microbial process.

The need to reduce residual CO to very low levels makes mass transfer the plausibly limiting factor in the design of such a process. This suggests the need for counter-current gas-liquid contacting systems, as in trickling filters used in waste treatment or plug-flow systems typical in commercial gas biofiltration processes (Andrew and Noah, 1995). Hollow fibers have been studied for application to such a process (Markov et al., 1996), and exhibit superior mass transfer rates (Grasso et al., 1995) but are likely too expensive in such applications.

Maximal (CO-saturated) conversion rates of 3.5 to 10 mmol CO/g dry weight biomass per hour have been extrapolated by Klasson et al. (1993), while Weaver et al. (1995) reported about an order of magnitude increase in such rates through a strain screening effort. Genetic strain improvements should be able to further increase the cell specific activity. The available data should allow at a least an initial estimation of the size of the required bioreactors and likely costs of such a process. Indeed, preliminary cost projections for wood gasification in combination with a microbial CO shift reaction report final H_2 costs within the target for merchant hydrogen (Mann, 1996). Thus, the microbial shift reaction may be a candidate for near-term biohydrogen process development.

4. BIOHYDROGEN FROM ORGANIC SUBSTRATES

Dark microbial H_2 fermentations of organic substrates are inherently more practical than photobiological conversion in that they do not require the large surface areas for photobioreactors demanded by even the most efficient solar conversion process.

Starches or sugars would be too expensive for H_2 production. Assuming a yield of 80% from fermentable starches or sugars on a higher heating value basis (almost 10 mol H_2 per mole of glucose), then the raw material costs alone would be about $14/GJ for a substrate cost of $200/ton. The U.S. corn starch to ethanol fermentation industry is only possible because of the large subsidies (>$10/GJ) available. Anyway, using foods for renewable energy production is not acceptable in the long-term.

One option is to combine H_2 fermentations with the production of higher value co-products. For examples, H_2 could be co-produced with acetic and other organic acids. Co-production of H_2 with bioplastics has also been suggested, although there would be competition for substrates between these products. Woodward et al. (1996) demonstrated an in vitro glucose to gluconic acid plus H_2 conversion. However, this reaction only produces relatively minor amounts of H_2 (about 1% of product value), with a maximum potential of a few hundred tons (Benemann, 1996). This applies to most such co-production processes, whose overall H_2 production potential is limited.

Of greater potential would be to combine H_2 fermentations with waste treatment processes. Many wastes, in particular waste water, contain relatively large amounts of readily fermentable materials, which present costly disposal problems. Currently, waste water treatment processes are aerobic (activated sludge, trickling filters, etc.) or anaerobic, producing biogas (methane + CO_2). Biogas is generally converted to electricity (Augenstein et al., 1994). Where a local market for H_2 exists, or if small-scale (<1 MW) fuel cell technology were to become competitive, H_2 waste fermentations could be favored over methane fermentations. However, the technical feasibility of such H_2 fermentations has not yet been demonstrated. As stated above, theoretical fermentative H_2 yields are maximally 4 mol of H_2 per mol glucose substrate, and in practice maximally 10–20% of the energy content in glucose is recoverable as H_2, compared to 80–90% for methane and ethanol fermentations. This is because of the limited metabolic energy derived from H_2 fermentations (Schink, 1997). However, there are several approaches to overcome this difficulty. Nutrient limitations can be used to arrest growth, reducing metabolic energy requirements. Operation at higher temperature would shift the thermodynamics in favor of H_2 production (Heijen, 1995). Metabolic engineering could be used to redirect bacterial metabolism toward H_2 production (see Keasling et al., 1998, these proceedings).

Clearly, the development of high-yield H_2 fermentations will require a significant R&D effort. Among the problems to be addressed are contamination with methane-producing and other microbes. However, for one example only, Ueno et al. (1995) demonstrated stable, prolonged, though low-yield (approximately 20% of stoichiometric) H_2 fermentations of sugar processing wastewaters at 60 °C. H_2 gas transfer may present further process limitations (Pauss et al., 1990). Clearly, more R&D is required. However, if developed, such H_2 fermentations would be of relatively low cost, similar to methane fermentations, as they would use similar mixed-tank reactors. Methane fermentations costs are in the range of $4–8/GJ. Thus, dark H_2 fermentations of wastes could be competitive with fossil-fuel derived H_2 and provide a plausible near- to mid-term approach to practical biohydrogen production.

However, most of the research on biohydrogen production from organic wastes over the past two decades has focused on the use of photosynthetic bacteria (see Sasikala, et al., 1993, for a review). The organic substrates could be produced by microalgae cultures, a concept already introduced some 20 years ago by Krampitz (1977; see also Weetall, 1977), and many others since, and most recently demonstrated in Japan (Ikuta et al., 1997, these proceedings). The attractiveness of using photosynthetic bacteria is the near-stoichiometric H_2 yields achievable. The major disadvantage is the rather low solar conversion efficiencies, typically not much higher than for algal biophotolysis systems. These are due to three factors: the use of the very inefficient nitrogenase enzyme, the very low light intensities at which bacterial photosynthesis saturates, and the inherently low efficiency of light conversion by photosynthetic bacteria. Although it can be argued that these problems could be overcome with more R&D, such as replacing nitrogenase with hydrogenase and reducing the antenna size of the photosynthetic apparatus of these bacteria, dark fermentations would be of much lower costs than photofermentations, even with less than stoichiometric yields. In conclusion, photosynthetic bacteria are not a promising approach to biohydrogen production, except, possibly, in the above described dark CO shift reaction.

In any event, biohydrogen from organic wastes is limited by the availability of wastes and by competition with alternative processes (methane or ethanol fermentations). H_2 fermentations of lignocellulosic wastes and biomass resources could increase this potential, but this is a very difficult problem and gasification of such biomass is a more likely process for H_2 production. Thus, for the near-term, readily degradable liquid wastes appear to be the most likely target for H_2 fermentations, even though they would be limited to niche markets. However, it is possible to currently operate methane fermentation (anaerobic digestion) processes (such as at sewage treatment plants) to produce some H_2, or $H_2:CH_4$ mixtures (see Benemann and Zaborsky, 1996, for discussion). This presents some near-term market opportunities for biohydrogen.

5. BIOPHOTOLYSIS PROCESSES

The practical production of H_2 from water and sunlight, biophotolysis, is the ultimate goal of biohydrogen R&D. The major limitations are the achievable overall solar conversion efficiency and the capital and operation costs of the solar converter.

With an annual average insolation in the U.S. Southwest of about 22 MJ/m^2/day, and assuming a 10% conversion into H$_2$ (higher heating value, HHV, at 142 MJ/kg), this yields 170 L H$_2$/m^2/day, or 0.8 GJ/m^2/year. With current U.S. retail costs of natural gas ($2.50/GJ), liquid fuels ($10/GJ), and electricity (at $0.08/kWh), such a H$_2$ output is worth about $2, $8, and $22/m^2/year, respectively, to the consumer. Biohydrogen would compete more closely with electricity. A producer cost goal of about $15/GJ, or some $11/m^2/year, would thus appear reasonable, allowing for H$_2$ distribution and some taxes. Assuming $2/m^2/year for operations and a total annual capital charge (return on investment, taxes, depreciation, insurance, etc.) of 25%, this would permit $36/m^2 of capital investment. Even if higher value H$_2$ or lower capital charges were allowed, total system costs, including photobioreactors, could not be above $50/m^2. Solar H$_2$ production is clearly a difficult goal, as pointed out, for one example only, by Myers (1977): "It is not my intent to demonstrate that biosolar conversion is impossible; only that it is difficult." However, future (year 2020) costs for photovoltaic-electrolyzer H$_2$, an already well-advanced technology, are projected optimistically at about $20/GJ (Block and Melody, 1992). Applied biophotolysis R&D can be justified if it could compete with such alternative processes.

For this, however, applied biohydrogen R&D must focus on the most likely processes and approaches and eliminate the more implausible. Perhaps the first such process investigated, using the heterocystous cyanobacteria, is also the first candidate for such elimination. This system is based on nitrogenase-driven hydrogen evolution. Nitrogenase is not only a very slow and unstable enzyme but also requires 4 ATP/H$_2$ produced, which would reduce any H$_2$ produced by roughly half compared to hydrogenase-based processes. Further, heterocysts themselves, the differentiated cells in which nitrogenase is located, have very high respiration rates and maintenance energy requirements (Turpin, 1985). Also problematic is that these produce O$_2$ and H$_2$ simultaneously, while still requiring CO$_2$ as an intermediate, making separation of these gases another cost issue. Thus, applied biophotolysis R&D using heterocystous algae, indeed using any nitrogen-fixing bacteria, should be given a low priority.

One alternative would be a direct biophotolysis system using reversible hydrogenase. Such a process would also produce O$_2$ and H$_2$ simultaneously, but could, at least theoretically, achieve higher efficiencies than heterocystous cyanobacteria. Indeed, sustained and highly efficient direct biophotolysis has been demonstrated with green algae in the laboratory, with an overall light conversion efficiency of more than 22% of visible radiation, corresponding to about 10% solar energy conversion (Greenbaum, 1988). However, these experiments were carried out under conditions of extremely low partial pressures of O$_2$ (and low light intensities). It is presently uncertain if it is possible to evolve H$_2$ in the presence of the high O$_2$ concentrations that would be present in any practical process. One argument against this approach is the lack of aerobic H$_2$ production reactions in nature. However, this is not too surprising, as such a reaction would likely serve no useful function. But where there is a need, nature seems to generally evolve a way. For example, the highly reducing CO$_2$ fixation reaction, carried out by all algae and higher plants, takes place at high dissolved O$_2$ tensions, even if with some difficulty. This suggests a possibility for O$_2$-resistant proton reduction, and genetic screening for algae with an O$_2$-stable hydrogenase is underway (Ghirardi et al., 1997). However, it must be pointed out that the issue is not hydrogenase stability to O$_2$, but, rather, the inhibition and inactivation of the hydrogenase reaction itself. Indeed, some hydrogenases are quite stable under O$_2$, and even labile hydrogenases can be stabilized by immobilization (Berenson et al., 1977). The mechanisms of O$_2$ inhibition and inactivation of the hydrogenase reaction and enzyme(s) must be a basic research priority to determine if a direct biophotolysis process is even possible before applied process development could be considered.

The alternative to a direct biophotolysis process is indirect biophotolysis processes in which CO_2 serves as the electron carrier between the O_2-producing water splitting and O_2-sensitive hydrogenase reactions. Thus, these reactions are temporally separated, or even spatially in different reactors. In such concepts, the algae would undergo a cycle of CO_2 fixation into storage carbohydrates (starch, glycogen) followed by its conversion to H_2 by endogenous metabolic processes, first in the dark and then in the light. The basic reactions are well-known, as is the storage of large amounts of carbohydrates by algae in response to nutrient limitations. However, dark anaerobic incubation of microalgae mainly results in little H_2 production, with mostly excreted fermentation products (Gibbs et al., 1986). Repeated cycles of algal CO_2 fixation into starch, followed by dark fermentation to produce some H_2 (and, mostly, organic acids), has been demonstrated (Miyamoto, 1994). What is not yet clear is how to manage such a process for efficient H_2 production. Indeed, when dark adapted (hydrogenase-induced) cells are exposed to light, photosynthetic O_2 production commences, inhibiting H_2 evolution. Thus, an important issue is how to delay the onset of PSII activity during light-driven H_2 production until all the substrate is exhausted. Such control is observed in nitrogen-fixing cyanobacteria, in particular non-heterocystous species that exhibit alternating cycles of O_2 evolution and anaerobic H_2 production (Weare and Benemann, 1974; see also Mitsui 1992). It should be possible to adapt similar metabolic controls to a hydrogenase-mediated indirect biophotolysis process.

A highly preliminary and conceptual economic analysis of an indirect biophotolysis process has been recently presented, which, based on many assumptions and caveats, arrived at a cost of \$10/GJ H_2 (Benemann, 1998). A one-million GJ/year scale plant was projected. The high carbohydrate (60% by dry weight) algal biomass was to be cultivated in large (approximately 10 ha) open ponds, at an efficiency of 10% solar energy conversion (CO_2 fixation into fermentable starches). The algal biomass was then concentrated (by settling) and transferred to a dark anaerobic fermenter (essentially a large covered pit). H_2, 4 mol/mol glucose, would be produced in this dark stage. The biomass would then be transferred to a tubular photobioreactor to complete the conversion of stored carbohydrates to H_2 in the light. The algal biomass would then be recycled back to the growth ponds. The costs of a 140-ha algal pond system were based on prior estimates for algal CO_2 fixation (Benemann, 1993). The gas handling/purification/storage costs were based on prior work for a conceptual solar chemical process (Copeland, 1991). The 14-ha tubular photobioreactor was simply assumed to cost \$140/m^2 (including engineering, contingencies, and other indirect costs) without further elaboration. The size of the photobioreactors assumed a single photon being required per H_2 produced. A total capital cost of \$43 million was estimated, including \$12 million for the 140 ha of ponds, almost \$19 million for 14 ha of photobioreactors, and the remainder for gas purification and handling and general supporting systems. Annual operating costs of \$1.5 million were projected. Assuming a capital charge of 25% per annum, total product costs would be \$10/GJ. The dominant cost in this analysis is the photobioreactors, whose costs were simply assumed. Tredici et al. (1998, these proceedings), present a cost estimate for a near-horizontal tubular photobioreactor that projects costs of only about \$50/m^2. With very favorable assumptions about capital charges (17% per annum), this may allow consideration of such reactors in single-stage biophotolysis processes.

Clearly, such rough and ready cost analyses and estimates can be critiqued, but they provide a focus for future analysis and for targeting R&D to key underlying assumptions, of which perhaps the most important is the solar conversion efficiency of photosynthesis. A 10% total solar efficiency has been generally assumed to be the maximum achievable (see Bolton, 1996, for a recent example). As already pointed out by the pioneers of microalgal solar energy conversion (e.g., Myers, 1957; Kok, 1973), and reminded occasionally since (Benemann, 1990), the key to achieving high efficiencies is to reduce the antenna sizes of the photosynthetic apparatus of the algae to match light saturation levels of photosynthesis with solar light intensities. Recent experiments by Melis et al. (1998, these proceedings), demonstrate the fundamental feasibility of this approach. Much more R&D is required to develop the highly efficient microalgal catalysts that can convert solar energy and water to hydrogen fuel.

6. CONCLUSIONS

This rather general and idiosyncratic overview suggests that biohydrogen R&D must evolve through a process of selection among contending concepts. It dismisses several major lines of past and current research as being of limited utility. Others, some not even discussed above, have not yet reached a level of basic science to allow their consideration in applied R&D. In this category, I place the biomimetic (artificial photosynthesis) systems, as there is no currently plausible mechanism that could substitute for natural photosynthesis (Bard and Fox, 1995). Also included are the cell-free (in vitro) systems, as even a relatively simple two-enzyme process, such as the conversion of glucose to gluconic acid and H_2 (Woodward et al., 1996), would be exceedingly difficult to scale to a practical process. That the complex photosynthetic apparatus, with scores of membrane-bound proteins (Melis, 1991), could be manufactured at low cost into highly stable solar conversion systems is beyond any foreseeable biotechnology. Biohydrogen processes must be based on self-reproducing microbial cultures that require only modest fabrication (e.g., inoculum production) efforts.

And, not all microbial systems are promising. Nitrogen-fixing cyanobacteria and photosynthetic bacteria are dismissed in the above analysis as being limited by the energetically inefficient nitrogenase enzyme. Even if the nitrogenase were replaced with hydrogenase, most of these systems still have significant limitations, compared to the alternatives, green algae, non-nitrogen-fixing cyanobacteria, and fermentative bacteria. Another approach using immobilized microbial systems, a focus of considerable R&D in biohydrogen over the years, also has limitations, that is, the costs of fabrication and lack of clear advantages over free cells, at least at this juncture. Such an almost wholesale dismissal of systems and concepts that have been the focus of most of the applied R&D in this field for over two decades will likely elicit comment. I therefore hasten to add that I do not mean to inhibit basic research into such topics. However, applied R&D has already advanced sufficiently to clearly demonstrate the relative limitations of most approaches. Indeed, a major objective of applied R&D must be to reduce options to the more promising, or at least still plausible, concepts.

I do not deny the argument that, for example, nitrogenase-based systems could be used to develop and demonstrate biophotolysis processes and photobioreactors. However, this should only be a stepping stone toward developing inherently more efficient and economical hydrogenase-based biophotolysis processes. Of course, photobioreactors are critical components in the economics of biophotolysis. Most current designs and concepts have significant engineering limitations (Weissman et al., 1987). For example, light diffusion devices are obviously unsuitable in low-cost applications. A major issue is if, indeed, photobioreactors systems could meet a cost goal of about $50/m^2, which would make single-stage biophotolysis processes feasible, at least in principle. Although, as discussed above, this may be possible, much more R&D is required. Low-cost photobioreactors could allow development of relatively small, possibly even "roof-top," systems, which, by providing fuel directly to the consumer, could be overall economically and practically more attractive than large-scale multi-hectare systems.

One emphasis in this overview has been the need for greater R&D into dark anaerobic processes, specifically fermentations of wastes and conversion CO. These have received little attention in the past, compared to photobiological processes, particularly considering their potential for developing practical processes in the near-term. Of course, these approaches have resource limitations, and in the longer-term biophotolysis processes will be required to help solve our environmental and sustainability problems. As pointed out above, two key R&D issues are solar conversion efficiencies and photobioreactor engineering and costs, in addition to H_2 evolution per se. And these issues transcend biophotolysis; they also apply to other microalgae processes, such as CO_2 mitigation or waste treatment.

Although the objective of applied biohydrogen R&D is a practical one, it should not be too constrained by cost goals or process specifics. Renewable energy sources may become more competitive if full externality costs were assessed to fossil fuels, accounting for their role in pollution and global warming. Neither should the R&D objectives be restricted to any one approach or system. Even with the narrowing of the applied R&D focus advocated herein, there are still many different concepts and a multitude of research needs and opportunities. Perhaps most important, applied biohydrogen R&D requires the support of basic R&D, directed into areas most likely to advance the scientific basis of this technology.

Technology development for mitigating climate change gases is moving to the forefront of the international agenda. This provides an opportunity to develop international cooperative R&D programs. Such an international effort in biohydrogen, moving beyond traditional information exchange activities to actual joint R&D activities, could help to greatly accelerate the development of this technology (Rossmeissl, 1997, these proceedings).

To summarize:

First ferment wastes in the dark we must do
Second convert carbon monoxide to h-two
Third, water split
with algae that sit
in the sun in ponds and tubes, it's true.

REFERENCES

Andrews, G.F., and Noah, K.S., 1995, Design of gas-treatment bioreactors, *Biotech. Prog.*, 11:498–509.
Augenstein, D.C., Benemann, J.R., and Hughes, E., 1994, Electricity from biogas, in *Proceedings of Second Interamerican Biomass Conference*, Reno, Nevada, October.
Bard, A.J., and Fox, M.A., 1995, Artificial photosynthesis: solar splitting of water to hydrogen and oxygen, *Accounts Chem. Res.*, 28:141–145.
Benemann, J.R., 1973, A model system for nitrogen fixation and hydrogen evolution by non-heterocystous bluegreen algae, *Fed. Proceed.*, 32:632.
Benemann, J.R., 1977, Hydrogen and methane production through microbial photosynthesis, in *Living Systems as Energy Converters*, Buvet, R. et al. (eds.), Elsevier/North-Holland Press, Amsterdam, pp. 285–298.
Benemann, J.R., 1990, The future of microalgae biotechnology, in *Algal Biotechnology*, Cresswell, R.C. et al. (eds.), Longman, London, pp. 317–337.
Benemann, J.R., 1993, Utilization of carbon dioxide from fossil fuel-burning power plants with biological systems, *Energy Cons. Mgmt.*, 34:999–1004.

Benemann, J.R., 1994, Feasibility analysis of photobiological hydrogen production, in *Hydrogen Energy Progress X, Proceedings of the 10th World Hydrogen Energy Conference*, Block, D.L., and Versiroglu, T.N. (eds.), Cocoa Beach, Florida, United States, June 20–24, 1994, pp. 931–940.

Benemann, J.R., 1996, Hydrogen biotechnology: progress and prospects, *Nature Biotechnology*, 14:1101–1103.

Benemann, J.R., 1998, Processes analysis and economics of biophotolysis: a preliminary assessment, Report to the International Energy Agency, Subtask B, Annex 10, Photoproduction of Hydrogen Program, in press.

Benemann, J.R., Berenson, J.A., Kaplan, N.O., and Kamen, M.D., 1973, Hydrogen evolution by a chloroplast-ferredoxin-hydrogenase system, in *Proceedings of the National Academy of Sciences (USA)*, 70:2317–2320.

Benemann, J.R., Miyamoto, K., and Hallenbeck, P.C., 1980, Bioengineering aspects of biophotolysis, *Enzyme and Microbial Technology*, 2:103–111.

Benemann, J.R., and Weare, N.M., 1974, Hydrogen evolution by nitrogen-fixing *Anabaena cylindrica* cultures, *Science*, 184:1917–175.

Benemann, J.R., and Zaborsky, O.R., 1996, Biohydrogen: market potential, in *Proceedings of the Annual Meeting of the National Hydrogen Association*, Washington D.C., April 1996.

Berenson, J.A., and Benemann, J.R., 1977, Immobilization of hydrogenase and ferrodoxins on glass beads, *FEBS Letters*, 76:105–107.

Block, D.L., and Melody, I., 1992, Efficiency and cost goals for photoenhanced hydrogen production processes, *Int. J. Hydrogen Energy*, 17:853–861.

Bolton, J.R., 1996, Solar photoproduction of hydrogen, *Report to the International Energy Agency*, under Agreement on the Production and Utilization of Hydrogen, IEA/H2/TR-96, September 1996.

Copeland, R.J., 1991, Low cost hydrogen systems, presented at U.S. Department of Energy/Solar Energy Research Institute Hydrogen Program Review, Washington, D.C., January 23–24.

Gaffron, H., and Rubin, J., 1942, Fermentative and photochemical production of hydrogen in algae, *J. Gen. Physiol.*, 26:219–240.

Gest, H., and Kamen, M.J., 1949, Photoproduction of molecular hydrogen by *Rhodospirillum rubrum*, *Science*, 109:558–559.

Ghirardi, M.L., Togasaki, R.K., and Seibert, M., 1997, Oxygen sensitivity of algal hydrogen production, *App. Biochem. Biotech.*, 63:141–151.

Gibbs, M., Hollaender, A., Kok, B., Krampitz, L.O., and San Pietro, A., 1973, in *Proceedings of the Workshop on Bio-Solar Hydrogen Conversion*, September 5–6, Bethesda Maryland, United States.

Gibbs, M., Gfeller, R.P., and Chen, C., 1986, Fermentative metabolism of *Chalmydomonas reinhardii*, *Plant Physiol.*, 82:160–166.

Grasso, D., Strevett, K., Fisher, R., 1995, Uncoupling mass transfer limitations of gaseous substrates in microbial systems, *Chemical Eng. J.*, 59:195–204.

Greenbaum, E., 1980, Simultaneous photoproduction of hydrogen and oxygen by photosynthesis, *Biotech. Bioeng. Symp.*, 10:1–13.

Greenbaum, E., 1988, Energetic efficiency of hydrogen photoevolution by algal water splitting, *Biophys. J.*, 54:365–368.

Hallenbeck, P.C., and Benemann, J.R., 1979, Hydrogen from algae, in *Photosynthesis in Relation to Model Systems*, Barber, J. (ed.), Elsevier/North-Holland Biomedical Press.

Happe, R.P, Roseboom, W., Pierik, A.J. and Bagley, K.A., 1997, Biological activation of hydrogen, *Nature*, 385:126.

Healy, F.P., 1970, The mechanism of hydrogen evolution by *Chlamydomonas moewusii*, *Plant Physiol.*, 45:153–159.

Heijnen, S.J., 1995, Thermodynamics of microbial growth and its implications for process design, *Trends in Biotechnology*, 12:483–492.

Hollaender, A., Monty, K.J., Paerlstein, R.M., Schidt-Bleek, F., Snyder, W.T., and Volkin, E., 1972, An inquiry into biological energy conversion, Workshop Report (October 12–14, 1972), Gatlinburg, Tennessee NSF-RANN, University of Knoxville, December 1972.

Jackson, D.D., and Ellms, J.W., 1986, On odors and tastes of surface waters with special reference to *Anabaena*, a microscopial organsim found in certain water supplies of Massachusetts, *Report to the Massachusetts State Board Health*, pp. 410–420.

Kerby, R.L., Hong, S.S., Ensign, S.A., Copoc, L.J., Ludden, P.W., and Roberts, G.P., 1992, Genetic and physiologiclal characterization of the *Rhodospirillum rubrum* carbon monoxide dehydrogenase system, *J. Bactriol.*, 174:5284–5294.

Klasson, K.T., Gupta, A., Claussen, E.C., and Gaddy, J.L., 1993, Evaluation of mass-transfer and kinetic parameters for *Rhodospirillum rubrum* in a continuous stirred tank reactor, *App. Biochem. Biotech.*, 39/40:549–557.

Kok, B., 1973, Photosynthesis, in *Proceedings of the Workshop on Bio-Solar Hydrogen Conversion*, Gibbs, M. et al. (eds.), September 5–6, Bethesda, Maryland, pp. 22–30.

Krampitz, L.O., 1977, Potentials of hydrogen production through biophotolysis, in *Symposium Papers: Clean Fuels from Biomass and Wastes*, January 25–28, Orlando, Florida, Institute of Gas Technology, Chicago, Illinois, pp. 141–151.

Lien, S. and San Pietro, A., 1975, An inquiry into biophotolysis of water to produce hydrogen, Report to the National Science Foundation.

Mann, M., 1996, Technical and economic assessment of renewable hydrogen production, presented at the Annual Review of the U.S. Department of Energy Hydrogen Program, Miami, Florida, April 29–May 3.

Markov, S.A., Weaver, R., and Seibert, M., 1996, Hydrogen production using microorganisms in hollow-fiber bioreactors, in *Proceedings of Hydrogen 96*, Suttgart, Germany.

McTavish, H., Sayavedra-Soto, L.A., and Arp, D.J., 1995, Substitution of *Azotobacter vinelandii* hydrogenase small subunit cysteines by serines can create insensitivity to inhibition by O_2 and preferentially damages H_2 oxidation over H_2 evolution, *J. Bact.*, 177:3960–3964.

Melis, A., 1991, Dynamics of photosynthetic membrane composition and function, *Biochim Biophys. Acta* (Reviews on Bioenergetics), 1058:87–106.

Mitsui, A., 1974, The utilization of solar energy for hydrogen production by cell free system of photosynthetic organisms, in *Hydrogen Energy, Part A*, Veziroglu, T.N. (ed.), pp. 309–316.

Mitsui, A., 1992, Biological hydrogen photoproduction (task B), in *Proceedings of 1992 U.S. Department of Energy/National Renewable Energy Laboratory Hydrogen Program Review*, May 6–7, Honolulu, Hawaii, NREL/CP-450-4972, pp. 129–156.

Miyamoto, K., 1994, Hydrogen production by photosynthetic bacteria and microalgae, in *Recombinant Microbes for Industrial and Agricultural Applications*, Murooka, Y., and Imanaka, T. (eds.), Marcel Dekker, New York, pp. 771–786.

Myers, J., 1957, Algal culture, *Encyclopedia of Chemical Technology*, Interscience, New York, pp. 649–680.

Myers, J., 1977, Bioengineering Approaches and Constraints, in *Biological Solar Energy Conversion*, Mitsui, A. et al. (eds.), Academic Press, New York, pp. 449–454.

Pauss, A., Andre, G., Perrier, M., and Guiot, S.R., 1990, Liquid-to-gas mass transfer in anaerobic processes: inevitable transfer limitations of methane and hydrogen in the biomethanation processes, *App. Env. Microbiol.*, 56:1636–1644.

Sasikala, K., Ramana, C.V., Rao, P.R., and Kovacs, K.L., 1993, Anoxygenic phototrophic bacteria: physiology and advances in hydrogen technology, *Adv. Appl. Microbio.*, 38:211–295.

Schink, B., 1997, Energetics of syntrophic cooperation in methanogenic degradation, *Microbiol. and Molecular Biol. Reviews*, 61:262–280.

Spruit, C.J.P., 1958, Simultaneous photoproduction of hydrogen and oxygen by *Chlorella, Mededel. Landbouwhogeschool Wageningen*, 58:1–17.

Thauer, R., 1976, Limitation of microbial hydrogen formation via fermentation, in *Microbial Energy Conversion*, Schlegel, H.G., and Barnea, J. (eds.), Erich Goltze, Gottingen, pp. 201–294.

Turpin, D.H., Layzell, D.B., and Elrifi, I.R., 1985, Modeling the carbon economy of *Anabaena flos-aquae*, *Plant Physiol.*, 78:746–752.

Ueno, Y., Morimoto, M., Ootsuka, S., Kawai, T., and Satou, S., 1995, Process for the production of hydrogen by microorganisms and for wastewater treatment, U.S. Patent, 5,464,539 (November 7, 1995).

Weare, N.M., and Benemann, J.R., 1974, Nitrogenase activity and photosynthesis by *Plectonema boryanum* 594, *J. Bacteriol.*, 119:258–268.

Weaver, P.F., Lien, S., and Seibert, M., 1980, Photobiological production of hydrogen, *Solar Energy*, 24:3–45.

Weaver, P., Maness, P.C., Fank, A., Li, S., and Toon, S., 1995, Biological water-gas shift activity, in *Proceedings of the Annual Review Meeting of the U.S. Department of Energy Office of Utility Technologies Hydrogen Program Review*, Miami, Florida.

Weetall, H.H., 1979, Biophotolysis of Water, U.S. Patent 4,148,690.

Weissman, J.C., and Benemann, J.R., 1977, Hydrogen production by nitrogen-fixing cultures of *Anabaena cylindrica*, *Appl. Env. Microbiol.*, 33:123–131.

Weissman, J.C., Goebel, R.P., and Benemann, J.R., 1988, Photobioreactor design: comparison of open ponds and tubular reactors, *Bioeng. Biotech.*, 31:336–344.

Woodward, J., Mattingly, S.M., Danson, M., Hough, D., Ward, N., and Adams, M., 1996, In vitro hydrogen production by glucose dehydrogenase and gydrogenase, *Nature Biotechnology*.

Zurrer, H., and Bachofen, R., 1979, Hydrogen production by the photosynthetic bacterium *Rhodospirillum rubrum*, *App. Envron. Microbiol.*, 37:789–793.

4

MARINE GENOMES

Tadashi Matsunaga and Haruko Takeyama

Department of Biotechnology
Tokyo University of Agriculture and Technology
Tokyo, Japan

1. SUMMARY

The diversified marine environment harbors a wide variety of organisms having many unique genes. From unique marine environments, photosynthetic microorganisms that are the largest primary producers of biomass are attracting attention as new gene resources. We report the results of screening of useful products from marine photosynthetic microorganisms, as well as the use of those organisms as a gene resource. Productivity of useful products has been improved by metabolic engineering. The survival strategies of novel marine cyanobacterial strains to environmental stress at the molecular level are described. Furthermore, a genome analysis of fish is introduced.

2. INTRODUCTION

Development of amino acid sequencing and protein electrophoresis techniques during the early 1960s has revealed that extensive protein polymorphism exists in natural populations. Subsequent introduction of new biochemical techniques, such as recombinant DNA and nucleotide sequence analysis, has tremendously accelerated the analysis of gene structure, organization, and function. Genome analysis by sequencing all DNA in the target organism has been carried out for several organisms (Fleischmann et al., 1995; Fraser et al., 1995; Bult et al., 1996; Kaneko et al., 1996; Bussey, 1997; Adams et al., 1995). These efforts will reveal not only the function of individual genes, but also the functional networks among the genes.

The marine environment is remarkably diverse. The marine environment harbors a wide variety of organisms that have acquired novel metabolic functions and corresponding genes through evolutionary adaptation. Hence, there may be as many unique genes as there are species of organisms. Marine organisms are potentially a treasure of gene resources. One of the largest gene resources in the marine environment consists of photosynthetic microorganisms. These are the largest primary producers of biomass. Fish is the most important food source for humans. With the development of marine biotechnology, photosynthetic microorganisms from unique and extreme marine environments are attracting attention as new gene resources. They have evolved to survive unique stressful environments. Elucidation of these survival mechanisms may lead to the isolation of novel products and/or genes.

In this paper, we report the results of screening of useful products from marine photosynthetic microorganisms and manipulating their production by metabolic engineering. We describe the survival strategy of novel marine cyanobacterial strains against environmental stress at the molecular level. Furthermore, a genome analysis in fish is introduced.

3. USEFUL PRODUCTS FROM MARINE PHOTOSYNTHETIC MICROORGANISMS

Most of the marine photosynthetic microorganisms maintained in our laboratory are isolates from Asian coastal areas and Micronesia. They have been tested for the production of useful chemicals and energy (Table 1).

3.1. Plant Growth Regulators

Efficient plant regeneration is a vital objective in plant tissue culture. Somatic embryos can be produced for various plant species using in vitro techniques. However, it remains difficult to develop somatic embryos into plantlets at a high frequency for most species.

Twenty-five strains of marine cyanobacteria were subjected to screening for their ability to promote carrot somatic embryogenesis. Extracts of *Synechococcus* sp. NKBG042902 promoted plantlet formation in carrot cells (Wake et al., 1991). A high molecular weight, non-dialyzing fraction from the extract promoted the formation of plantlets. Plantlet-formation frequency was 60 and 28%, with and without the non-dialysate, respectively (Wake et al., 1992a). This non-dialysate was applied to improving an artificial seed system (Wake et al., 1992b). A high frequency of seed germination was obtained in the presence of the non-dialysate (100 mg/L). The active substances enhancing induction, maturation, and germination appear to be lipopolysaccharides and/or peptides.

3.2. Fatty Acid Production

Recent pharmaceutical interest in unsaturated fatty acids has triggered a search for sources of these valuable compounds. Monounsaturated fatty acids such as palmitoleic acid (C16:1), undecylenic acid (C11:1), and tridecenic acid (C13:1) have the potential for preventing several diseases. Two marine cyanobacteria, *Phormidium* sp. NKBG 041105 and *Oscillatoria* sp. NKBG 091600, showed a high content of *cis*-palmitoleic acid (54.5 and 54.4% of total fatty acid, respectively) (Matsunaga et al., 1995). Furthermore, the content of *cis*-palmitoleic acid was found to remain constant with varying temperature.

Spirulina is well known to have a high content of γ-linoleic acid (C18:3, GLA). However, the major form of linoleic acid found in microalgae is α-linoleic acid. The marine microalga *Chlorella* sp. NKG 042401 was found to contain 10% GLA. This was mainly found in the form of galactolipid (Hirano et al., 1990; Miura et al., 1993a).

Several eukaryotic microalgae are known to produce highly unsaturated fatty acids such as eicosapentaenoic acid (EPA, C20:5) and docosahexaenoic acid (DHA, C22:6), which

Table 1. Production of useful materials by marine photosynthetic microorganisms

Product	Strain	Amount (mg or unit* /g dry wt.)	Ref.
Coccolith	Emiliania huxleyi Pleurochrysis carterae	740	Takano et al. (1993,1994,1995a)
γ-Linolenic acid	Chlorella.sp NKG042401	9.5	Hirano et al. (1990) Miura et al. (1993b)
Palmitoleic acid	Phormidium sp. NKBG 041105	47	Matsunaga et al. (1995)
Docosahexaenoic acid (DHA)	Isochrysis galbana UTEX LB 2307	15.7	Burgess et al. (1993)
Eicosapentaenoic acid (EPA)	Synechococcus sp, NKBG042902	0.64	Takeyama et al. (1997)
β-Carotene	Rhodovulum suifidophilum NKPB 160471R	0.02	Takeyama et al. (1997b)
Polysaccharide	Aphanocapsa halophytia	45	Sudo et al. (1994)
Glutamate	Synechococcus sp. NKBG040607	15.4	Matsunaga et al.(1988,1991)
Phycocyanin	Synechococcus sp. NKBG042902	150	Takano et al. (1995b)
UV-A absorbing biopteringlucoside	Oscillatoria sp. NKBG 091600	0.2	Matsunaga et al. (1993) Wachi et al. (1995a)
Antimicrobial compound	Chlorella sp. NKG 0111	-	Miura et al.(1993b)
	Chromatium purpuratum NKPB 031704	-	Burgess et al. (1991)
Plant growth regulator	Synechococcus sp. NKBG042902	-	Wake et al. (1991, 1992a, 1992b)
Lactic acid bacteria growth regulator	Synechococcus sp. NKBG040607	-	
Tyrosinase inhibitor	Synechococcus sp. NKBG15041c	107*	Wachi et al. (1995a, 1995b)
SOD	Synechococcus sp. NKBG042902	-	Wachi et al. (1995c)
Hydrogen	Cromatium sp. NKPB 0021	0.96 μ mol/mg(dry wt.)/h	Yamada et al. (1996)
	Rhodovulum suifidophilum NKPB 160471R	9.4 μ mol/mg(dry wt.)/h	

are valuable dietary components (Burgess et al., 1993; Takeyama et al., 1996a). They may be involved in preventing several human diseases (Terano et al., 1984; Abbey et al., 1990). However, wild-type cyanobacteria do not possess the biosynthetic pathway for these fatty acids. The EPA synthesis gene-cluster (ca. 38 kbp) from the marine bacterium *Shewanella putrefaciens* SCRC-2738 was cloned into the marine cyanobacterium *Synechococcus* sp. using a broad-host cosmid vector, pJRD215 (10.2 kbp, Sm^r Km^r). The constructed plasmid (pJRDEPA, ca. 48 kbp) could be transferred using transconjugation to the cyanobacterial host at a frequency of 2.2×10^{-7}. Cyanobacterial transconjugants grown at 29 °C produced only 0.12 mg EPA/g dry weight cell, whereas those grown at 23 °C produced 0.56 mg/g dry weight cell. Furthermore, the content of EPA in cells grown at 23 °C increased by 0.64 mg/g dry weight cell after pre-incubation for one day at 17 °C (Takeyama et al., 1997a).

3.3. β-Carotene Production

The antioxidant property of β-carotene has received attention due to its application in clinical and nutritional fields. It may function in preventing several diseases possibly caused by oxidative damage. Several microorganisms produce these antioxidants and are expected to produce a more stable supply than chemical synthesis or extraction from higher plants. β-carotene is produced by photosynthetic microorganisms to protect chlorophyll and the thylakoid membrane against photo-oxidative damage (Takeyama et al., 1997b).

The genes encoding enzymes involved in carotenoid biosynthesis, *crtI* and *crtY*, were isolated from *Erythrobacter longus* OCh101 (Matsumura et al., 1997) and cloned into a marine photosynthetic bacterium, *Rv. sulfidophilum* NKBG190471R, producing sphcroidene as a major carotenoid, not β-carotene. β-carotene synthesis (20 µg/g dry cells) was obtained by re-routing host carotenoid bio-synthesis (Takeyama et al., 1996b). Furthermore, higher β-carotene production was obtained using mutants deficient for carotenoid biosynthesis through chemical mutagenesis.

4. MOLECULAR ADAPTATION TO STRESS ENVIRONMENTS BY MARINE CYANOBACTERIA

4.1. Analysis of a UV-A Stress Responsive Gene in Marine Cyanobacteria

Ultraviolet light (320–400 nm, UV-A and B) in solar radiation penetrates the earth's surface. Shallow-water, reef-building corals are normally exposed to high levels of near-UV light. Some microalgae associated with these corals were found to produce UV-absorbing compounds such as mycosporine-like amino acids (Shibata, 1969; Dunlap et al., 1986).

Among 89 isolates of marine cyanobacteria, *Oscillatoria* sp. NKBG 091600 was found to have resistance to UV-A irradiation and to produce high levels of a UV-A absorbing compound, biopterin glucoside (BG, Figure 1A) in response to UV-A irradiation (Matsunaga et al., 1993; Wachi et al., 1995a). Intracellular levels of BG increased after UV-A irradiation of approximately 8 h (Figure 1B). It increased with increasing intensity of UV-A irradiation. These results suggest that BG might work as a UV-A absorbing substance for protecting cells.

Intracellular protein analysis by SDS-PAGE revealed the induction of a specific protein under UV-A irradiation. A 60-kDa protein was simultaneously induced under UV-A ir-

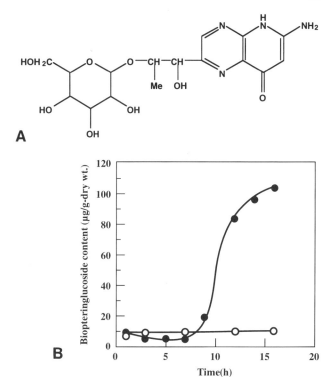

Figure 1. (A) Structure of biopterin glucoside. (B) Effect of UV-A light irradiation (900 μW/cm^2) on intracellular production of biopterin glucoside with time by *Oscillatoria* sp. NKBG091600. Dark circles = UV plus fluorescent white light; open circles = fluorescent white light only (200 μW/cm^2).

radiation alongside BG. Amino acid sequence analysis indicated that this protein was the heat shock protein GroEL. It is well known that GroEL (bacterial chaperonin) works to reactivate denatured enzymes. Induction of GroEL by heat shock was reported to be observed in 60 min in cyanobacteria (Webb et al., 1990). The mechanism for regulation of the *groEL* promoter by UV-A and heat shock may be different in this strain. GroEL may have a role in protecting enzymes required for BG synthesis from UV-A damage.

4.2. Analysis of Salinity Stress Responsive Gene in Marine Cyanobacteria

Many cyanobacteria harboring endogenous plasmids have been reported. Some functional genes were found to be coded on fresh water cyanobacterial plasmids (Muro-Pastor et al., 1994), and replication regions of several fresh water cyanobacterial plasmids have been determined (Van der Plas et al., 1992; Yang and McFadden, 1993). However, most cyanobacterial plasmids are cryptic.

The marine cyanobacterium *Synechococcus* sp. NKBG 042902 harboring the plasmid pSY10 grows well in the presence of 0–5% NaCl. The copy number of the marine plasmid pSY10 was found to be controlled by changing the NaCl concentration of the culture medium (Takeyama et al., 1991). The pSY10 was maintained at a high copy number under sea water conditions, and at a low copy number under fresh water conditions. The RepA protein sequence was found on pSY10 and RepA protein revealed to participate in pSY10 replication (Kawaguchi et al., 1994). The mechanism of pSY10 replication involving the RepA protein and regulation of *repA* gene expression has been elucidated under different salinities. A novel replication mechanism has been proposed for this marine plasmid in which a stress-respon-

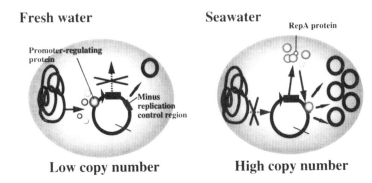

Figure 2. The novel mechanism for replication of marine plasmid pSY10 in *Synechococcus sp. NKBG042902*.

sive protein encoded by the chromosome is expressed under fresh water conditions in *Synechococcus* sp. NKBG042902. This protein depresses expression of RepA protein resulting in a low copy number of pSY10. This is a novel plasmid replication mechanism involving chromosome control (Figure 2). The function of this plasmid remains unclear. Further investigation of the replication mechanism controlled by chromosomally encoded proteins may allow further understanding of the role of this plasmid in the host cell under salt stress.

5. SCREENING OF HYDROGEN-PRODUCING MARINE PHOTOSYNTHETIC BACTERIA USING THE *nifH* GENE

Nitrogen fixation mediated by nitrogenase is widely observed among eubacteria and archaebacteria (Young, 1992). The phylogeny of one component gene of nitrogenase, the dinitrogenase reductase gene *nifH*, is consistent with that of 16S rRNA phylogenies. Rapid detection of nitrogen-fixing bacteria in sample of sea water is easily carried out by using *nifH* gene amplifications.

A novel phototrophic purple non-sulfur bacterium was isolated from sea water sediment using the *nifH* gene amplification. This strain (NKPB 160471) was identified as *Rhodovulum sulfidophilum* by phylogenetic analysis of the *nifH* gene sequence. A 16S rDNA sequence analysis further identified this bacterium as *R. sulfidophilum*. This strain produced hydrogen at 157 nmol/mg dry cell at a light intensity of 1770 W/m^2 (Yamada et al., 1996).

6. GENOME ANALYSIS IN FISH

A sex-linked DNA fragment of medaka (*Oryzias latipes*) was previously reported (Matsuda et al., 1997). We have investigated the nucleotide sequences of the region adjacent to this DNA fragment and found two tRNA-related sequences. The 3' and 5' ends of these tRNA-related sequences were found to be flanked by direct and inverted repeats. Several domains in these fragments were observed to have homology with short interspersed repetitive elements (SINEs) of salmon, squid, and proviral MPMV (Simian Mason-Pfizer D-type retrovirus). The tRNA-related sequences observed are proposed to be transposable elements (Tes). Insertion may occur with or without an RNA intermediate. It is suggested that the repetitive elements in tRNA-related sequences from this species were transposed directly through DNA copies from distant species (rat and *D. melanogaster*) by horizontal transfer.

REFERENCES

Abbey, M., Clifton, P., Kestin, M., Belling, B., and Nestel, P., 1990, Effect of fish oil on lipoproteins, lecithin, cholesterol acyltransferase, and lipid transfer protein activity in humans, *Arteriosclerosis*, 10:85–94.

Adams, M.D., 1995, Initial assessment of human gene diversity and expression patterns based upon 83 million nucleotides of cDNA sequence, *Nature*, 377:3–17.

Bult, C.J., et al., 1996, Complete genome sequence of the methanogenic archeon, *Methanococcus jannaschii*, *Science*, 273:1058–1073.

Bussey, H., 1997, The nucleotide sequence of *Saccharomyces cerevisiae* chromosome XVI, *Nature*, 387(6632):103–105.

Burgess, J.G., Iwamoto, K., Miura, Y., Takano, H., and Matsunaga, T., 1993, An optical fiber photobioreactor for enhanced production of the marine unicellular alga *Isochrysis* aff. *galbana* T-Iso (UTEX LB 2307) rich in docosahexaenoic acid, *Appl. Microbiol. Biotechnol.*, 39:456–459.

Burgess, J.G., Miyashita, H., Sudo, H., and Matsunaga, T., 1991, Antibiotic production by the marine photosynthetic bacterium *Chromatium purpuratum* NKPB 031704: localization of activity to the chromatophores, *FEME Microbiol. Lett.*, 84:301–306.

Dunlap, W.C., and Chalker, B.E., 1986, Identification and quantitation of near-UV absorbing compounds (S-320) in a hermatypic scleractinian, *Coral Reefs*, 5:155–159.

Fleischmann, R.D., et al., 1995, Whole-genome randam sequencing and assembly of *Haemophilus influenza* RD, *Science*, 269:496–512.

Fraser, C.M., et al., 1995, The minimal gene complement of *Mycoplasma genitalium*, *Science*, 270:397–403.

Hirano, M., Mori, H., Miura, Y., Matsunaga, N., Nakamura, N., and Matsunaga, T., 1990, Gamma linolenic acid production by microalgae, *Appl. Biochem. Biotechnol.*, 24/25:183–191.

Kaneko, T., et al., 1996, Sequence analysis of the genome of the unicellular cyanobacterium *Synechocystis* sp. strain PCC6803. II; sequence determination of the entire genome and assignment of potential protein-coding regions, *DNA Res.*, 3:109–136.

Kawaguchi, R., Nagaoka, T., Burgess, J.G., Takeyama, H., and Matsunaga, T., 1994, Sequence of a 2.6-kb cryptic plasmid from a marine cyanobacterium *Synechococcus* sp., *Plasmid*, 32:245–253.

Matsuda, N., Kusama, T., Oshiro, T., Kurihara, Y., Hamaguchi, S., and Sakaizumi, M., 1997, Isolation of a sex chromosome-specific DNA sequence in the medaka, *Oryzias latipes*, *Gen. Genet. Syst.*, 72:263–268.

Matsumura, H., Takeyama, H., Kusakabe, E., Burgess, J.B., and Matsunaga, T., 1997, Cloning, sequencing and expressing the carotenoid biosynthesis genes, lycopen cyclase and phytoene desaturase, from the aerobic photosynthetic bacterium *Erythrobacter longus* sp. strain Och101 in *Escherichia coli*, *Gene*, 189:169–174.

Matsunaga, T., Burgess, J.G., Yamada, N., Komatsu, K., Yoshida, S., and Wachi, Y., 1993, An ultraviolet (UV-A) absorbing biopterin glucoside from the marine planktonic cyanobacterium *Oscillatoria* sp., *Appl. Microbiol. Biotechnol.*, 39:250–253.

Matsunaga, T., Nakamura, N., Tsuzaki, N., and Takeda, N., 1988, Selective production of glutamate by immobilized blue-green alga *Synechococcus* sp., *Appl. Microbiol. Biotechnol.*, 28:373–376.

Matsunaga, T., Takeyama, H., Miura, Y., Yamazaki, T., Furuya, H., and Sode, K., 1995, Screening of marine cyanobacteria for high palmitoleic acid production, *FEMS Microbiol. Lett.*, 133:137–141.

Matsunaga, T., Takeyama, H., Sudo, H., Oyama, N., Ariura, S., Takano, H., Hirano, M., Burgess, J.G., Sode, K., and Nakamura, N., 1991, Glutamate production from CO_2 by marine cyanobacterium *Synechococcus* sp. using a novel biosolar reactor employing light-diffusing optical fibers, *Appl. Biochem. Biotechnol.*, 28/29:157–167.

Miura, Y., Sode, K., Narasaki, Y., and Matsunaga, T., 1993b, Light-induced antimicrobial activity of extracts from marine *Chlorella*, *J. Mar. Biotechnol.*, 1:143–146.

Miura, Y., Sode, K., Narasaki, Y., and Matsunaga, T., 1993a, Production of g-linolenic acid from the marine green alga *Chlorella* sp. NKG 042401, *FEMS Microbiol. Lett.*, 107:163–168.

Muro-Pastor, A.M., Kuritz, T., Flores, E., Herrero, A., and Wolk, C.P., 1994, Transfer of a genetic marker from a megaplasmid of *Anabaena* sp. strain PCC 7120 to a megaplasmid of a different *Anabaena* strain, *J. Bacteriol.*, 176:1093–1098.

Shibata, K., 1969, Pigments and a UV-absorbing substance in corals and a blue green alga living in the Great Barrier Reef, *Plant Cell Physiol.*, 10:325–335.

Sudo, H., Burgess, J.B., Takemasa, H., Nakamura, N., and Matsunaga, T., 1994, Sulfated exopolysaccharide production by the halophilic cyanobacterium *Aphanocapsa halophytia*, *Curr. Microbiol.*, 30:1–4.

Takano, H., Arai, T., Hirano, M., and Matsunaga, T., 1995b, Effect of intensity and quality of light on phycocyanin production by a marine cyanobacterium *Synechococcus* sp. NKBG042902, *Appl. Microbiol. Biotechnol.*, 43:1014–1018.

Takano, H., Jeon, J., Burgess, J.G., Manabe, E., Izumi, Y., Okazaki, M., and Matsunaga, T., 1994, Continuous production of extracellular ultrafine calcite particles by the marine coccolithophorid alga *Pleurochrysis carterae, Appl. Microbiol. Biotechnol.*, 40:946–950.

Takano, H., Manabe, E., Hirano, M., Okazaki, M., Burgess, J.G., Nakamura, N., and Matsunaga, T., 1993, Development of a rapid isolation procedure for coccolith ultrafine particles produced by coccolithophorid algae, *Appl. Biochem. Biotechnol.*, 39/40:239–247.

Takano, H., Takei, R., Manabe, E., Burgess, J.G., Hirano, M., and Matsunaga, T., 1995a, Increased coccolith production by *Emiliania huxleyi* cultures enriched with dissolved inorganic carbon, *Appl. Microbiol. Biotechnol.*, 43:460–465.

Takeyama, H., Burgess, J.G., Sode, K., and Matsunaga, T., 1991, Salinity dependent copy number increase of a marine cyanobacterial endogenous plasmid, *FEMS Microbiol. Lett.*, 90:95–98.

Takeyama, H., Iwamoto, K., Hata, S., Takano, H., and Matsunaga, T., 1996a, DHA enrichment of rotifers; a simple two-step culture using the unicellular algae *Chlorella regularis* and *Isochrysis galbana, J. Mar. Biotechnol.*, 3:244–247.

Takeyama, H., Sunarjo, J., Yamada, A., Matsumura, H., Kusakabe, E., and Matsunaga, T., 1996b, β-carotene production in a novel hydrogen-producing marine photosynthetic bacterium *Rhodovulum sulfidophilum* expressing the *Erythrobacter longus* OCh101 *crtI* and *crtY* genes, *J. Mar. Biotechnol.*, 4:224–229.

Takeyama, H., Kanamaru, A., Yoshino, Y., Kakuta, H., Kawamura, Y., and Matsunaga, T., 1997b, Production of antioxidant vitamins, β-carotene, vitamin C, and vitamin E by two-step culture of *Euglene gracilis* Z, *Biotechnol. Bioeng.*, 53:185–190,

Takeyama, H., Takeda, D., Yazawa, K., Yamada, A., and Matsunaga, T., 1997a, Expression of the eicosapentaenoic acid synthesis gene cluster from *Shewanella* sp. in a *Synechococcus* sp., *Microbiology*, 143:2725–2731.

Terano, T., Salmon, J.A., and Moncada, S., 1984, Effect of orally administered eicosapentaenoic acid (EPA) on the formation of leukotriene B_4 and leukotrien B_5 by rat leukocytes, *Prostaglandins*, 27:217–232.

van der Plas, J., Oosterhoff-Teertstra, R., Borrias, M., and Weisbeek, P., 1992, Identification of replication and stability functions in the complete nucleotide sequence of plasmid pUH24 from the cyanobacterium *Synechococcus* sp. PCC7942, *Mol. Microbiol.*, 6:653–664.

Wachi, Y., Burgess, J.B., Iwamoto, K., Yamada, N., Nakamura, N., and Matsunaga, T., 1995a, Effect of ultraviolet-A (UV-A) light on growth, photosynthetic activity, and production of biopterin glucoside by the marine UV-A resistant cyanobacterium *Oscillatoria* sp, *Biochim. Biophys. Acta*, 1244:165–168.

Wachi, Y., Burgess, J.B., Takahashi, J., Matsunaga, T., and Nakamura, N., 1995d, Production of superoxide dismutase by marine cyanobacteria, *J. Mar. Biotechnol.*, 3:258–261.

Wachi, Y., Burgess, J.G., Takahashi, J., Nakamura, N., and Matsunaga, T., 1995b, Tyrosinase inhibition by the water-soluble fraction of marine microalgae, *J. Mar. Biotechnol.*, 2:210–213.

Wachi, Y., Sode, K., Horikoshi, K., Takeyama, H., and Matsunaga, T., 1995c, Screening of melanin biosynthesis inhibitors from marine microalgae using *Streptomyces bikiniensis* bioassay, *Biotechnol. Techniques*, 9:633–636.

Wake, H., Akasaka, A., Umetsu, H., Ozeki, Y., Shimomura, K., and Matsunaga, T., 1992b, Enhanced germination of artificial seeds by marine cyanobacterial extract, *Appl. Microbiol. Biotechnol.*, 36:684–688.

Wake, H., Akasaka, A., Umetsu, H., Ozeki, Y., Shimomura, K., and Matsunaga, T., 1992a, Promotion of plantlet formation from somatic embryos of carrot treated with a high molecular weight extract from a marine cyanobacterium, *Plant Cell Reports*, 11:62–65.

Wake, H., Umetsu, H., Shimomura, K., and Matsunaga, T., 1991, Extracts of marine cyanobacteria stimulated somatic embryogenesis of *Daucus carota* L, *Plant Cell Reports*, 9:655–658.

Webb, R., Reddy, K.J., and Shweman, L.A., 1990, Regulation and sequence of the *Synechococcus* sp. Strain PCC 7942 *groEL* operon, encoding a cyanobacterial chaperonin.

Yamada, A., Takano, H., Burgess, J.G., and Matsunaga T., 1996, Enhanced hydrogen production by a marine photosynthetic bacterium, *Rhodobacter marinus*, immobilized onto light-diffusing optical fibers, *J. Mar. Biotechnol.*, 4:23–27.

Yang, X., and McFadden, B.A., 1993, A small plasmid, pCA2.4, from the cyanobacterium *Synechocystis* sp. strain PCC 6803 encodes a Rep protein and replicates by a rolling circle mechanism, *J. Bacteriol.*, 175:3981–3991.

Young, J.P.W., 1992, Phylogenetic classification of nitrogen-fixing organisms, in *Biological Nitrogen Fixation*, Stacey, G. et al. (eds.), Chapman and Gall Press, New York, pp. 43–86.

5

COMMENCEMENT CHALLENGE

Patrick Takahashi

Hawaii Natural Energy Institute
University of Hawaii
2540 Dole Street, Holmes Hall 246
Honolulu, Hawaii 96822

It is a curious thing that in the United States, at the completion of an educational career, the graduation ceremonies are referred to as a "commencement." It is, thus, entirely appropriate that as the final speaker after four days of intense information exchange, I provide a commencement challenge, for what we have accomplished during the past week is to establish the foundation for the future. We are not at an end, but in the process of taking the next step—we are continuing what will be a long journey that will hopefully lead to cost-competitive biohydrogen in harmony with the environment.

We seem to be agreeing on a course of action. First, as Neil Rossmeissl reiterated, we need *true* international collaboration. You heard him yesterday and again today, so I do not need to elaborate.

Second, we should link biohydrogen to the environment. Thus, our next phase of activity might well be thought of as biohydrogen-dot-carbon-dioxide, or "$BioH_2 \cdot CO_2$." The combination of the ultimate fuel, hydrogen, with a real concern for planet Earth, carbon dioxide, and the potential for global climate warming, could provide just the right marriage for more substantive funding. During the mid-'80s, Hawaii hosted three workshops on the photobiological production of hydrogen. This was, thus, the fifth in the series. In comparing those proceedings with the abstracts from BioHydrogen '97, I can be kind by characterizing our progress as modest. In consideration of how fields such as computers and the Internet have advanced, one conclusion is that the billions of defense dollars that went into the development of these fields made a difference. While it is unreasonable to expect hydrogen to garner similar support, a good next step for expanded funding to accelerate progress might best be attained through a coupling with the greenhouse effect.

Third, as we look into the future, consensus is forming for a Year 2000 gathering of this group in Europe. I trust that our Department of Energy will interact with the RITE leadership and their European counterparts to plan for this upcoming BioHydrogen • Carbon Dioxide gathering. David Hall and Grant Burgess have expressed interest in together working toward such a session, perhaps a day or two in London, followed by the actual conference in Edinburgh. Surely, then, under these conditions, field trips should be offered to the University of Florence (Italy) and Potsdam (located close to Berlin) to visit their state-of-the-art bioreactor facilities.

Finally, it is my understanding that Japan will be hosting a major marine biotechnology congress in 2003. It would make sense for biohydrogen to piggy-back a symposium to this conference, particularly as the ultimate hydrogen/carbon dioxide system shows promise for having a marine setting. Should this direction be taken, I would like to *challenge*[*] some of you to look closely at the ultimate bioreactor—the ocean. The utilization of an induced, upwelled, open-ocean system to produce biohydrogen while converting atmospheric carbon dioxide into compounds that can be stored at the bottom of the ocean, as an added value "environmental" component to next-generation fisheries and marine biomass plantations, would be an inspired unification merging sustainable energy with economic productivity toward a more healthful environment.

In closing, let us again acknowledge the truly fine efforts of Oskar and his team—Mary Kamiya, Pam Walton, Tracie Nagao, and Brandon Yoza, plus Eileen Kalim of the National Renewable Energy Laboratory. This has been a most rewarding and enjoyable experience.

REFERENCES

Takahashi, P.K., McKinley, K.R., Phillips, V.D., Magaard, L., and Koske, P., 1993, Marine macrobiotechnology systems, *J. Marine Biotech.*, 1:9–15. (Condensed, Japanese language version of the above can be found in the Spring 1995 issue of *RITE NOW*, pp. 2–3, while a summary of the 1994 biological hydrogen production workshop held in Tokyo is provided on pp. 4–9.)

Takahashi, P.K., 1996, Project blue revolution, *Journal of Energy Engineering*, 122(3):114–124.

Takahashi, P.K., 1997, Artificial upwelling for environmental enhancement, In *Proc. of Oceanology International (IOA) '97*, Singapore, May 11–14, 1997.

---

[*] Note: The following references can be consulted for information on this proposed marine biosystem.

6

MAXIMIZING PHOTOSYNTHETIC PRODUCTIVITY AND LIGHT UTILIZATION IN MICROALGAE BY MINIMIZING THE LIGHT-HARVESTING CHLOROPHYLL ANTENNA SIZE OF THE PHOTOSYSTEMS[*]

Anastasios Melis, John Neidhardt, Irene Baroli, and John R. Benemann

Department of Plant and Microbial Biology
University of California
411 Koshland Hall
Berkeley, California 94720-3102

Key Words

chlorophyll antenna size, chloroplast acclimation, *Dunaliella salina*, photosynthetic productivity, light-saturation curve of photosynthesis, P_{max}, quantum yield of photosynthesis

1. SUMMARY

The photosynthetic characteristics of the green alga *Dunaliella salina* were analyzed in growth under low or high irradiance and in the presence of $NaHCO_3$ or supplemental CO_2 as the inorganic carbon source. High-light $NaHCO_3$-grown cells exhibited signs of chronic photoinhibition, characterized by a lower pigment content, a highly truncated chlorophyll antenna size for the photosystems, and accumulation of photodamaged photosystem-II reaction centers in the chloroplast thylakoids. In spite of these deficiencies,

---

[*] Abbreviations: Chl, chlorophyll; PS, photosystem; P700, the photochemical reaction center chlorophyll of PSI; Q_A, the primary quinone acceptor of PSII; D1, the 32-kD reaction center protein of PSII coded by the chloroplast *psbA* gene; D2, the 34-kDa reaction center protein of PSII coded by the chloroplast *psbD* gene; I_s, light intensity at which photosynthesis is saturated; LHC, the chlorophyll *a-b*-binding light-harvesting complex; RC, reaction center; PQ, plastoquinone; P_{max}, the light-saturated rate of photosynthesis; Φ, quantum efficiency of photosynthesis.

BioHydrogen, edited by Zaborsky *et al.*
Plenum Press, New York, 1998

high-light $NaHCO_3$-grown cells showed photosynthetic productivity on a per chlorophyll basis (300 mmol O_2 mol^{-1} Chl s^{-1}) that was ≈3 times greater than that in the normally pigmented LL-grown cells (≈100 mmol O_2 mol^{-1} Chl s^{-1}). Repair of photodamaged centers in the high-light $NaHCO_3$-grown cells, occurring in the absence of a light-harvesting chlorophyll antenna size enlargement, increased photosynthetic productivity further to ≈650 mmol O_2 mol^{-1} Chl s^{-1} in these cells. From the analysis of the results, it was inferred that *Dunaliella salina* with a highly truncated chlorophyll antenna size, when grown under supplemental CO_2 rather than $NaHCO_3$, will display light-saturated rates of photosynthesis approaching 2000 mmol O_2 mol^{-1} Chl s^{-1}. Microalgae with such superior photosynthetic productivity and light utilization efficiency are ideal for commercial applications in CO_2 mitigation, and novel biochemical, biomass, or hydrogen production.

2. INTRODUCTION

Microalgal cultures growing under full sunlight have lower light-to-biomass energy conversion efficiencies than those growing under low light intensities. The reason for this inefficiency is that, at high photon flux densities, the rate of photon absorption by the antenna chlorophylls of the first few layers of cells in the culture exceeds the rate at which photosynthesis can utilize them, resulting in dissipation of the excess energy as fluorescence or heat. Up to 80% of absorbed photons could thus be wasted, reducing solar conversion efficiencies and cellular productivity to unacceptably low levels. To make the situation worse, cells deeper in the culture are deprived of much-needed sunlight as it is strongly attenuated due to filtering by the first layer of cells in the culture container (Naus and Melis, 1991; Neidhardt et al., 1998).

A truncated chlorophyll antenna size of the photosystems in the chloroplast of microalgae could alleviate this difficulty because it will minimize absorbance of bright incident sunlight by the first layer of chloroplasts in a culture, diminish mutual cell shading and wasteful dissipation of excitation energy, permit greater transmittance of light through the culture, and, thus, result in more uniform illumination of the cells. This, in turn, would result in greater photosynthetic productivity and light utilization efficiency in mass microalgal cultures. It is predicted that a smaller chlorophyll antenna size will result in a relatively higher light intensity for the saturation of photosynthesis in individual cells but, concomitantly, in much greater cellular productivity on a per chlorophyll basis. The experimental demonstration of this concept would be to take algal cells that possess a small chlorophyll antenna size for their photosystems and measure the light-saturation curve of photosynthesis.

Long-term exposure of plants, algae, or cyanobacteria to high levels of growth irradiance brings about reversible structural and functional adjustments in their photosynthetic apparatus (Powles, 1984; Anderson, 1986; Melis, 1991; Barber and Andersson, 1992). Previous work has demonstrated that the chlorophyll antenna size of green algae, such as *Chlorella vulgaris* (Ley and Mauzerall, 1982), *Dunaliella tertiolecta* (Sukenik et al., 1988), *Dunaliella salina* (Smith et al., 1990), and *Chlamydomonas reinhardtii* (Neale and Melis, 1986; Melis et al., 1996), is unusually plastic, and that at high light intensities the chlorophyll antenna size and the concentration of the photosystems are significantly altered.

When grown under irradiance stress, that is, high light conditions (HL = 2000 μmol photons m^{-2} s^{-1}) in the presence of $NaHCO_3$ as the sole carbon source, *D. salina* chloroplasts assemble ≈7% of the PSI complexes and ≈65% of the PSII complexes compared to low light (LL = 100 μmol photons m^{-2} s^{-1}) controls. However, of the PSII present in the thylakoid of HL-grown cells, only about 20–25% was photochemically competent, the rest occurring as photodamaged centers containing an irreversibly inactive PSII reaction center (D1) protein (Vasilikiotis and Melis, 1994). Thus, in HL-acclimated, $NaHCO_3$-grown *D. salina*, photosynthesis and growth depend solely on ≈7% of the PSI and ≈15% of the PSII centers that are operational in LL-grown cells.

With the caveat of the photodamaged PSII centers and the lower amount of PSI centers in the HL-acclimated *D. salina*, it appears that the latter may be a good experimental material in which to measure the productivity of photosynthesis *on a per chlorophyll basis* and, thus, to test for biomass production in microalgal cultures under bright light intensities. The work in this paper extends recent studies in this laboratory (Smith et al., 1990; Kim et al., 1993; Neidhardt et al., 1997) and presents a comparative analysis of the photosynthetic performance of *D. salina* grown under low or high irradiance and in the presence of $NaHCO_3$ or supplemental CO_2 as the inorganic carbon source. The results support the notion that, on a per chlorophyll basis, cells with a highly truncated chlorophyll antenna size and abundant inorganic carbon will exhibit superior photosynthetic productivity and light-utilization efficiency compared to the normally pigmented control cells.

3. EXPERIMENTAL

3.1. Growth of *Dunaliella salina* Cultures

The unicellular green alga *Dunaliella salina* was grown in a hypersaline medium containing 1.5 M NaCl, 0.2 M Tris-HCl (pH 7.5), 0.1 M KNO_3, 0.1 M $MgSO_4$, 6 mM $CaCl_2$, 2 mM KH_2PO_4, and 40 μM $FeCl_3$ dissolved in 400 μM EDTA (Pick et al., 1986). Bicarbonate was added to the medium as the ever-present carbon source to a concentration of 25 mM. Supplemental CO_2 was provided by bubbling the culture with a mixture of 3% CO_2 in air. The medium also contained a mixture of micronutrients in the following concentrations: 150 μM H_3BO3, 10 μM $MnCl_2$, 2 μM Na_2MoO_4, 2 μM $NaVO_3$, 0.8 μM $ZnCl_2$, 0.3 μM $CuCl_2$, and 0.2 μM $CoCl_2$.

Growth media were inoculated with several milliliters of a stock suspension of *D. salina* cells and cultivated in flat bottles (about 4 cm thick) at a temperature between 26 and 29 °C. The cells grew exponentially in the density range of $0.15–1.5 \times 10^6$ cells/mL (Naus and Melis, 1991). Measurements were performed with cultures having a cell density between 0.8 and 1.3×10^6 cells/mL.

The cultures were grown under either low light (incident irradiance of ≈100 μmol photons m^{-2} s^{-1}) or high light conditions (irradiance of ≈2000 μmol photons m^{-2} s^{-1}). The incident irradiance was measured with a LI-COR Model LI-185B radiometer. Shaking of the cultures along with the use of light reflectors ensured uniform illumination of the cells.

3.2. Cell Count and Chlorophyll Quantitation

The cell density in the cultures was obtained by counting with a Hemacytometer (improved Neubauer chamber) and use of an Olympus BH-2 compound microscope at a magnification of ×100. For the counting, cells were immobilized and stained by the addition of several microliters of Utermoehl oil to 0.25–1.0 mL aliquot of the culture.

Chlorophyll concentrations were measured upon pigment extraction in 80% acetone after removal of cell debris by centrifugation and by measuring the absorbance of the solutions at 663 and 645 nm. The amount of chlorophyll was calculated by use of Arnon's equations (1949).

3.3. Photosynthesis Measurements

Photosynthetic activity of the cells was measured by a Clark-type oxygen electrode. An aliquot of 5 mL cell suspension was applied to the oxygen electrode chamber. In order to compare the relative quantum yield of photosynthesis between the different samples, about the same chlorophyll concentration (2–3 µM) was loaded in the oxygen electrode. To ensure that oxygen evolution was not limited by the carbon source available to the cells, 100 µL of a 0.5 M sodium bicarbonate solution (pH 7.4) was added prior to the oxygen evolution measurements. Samples were illuminated with increasing light intensities under stirring and at a temperature of 25 °C. The rate of oxygen evolution under each of these light intensities was recorded continuously for a period of 2.5 min. The results were plotted to show the light saturation curves of photosynthesis either on a per chlorophyll or per cell basis.

The concentration of the photosystems in thylakoid membranes was estimated spectrophotometrically from the amplitude of the light *minus* dark difference at 700 nm (P700) for PSI and 320 nm (Q_A) for PSII (Melis, 1989). The light-harvesting chlorophyll antenna size of PSI and PSII was measured from the kinetics of P700 photo-oxidation and Q_A photoreduction, respectively (Melis, 1989).

4. RESULTS AND CONCLUSIONS

Information about the efficiency and productivity of photosynthesis on a per chlorophyll basis can be obtained from the light-saturation curve of photosynthesis (the so-called "P vs. I" curve), in which the rate of O_2 evolution, or CO_2 assimilation, is measured and plotted as a function of the actinic light intensity. In such a presentation, the rate of photosynthesis first increases linearly with light intensity and then levels off as the saturating light intensity (I_s) is approached. The slope of the initial linear increase defines the quantum efficiency of photosynthesis (Φ, O_2 evolved per photon absorbed) (Björkman and Demmig, 1987). The rate of photosynthesis is saturated at light intensities higher than I_s. This light-saturated rate (P_{max}) provides a measure of the capacity of photosynthesis for the particular sample (Powles and Critchley, 1980). The three parameters (Φ, I_s, and P_{max}) measured with dilute cultures under conditions of little mutual shading define the photosynthetic properties and capabilities of the algal culture.

It was of interest to compare the performance characteristics of low- and high-light acclimated *Dunaliella salina*, grown in the presence of $NaHCO_3$ as the sole inorganic carbon source, or in the presence of saturating amounts of CO_2, provided upon bubbling the growth medium with 3% CO_2 in air. The objective of this experiment was to assess the organization and function of the photosynthetic apparatus and test the hypothesis that a truncated chlorophyll antenna size would actually help cells to achieve a higher *per chlorophyll* capacity of photosynthesis under a variety of culture conditions.

4.1. Photosynthetic Efficiency and Productivity in *Dunaliella salina* Grown in the Presence of $NaHCO_3$ as the Sole Carbon Source

Cells grown at low light (100 µmol photons m^{-2} s^{-1}) or at high light (2000 µmol photons m^{-2} s^{-1}) in the presence of 25 mM $NaHCO_3$ had similar doubling times (8–8.5 h, Table 1). However, they had substantially different photosynthetic apparatus organization. Compared to the LL-grown, HL cells had only 25% the cellular Chl content, a much

Table 1. Effect of carbon source and growth irradiance on pigment content, photosynthetic apparatus organization, and rate of photosynthesis in *Dunaliella salina*

	Low-light NaHCO$_3$-grown	High-light NaHCO$_3$-grown	Low-light CO$_2$-bubbled	High-light CO$_2$-bubbled
Cell doubling time (h)	8.0	8.5	8.0	6.0
Chl/cell (molecules/cell)	0.41·10^9	0.10·10^9	1.13·10^9	0.37·10^9
Chl a/Chl b (mol:mol)	4.5:1	12:1	4.5:1	6:1
N$_{PSI}$	230	105	250	150
N$_\alpha$	560 (65%)	—	530 (75%)	500 (26%)
N$_\beta$	140 (35%)	130 (5%)	140 (25%)	120 (74%)
N$_{core}$	—	60 (95%)	—	—
P$_{max}$ (mmol O$_2$ mol^{-1} Chl s^{-1})	100	300	150	300
Quantum yield (Φ, rel. units)	1.0	0.37	1.0	0.56

Cells were grown at low light (100 µmol photons m^{-2} s^{-1}) or at high light (2000 µmol photons m^{-2} s^{-1}) in the presence of 25 mM NaHCO$_3$, with or without supplemental CO$_2$.

higher Chl a/Chl b ratio, and substantially truncated Chl antenna size for both PSI and PSII in their thylakoid membranes (Table 1). The number of Chl (a and b) molecules specifically associated with PSI (N$_{PSI}$) was lowered from 230 in LL to 105 in HL. In LL cells, about 65% of the functional PSII centers were PSII$_\alpha$ with an antenna size N$_\alpha$ of approximately 500 Chl (a and b) molecules. The remaining 35% of the functional PSII were of the PSII$_\beta$ type with an antenna size N$_\beta$ of ≈140 Chl (a and b) molecules. This well-known PSII α-β antenna heterogeneity (Melis, 1991) was essentially absent in the HL cells. In the latter, 95% of all functional PSII centers possessed a small antenna consisting of ≈60 Chl molecules. These results are consistent with the notion that HL cells grown in the presence of NaHCO$_3$ exist in a state of chronic irradiance stress as they are limited in terms of photosynthesis because of the slow carbon supply by NaHCO$_3$ (Smith et al., 1990; Vasilikiotis and Melis, 1994; Baroli and Melis, 1996). The effect of the truncated Chl antenna size and chronic photoinhibition status on the rate and quantum yield of photosynthesis was assessed.

Figure 1A shows the light-saturation curves of LL and HL NaHCO$_3$-grown *D. salina*. It is evident that LL-grown cells, in which the rate of light absorption limits photosynthesis, have a light-saturated rate (P$_{max}$) of ≈100 mmol O$_2$ mol^{-1} Chl s^{-1}. The HL cells with the truncated Chl antenna size and chronic photoinhibition reached a light-saturated rate of photosynthesis (P$_{max}$ = ≈320 mmol O$_2$ mol^{-1} Chl s^{-1}) that was ≈3 times greater than that of the LL-acclimated cells. This difference is attributed to the much smaller Chl antenna size for the HL-grown cells, translating into higher per Chl productivity. Consistent with this interpretation is the difference in the I$_s$ values, which is 8–10 times greater for the HL-grown than for the LL-grown cells, suggesting an 8–10 times greater Chl antenna size for the LL- than for the HL-grown cells (Table 1; Herron and Mauzerall, 1972).

The same results, plotted on a per cell basis (Figure 1B), show a greater cellular productivity for the LL-grown cells (≈75 pmol O$_2$ 10^{-6} cells s^{-1}), compared with the HL-

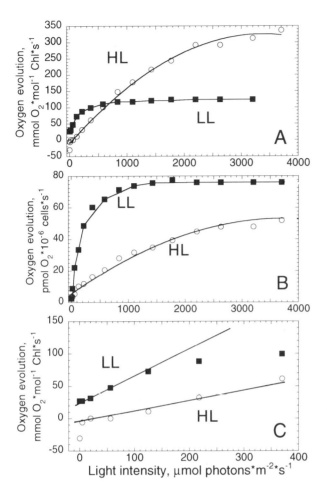

Figure 1. Light saturation curve of photosynthesis in $NaHCO_3$-grown *Dunaliella salina*. Cells were grown either at ≈100 μmol photons m^{-2} s^{-1} (LL) or at ≈2000 μmol photons m^{-2} s^{-1} (HL). (A) Rates of oxygen evolution on a *per chlorophyll* basis measured as a function of incident intensity to the cell suspension. (B) Rates of oxygen evolution on a *per cell* basis measured as a function of incident intensity to the cell suspension. (C) Initial slope of the rate of photosynthesis versus irradiance (relative quantum yield on a Chl basis) in the LL- and HL-grown cells.

grown cells in which the cell productivity was at ≈55 pmol O_2 10^{-6} cells s^{-1}. Again, this difference underscores the chronic photoinhibition status of the HL-grown cells where, in addition to the truncated Chl antenna size, a significant fraction of PSII centers are photochemically inert due to photodamage (Vasilikiotis and Melis, 1994). This configuration of the photosynthetic apparatus in the HL cells resulted in similar growth rates with the LL cells (Table 1). Clearly, however, both rates of growth are below those that can be achieved under optimal growth conditions (Baroli and Melis, 1996).

It is also evident from the results of Figure 1A that the initial slopes of the light-saturation curves (which provide a measure of the quantum yield of photosynthesis, Φ) are different for the two samples (Figure 1C), with that of the LL-grown cells (Φ = 0.42 rel. units) being steeper than that of the HL-grown cells (Φ = 0.15 rel. units). This difference reflects the fact that not all Chl molecules are photochemically competent in the HL-grown cells due to the chronic photoinhibition of photosynthesis that prevails in these cells (Smith et al., 1990; Kim et al., 1993; Baroli and Melis, 1996). On the basis of the relative quantum yield in these two measurements (Table 1), it would appear that only about 37% of the Chl molecules are photochemically competent in the HL, the rest being photochemically inert due to the accumulation of photodamaged and, therefore, inactive PSII centers in the HL thylakoids. In principle, then, the P_{max} = ≈300 mmol O_2 mol^{-1} Chl s^{-1} and the cellular productivity

4.2. Photosynthetic Efficiency and Productivity in *Dunaliella salina* Grown in the Presence of Supplemental CO_2

Cells grown at low light (100 µmol photons m^{-2} s^{-1}) or at high light (2000 µmol photons m^{-2} s^{-1}) under supplemental CO_2 had different doubling times (8 h for the LL and 6 h for the HL, Table 1). On average, supplemental CO_2-induced cells tend to accumulate more Chl relative to that of the $NaHCO_3$-grown cells (Table 1). Low-light and HL CO_2-bubbled cells had a different photosynthetic apparatus organization. Compared to the LL-grown, HL cells had a slightly higher Chl a/Chl b ratio and a somewhat smaller Chl antenna size for both PSI and PSII in their thylakoid membranes (Table 1). The number of Chl (a and b) molecules specifically associated with PSI was 250 in LL and 150 in HL. In LL cells, about 75% of the functional PSII centers were $PSII_\alpha$ with an antenna size of approximately 530 Chl (a and b) molecules. The remaining 25% of the functional PSII were of the $PSII_\beta$ type with an antenna size of ≈140 Chl (a and b) molecules. The PSII α-β antenna heterogeneity was present but distorted in the HL cells with 26% $PSII_\alpha$ and 74% $PSII_\beta$ centers. These results are consistent with the notion that supplemental CO_2 alleviates to a significant extent the irradiance stress of the HL $NaHCO_3$-grown cells.

Figure 2A shows the light-saturation curves of CO_2 supplemented LL- and HL-grown *D. salina*. The LL-grown cells had a P_{max} of ≈150 mmol O_2 mol^{-1} Chl s^{-1}. The HL cells with the partially truncated Chl antenna size reached a greater light-saturated rate of photosynthesis (P_{max} = ≈270 mmol O_2 mol^{-1} Chl s^{-1}). This difference is again attributed to the smaller Chl antenna size for the HL-grown cells, translating into higher per Chl productivity. Consistent with this interpretation is also the difference in the I_s values, which is ≈2 times greater for the HL-grown than for the LL-grown cells, suggesting an average ≈2 times greater Chl antenna size for the LL- than for the HL-grown cells (Table 1).

The same results, plotted on a per cell basis (Figure 2B), showed greater cellular productivity for the LL-grown cells (≈320 pmol O_2 10^{-6} cells s^{-1}), compared with the HL-grown cells in which the cell productivity was at ≈170 pmol O_2 10^{-6} cells s^{-1}. Again, this difference points to the occurrence of some photoinhibition in the HL-grown cells in which a fraction of PSII centers must be photochemically inert due to photodamage (results not shown). A comparison of the results in Figure 2B (LL) with the corresponding data in Figure 1B (LL) shows that supplemental CO_2 increased the light-saturated cellular productivity by about fourfold. Similarly, a comparison of the results in Figure 2B (HL) with the corresponding data in Figure 1B (HL) shows that supplemental CO_2 increased cellular productivity by about threefold.

It is evident from the results shown in Figure 2A that the initial slopes of the light-saturation curves (which provide a measure of the quantum yield of photosynthesis, Φ) are different for the two samples (Figure 2C), with that of the LL-grown cells (Φ = 0.43 rel. units) being steeper than that of the HL-grown cells (Φ = 0.24 rel. units). This difference in Φ further reflects on the fact that not all Chl molecules are photochemically competent in the HL-grown cells due to photoinhibition. On the basis of the relative quantum yield in these two measurements (Table 1), it would appear that about 56% of the Chl molecules are photochemically competent in the HL, the rest being photochemically inert due to the accumulation of photodamaged and, therefore, inactive PSII centers.

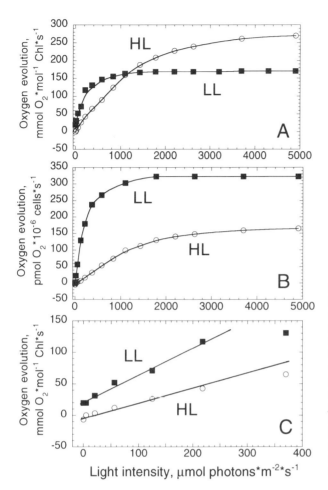

Figure 2. Light saturation curve of photosynthesis in CO_2-supplemented *Dunaliella salina*. Cells were grown either at ≈100 µmol photons m^{-2} s^{-1} (LL) or at ≈2000 µmol photons m^{-2} s^{-1} (HL). (A) Rates of oxygen evolution on a *per chlorophyll* basis measured as a function of incident intensity to the cell suspension. (B) Rates of oxygen evolution on a *per cell* basis measured as a function of incident intensity to the cell suspension. (C) Initial slope of the rate of photosynthesis versus irradiance (relative quantum yield on a Chl basis) in the LL- and HL-grown cells.

These results suggest that a truncated Chl antenna size for the photosystems and elevated concentrations of inorganic carbon substrate (provided as supplemental CO_2) will enhance the light utilization efficiency and productivity of photosynthesis. To further illustrate the effect of a truncated Chl antenna size, it was important to correct for the effect of chronic photoinhibition on the cellular photosynthesis. To this end, we devised an experimental approach that would promote the repair of photodamaged PSII centers without the induction of a concomitant, significant increase in the Chl antenna size of the photosystems. We performed "light shift" experiments in which HL $NaHCO_3$-grown cultures, with cells in the exponential phase of growth, were shifted to LL growth conditions. We reasoned that upon a HL→LL transition, both the repair of the photodamaged PSII centers and an increase in the chlorophyll antenna size would occur. However, the PSII repair reportedly occurs with a half-time of about 60 min (Vasilikiotis and Melis, 1994; Baroli and Melis, 1996), whereas the increase in the Chl antenna size of PSII occurs with slower kinetics having a half-time of ≈4 h. Thus, in the early stages of a HL→LL shift, one would encounter a situation where a significant fraction of PSII centers would have been repaired with only a minimal increase in the Chl antenna size of the photosystems.

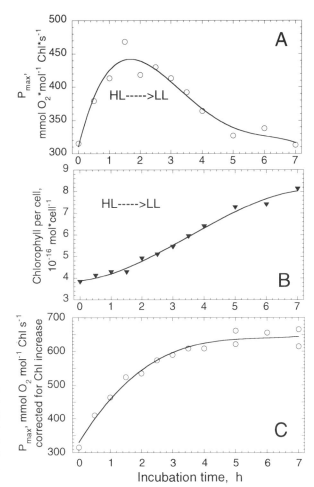

Figure 3. Changes in cellular photosynthesis and chlorophyll content in NaHCO$_3$-grown *D. salina* following a switch of HL-grown cells to LL-growth conditions. The switch in growth irradiance occurred at zero time. (A) Changes in the light-saturated rate of photosynthesis (P_{max}) following a HL→LL transition. (B) Changes in the Chl/cell ratio following the HL→LL transition. (C) Rates of photosynthesis (P_{max}), corrected for the change in the Chl content of the cells, reflecting the repair of photodamaged PSII centers as a function of time in LL.

4.3. Light Shift Experiments

Figure 3A shows the adjustment of the light-saturated rate of photosynthesis (P_{max}) in NaHCO$_3$-grown cells following a HL→LL transition. It is evident that P_{max} increases promptly as a function of time upon the HL→LL transition, from ≈310 mmol O$_2$ mol^{-1} Chl s^{-1}, measured at zero time, to a transient maximum of ≈475 mmol O$_2$ mol^{-1} Chl s^{-1}, attained within the first 2 h under LL conditions. This change reflects the repair of PSII centers and the de novo biosynthesis/assembly of PSI centers, both of which bring about a greater capacity for photosynthetic electron transport in the thylakoid membranes (Neidhardt et al., 1998). Subsequent incubation under LL conditions caused a gradual decline in the value of P_{max}, reflecting the significant accumulation of Chl in the chloroplasts, and an increase in the light-harvesting Chl antenna size, which resulted in a lower per Chl P_{max} value for the cells.

Figure 3B shows the change in the Chl/cell ratio following a HL→LL transition. Within 7 h, the Chl/cell ratio increased from less than 4 to about 9 × 10^{-16} mol cell^{-1}. Concomitantly, the Chl *a*/Chl *b* ratio of the cells decreased from ≈12/1 to a low value of ≈6/1 over the same period (not shown). The lowering of the Chl *a*/Chl *b* ratio reflects accumulation of Chl b and the ensuing increase in the auxiliary light-harvesting chlorophyll antenna size of the photosystems. Both changes are consistent with earlier measurements on the Chl antenna size increase upon a HL→LL transition (Kim et al., 1993).

Figure 3C shows estimated values of P_{max} as a function of incubation time following a HL→LL transition. This presentation depicts the P_{max} values that would have been attained by repair of photodamaged PSII in the absence of a Chl antenna size increase. Results in this plot were obtained by multiplication of the corresponding values of P_{max} (Figure 3A) and Chl/cell (Figure 3B). Figure 3C shows that, following repair of PSII, P_{max} in $D.\ salina$ cells would have increased from ≈ 300 to over 650 mmol O_2 mol^{-1} Chl s^{-1}. The exponential increase in the value of P_{max} following the HL→LL transition reflects the kinetics of the repair of photodamaged PSII centers. The measured half-time of ≈ 1 h is consistent with earlier findings on the half-time of the PSII repair from photodamage (Vasilikiotis and Melis, 1994; Baroli and Melis, 1996).

A light-saturated rate of 650 mmol O_2 mol^{-1} Chl s^{-1} would have been attained upon the repair of PSII under $NaHCO_3$-growth conditions. On the basis of the results in Figures 1B and 2B, it could be argued that supplemental CO_2 would further enhance the light-saturated rate of photosynthesis by a factor of ≈ 3, thereby bringing the productivity of microalgae from ≈ 650–2000 mmol O_2 mol^{-1} Chl s^{-1}.

The biotechnology of microalgal cultures has developed over the past decades into a commercially viable industry, with both fermentation and open pond culture processes (Benemann, 1990). Although current products are of relatively high monetary value, and production scales are relatively small (<100 hectare total pond area; <1000 tons/year total annual production), future technological improvements could expand this industry into commodity-scale products and even chemicals and fuels production. Of the latter, CO_2 mitigation efforts (Benemann, 1993; Brown and Zeiler, 1993; Mulloney, 1993; Nakicenovic, 1993), algal biomass (Vazquezduhalt, 1991; Westermeier and Gomez, 1996), and hydrogen production (Greenbaum, 1984; Cinco, et al., 1993; Greenbaum, et al., 1995; Ghirardi, et al., 1997) have received considerable attention for the past two decades. In any algal production system, however, critical for low-cost generation of biomass, hydrogen, or CO_2 mitigation, the achievable photosynthetic productivity and light utilization efficiency is the single most important factor.

Results of this work demonstrate a novel method for maximizing photosynthetic efficiencies and light utilization in microalgae. They suggest that a valid experimental approach by which to achieve this goal in mass cultures would be to minimize the number of light-harvesting pigments of the photosystems. A truncated light-harvesting chlorophyll antenna size in the chloroplast would permit more uniform illumination of the cells and higher (per chlorophyll) cellular productivity because it would minimize mutual cell shading and wasteful dissipation of bright incident sunlight. It is evident from the results of this work that whenever the light-harvesting antenna size of photosynthesis is small, light absorption by individual cells will be lowered, permitting light to penetrate deeper into the culture and resulting in overall greater per chlorophyll photosynthesis and culture biomass production.

It is also evident from this work that a combination of a truncated Chl antenna size and provision of high concentrations of inorganic carbon, ideally provided as supplemental CO_2 to the growth medium, will result in truly superior rates of photosynthesis with light-saturated values reaching 2000 mmol O_2 mol^{-1} Chl s^{-1}. Such microalgae will be ideal for commercial applications including CO_2 mitigation and novel biochemical, biomass, or hydrogen production.

ACKNOWLEDGMENT

The work was supported by U.S. Department of Energy Cooperative Agreement Number DE-FC36-98GO10278.

REFERENCES

Anderson, J.M., 1986, Photoregulation of the composition, function, and structure of thylakoid membranes, *Annu. Rev. Plant. Physiol.*, 37:93–136.
Arnon, D., 1949, Copper enzymes in isolated chloroplasts; polyphenol oxidase in *Beta vulgaris*, *Plant Physiol.*, 24:1–15.
Barber, J., and Andersson, B., 1992, Too much of a good thing: light can be bad for photosynthesis, *Trends Biochem. Sci.*, 17:61–66.
Baroli, I., and Melis, A., 1996, Photoinhibition and repair in *Dunaliella salina* acclimated to different growth irradiances, *Planta*, 198:640–646.
Benemann, J.R., 1990, The future of microalgae biotechnology, in *Algal Biotechnology*, Cresswell, R.C., Rees, T.A.V., and Shah, N. (eds.), pp. 317–337, Longman, London.
Benemann, J.R., 1993, Utilization of carbon dioxide from fossil fuel-burning power plants with biological systems, *Energy Conversion and Management*, 34:999–1004.
Björkman, O., and Demmig, B., 1987, Photon yield of O_2 evolution and chlorophyll fluorescence characteristics at 77 K among vascular plants of diverse origins, *Planta*, 170:489–504.
Brown, L.M., and Zeiler, K.G., 1993, Aquatic biomass and carbon dioxide trapping, *Energy Conversion and Management*, 34:1005–1013.
Cinco, R.M., Macinnis, J.M., and Greenbaum, E., 1993, The role of carbon dioxide in light-activated hydrogen production by *Chlamydomonas reinhardtii*, *Photosynth. Res.*, 38:27–33.
Ghirardi, M.L., Togasaki, R.K., and Seibert, M., 1997, Oxygen sensitivity of algal hydrogen production, *Appl. Biochem. Biotech.*, in press.
Greenbaum, E., 1984, Biophotolysis of water: the light saturation curves, *Photobiochem. Photobiophys.*, 8:323–332.
Greenbaum, E., Lee, J.W., Tevault, C.V., Blankinship, S.L., and Mets, L.J., 1995, CO_2 fixation and photoevolution of H_2 and O_2 in a mutant of *Chlamydomonas* lacking photosystem I, *Nature*, 376: 438–441.
Herron, H.A., and Mauzerall, D., 1972, The development of photosynthesis in a greening mutant of *Chlorella* and an analysis of the light saturation curve, *Plant Physiol.*, 50:141–148.
Kim, J.H., Nemson, J.A., and Melis, A., 1993, Photosystem II reaction center damage and repair in *Dunaliella salina* (green alga): analysis under physiological and irradiance-stress conditions, *Plant Physiol.*, 103:181–189.
Ley, A.C., and Mauzerall, D.C., 1982, Absolute absorption cross-sections for photosystem II and the minimum quantum requirement for photosynthesis in *Chlorella vulgaris*, *Biochim. Biophys. Acta*, 680:95–106.
Melis, A., 1989, Spectroscopic methods in photosynthesis: photosystem stoichiometry and chlorophyll antenna size, *Phil. Trans. R. Soc. Lond.*, B323:397–409.
Melis, A., 1991, Dynamics of photosynthetic membrane composition and function, *Biochim. Biophys. Acta* (Reviews on Bioenergetics), 1058:87–106.
Melis, A., Murakami, A. Nemson, J.A., Aizawa,K., Ohki, K., and Fujita, Y., 1996, Chromatic regulation in *Chlamydomonas reinhardtii* alters photosystem stoichiometry and improves the quantum efficiency of photosynthesis, *Photosynth. Res.*, 47:253–265.
Mulloney, J.A., 1993, Mitigation of carbon dioxide releases from power production via sustainable agri-power; the synergistic combination of controlled environmental agriculture (large commercial greenhouses) and disbursed fuel cell, *Energy Conversion and Management*, 34:913–920.
Nakicenovic, N., 1993, Carbon dioxide mitigation measures and options, *Environmental Science and Technology*, 27:1986–1989.
Naus, J., and Melis, A., 1991, Changes of photosystem stoichiometry during cell growth in *Dunaliella salina* cultures, *Plant Cell Physiol.*, 32:569–575.
Neale, P.J., and Melis, A., 1986, Algal photosynthetic membrane complexes and the photosynthesis-irradiance curve: a comparison of light-adaptation responses in *Chlamydomonas reinhardtii*, *J. Phycol.*, 22:531–538.

Neidhardt, J., Benemann, J.R., Zhang, L., and Melis, A., 1998, Photosystem II repair and chloroplast recovery from irradiance stress: relationship between chronic photoinhibition, light-harvesting chlorophyll antenna size, and photosynthetic productivity in *Dunaliella salina* (green algae), *Photosynth. Res.*, in press.

Pick, U., Karni, L., and Avron, M., 1986, Determination of ion content and ion fluxes in the halotolerant alga *Dunaliella salina*, *Plant Physiol.*, 81:92–96.

Powles, S.B., 1984, Photoinhibition of photosynthesis induced by visible light, *Annu. Rev. Plant Physiol.*, 35:15–44.

Powles, S.B., and Critchley, C., 1980, Effect of light intensity during gast acclimation strategy to irradiance stress, in *Proceedings of the National Academy of Sciences*, 91:7222–7226.

Vazquez-Duhalt, R., 1991, Light-effect on neutral lipids accumulation and biomass composition of *Botryococcus sudeticus* (Chlorophyceae), *Cryptogamie Algologie*, 12:109–119.

Westermeier, R., and Gomez, I., 1996, Biomass, energy contents, and major organic compounds in the brown alga *Lessonia nigrescens* (Laminariales, Phaeophyceae) from Mehuin, south Chile, *Botanica Marina*, 39:553–559.

7

Nostoc PCC 73102 AND H_2

Knowledge, Research, and Biotechnological Challenges

Peter Lindblad, Alfred Hansel, Fredrik Oxelfelt, Paula Tamagnini, and Olga Troshina

Department of Physiological Botany
Uppsala University
Villavägen 6
S-752 36 Uppsala, Sweden

Key Words

cyanobacteria, *Nostoc*, uptake hydrogenase, bidirectional hydrogenase, *hup*-genes, *hox*-genes, *hyp*-genes

1. SUMMARY

In N_2-fixing cyanobacteria, at least three enzymes may be involved in H_2 metabolism: (a) nitrogenase, evolving H_2 during N_2 fixation; (b) an uptake hydrogenase, reutilizing this H_2; and (c) a bidirectional (reversible) hydrogenase. Our studies focus on the hydrogen metabolism of the filamentous, heterocystous cyanobacterium *Nostoc* PCC 73102, a free-living strain originally isolated from coralloid roots of the Australian cycad, *Macrozamia*. Immunological studies using polyclonal antisera directed against hydrogenases purified from several different microorganisms indicated the presence of two native enzymes/forms in N_2-fixing cells of *Nostoc* PCC 73102, with at least one common subunit of approximately 58 kDa. Moreover, a cellular localization in both the N_2-fixing heterocysts and the vegetative cells was observed. Measurements with an H_2 electrode revealed the presence of a light-stimulated in vivo H_2 uptake in N_2-fixing cells of *Nostoc* PCC 73102. We have identified and sequenced genes encoding an uptake hydrogenase (*hupSL*) in *Nostoc* PCC 73102. They are highly homologous to corresponding genes in *Anabaena* PCC 7120. However, there is no rearrangement within *hupL*. Using both molecular and physiological techniques, it was not possible to demonstrate any evidence of the presence of either *hox* genes or corresponding bidirectional enzyme activities in *Nos-*

toc PCC 73102. This fact, together with the availability of the genes encoding an uptake hydrogenase, makes *Nostoc* PCC 73102 an interesting candidate for further molecular studies of its hydrogen metabolism.

2. INTRODUCTION

Molecular hydrogen is an environmentally clean source of energy that may be a valuable alternative to the limited fossil fuel resources of today. For photobiological H_2 production, cyanobacteria are among the ideal candidates because they have the simplest nutritional requirements: they can grow in air (N_2 and CO_2), water (electrons and reductant), and simple mineral salts, with light as the only source of energy. Cyanobacteria are Gram-negative bacteria performing plant-type photosynthesis. They are common organisms in fresh and brackish waters and marine environments, as well as in different soils and ecosystems. Moreover, they have the capacity to establish symbioses with a number of "higher organisms": algae (a few genera), fungi (lichens, ~8% of the approximately 18000 described lichen species contain cyanobacteria), bryophytes (a few liverwort genera), pteridophytes (*Azolla*, 7 spp), gymnosperms (cycads), and angiosperms (*Gunnera*, ~50 spp) (Rai, 1990). In N_2-fixing cyanobacteria, H_2 is mainly produced by nitrogenases, but its partial consumption is quickly catalyzed by a unidirectional uptake hydrogenase. In addition, a bidirectional enzyme may also oxidize some of the molecular hydrogen (for general reviews see Smith, 1990; Rao and Hall, 1996; Schulz, 1996). Some filamentous genera, for example, *Nostoc* and *Anabaena*, may differentiate three cell types: vegetative cells (photosynthesis), heterocysts (N_2 fixation), and akinetes ("spores"). Structurally, the vegetative cells contain thylakoids, with distinct phycobilisomes harboring the light-harvesting phycobiliproteins. Heterocysts develop from vegetative cells in response to a low level of available nitrogen. They serve as N_2-fixing entities that receive fixed carbon from the vegetative cells and return fixed nitrogen in the form of glutamine. Heterocysts possess thick cell walls. Their thylakoid membranes extend throughout the cytoplasm. Photosystem II is not functional in mature heterocysts, thus providing a more or less oxygen-free environment for nitrogenase activity. Filamentous cyanobacteria have been used in bioreactors for the photobiological conversion of water to hydrogen (Rao and Hall, 1996). However, the conversion efficiencies achieved are low because the net H_2 production is the result of H_2 evolution via a nitrogenase and H_2 consumption mainly via an uptake hydrogenase. Consequently, improvements in the conversion efficiency might be achieved through the optimization of the conditions for H_2 evolution by nitrogenase and through the production of mutants deficient in H_2 uptake activity. Symbiotic cells are of fundamental interest because they in situ "function as a bioreactor," that is, high metabolism and transfer of metabolite(s) from symbiont to host, but almost no growth. The potential, problems, and prospects of H_2 production by cyanobacteria/hydrogen biotechnology have been reviewed (Benemann, 1996; Hansel and Lindblad, 1998; Rao and Hall, 1996; Schulz, 1996).

The structural genes encoding hydrogenases have been sequenced and characterized in many microorganisms representing several different taxonomic groups (e.g., Vignais and Toussaint, 1994). However, molecular studies concerning hydrogenases of cyanobacteria are rare. In 1995 (Carrasco et al.), a developmental genome rearrangement for *Anabaena* PCC 7120 was described. It is present in addition to the known *nifD* (Golden et al., 1985) and *fdxN* (Golden et al., 1991) rearrangements taking place during the differentiation of a photosynthesizing vegetative cell into a nitrogen-fixing heterocyst. This latter rearrangement occurs within a gene (*hupL*) that exhibits homology to genes coding for the large subunits of membrane-bound uptake hydrogenases. A 10.5-kb element is excised late in the heterocyst differentiation process, indicating that the gene coding for HupL in *A.* PCC 7120 is expressed only in heterocysts (Carrasco et al., 1995).

Bidirectional/reversible hydrogenase catalyzes both H_2 production and H_2 consumption (Smith, 1990). It is believed to be a common cyanobacterial enzyme, and its presence is not linked to nitrogenase. The structural genes (*hox*) coding for a bidirectional hydrogenase have been sequenced in *Anabaena variabilis* (Schmitz et al., 1995) and in the unicellular non-N_2-fixing *Anacystis nidulans* (Boison et al., 1996). Nucleotide sequence comparisons showed that there is a high degree of homology between the *hox* genes of cyanobacteria and genes coding for the NAD^+-reducing hydrogenase from the chemolithotrophic H_2-metabolizing bacterium *Alcaligenes eutrophus,* as well as methyl viologen-reducing hydrogenases from species of the archaebacterial genera *Methanobacterium*, *Methanococcus*, and *Methanothermus*. Moreover, the sequence of a NAD(P)-reducing hydrogenase operon of the unicellular cyanobacterium *Synechocystis* PCC 6803 has been determined (Appel and Schulz, 1996).

3. EXPERIMENTAL

At present, we are concentrating our studies on the filamentous, heterocystous cyanobacterium *Nostoc* PCC 73102, a free-living strain originally isolated from coralloid roots of the Australian cycad *Macrozamia*. *Nostoc* PCC 73102 has been proposed as the type strain of the species *Nostoc punctiforme* in the Pasteur Culture Collection (Paris, France). It is very important to describe and characterize all hydrogenases/H_2-metabolism in a particular cyanobacterial strain in detail. With this knowledge, further molecular experiments, for example, the construction of specific mutants with desired hydrogen metabolism, can be performed and correctly evaluated. For information about additional organisms and growth conditions, see Tamagnini et al. (1997).

Immunological, physiological, and molecular experiments were all carried out as described previously (Oxelfelt et al., 1995; Oxelfelt et al., 1998; Tamagnini et al., 1995; Tamagnini et al., 1997).

4. RESULTS AND CONCLUSIONS

4.1. Presence of Hydrogenases (Tamagnini et al., 1995)

Immunological studies using polyclonal antisera directed against hydrogenases purified from *Bradyrhizobium japonicum, Azotobacter vinelandii, Methanosarcina barkeri,* and *Thiocapsa roseopersicina* demonstrated the presence of two native enzymes/forms in N_2-fixing cells of *Nostoc* PCC 73102 (Figure 1a), with at least one common subunit of approximately 58 kDa (Figure 1b). Moreover, two additional polypeptides with molecular masses of about 34 and 70 kDa were recognized with some of the antisera used (Figure 1b). The antigens were localized in both the N_2-fixing heterocysts and the photosynthetic vegetative cells, with considerable higher antigen content in the latter cell type.

4.2. Uptake Hydrogenase (Oxelfelt et al., 1995; Oxelfelt et al., 1998)

Nostoc PCC 73102 has the capacity to take up atmospheric hydrogen (Figure 2). This uptake is stimulated by light and positively regulated by the substrate H_2 (added

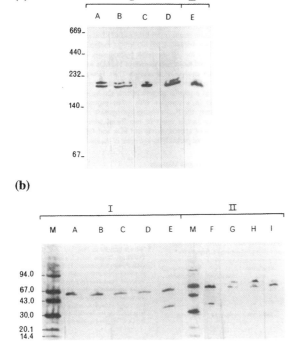

Figure 1. Native (a) and SDS (b) PAGE/Western immunoblots demonstrating the presence of imunologically recognizable proteins in *Nostoc* PCC 73102. The cell extracts were electroblotted onto two nitrocellulsoe membranes (I and II) that were cut and incubated with, in (a), the holoenzyme purified from *Azotobacter vinelandii* (A), the large subunit purified from *Bradyrhizobium japonicum* (B), the small subunit purified from *Bradyrhizobium japonicum* (C), the F_{420}-reducing hydrogenase purified from *Methanosarcina barkeri* (D), a hydrogenase purified from *Thiocapsa roseopersiocina* (E), and in (b), the small subunit purified from *B. japonicum* (A and B), a hydrogenase purified from *T. roseopersiocina* (E and F), the large subunit purified from *B. japonicum* (C and D), the F_{420}-reducing hydrogenase purified from *M. barkeri* (G and H), and the holoenzyme purified from *A. vinelandii* (I). The positions of standard proteins, in kDa, are shown to the left of each membrane (Tamagnini et al., 1995).

either directly as H_2 from a cylinder or indirectly through the action of nitrogenase). Additionally, the in vivo nitrogenase and hydrogen uptake activities appear to be co-regulated when exposing nitrogen-fixing cells to either combined-N or organic carbon sources (Figures 3 and 4). We have identified, cloned, and sequenced genes coding for an uptake hydrogenase (*hupSL*) in *N.* PCC 73102. They are highly homologous to corresponding genes in *Anabaena* PCC 7120, but differ in showing no rearrangement in *hupL*. When performing low-stringency Southern hybridizations using cloned *Nostoc hupSL* as a probe and genomic DNA digested with *Eco*RI, two additional bands (in addition to the probe itself) were observed (Figure 5). They might represent either an additional copy of *hupSL*; other *hup*-genes sharing homology to *hupSL*, for example *hupUV*; or (3) totally unrelated genes/sequences. The genes *hupUV*, putatively involved in sensing (e.g., H_2) have been described in *Rhodobacter capsulatus* (Colbeau and Vignais, 1992; Elsen et al., 1993; Elsen et al., 1996) where they, together with the small regulatory *hupT*, form the suggested

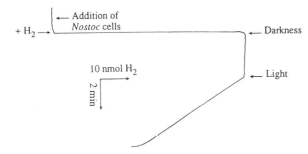

Figure 2. Example of a graph from the H_2 electrode demonstrating an in vivo, light-dependent, hydrogen uptake in N_2-fixing cells of *Nostoc* PCC 73102 (Oxelfelt et al., 1995).

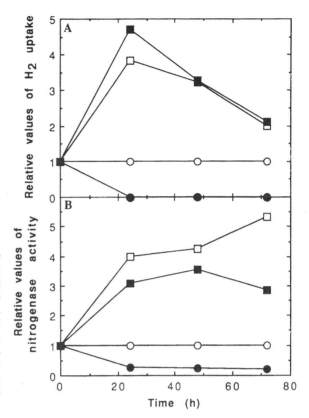

Figure 3. Effect of heterotrophic growth/metabolism on the in vivo H_2 uptake (A) and nitrogenase (B) activities by *Nostoc* PCC 73102. (○) = light, (●) = darkness, (□) = light with 30 mM glucose and 30 mM fructose, (■) = darkness with 30 mM glucose and 30 mM fructose. The control culture is standardized to 1.0, which at time 0 corresponds to: (A) hydrogen uptake activity of 11 nmol H_2 h^{-1} μg^{-1} Chl$a$, and in (B) a nitrogenase activity of 30 nmol C_2H_4 h^{-1} μg^{-1} Chl$a$ (Oxelfelt et al., 1995).

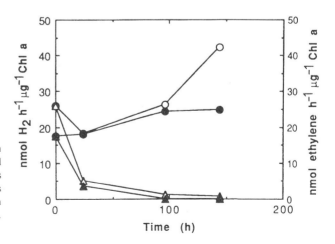

Figure 4. Effect of ammonia on the in vivo H_2 uptake (black symbols) and nitrogenase (white symbols) activities by *Nostoc* PCC 73102. The cultures were grown without (○,●) or with (△,▲) the addition of 5 mM NH_4Cl. (Oxelfelt et al., 1995).

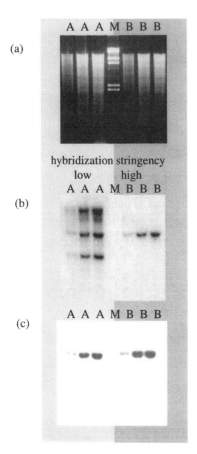

Figure 5. Southern hybridization using cloned *Nostoc* PCC 73102 *hupSL* as a probe against genomic, digested (*Eco*RI) *Nostoc* PCC 73102 DNA. Two identical membranes, I (lanes A, different amounts of DNA) and II (lanes B, different amounts of DNA), were hybridized against the identical probe but at different temperatures, I at 55 °C, II at 65 °C. After washing at 55 °C (and 3x SSPE), three bands appeared in membrane I, whereas only one in membrane II (b). After a subsequent wash at 57 °C (and 0.5x SSPE), only the band at 3.2 kb remained (c). The 3.2-kb DNA fragment contains the cloned *hupSL*.

hupTUV operon. *hupUV* reveals considerable homology to *hupSL* in this organism (Elsen et al., 1996).

4.3. Bidirectional Hydrogenase (Tamagnini et al., 1997)

We designed oligonucleotide primers from conserved sequences within the *hoxY* and *hoxH* genes of *Anabaena variabilis* ATCC 29413. These primers were used in PCRs with genomic DNA from *A. variabilis* as the template. PCR products of the expected sizes were obtained and their respective identities were established by sequencing. Genomic DNA of *A. variabilis*, *Nostoc* PCC 73102, *A.* PCC 7120, and *N. muscorum* CCAP 1453/12 was digested with the restriction endonucleases *Eco*RI and/or *Hin*dIII. Low-stringency Southern hybridizations with the ^{32}P-labeled *hox* probes revealed the occurrence of corresponding sequences in *A. variabilis* (control), *A.* PCC 7120, and *N. muscorum,* but not in *N.* PCC 73102 (Figure 6). Moreover, using PCR, genomic DNA of *A. variabilis*, *A.* PCC 7120, or *N. muscorum* and primer pairs within *hoxY* and *hoxH* resulted in DNA fragments of expected sizes and homologous sequences. The primer pair OL6A and OL8B (*hoxH*), for example, resulted in 89 and 79% identical DNA fragments when comparing the PCR-generated DNA fragment from *A. variabilis* with those generated from *A.* PCC 7120 and *N. muscorum*, respectively (Figure 7). However, PCRs using genomic DNA of *N.* PCC 73102 resulted in no amplification at all. Cells/cell extracts of *N.* PCC 73102, *N.*

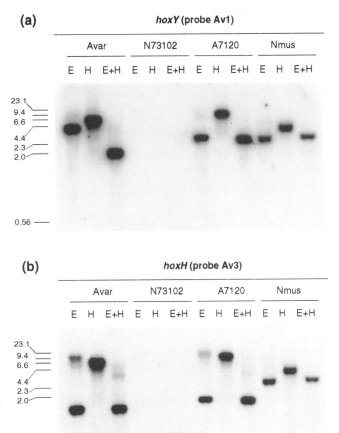

Figure 6. Autoradiographs demonstrating the presence of sequences homologous to the structural genes *hoxY* (a) and *hoxH* (b) in four filamentous, heterocystous cyanobacteria. Total genomic DNA of *Anabaena variabilis* (Avar), *Nostoc* PCC 73102 (N73102), *Anabaena* PCC 7120 (A7120), or *Nostoc muscorum* (Nmus) was digested with *Eco*RI (E), *Hin*dIII (H), or a combination of *Eco*RI and *Hin*dIII (E+H), separated by agarose gel electrophoresis, transferred to a nylon membrane, and hybridized with a ^{32}P-labeled part of the *Anabaena variabilis hoxH* as probe (b), before being stripped and rehybridized with part of *A. variabilis hoxY* as probe (a). The positions of the molecular length markers (in kB) are shown to the left (Tamagnini et al., 1997).

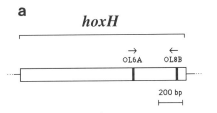

Figure 7. (a) Schematic presentation of *Anabaena variabilis* ATCC 29413 *hoxH* gene (Schmitz et al., 1995) with the conserved regions 6 and 8 indicated. → sense primer (OL6A), ← antisense primer (OL8B) (see also Tamagnini et al., 1997). (b) Identities between the PCR-generated DNA fragments/sequences (*Anabaena* PCC 7120 and *Nostoc muscorum*) and the corresponding sequence in *Anabaena variabilis*.

PCR-generated DNA fragment using genomic DNA from	Identity comparing to the corresponding sequence in *Anabaena variabilis* (Schmitz et al., 1995)
Anabaena variabilits	100%
Anabaena PCC 7120	89%
Nostoc muscorum	79%

Table 1. (a) *In vivo* (top) and (b) *in vitro* (bottom) bidirectional hydrogenase activities in four filamentous heterocystous cyanobacteria

Organism	Growth media*		
	A[1]	B[1]	C[1]
Anabaena variabilis	5.5 ± 0.2	6.3 ± 0.3	18.5 ± 0.7
Nostoc PCC 73102	non-detectable	non-detectable	non-detectable
Anabaena PCC 7120	7.2 ± 0.1	15.9 ± 0.4	26.2 ± 1.6
Nostoc muscorum	2.2 ± 0.2	2.3 ± 0.1	6.2 ± 0.2
		B[2]	C[2]
Anabaena variabilis		6.4 ± 0.7	11.3 ± 0.8
Nostoc PCC 73102		non-detectable	non-detectable
Anabaena PCC 7120		8.6 ± 0.8	18.0 ± 1.6
Nostoc muscorum		2.3 ± 0.2	4.0 ± 0.3

(Tamagnini et al., 1997)

* A - $BG11_0$, B - $BG11_0$ with the exclusion of Na_2CO_3, C - 1/4 Allen and Arnon medium + A5 micronutrient solution. All growth media were supplemented with 5 mM NH_4Cl, 1 μM $NiSO_4$, and 10 mM HEPES (pH 7.5).
[1](nmoles of H_2 produced h^{-1} μg^{-1} chl *a*; avg ± SD, n = 3-7).
[2](nmoles of H_2 produced min^{-1} mg^{-1} protein; avg ± SD, n = 3-7).

muscorum, *A.* PCC 7120, and *A. variabilis* were assayed for the presence of both an in vivo and in vitro active bidirectional hydrogenase. *N. muscorum*, *A.* PCC 7120, and *A. variabilis* in all experiments exhibited detectable enzymes/enzyme activities, whereas *N.* PCC 73102 consistently lacked any detectable activity (Table 1). Native PAGE/in situ hydrogenase activity staining was used to further demonstrate the presence or absence of a functional enzyme in partially purified extracts from the four cyanobacteria. Again, bands corresponding to hydrogenase activities were observed for *N. muscorum*, *A.* PCC 7120, and *A. variabilis*. However, no bands were detected for *N.* PCC 73102 (Figure 8).

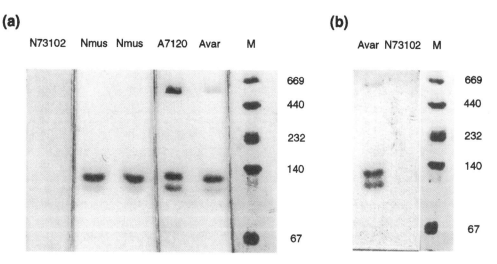

Figure 8. Native PAGE/in situ hydrogenase activity of partially purified extracts of *Nostoc* PCC 73102 (N73102), *Nostoc muscorum* (Nmus), *Anabaena* PCC 7120 (A7120), and *Anabaena variabilis* (Avar). (a) and (b) represent two different growth media (see footnote in Table 1). Standard proteins (M, in kDa) are shown to the right in each gel. (Tamagnini et al., 1997).

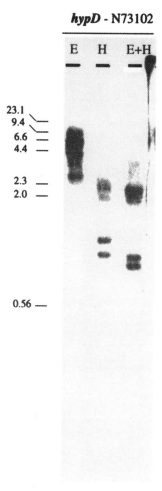

Figure 9. Autoradiograph demonstrating the presence of a sequence homologous to *hypD* in *Nostoc* PCC 73102. Total genomic DNA of *N.* PCC 73102 was digested (E - *Eco*RI, H - *Hin*dIII, E+H - *Hin*dIII & *Eco*RI), separated by agarose gel electrophoresis, transferred to a nylon membrane, and hybridized with a ^{32}P-labeled part of the *Synechocystis* PCC 6803 *hypD* as probe. The positions of the molecular length markers (in kb) are shown to the left.

4.4. *hyp*-Genes

Hydrogenases contain nickel in their active site, and the maturation of the individual subunits are believed to be nickel-dependent. In *Nostoc* PCC 73102, we could demonstrate a stimulatory effect upon addition of nickel ions on the in vivo, light-dependent, H_2 uptake activity (Oxelfelt et al., 1995). Several genes involved in nickel-insertion/maturation (*hyp*-genes) have been characterized. In *Azotobacter, Rhodobacter, Rhizobium, Bradyrhizobium,* and *Alcaligenes*, up to six conserved *hyp*-genes (*hypABCDEF*) have been identified (Chen and Mortensson, 1992; Colbeau et al., 1993; Du et al., 1992; Dernedde et al., 1993; van Soom et al., 1993; Brito et al., 1994; Hernando et al., 1995; Dernedde et al., 1996; Rey et al., 1996). Mutants in these genes produce immature and nickel-free large subunits that lack any detectable catalytic activity. We have identified conserved regions within several *hyp* genes, constructed primers, and performed PCR using genomic DNA from the unicellular cyanobacterium *Synechocystis* PCC 6803 (CyanoBase). The identities of the obtained DNA fragments were confirmed by sequencing. Initial Southern hybridizations using part of the *Synechocystis* PCC 6803 *hypD* as the probe indicate the presence of a corresponding sequence in *Nostoc* PCC 73102 (Figure 9).

4.5. Biotechnological Potential

Nostoc PCC 73102 seems to be an unusual cyanobacterium. Both a nitrogenase (Martel et al., 1993; Oxelfelt et al., 1995) and an uptake hydrogenase (Oxelfelt et al., 1995; Oxelfelt et al., 1998) are clearly present in the cells, whereas there is no evidence of the presence of a bidirectional enzyme (Tamagnini et al., 1997). Additionally, this strain in situ exhibits high metabolic activity with almost no growth, a natural bioreactor. A *hupL* deficient mutant has been described in *Rhodobacter* (Jahn et al., 1994), and *hup$^-$* mutants used in biotechnological experiments (Willison et al., 1984; Krahn et al., 1996; Zorin et al., 1996). We are presently constructing mutants lacking a functional *hupL* gene (Hansel et al.; see article in this publication). Such a mutant, in comparison with the wild type, could give an insight into the role of the uptake hydrogenase in *N.* PCC 73102. From a biotechnological point of view, the mutant(s) should produce H_2 through the action of nitrogenase, have no functional uptake hydrogenase, and reveal no potential uptake or regulatory effect(s) by a bidirectional enzyme.

ACKNOWLEDGMENTS

This work was financially supported by the Swedish Institute, Uppsala University, the Swedish Natural Science Research Council, Swedish National Board for Industrial and Technical Development, PRAXIS XXI, and the Carl Tryggers and OK Environment Foundations.

REFERENCES

Appel, J., and Schulz, R., 1996, Sequence analysis of an operon of a NAD(P)-reducing nickel hydrogenase from the cyanobacterium *Synechocystis* sp. PCC 6803 gives additional evidence for direct coupling of the enzyme to NAD(P)H-dehydrogenase (complex I), *Biochim. Biophys. Acta*, 1298:141–147.

Benemann, J., 1996, Hydrogen biotechnology: progress and prospects, *Nature Biotech.*, 14:1101–1103.

Boison, G., Schmitz, O., Mikheeva, L., Shestakov,S., and Bothe, H., 1996, Cloning, molecular analysis, and insertional mutagenesis of the bidirectional hydrogenase genes from the cyanobacterium *Anacystis nidulans*, *FEBS Lett.*, 394:153–158.

Brito, B., Palacios, J., Hidalgo, E., Imperial, J., and Ruiz-Argueso, T., 1994, Nickel availability to pea (*Pisum sativum* L.) plants limits hydrogenase activity of *Rhizobium leguminosarum* bv. *viciae* bacterioides by affecting the processing of the hydrogenase structural subunits, *J. Bacteriol.*, 176:5297–5303.

Carrasco, C.D., Buettner, J.A., and Golden, J.W., 1995, Programmed DNA rearrangement of a cyanobacterial *hupL* gene in heterocysts, in *Proceedings of the National Academy of Sciences (USA)*, 92:791–795.

Chen, J.C., and Mortenson, L.E., 1992, Identification of six open reading frames from a region of the *Azotobacter vinelandii* genome likely involved in dihydrogen metabolism, *Biochim. Biophys. Acta*, 1131:199–202.

Colbeau, A., and Viganis, P.M., 1992, Use of *hupS:lacZ* gene fusion to study regulation of hydrogenase expression in *Rhodobacter capsulatus*: stimulation by H_2, *J. Bacteriol.*, 174:4528–4264.

Colbeau, A., Richaud, P., Toussaint, B., Caballero, F.J., Elster, C., Dephin, C., Smith, R.L., Chabert, J., and Vignais, P.M., 1993, Organization of the genes necessary for hydrogenase expression in *Rhodobacter capsulatus*; sequence analysis and identification of two *hyp* regulatory mutants, *Mol. Microbiol.*, 8:15–29.

Dernedde, J., Eitinger. M., and Friedrich, B., 1993, Analysis of a pleotropic gene region involved in formation of catalytically active hydrogenase in *Alcaligenes eutrophus* H16, *Arch. Microbiol.*, 159:545–553.

Dernedde, J., Eitinger, M., Patange, N., and Friedrich, B., 1996, *hyp* gene products in *Alcaligenes eutrophus* are part of a membrane-maturation system, *Eur. J. Biochem.*, 235:351–358.

Du, L., Stejskal, and Tibelius, K.H., 1992, Characterization of two genes (*hupD* and *hupE*) required for hydrogenase activity in *Azotobacter chrococcum*, *FEMS Microbiol. Lett.*, 75:93–101.

Elsen, S., Colbeau, A., Chabert, J., and Vignais, P.M., 1996, The *hupTUV* operon is involved in negative control of hydrogenase synthesis in *Rhodobacter capsulatus*, *J. Bacteriol.*, 178:5174–5181.

Elsen, S., Richaud, P., Colbeau, A., and Vignais, P.M., 1993, Sequence analysis and interposon mutagenesis of the *hupT* gene, which encodes a sensor protein involved in repression of hydrogenase synthesis in *Rhodobacter capsulatus*, *J. Bacteriol.*, 175:7404–7412.

Golden, J.W., Robinson, S.J., and Haselkorn, R., 1985, Rearrangement of nitrogen-fixation genes during heterocyst differentiation in the cyanobacterium *Anabaena*, *Nature*, 314:419–423.

Golden, J.W., Whorff, L.L., and Wiest, D.R., 1991, Independent regulation of *nif*HDK operon transcription and DNA rearrangement during heterocyst differentiation in the cyanobacterium *Anabaena* sp. strain PCC 7120, *J. Bacteriol.*, 173:7098–7105.

Hansel, A., and Lindblad, P., 1998, Towards optimization of cyanobacteria as biotechnologically relevant producers of molecular hydrogen, a clean and renewable energy source, *Appl. Microbiol. Biotechnol.*, in press.

Hernando, Y., Palacios, J.M., Imperial, J., and Ruiz-Argueso, T., 1995, The *hyp*BFCDE operon from *Rhizobium leguminosarum* biovar *viciae* is expressed from a promotor that escapes mutagenesis of the *fnr*N gene, *J. Bacteriol.*, 177:5661–5669.

Jahn, A., Keuntje, B., Dörffler, M., Klipp, W., and Oelze, J., 1994, Optimizing photoheterotrophic H_2 production by *Rhodobacter capsulatus* upon interposon mutagenesis in the *hupL* gene, *Appl. Microbiol. Biotechnol.*, 40:687–690.

Krahn, E., Scneider, K., and Müller, A., 1996, Comparative characterization of H_2 production by the convential Mo nitrogenase and the alternative "iron-only" nitrogenase of *Rhodobacter capsulatus hup⁻* mutants, *Appl. Microbiol. Biotechnol.*, 46:285–290.

Martel, A. Jansson, E., Garcia-Reina, G., and Lindblad, P., 1993, Ornithine cycle in *Nostoc* PCC 73102. arginase, OCT and arginine deiminase and the effects of addition of external arginine, ornithine, or citrulline, *Arch. Microbiol.*, 159:506–511.

Oxelfelt, F., Tamagnini, P., Salema, R., and Lindblad, P., 1995, Hydrogen uptake in *Nostoc* strain PCC 73102: effects of nickel, hydrogen, carbon, and nitrogen, *Plant Physiol. Biochem.*, 33:617–623.

Oxelfelt, F., Tamagnini, P., and Lindblad, P., 1998, Hydrogen uptake in *Nostoc* PCC 73102; cloning and characterization of a *hupSL* homologue, *Arch. Microbiol.*, 169:267–274

Rai, A.N., (ed.), 1990, *Handbook of Symbiotic Cyanobacteria*, CRC Press.

Rao, K.K., and Hall, D.O., 1996, Hydrogen production by cyanobacteria: potential, problems, and prospects, *J. Mar. Biotechnol.*, 4:10–15.

Rey, L., Fernandez, D., Brito, B., Hernando, Y., Palacios, J.M., Imperial, J., and Ruiz-Argueso, T., 1996, The hydrogenase gene clusters of *Rhizobium leguminosarum* bv. *viciae* contains an additional gene (*hypX*), which encodes a protein with sequence similarity to the N10-formyltetrahydrofolate-dependent enzyme family and is required for nickel-dependent hydrogenase processing and activity, *Mol. Gen. Genet.*, 252:237–248.

Schmitz, O., Boison, G., Hilscher, R., Hundeshagen, B., Zimmer, W., Lottspeich, F., and Bothe, H., 1995, Molecular biological analysis of a bidirectional hydrogenase from cyanobacteria, *Eur. J. Biochem.*, 233:266–276.

Schulz, R., 1996, Hydrogenases and hydrogen production in eukaryotic organisms and cyanobacteria, *J. Mar. Biotechnol.*, 4:16–22.

Smith, G.D., 1990, Hydrogen metabolism in cyanobacteria, in *Phycotalk 2*, Kumar, H.O. (ed.), Rastogi and Company, Meerut, India, pp. 131–143.

Tamagnini, P., Oxelfelt, F., Salema, R., and Lindblad, P., 1995, Immunological characterization of hydrogenases in the nitrogen-fixing cyanobacterium *Nostoc* sp. strain PCC 73102, *Current Microbiol.*, 31:102–107.

Tamagnini, P., Troshina, O., Oxelfelt, F., Salema, R., and Lindblad, P. 1997, Hydrogenases in *Nostoc* sp. strain PCC 73102, a strain lacking a bidirectional enzyme, *Appl. Environ. Microbiol.*, 63:1801–1807.

van Soom, C., Verreth, C., Sampaio, M.J., and vanderLeyden, J., 1993, Identification of a potential transcriptional regulator of hydrogenase activity in free-living *Bradyrhizobium japonicum* strains, *Mol. Gen. Genet.*, 239:235–240.

Vignais, P.M., and Toussaint, B., 1994, Molecular biology of membrane-bound H_2 uptake hydrogenases, *Arch. Microbiol.*, 161:1–10.

Willison, J.C., Madern, D., and Vignais, P.M., 1984, Increased photoproduction of hydrogen by non-autotrophic mutants of *Rhodopseudomonas capsulata*, *Biochem. J.*, 219:593–600.

Zorin, N.A., Lissolo, T., Colbeau, A., and Vignais, P.M., 1996, Increased hydrogen production by *Rhodobacter capsulatus* strains deficient in uptake hydrogenase, *J. Mar. Biotechnol.*, 4:28–33.

8

MOLECULAR BIOLOGY OF HYDROGENASES

Claudio Tosi, Elisabetta Franchi, Francesco Rodriguez, Alessandro Selvaggi, and Paola Pedroni

Eniricerche S.p.A.
Environmental Technologies
Via F. Maritano, 26-20097 San Donato
Milanese, Italy

Key Words

hydrogenase, [Fe-S] cluster, hydrogenase gene cluster, *Pyrococcus furiosus*, *Acetomicrobium flavidum*, *Rhodobacter sphaeroides* RV

1. SUMMARY

Hydrogenases, found in a wide variety of organisms, catalyze either the consumption or the production of H_2 in response to different physiological conditions. In recent years, a large body of biochemical and genetic data on enzymes isolated from different sources has contributed to elucidating fundamental aspects about their catalytic properties and gene organization. In addition, the recently obtained crystal structure of a [Ni-Fe] hydrogenase sheds light on structure-function relationships in these enzymes. With the ultimate goal of engineering photosynthetic strains to improve their light-dependent hydrogen evolution capacity, we have characterized at the molecular level hydrogenases from different microbial sources. In particular, we have focused our attention on the H_2-evolving hydrogenases from *Pyrococcus furiosus* and *Acetomicrobium flavidum* and on the uptake hydrogenase system from *Rhodobacter sphaeroides* RV. This microorganism represents the species selected for use in a photobioreactor, as planned in the Japanese hydrogen production project in which we are involved.

2. INTRODUCTION

Hydrogenases are a heterogeneous group of redox enzymes that reversibly catalyze the production or consumption of hydrogen according to the equation: $H_2 \leftrightarrow 2H^+ + 2e^-$.

This reaction is coupled with the reduction or oxidation of an associated electron carrier specifically interacting with the enzyme. Hydrogenases are widespread in nature, having been isolated from organisms belonging to the three different domains of life (archaea, bacteria, and eucarya) and their activity enables the use of H_2 either as a source of energy or as a final electron acceptor. Functionally, hydrogenases may be differentiated into H_2-evolution enzymes when their physiological function is the production of hydrogen, or H_2-uptake when they catalyze the hydrogen consumption.

During the last three decades, hydrogenases have attracted considerable interest both from a fundamental point of view and for potential applications. These enzymes have been purified from a wide variety of microorganisms and extensively characterized with regard to structure, catalytic properties, and genetic organization (Adams, 1990; Reeve and Beckler, 1990; Przybyla et al., 1992; Wu and Mandrand, 1993; Friedrich and Schwartz, 1993; Albracht, 1994; Vignais and Toussaint, 1994). With the exception of a novel hydrogenase isolated from methanogenic archaea (Zirngibl et al., 1992; Hartmann et al., 1996), all hydrogenases characterized to date are iron-sulfur proteins that contain Fe atoms arranged in Fe-S clusters. They differ, however, in cellular location, subunit composition, electron carrier specificity, and sensitivity to O_2 inactivation. On the basis of their metal content, they are usually divided into two major groups: [Fe] hydrogenases containing only iron-sulfur clusters, one of which, the so-called H cluster, is the hydrogen-binding site (Adams, 1990); and [NiFe] hydrogenases, which contain nickel in addition to iron. Spectroscopic data indicate that the Ni metallocenter is directly involved in H_2 activation (Albracht, 1994). The latter group also includes the subset of [NiFeSe] hydrogenases, in which one of the sulfur atoms is replaced by selenium in the form of selenocysteine. To date, [Fe] hydrogenases seem to be rather limited in distribution because they have been detected only in strictly anaerobic bacteria. On the other hand, [NiFe] variants appear to be the most common, being present in anaerobes, facultative anaerobes, and aerobes. As a result, Ni-containing enzymes have been the most intensively studied, and in recent years, significant advances in their characterization have been made.

Structurally, the minimal functional unit of [NiFe] hydrogenases consists of two moieties: a large subunit, with molecular mass in the range of 45–65 kDa, carrying the nickel-containing active site, and a small subunit, with a molecular mass ranging between 28 and 35 kDa, which includes Fe-S clusters and has an electron-transfer function between redox proteins and the large subunit. Immunological data (Kovacs et al., 1989) provide evidence that the structure of large subunits is usually very well conserved even in distantly related organisms, whereas small subunits show a lower degree of similarity. According to amino acid sequence homologies, metal content, and physiological role, [NiFe] hydrogenases have been grouped into four subclasses which take into account the evolutionary relationships and motif compositions of the different enzymes (Wu and Mandrand, 1993). Besides the hydrogenase dimer, which constitutes the sole structural component of heterodimeric hydrogenases, one or two extra subunits, responsible for the interaction with the specific electron acceptor or donor, are also present in multimeric hydrogenases.

Recently, the first crystal structure of a [NiFe] hydrogenase, isolated from *Desulfovibrio gigas*, was determined (Volbeda et al., 1995), elucidating fundamental details about the catalytic site of the enzyme, the pathways of electron transfer, and the positions of the three iron-sulfur clusters. The functional model presented provides the first structural basis for understanding, at the atomic level, how molecular hydrogen is consumed or evolved by microorganisms.

As far as molecular characterization is concerned, the cloning and sequencing of a significant number of [NiFe] hydrogenase structural and related genes led to the conclusion that they are clustered and organized in a very conservative way. In particular, a surprisingly large number of accessory genes, linked to those encoding the structural subunits, were found in operons encoding heterodimeric membrane-bound hydrogenases. These additional genes are arranged in functional units and proved to be involved in the formation of the enzymatically active hydrogenase and in its regulation. The structure of membrane-bound hydrogenase operons is highly similar and generally heterodimeric hydrogenase sequences can be very easily aligned. With the only exception of the genes encoding *E. coli* hydrogenase isoenzyme-2, small subunit genes always precede large subunit ones. Auxiliary genes may specifically act on a single hydrogenase, or if multiple isoenzymes are present in the same organism, may exert a pleiotropic effect on all hydrogenase activities. On the other hand, the physical organization of genes coding for cytoplasmic or membrane-associated multimeric hydrogenases is less constant and the degree of sequence homology appears to be lower. In this group of enzymes, the variability of the extra subunits is of fundamental importance because they represent the electrochemical interface with the metabolism and determine the functional specificity of the hydrogenase involved.

The ability of biological systems to evolve hydrogen has been intensively investigated for the development of energy production processes based on environmentally friendly sources. With the final objective of producing hydrogen using microorganisms, particularly through the engineering of the photosynthetic bacterium *Rhodobacter sphaeroides* RV, we have biochemically and genetically characterized several hydrogenase systems from archaeal and bacterial origins. In this communication, we present our most recent work on the molecular characterization of different [NiFe] enzymes and their applications within the biological hydrogen production project, supported by the Japanese New Energy and Industrial Technology Development Organization (NEDO).

3. *Pyrococcus furiosus* SULFHYDROGENASE

P. furiosus is an anaerobic hyperthermophilic archaebacterium that grows at temperatures up to 100 °C (Fiala and Stetter, 1986). It exhibits a fermentative-type metabolism and utilizes simple and complex carbohydrates liberating H_2 and CO_2 in the absence of S°. When elemental sulfur is present, H_2S is produced. Its hydrogenase, physiologically H_2-evolving, is an enzyme endowed with unique features particularly attractive for applicative purposes. It is highly active, soluble, and besides being thermostable, is remarkably resistant to chemical and O_2 inactivation. The enzyme, purified and characterized from a biochemical point of view (Bryant and Adams, 1989; Mura et al., 1991), consists of four different subunits organized in an $\alpha\beta\gamma\delta$ tetrameric structure. The corresponding gene cluster, characterized using primers designed on the basis of the subunit N-terminal sequences (Pedroni et al., 1995), is shown in Figure 1. In the *P. furiosus* enzyme, two structurally different dimers can be distinguished on the basis of amino acid sequence comparisons: the $\delta\alpha$ dimer, related to small and large hydrogenase subunits of class IV, and the $\beta\gamma$ dimer, homologous to the AsrA and AsrB subunits of *Salmonella typhimurium* sulfite reductase. Indeed, the bifunctional nature of the *P. furiosus* enzyme, now referred to as sulfhydrogenase, has been biochemically demonstrated (Ma et al., 1993); in addition to proton reduction, the enzyme is able to act as a sulfur reductase, catalyzing S° or polysulfide reduction to H_2S. The possibility of dissociating the two structural components without losing the individual catalytic functions remains to be demonstrated. These findings might also suggest the evolutionary pathway followed by ancestral hydrogenases because archaea are considered to be closely related to ancient organisms.

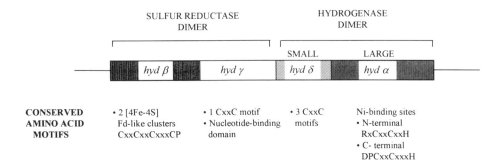

Figure 1. Molecular organization of the *P. furiosus* sulfhydrogenase gene cluster and schematic representation of the conserved motifs in the amino acid sequences.

The inherent properties of *P. furiosus* sulfhydrogenase prompted us to test this enzyme as a catalyst in a light-mediated in vitro system (Figure 2) (Mura et al., 1996; Pedroni et al., 1996). Photoinduced hydrogen production was performed by coupling the *P. furiosus* sulfhydrogenase to the semiconductor titanium dioxide in the presence of methylviologen as an electron carrier and Tris-HCl as an electron donor. Under the experimental conditions used, the maximum yield of hydrogen evolved was 62 μmol/mL and the rate was as high as 3 μmol/min. Future experiments will investigate the possibility of using waste products as electron donors to make the system both economically and environmentally attractive.

4. *Acetomicrobium flavidum* HYDROGENASE

A. flavidum is an anaerobic thermophilic bacterium optimally growing at 60 °C, which ferments 1 mol glucose to 2 mol acetate and CO_2 and 4 mol of H_2 (Soutschek et al., 1984). Its H_2-evolving hydrogenase exhibits the same biochemical properties already described for the *P. furiosus* enzyme, with the exception of the sulfur reductase activity and temperature optimum. Biochemical characterization revealed that it is a heterodimeric hydrogenase composed of two different subunits arranged as an $\alpha_2\beta_2$ tetramer. The corre-

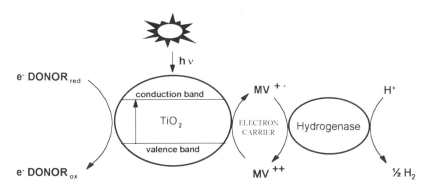

Figure 2. Reaction scheme for light-mediated in vitro H_2 production. TiO_2: titanium dioxide; MV: methylviologen in the oxidized form (MV^{++}) and in the reduced form ($MV^{+\bullet}$).

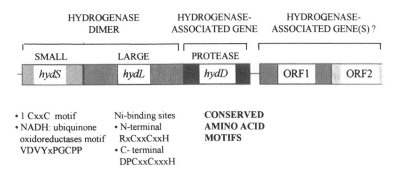

Figure 3. Molecular organization of the *A. flavidum* hydrogenase gene cluster and schematic representation of the conserved motifs in the amino acid sequences.

sponding gene cluster has been isolated using the same strategy successfully adopted for the cloning of the *P. furiosus* hydrogenase operon (Figure 3) (Mura et al., 1996b). Comparative analysis of the subunit sequences and the presence of conserved amino acid motifs suggest that the enzyme can be classified as a [NiFe] hydrogenase of class IV. Additional ORFs are present downstream from the *hydSL* structural genes: the first gene, designated *hydD*, encodes a polypeptide showing homology to hydrogenase-associated proteins from different bacteria, whereas the products of the other two putative ORFs exhibit no homology to protein sequences stored in data libraries. On the basis of sequence similarity and *hydD* gene location, we assume that HydD is the protease which specifically processes the hydrogenase large subunit at its C-terminal end. This proteolytic cleavage, which occurs after nickel incorporation into the catalytic subunit, represents one of the posttranslational maturation steps fundamental to the formation of enzymatically active [NiFe] hydrogenases (Maier and Böck, 1996). Experiments aimed at the co-expression in *E. coli* of the three *hydSLD* genes in a stable form are underway to verify the large subunit processing reaction as mediated by the specific protease in vivo.

5. *Rhodobacter spaeroides* RV UPTAKE HYDROGENASE

The purple non-sulfur bacterium *R. sphaeroides* RV, a H_2-overproducing strain isolated in Japan (Mao et al., 1986), is the microorganism selected for the photobioreactor construction in our biological hydrogen production project. In purple non-sulfur bacteria, at least two different enzymes are involved in hydrogen metabolism: the nitrogenase complex, responsible for H_2 production, and a membrane-bound uptake hydrogenase that mediates H_2 consumption. It has been suggested that the latter enzyme recycles the H_2 produced by the nitrogenase, thereby decreasing the rate and yield of H_2 photoproduction. This hypothesis was experimentally confirmed with *R. capsulatus* and *R. rubrum*; inactivation of the structural and/or regulatory *hup* genes significantly increased the efficiency of H_2 production (Jahn et al., 1994; Kern et al., 1994).

To further enhance the hydrogen evolution capability of *R. sphaeroides* RV, we have undertaken experiments aimed either at the construction of mutants impaired in the uptake hydrogenase activity or at the homologous overexpression of the regulatory gene(s) involved in the repression of the uptake system. In particular, we plan to obtain uptake hydrogenase mutants through inactivation of the corresponding structural gene(s) by

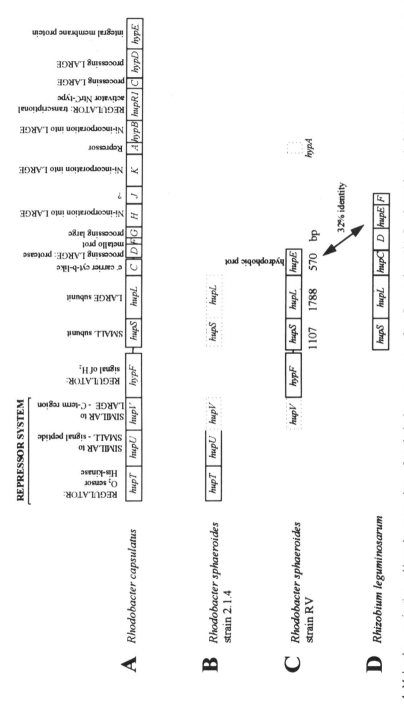

Figure 4. Molecular organization and homologous products of uptake hydrogenase operons from *R. capsulatus* (A), *R. sphaeroides* strain 2.4.1 (B), *R. sphaeroides* strain RV (C), and *R. leguminosarum* (D). Features or functions of the gene products are also indicated.

interposon mutagenesis and to overproduce the negative regulator(s) using expression vectors suitable for photosynthetic bacteria. Thus, the molecular characterization of the uptake hydrogenase operon from this strain was a prerequisite for both strategies.

The phylogenetically closely related bacterium *R. capsulatus* has been extensively characterized as far as the uptake hydrogenase gene cluster is concerned (Figure 4A). As already mentioned, genes encoding membrane-bound hydrogenases are clustered in a very conservative way and are organized in distinct functional units. Up to 20 accessory genes were found to be linked to *hupS* and *hupL*, whose products are not a constitutive part of the mature enzyme, but rather function as potential electron carriers channelling electrons from the enzyme to the natural electron acceptor, as positive or negative regulators, as proteins taking part in nickel metabolism, and as proteins involved in hydrogenase activation steps (i.e., nickel insertion, subunit processing, assembling of the holoenzyme, and membrane attachment).

Unlike *R. capsulatus*, the *R. sphaeroides* uptake hydrogenase operon has been so far poorly characterized. The only DNA sequences completely determined were the *hupTU* genes from *R. sphaeroides* strain 2.4.1 (Gomelsky and Kaplan, 1995) (Figure 4B), whose products, together with those of *hupV* and *hypA* genes, are thought to be involved in the repression of the uptake hydrogenase activity. By PCR and IPCR amplifications, we have isolated from the strain *R. sphaeroides* RV several ORFs belonging to the uptake hydrogenase operon (Figure 4C). In particular, the structural genes *hupSL*, the *hupE* gene, potentially encoding a hydrophobic protein of unknown function, and the *hypF* gene, coding a regulatory zinc-finger protein, have been fully sequenced, whereas PCR products of the size expected for *hupV* and *hypA* are presently being characterized. It is worth noting that in *R. sphaeroides* RV, the gene organization within the uptake hydrogenase cluster differs from what has previously been found in highly related photosynthetic bacteria and a homology search revealed that a HupE-like protein is present only in the hydrogenase operon from *Rhizobium leguminosarum* (Hidalgo et al., 1992) (Figure 4D).

ACKNOWLEDGMENTS

This work was partially supported by Eniricerche S.p.A. and performed under the management of the Research Institute of Innovative Technology for the Earth (RITE) as a part of the Research and Development Project on Environmentally Friendly Technology for the Production of Hydrogen, supported by the New Energy and Industrial Technology Development Organization (NEDO), Japan.

REFERENCES

Adams, M.W.W., 1990, The structure and mechanism of iron-hydrogenases, *Biochim. Biophys. Acta*, 1020:115–145.
Albracht, S.P.J., 1994, Nickel hydrogenases: in search of the active site, *Biochim. Biophys. Acta*, 1188:167–204.
Bryant, F.O., and Adams, M.W.W., 1989, Characterization of hydrogenase from the hyperthermophilic archaebacterium *Pyrococcus furiosus*, *J. Biol. Chem.*, 246:5070–5079.
Fiala, G., and Stetter, K.O., 1986, *Pyrococcus furiosus* sp. nov. represents a novel genus of marine heterotrophic archaebacteria growing optimally at 100 °C, *Arch. Microbiol.*, 145:56–61.
Friedrich, B., and Schwartz, E., 1993, Molecular biology of hydrogen utilization in aerobic chemolithotrophs, *Annu. Rev. Microbiol.*, 47:351–383.

Gomelsky, M., and Kaplan, S., 1995, Isolation of regulatory mutants in photosynthesis gene expression in *Rhodobacter sphaeroides* 2.4.1 and partial complementation of a PrrB mutant by the HupT histidine kinase, *Microbiology*, 141:1805–1819.

Hartmann, G.C., Klein, A.R., Linder, M., and Thauer, R.K., 1996, Purification, properties, and primary structure of H_2-forming N^5,N^{10}-methylenetetrahydromethanopterin dehydrogenase from *Methanococcus thermolithotrophicus*, *Arch. Microbiol.*, 165:187–193.

Hidalgo, E., Palacios, J.M., Murillo, J., and Ruiz-Argüeso, T., 1992, Nucleotide sequence and characterization of four additional genes of the hydrogenase structural operon from *Rhizobium leguminosarum* bv. viciae, *J. Bacteriol.*, 174:4130–4139.

Jahn, A., Keuntje, B., Dörffler, M., Klipp, W., and Oelze, J., 1994, Optimizing photoheterotrophic H_2 production by *Rhodobacter capsulatus* upon interposon mutagenesis in the *hupL* gene, *Appl. Microbiol. Biotechnol.*, 40:687–690.

Kern, M., Klipp, W., and Klemme, J.H., 1994, Increased nitrogenase-dependent H_2 photoproduction by *hup* mutants of *Rhodospirillum rubrum*, *Appl. Environmental Microbiol.*, 60:1768–1774.

Kovacs, K.L., Seefeld, L.C., Tigyi, G., Dole, C.M., Mortenson, L.E., and Arp, D.J., 1989, Immunological relationship among hydrogenases, *J. Bacteriol.*, 171:430–435.

Ma, K., Schicho, R.N., Kelly, R.M., and Adams, M.W.W., 1993, Hydrogenase of the hyperthermophile *Pyrococcus furiosus* is an elemental sulfur reductase or sulfhydrogenase: evidence for a sulfur-reducing hydrogenase ancestor, in *Proceedings of the National Academy of Sciences (USA)*, 90:5341–5344.

Maier, T., and Böck, A., 1996, Nickel incorporation into hydrogenases, in *Mechanisms of Metallocenter Assembly*, Hausinger, R.P., Eichhorn, G.L., and Marzilli, L.G., (eds.), VCH Publishers, Inc., New York, pp. 173–192.

Mao, X.Y., Miyake, J., and Kawamura, S., 1986, Screening photosynthetic bacteria for hydrogen production from organic acids, *J. Ferment. Technol.*, 64:245–249.

Mura, G.M., Pedroni, P., Branduzzi, P., Grandi, G., Park, J.B., Adams, M.W.W., and Galli, G., 1991, Characterization of *Pyrococcus furiosus* hydrogenase, in *Proceedings of the 15th International Congress of Biochemistry*, Jerusalem, p. 313.

Mura, G.M., Galli, G., Pratesi, C., Frascotti, G., Serbolisca, L., Pedroni, P., and Grandi, G., 1996, Photoinduced hydrogen production using titanium dioxide coupled to thermostable hydrogenases, *J. Marine Biotechnol.*, 4:68–74.

Mura, G.M., Pedroni, P., Pratesi, C., Galli, G., Serbolisca, L., and Grandi, G., 1996b, The [Ni-Fe] hydrogenase from the thermophilic bacterium *Acetomicrobium flavidum*, *Microbiology*, 142:829–836.

Pedroni, P., Della Volpe, A., Galli, G., Mura, G.M., Pratesi, C., and Grandi, G., 1995, Characterization of the locus encoding the [Ni-Fe] sulfhydrogenase from the archaeon *Pyrococcus furiosus*: evidence for a relationship to bacterial sulfite reductases, *Microbiology*, 141:449–458.

Pedroni, P., Mura, G.M., Galli, G., Pratesi, C., Serbolisca, L., and Grandi, G., 1996, The hydrogenase from the hyperthermophilic archaeon *Pyrococcus furiosus*: from basic research to possible future applications, *Int. J. Hydrogen Energy*, 21:853–858.

Przybyla, A.E., Robbins, J., Menon, N., and Peck, H.D., 1992, Structure-function relationship among the nickel-containing hydrogenases, *FEMS Microbiol. Rev.*, 88:109–136.

Reeve, J.N., and Beckler, G.S., 1990, Conservation of primary structure in prokaryotic hydrogenases, *FEMS Microbiol. Rev.*, 87:419–424.

Soutschek, E., Winter, J., Scindler, F., and Kandler, O., 1984, *Acetomicrobium flavidum*, gen. nov., sp. nov., a thermophilic, anaerobic bacterium from sewage sludge, forming acetate, CO_2 and H_2 from glucose, *Syst. Appl. Microbiol.*, 5:377–390.

Vignais, P.M., and Toussaint, B., 1994, Molecular biology of membrane-bound H_2 uptake hydrogenases, *Arch. Microbiol.*, 161:1–10.

Volbeda, A., Charon, M.H., Piras, C., Hatchikian, E.H., Frey, M., and Fontecilla-Camps, J.C., 1995, Crystal structure of the nickel-iron hydrogenase from *Desulfovibrio gigas*, *Nature*, 373:580–587.

Wu, L.F., and Mandrand, M.A., 1993, Microbial hydrogenases: primary structure, classification, signature, and phylogeny, *FEMS Microbiol. Rev.*, 104:243–270.

Zirngibl, C., van Dongen, W., Schwörer, B., von Bünau, R., Richter, M., Klein, A., and Thauer, R.K., 1992, Hydrogen-forming methylenetetrahydromethanopterin dehydrogenase, a novel type of hydrogenase without iron-sulfur clusters in methanogenic archaea, *Eur. J. Biochem.*, 208:511–520.

9

EFFECT OF HYDROGENASE 3 OVER-EXPRESSION AND DISRUPTION OF NITRATE REDUCTASE ON FERMAMENTIVE HYDROGEN PRODUCTION IN *Escherichia coli*

A Metabolic Engineering Approach

Koji Sode,[1] Mika Watanabe,[1] Hiroshi Makimoto,[1] and Masamitsu Tomiyama[2]

[1]Department of Biotechnology
Tokyo University of Agriculture and Technology
2-24-16 Nakamachi
Koganei, Tokyo 184
[2]National Institute of Agrobiological Resources
2-1-2 Kannondai
Tsukuba City, Ibaraki 305, Japan

Key Words

metabolic engineering, formate hydrogenlyase system, hydrogenase, nitrate reductase, *Escherichia coli*

1. SUMMARY

Based on available molecular genetics data on the regulations and enzymes in the *E. coli* anaerobic metabolism, we are examining approaches for the enhancement of hydrogen production by *E. coli* through metabolic engineering. A regulation mutant, HD701, in which the repressor gene *hyc*A for the terminal enzyme for hydrogen production (hydrogenase 3) was disrupted, increased the level of expression hydrogenase 3. However, this mutation did not result in the enhancement of hydrogen production efficiency. Subsequently, the *E. coli* mutant strain RK5265, of which *nar*G, a gene encoding the α-subunit of nitrate reductase, was disrupted, was examined for hydrogen production. The elimination of nitrate reductase resulted in hydrogen production in the presence of nitrate,

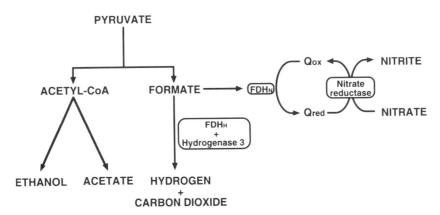

Figure 1. Fermentative pathway of *E. coli* in bacterial hydrogen production. FDH-N, formate dehydrogenase nitrate reductase linking; FDH-H, formate dehydrogenase formate hydrogenlyase linking.

although the nitrate reductase active parent strain (RK4353) could not produce hydrogen under the same conditions because of the repression of the formate hydrogenlyase system. These results revealed that the elimination of the branch reaction in formate utilization resulted in the construction of a versatile strain for practical hydrogen production.

2. INTRODUCTION

Bacterial hydrogen production based on anaerobic metabolism has been studied as a means of efficiently using waste biomass, such as molasses from sugar manufacturing. Anaerobic processes can decompose saccharides, recovering molecular hydrogen gas with high efficiency. Simultaneously produced organic acids can be further utilized for photosynthetic bacterial hydrogen production.

Molecular biology studies on the anaerobic metabolism of *Escherichia coli* have identified the majority of enzymes and genes responsible for fermentative hydrogen production (Böck et al., 1996). On the basis of the genomic information, *E. coli* was shown to be an attractive target microorganism for bacterial hydrogen production by redesigning the metabolic pathway. *E. coli* produces hydrogen by mixed acid fermentation, mainly from glucose. In *E. coli*, hydrogen evolves by a formate hydrogenlyase system (FHL) containing the formate dehydrogenase-H (FDH-H, hydrogenase linked), electron carrier intermediate(s), and hydrogenase 3 (Sawers et al., 1985). In evaluating a metabolic engineering strategy to improve bacterial hydrogen production, fermentative hydrogen production must be investigated using adequate mutant strains that show unique properties on the branch point and rate-limiting steps on carbon and/or electron flux from glucose to FHL, for example, the rate of glucose or other saccharide uptake and preference, the utilization of pyruvate in the competition between lactate dehydrogenase and pyruvate formatelyase, the utilization of formate in the competition between FHL and formate dehydrogenase-N coupling with nitrate reductase system, level of FHL expression, contribution of uptake hydrogenases to the rate of hydrogen production, and so on.

In this study, we focused on the following effects on hydrogen production: the expression level of hydrogenase 3 and the mutation of a branch pathway competing for the carbon flux from glucose to hydrogen production, formate dehydrogenase-N (FDH-N), which is coupled with a nitrate reductase system (Figure 1).

3. EXPERIMENTAL

3.1. Bacterial Strains and Cultivation

Bacterial strains used in this study are listed in Table 1. These strains were grown anaerobically in an argon atmosphere, at 37 °C for 5 h in 10 mL of a complex medium (1% tryptone, 0.5% yeast extract, 0.4% glucose, 100 mM potassium phosphate, 15 mM sodium formate, 1 µM sodium molybdate, and 5 µM nickel chloride, pH 6.5). Ten mg mL^{-1} tetracycline were added for the cultivation of the strain RK5265 possessing the Tn10 insertion.

3.2. Bacterial Hydrogen Production

Each *E. coli* strain was cultivated as described above, and cells were harvested by centrifugation (3000 x g, 4 °C), washed three times with a 100 mM potassium phosphate buffer, pH 7.0, and resuspended in 41 mL of the same buffer in a 83 mL Schlenk's tube. These prepared cell suspensions were treated by repeated argon gas purging and degassing in order to establish an argon atmospheric anaerobic condition in the reaction tube. The reactions for the hydrogen production were initiated by the addition of defined concentrations of either glucose or formate at 37 °C. The phase of the reaction tube was analyzed every hour using TCD type gas chromatography (GC-8AIT, Shimadzu Corp., Japan), equipped with a 1-mL sample loop. For the measurement of hydrogenase activity, methyl viologen (MV) was used as the substrate. This reaction mixture consisted of 40 mL of 16.5 mM MV solution containing 2.8 mM sodium phosphinate, 1.0 mM $MgCl_2$, and 9.6 mM sodium hydrosulfate in 0.1 M potassium phosphate buffer, pH 6.5. By injecting anaerobically prepared suspended *E. coli* cells into the reaction tube, hydrogen production was initiated. Every 20 min, the amount of hydrogen produced was determined as noted above and the rate of hydrogen evolution was used as the value for hydrogenase activity.

4. RESULTS AND DISCUSSION

4.1. Effect of Over-Expression of Hydrogenase 3 on Hydrogen Production

Three different hydrogenase isoenzymes have been reported in *E. coli*: hydrogenase 1 (Sawers et al., 1986), hydrogenase 2 (Ballantine et al., 1986), and hydrogenase 3 (Sawers et al., 1985), and each isoenzyme was encoded in *hya* (Menon et al., 1990), *hyb*

Table 1. Bacterial strains used

Strains	Genotype	Reference or source
MC4100	F$^-$ *araD139* Δ(*argF-lac*) *U 169 deoC1 flhD5301 ptsF25 relA1 rpsL150*	National Institute of Genetics (Japan)
HD701	MC4100, Δ*hycA*	M. Sauter et al. (1994)
RK4353	As MC4100 but *gryA219 non-9*	Stewart and MacGregor (1982)
RK5265	As RK4353 but *narG202::Tn10*	Stewart and MacGregor (1982)

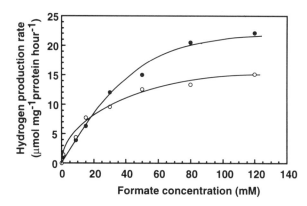

Figure 2. Correlation between formate concentration and hydrogen production rate by *E. coli* strains HD701 (open circles) and MC4100 (closed circles).

(Menon et al., 1994), and *hyc* (Böhm et al., 1990) operons, respectively. Hydrogenases 1 and 2 have been reported to act as hydrogen oxidation catalysts, so-called uptake hydrogenase, whereas hydrogenase 3 is a constituent of the FHL system, and therefore catalyzes the terminal reactions in hydrogen production. Recently, the existence of a fourth hydrogenase isoenzyme and its operon was reported, Hyf (hydrogenase 4) and *hyf*, although the role of this isoenzyme is still unknown (Andrews et al., 1997). The *hyc* operon, encoding hydrogenase 3, is composed of 8 genes, *hyc*A to *hyc*H (Böhm et al., 1990). The structural gene of the large subunit of hydrogenase 3 may be encoded in *hyc*E according to homology analyses. *hyc*A encodes the repressor gene of the FHL system, including *hyc*. It has been reported that the deletion of *hyc*A results in an increase in hydrogenase activity (Sauter et al., 1992). Therefore, we tested the potential of hydrogen evolution of a mutant *E. coli* strain deficient in the *hyc*A gene, in order to evaluate whether the hydrogenase 3 expression level might be the rate-limiting step in an *E. coli* hydrogen production system.

Table 2 summarizes hydrogenase activity and hydrogen production from glucose of the *hyc*A mutant strain (HD701) (Sauter et al., 1992) and its parent strain (MC4100). The HD701 strain showed twofold greater hydrogenase activity compared to MC4100. This was due to the overexpression of hydrogenase 3 by the deletion of *hyc*A. However, the hydrogen production rate from glucose was not affected by this mutation. This indicated that the enhancement of the terminal enzyme expression level will not result in improved hydrogen production efficiency.

Figure 2 shows the correlation between formate concentration and the hydrogen production rate of each strain. At formate concentrations higher than 10 mM, the mutant strain showed a lower hydrogen production rate than its parent strain, even though it possessed higher hydrogenase activity. Assuming that this correlation was a Michaelis-Menten-type saturation curve, the apparent K_m and V_{max} were obtained for each strain. The apparent V_{max} and K_m values of the HD701 strain were 18 µmol mg^{-1} protein h^{-1} and

Table 2. Hydrogen production rates from reduced MV and glucose

Substrate	Hydrogen production rates (µmol mg^{-1} protein h^{-1})	
	MC4100	HD701(ΔhycA)
Reduced MV	0.8	1.5
Glucose	1.8	1.8

5.6 mM glucose and 16.5 mM reduced MV were used in this study.

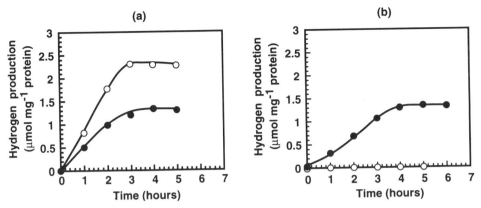

Figure 3. Time courses of hydrogen production from 5.6 mM glucose by *E. coli* strains RK5265 (closed circles) and RK4353 (open circles) in the (a) absence and (b) presence of 40 mM nitrate.

24 mM, respectively, whereas those of the MC4100 strains were 36 μmol mg^{-1} protein h^{-1} and 67 mM, respectively. Therefore, by the overexpression of hydrogenase 3, a drastic decrease in V_{max} values was observed. The FHL system is assumed to constitute a multiprotein complex located on the inner aspect of the cytoplasmic membrane (Sauter et al., 1992). Therefore, the overexpression of hydrogenase 3 may cause the partial dessociation of the complex, consequently resulting in the imbalance of formate utilization at a high concentration of substrate.

4.2. Hydrogen Production by a Mutant *E. coli* Strain Deficient in Nitrate Reductase

In the presence of nitrate, FDH-N and nitrate reductase are induced (Stewart, 1994). As a result, the formate is consumed preferentially by FDH-N due to its inherent high affinity for formate versus FDH-H. The K_m value for the formate of FDH-N is 0.12 mM, while that of FDH-H is 26 mM. Inasmuch as the FHL system is induced by the presence of formate, formate consumption by an FDH-N nitrate reductase system depresses the expression level of the FHL system. Consequently, both the expression of FHL and hydrogen production are repressed by the presence of nitrate in the medium. The removal of nitrate from waste water is not a practical process; therefore, a bacterial strain capable of hydrogen production even in the presence of nitrate is favorable. The subunits of nitrate reductase are encoded in the *fdn* operon. The α subunit of nitrate reductase is encoded in the *narG* locus. Therefore, we tested the potential of nitrate reductase mutant for hydrogen production in the presence of nitrate.

Figures 3a and 3b show the time courses of hydrogen production from glucose in the absence or presence of nitrate by the strain deficient in the *narG* locus (RK5265) and its parent strain (RK4353).

The total hydrogen produced by the parent strain RK4353 was 2.3 μmol mg^{-1} protein, in the absence of nitrate. The *E. coli* strain RK5265, in which *narG*, a gene encoding for an α subunit of nitrate reductase was disrupted by a Tn 10 insertion (Stewart et al., 1982), was then investigated for hydrogen production. Hydrogen production by RK5265 in the absence of nitrate was about 60% (1.3 μmol mg^{-1} protein) of that achieved by the parent strain. This would be due to the effect of the Tn10 insertion mutagenesis, which might cause the change in the ratio of total proteins, including enzymes relating to hydrogen production.

In the presence of 40 mM nitrate, however, hydrogen production by the strain RK4353 was greatly repressed, and no detectable hydrogen was produced. This indicated that cultivation in a rich medium utilized for pre-cultivation resulted in the induction of a nitrate reductase system. Therefore, the strain with a nitrate reductase system cannot be applied to bacterial hydrogen production in the presence of nitrate. In contrast, the strain RK5365 produced hydrogen in the presence of 40 mM nitrate. Hydrogen production was the same as the condition without nitrate (1.3 µmol mg^{-1} protein). Considering that the nitrate reductase system of parent strain RK4353 was induced by pre-cultivation in a rich medium, the FDH-N in RK5265 would be induced. However, the absence of the nitrate reductase due to the state of the gene locus *narG* achieved hydrogen production even in the presence of the 40 mM nitrate.

These results clearly indicate that by disrupting the nitrate reductase system, an *E. coli* strain suitable for practical hydrogen production can be developed and will not be affected by the contamination of nitrate in the starting substrate. The availability of a mutant strain enabled us to demonstrate the application of a mutant strain deficient in the *narG* locus. However, the existence of a constitutively synthesizing nitrate reductase was reported, a nitrate reductase Z (Bonnetoy et al., 1994). Indeed, by the cultivation of the strain RK5265 in the presence of 100 mM of nitrate, no FHL activity was observed (results not shown). Therefore, the ideal *E. coli* strain will be constructed by introducing mutation in the structural gene for FDH-N.

5. CONCLUSION

We have evaluated a strategy for the enhancement of hydrogen production by *E. coli* via metabolic engineering. The overexpression of a terminal enzyme responsible for hydrogen production, hydrogenase 3, did not result in the enhancement of hydrogen production efficiency. The elimination of nitrate reductase resulted in hydrogen production in the presence of nitrate, using cells precultivated in a rich medium. This result revealed that the elimination of the branch reaction in formate utilization resulted in the construction of a versatile strain for practical hydrogen production. Further engineering approaches are being conducted on the various branch points in the anaerobic pathway.

ACKNOWLEDGMENTS

The authors gratefully acknowledge Dr. A. Böck, University of München, and Dr. V. Stewart, Cornell University, for providing the mutant *E. coli* strains.

REFERENCES

Andrews, S.C., Quail, M.A., Golby, P., Guest, J.R., McClay, J., Amber, A., and Berks, B.C., 1997, A new 12-gene operon (*hyf*) encoding a fourth hydrogenase in *E. coli*, in Abstracts of the International Meeting of Anaerobic Metabolism, Electron Transfer Systems and Regulation, Marseille-Carry-Le-Rouet.

Ballantine, S.P., and Boxer, D.H., 1986, Isolation and characterization of a soluble active fragment of hydrogenase isoenzyme 2 from the membranes of anaerobically grown *Escherichia coli*, *Eur. J. Biochem.*, 156:277–284.

Böck, A., and Sawers, G., 1996, Fermentation, in *Escherichia coli and Salmonella*, 2nd ed., Neidhardt, F.C. et al. (eds.), ASM Press, Washington, D.C., pp. 262–282.

Böhm, R., Sauter, M., and Böck, A., 1990, Nucleotide sequence and expression of an operon in *Escherichia coli* coding for formate hydrogenlyase components, *Mol. Microbiol.*, 4:231–243.

Bonnefoy, V., and Demoss, J.A., 1994, Nitrate reductases in *Escherichia coli*, *Antonie van Leeuwenhoek*, 66:47–56.

Menon, N.K., Chatelus, C.Y., Dervartanian, M., Wendt, J.C., Shanmugam, K.T., Peck, H.D., Jr., and Pryzybyla, A.E., 1994, Cloning, sequencing, and mutational analysis of the *hyb* operon encoding *Escherichia coli* hydrogenase 2, *J. Baceriol.*, 176:4416–4423.

Menon, N.K., Robbins, J., Peck, H.D., Jr., Chatelus, C.Y., Choi, E. S., and Przybyla, A.E., 1990, Cloning and sequencing of a putative *Escherichia coli* [NiFe] hydrogenase operon containing six open reading frames, *J. Bacteriol.*, 172:1969–1977.

Sauter, M., Böhm, R., and Böck, A., 1992, Mutational analysis of the operon (*hyc*) determining hydrogenase 3 formation in *Escherichia coli*, *Mol. Microbiol.*, 6:1523–1532.

Sawers, R.G., Ballantine, S.P., and Boxer, D.H., 1985, Differential expression of hydrogenase isoenzymes in *Escherichia coli* K-12: evidence for a third isoenzyme, *J. Bacteriol.*, 164:1324–1331.

Sawers, R.G., and Boxer, D.H., 1986, Purification and properties of membrane-bound hydrogenase isoenzyme 1 from anaerobically grown *Escherichia coli* K-12, *Eur. J. Biochem.*, 156:265–275.

Stewart, V., 1994, Regulation of nitrate and nitrite reductase synthesis in enterobacteria, *Antonie van Leeuwenhoek*, 66:37–45.

Stewart, V., and MacGregor, C.H., 1982, Nitrate reductase in *Escherichia coli* K-12: involvement of *chl*C, *chl*E, and *chl*G loci, *J. Bacteriol.*, 151:1320–1325.

10

IMPROVEMENT OF BACTERIAL LIGHT-DEPENDENT HYDROGEN PRODUCTION BY ALTERING THE PHOTOSYNTHETIC PIGMENT RATIO

Masato Miyake,[1] Makoto Sekine,[1] Lyudmila G. Vasilieva,[2] Eiju Nakada,[2] Tatsuki Wakayama,[2] Yasuo Asada,[1] and Jun Miyake[3]

[1]National Institute of Bioscience and Human-Technology
Higashi 1-1, Tsukuba, Ibaraki 305
[2]Research Institute of Innovative Technology for the Earth
[3]National Institute for Advanced Interdisciplinary Research
Higashi 1-1-4
Tsukuba, Ibaraki 305, Japan

Key Words

light-harvesting complex, mutant, hydrogen production, photosynthetic bacteria, *puf* operon

1. SUMMARY

A method has been proposed for the enhancement of bacterial light-dependent hydrogen production by altering the LH1/LH2 ratio of *Rhodobacter sphaeroides* RV. A model mutant strain, P3, was obtained by UV irradiation from the wild strain (RV). The amount of bacteriochlorophyll in LH1 was reduced to 35% and that in LH2 was enhanced to 140% in P3 compared to RV. P3 showed 1.4 times higher hydrogen production rates than RV at 850 nm (an absorption maxima for LH2). At 875 nm (absorption maximum for LH1), P3 showed a similar hydrogen production rate to RV, despite the decreased level of LH1. P3 was restored to the wild-type LH1/LH2 ratio by genetic transformation with a plasmid carrying genes encoding LH1 pigment-binding proteins. The hydrogen production of the transformed P3 was also restored to that of the wild type. These results suggest that alteration of the LH1/LH2 ratio can enhance light-dependent hydrogen production.

2. INTRODUCTION

Light-dependent hydrogen production by photosynthetic bacteria has been extensively studied for the establishment of environmentally acceptable energy production.

BioHydrogen, edited by Zaborsky *et al.*
Plenum Press, New York, 1998

Many strong hydrogen producers have been screened and evaluated for hydrogen production, as described by Miyamoto (1994).

The purple bacterium *Rhodobacter sphaeroides* RV (Miyake and Kawamura, 1987) is a powerful hydrogen producer showing a 7% energy conversion efficiency in the presence of lactate and glutamate (Miyake and Kawamura, 1987; Miyake et al., 1989). The high rate of hydrogen production makes it suitable for photobioreactor applications (Tsygankov et al., 1994; Nakada et al., 1996).

Light absorption by the bacterial cells reduces the penetration of light into the photobioreactor and sometimes suppresses the hydrogen production rate of the photobioreactor (Nakada et al., 1995). Two different light-harvesting (LH) complexes are involved in the absorption of light quanta and transfer of excitation energy to the photosynthetic reaction center (RC) (Richter and Drews, 1991) in photosynthetic bacteria. *R. sphaeroides* shows absorption peaks at 800, 850, and 875 nm, corresponding to the LH2-bacteriochlorophyll B800, B850, and LH1-bacteriochlorophyll, respectively (Nagamine et al., 1996).

We have proposed a new genetic strategy to improve the hydrogen production rate of *R. sphaeroides* RV photobioreactor by altering the LH1/LH2 ratio, which affects the intracellular light energy transfer and light penetration into the photobioreactor.

3. EXPERIMENTAL

3.1. Bacteria and Cultures

Rhodobacter sphaeroides RV (Miyake and Kawamura, 1987) and its UV-mutant (Vasilieva et al., 1997) were cultivated in aSy medium (pH 6.8) (Miyake and Kawamura, 1987) containing 0.1% yeast extract, 10 mM ammonium sulfate, and 30 mM sodium succinate. The pH was adjusted with NaOH. The cells were cultivated with a 5% inoculum at 30 °C for 24 h under anaerobic conditions with tungsten lamps (33 W/m^2 lux). The 50-mL culture was transferred to a 200-mL bottle. The container was then filled with gL medium (pH 6.8) containing 50 mM sodium lactate and 10 mM sodium glutamate for hydrogen production. The culture was incubated at 30 °C for 18 h under anaerobic conditions with tungsten lamps (65 W/m^2). Only the cultures evolving hydrogen at 10–15 mL/h were used for further experiments.

3.2. Measurement of Hydrogen Production

Hydrogen production by RV or P3 was measured by using a hydrogen electrode system (Biott) (Figure 1). The reactor volume was 2.5 mL. The illumination area was 1.77 cm^2. A halogen lamp (Bromo Lamp Delux, LPL) was used as the light source. A water layer for cooling and band-path filters (Nippon Shinku Kougaku) for monochromatic light conditions were placed between the light source and the reactor. The light intensity was determined by a radiometer (Model 4090, SJI).

3.3. Genetic Manipulations

Standard cloning procedures were performed as described by Sambrook et al. (1989). Transconjugation of *R. sphaeroides* RV and the mutants was done according to Nagamine et al. (1996).

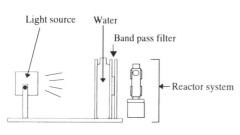
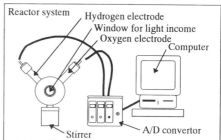

Figure 1. Schematic of the hydrogen electrode system. The reactor was made of stainless steel, the window made of crystal.

4. RESULTS AND CONCLUSIONS

4.1. P3 and the Complement

By spectroscopic analysis described by Vasilieva et al. (1998), the amount of bacteriochlorophylls in LH1 was reduced to 35% and that of LH2 was enhanced to 140% in P3, compared to RV. The absorption spectra of RV and P3 are indicated in Figure 2. The absorbance at 875 nm in P3 was lower than that in RV.

P3 was transformed with the plasmid pRKM-QBA (Figure 3a), which carried the *pufQBA* fragment (1.1 kbp) containing *puf* promoter Puf Q, LH1 alpha, and beta subunit structural genes. The absorption spectra of the transformants (strain P3-QBA) shows that the plasmid complemented the absorbance at 875 nm (Figure 3b). The amount of LH2 also reduced to the wild type.

4.2. Effect of LH1/LH2 Ratio on Hydrogen Production

The contribution of each light-harvesting complex (LH1 and LH2) to hydrogen production was examined under monochromatic light (Figures 4a and 4b). Under 850-nm light conditions, the maximum production rates of P3 were 1.4 times higher than those of RV. Moreover, P3 showed higher hydrogen production than RV even under the light-un-

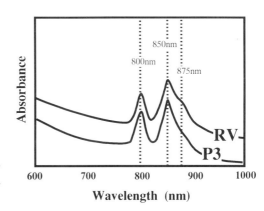

Figure 2. Absorption spectra of *R. sphaeroides* RV and mutant P3. Cell concentration used for measurement was 0.16 mg of dry weight per mL.

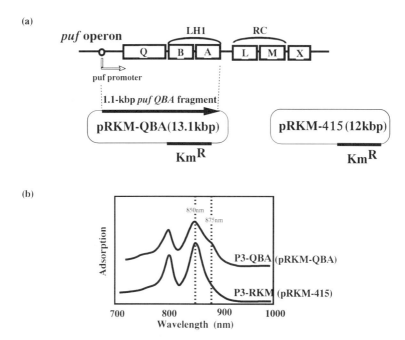

Figure 3. Complement of LH1 in P3: (a) schematic of the plasmids used for transformation of P3 and (b) absorbtion spectra of P3 transformants.

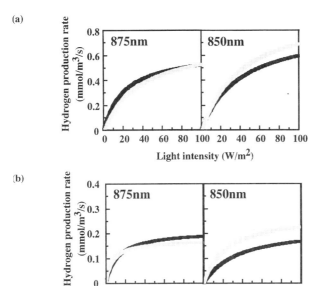

Figure 4. Hydrogen production rate of *R. sphaeroides* under monochromatic light conditions. Data sets approximated to the Michaelis-Menten equation with use of least-squares method. (a) Black and half-tone lines indicate *R. sphaeroides* RV and P3; (b) black and half-tone lines indicate P3-QBA and P3-RKM. Each cell concentration was 1.0 mg of dry weight per mL.

saturated conditions. Under 875-nm light conditions, the hydrogen production rates of P3 were similar to those of RV.

The P3 transformants showed lower hydrogen production rates than the wild type and P3, which was probably due to the kanamycin resistance (Nagamine et al., 1996). The hydrogen production rates of P3-QBA were lower than those of the P3 transformants harboring pRKM-415 (strain P3-RKM) under 850-nm light conditions (Figure 4b). Under 875-nm light conditions, P3-QBA showed similar hydrogen production rates as P3-RKM. The results strongly support the conclusion that the higher hydrogen production rates of P3 were caused by the alteration of the LH1/LH2 ratio.

We concluded that the alteration of the LH1/LH2 ratio induced the change in the pathway of energy from light-harvesting pigments to the photosynthetic reaction center. The changes in the amount of bacteriochlorophyll in LH2 and LH1 resulted in a change in the penetration of monochromatic light (850 and 875 nm) in the photobioreactor, which also affected hydrogen production.

Reduction of uptake hydrogenase in photosynthetic bacteria improves the net volume of hydrogen from a bioreactor but not the hydrogen production rate (Warthmann et al., 1993; Kern et al., 1994). Alteration of the LH2/LH1 ratio in this work was the first possible method for improving the hydrogen production rate in photosynthetic bacteria. Further optimization of pigment composition by genetic engineering could improve hydrogen production by photosynthetic bacteria.

ACKNOWLEDGMENTS

We thank Mr. Hiroshi Okano for his technical support. We also thank the Research Institute of Innovative Technology for the Earth for its cooperation.

REFERENCES

Kern, M., Klipp, W., and Klemme, J.H., 1994, Increased nitrogenase-dependent H_2 photoproduction by hup mutants of *Rhodospirillum rubrum*, *Appl. Env. Microbiol.*, 60:1768–1774.

Miyake, J., Asada, Y., and Kawamura, S., 1989, Nitrogenase, *Biomass Handbook*, Hall, C.W., and Kitani, O., (eds.), Gordon and Breech Scientific Publishers, New York, pp. 362–370.

Miyake, J., and Kawamura, S., 1987, Efficiency of light energy conversion to hydrogen by the photosynthetic bacterium *Rhodobacter sphaeroides*, *Int. J. Hydrogen Energy*, 12:147–149.

Miyamoto, K., 1994, Hydrogen production by photosynthetic bacteria and microalgae, *Recombinant Microbes for Industrial and Agricultural Applications*, Murooka, Y., and Imanaka, T. (eds.), Marcel Dekker, New York, pp. 771–786.

Nagamine, Y., Kawasugi, T., Miyake, M., Asada, Y., and Miyake, J., 1996, Characterization of photosynthetic bacterium *Rhodobacter sphaeroides* RV for hydrogen production, *J. Mar. Biotechnol.*, 4:34–37.

Nakada, E., Asada, Y., Arai, T., and Miyake, J., 1995, Light penetration into cell suspensions of photosynthetic bacteria and relation to hydrogen production, *J. Ferment. Bioeng.*, 80:53–57.

Nakada, E., Nishikata, S., Asada, Y., and Miyake, J., 1996, Hydrogen production by gel-immobilized cells of *Rhodobacter sphaeroides*—distribution of cells, pigments, and hydrogen evolution, *J. Mar. Biotechnol.*, 4:38–42.

Richter, P., and Drews, G., 1991, Incorporation of light-harvesting complex Iα and β polypeptides into the intracytoplasmic membrane of *Rhodobacter capsulatus*, *J. Bacteriol.*, 173:5336–5345.

Sambrook, J., Fritsch, E.F., and Maniatis, T., 1989, *Molecular Cloning: A Laboratory Manual*, 2nd ed. Cold Spring Harbor Lab Press, New York.

Tsygankov, A.A., Hirata, Y., Miyake, M., Asada, Y., and Miyake, J., 1994, Photobioreactor with photosynthetic bacteria immobilized on porous glass for hydrogen photoproduction, *J. Ferment. Bioeng.*, 77:575–578.

Vasilyeva, L.G., Miyake, M., Nakada, E., Asada, Y., and Miyake, J., 1998, Regulation of Bchl level in *Rb. sphaeroides* for optimization of hydrogen production, genetic approach, in *Proceedings of BioHydrogen 97*, in press.

Warthmann, R., Pfenning, N., and Cypionka, H., 1993, The quantum requirement for H_2 production by anoxygenic phototrophic bacteria, *Appl. Microbiol. Biotechnol.*, 39:358–362.

11

A TOOLKIT FOR METABOLIC ENGINEERING OF BACTERIA

Application to Hydrogen Production

J. D. Keasling, John Benemann, Jaya Pramanik, Trent A. Carrier, Kristala L. Jones, and Stephen J. Van Dien

Department of Chemical Engineering
University of California
Berkeley, California 94720-1462

Key Words

metabolic engineering, *Escherichia coli*, flux models, low-copy plasmids, hydrogen fermentations

1. SUMMARY

Dark, anaerobic hydrogen fermentations of starches and sugars have generally produced less than 20% of the stoichiometrically possible yield, compared to 80–90% yields in the case of ethanol and methane fermentations. Indeed, a 33% yield from glucose, with the remaining electrons secreted as acetate, is generally considered the maximum achievable hydrogen production. However, this restriction would not apply to non-growing cultures; in principle, non-growing cells could recycle much of the metabolic energy derived from the initial hydrogen/acetate fermentation into driving additional hydrogen production. Such improvements in hydrogen production may be possible using directed strategies to manipulate cellular metabolism—metabolic engineering.

Metabolic engineering for hydrogen production may necessitate the introduction of novel genes and coordinated expression of those genes. Once introduced into the cell, it will be necessary to optimize expression of these genes during the hydrogen production (slow- or non-growth) stage of the culture. To enable controlled and reproducible expression of the genes for a redirected metabolic pathway, a metabolic engineering toolkit has been developed for *Escherichia coli*. The first tool is a low-copy plasmid that exists at 1–2 copies/cell and is segregationally stable. The second tool is a strategy to introduce hairpins into the 5'-end of mRNA to control its stability. The third tool is a host organism that is unable to metabolize intracellular polyphosphates and, as a result, has improved control over gene expression from phosphate-starvation promoters. The fourth tool is a mathematical method to predict the fluxes through the metabolic pathways and to tune the expression level of the genes in the introduced pathways. Such tools should find important application in metabolic engineering for high-yield hydrogen production from organic substrates.

2. INTRODUCTION

Fermentative, dark, bacterial hydrogen formation has been studied extensively by microbiologists for decades, as hydrogen is a major product of anaerobic intermediate metabolism and nitrogen fixation. Most important, hydrogen is a crucial link between the breakdown of organic material and microbial methanogenesis, a major process in nature and in the biotechnology of waste treatment (Ferris, 1993). However, in such systems only very low hydrogen concentrations are typically encountered (2 to 200 nM, versus 600 µM for dissolved H_2 at 1 atm, [Dolfing, 1992]). The reason for this low concentration is that microbiologically formed hydrogen is quickly utilized by methanogenic (and other H_2-consuming) bacteria.

The stoichiometric decomposition of glucose to hydrogen (12 mol H_2/mol glucose) under standard conditions (pH 7, 1 atm H_2 pressure) has a Gibbs free energy of only -9.5 Kcal/mol glucose, not enough to allow even one adenosine triphosphate (ATP) to be generated per mole of glucose. Thus, it was suggested that hydrogen formation would be limited to about 33% yield, with the remaining electrons excreted in the form of acetate, yielding −51.6 Kcal/mol (Thauer, 1976; Thauer et al., 1977), theoretically sufficient for the production of several ATPs. Indeed, fermentative hydrogen yields (in the absence of methanogenesis or other hydrogen-consuming reactions) are low, typically less than 10% of stoichiometric. In all work published thus far, hydrogen is a minor by-product of intermediate metabolism, compared to methane or ethanol fermentations, where yields of between 80 and 90% of theoretical are routinely obtained. This discouraging state of affairs led Archer and Thompson (1987) to conclude "hydrogen production from wastes via fermentation is not at the present time feasible on an industrial scale." This, perhaps coupled to the historical accident that biological hydrogen production research and development was started by researchers of photosynthetic systems (Benemann, 1977; Benemann et al., 1973), accounts for the relative neglect of dark anaerobic fermentations for hydrogen production.

Here we argue that the limitations previously thought to exist can be overcome with a combination of both classical and modern approaches to industrial microbiology. First, the argument that substrate conversion reactions yielding less than 1 ATP/mol cannot support growth can be rejected, as it has been demonstrated that some bacteria can survive and grow on less than one ATP produced per mole of substrate converted (Schink, 1997). Moreover, growth is not required for hydrogen production, as such processes would be best carried out with cultures in which cell growth is arrested by a nutrient limitation (e.g., phosphate or nitrogen), as in many industrial fermentations. Further, operation at thermophilic temperatures, advantageous for a number of technical and economic reasons (including heat dissipation), greatly improves the thermodynamics of this process, making hydrogen production more feasible.

Finally, a 100% stoichiometric conversion efficiency is not necessary. Even a doubling of what is currently perceived by some as the maximum possible, to give a 66% yield, would be sufficient for economical hydrogen fermentation in many circumstances. This assumes similar fermentation processes and facilities as currently used in ethanol or methanol fermentations, which currently produce fuels at costs below typical prices for commercial hydrogen. Hydrogen fermentations could not only be a viable local source of hydrogen, but provide an additional waste treatment and re-use option.

The conversion of wastes to hydrogen by bacterial fermentations has also been the goal of the many projects carried out for the past 20 years with photosynthetic bacteria, including work in Europe, the U.S., and most recently, a very active research and development program in Japan. However, dark anaerobic fermentations are preferable to photosynthetic processes, even if efficiencies for the latter were increased by an order of magnitude more than 100% of solar, as initially speculated to be possible (Benemann, 1977). (It should be noted that these efficiencies do not include the energy contained in the substrate, which, if considered, would change the thermodynamic and entropy arguments in favor of these systems.)

Certainly, it must be recognized that dark anaerobic H_2 production, like H_2 production by photosynthetic bacteria, does not take place in nature to any significant extent. The reasons, in the case of dark fermentative bacteria, is that the transfer of electrons to other substrates—resulting in ethanol, lactate, methane, and others—are thermodynamically more favorable and provide greater growth potential for the bacteria. However, inspection of intermediate metabolic pathways in fermentative bacteria, such as the tricarboxylic acid (TCA) cycle and the pentose phosphate pathways, clearly demonstrates the potential for hydrogen production; however, these pathways are tightly regulated and alternatives (e.g., reduction of acetate) are preferred by the metabolic machinery.

We propose that the novel techniques of metabolic engineering provide a new approach to alter this fundamental pathway of hydrogen fermentations. The overall approach would be to develop such a process in a convenient bacterium, such as *Escherichia coli*, and then adapt and even transfer this genetic information to other fermentative bacteria that have been successfully used in long-term, though low-yield, hydrogen fermentations (Tanisho, 1995). Here we review metabolic engineering approaches and tools developed in our laboratory, and propose their application to metabolic engineering for dark hydrogen fermentations.

3. METABOLIC ENGINEERING OF HYDROGEN PRODUCTION

Escherichia coli is the organism of choice for demonstrating metabolic engineering processes, particularly in the early stages of biotechnology development. Genetic engineering of *E. coli* specifically for hydrogen production has been carried out by Karube's group in Japan (Kanayama et al., 1988; Kanayama et al., 1988), who cloned into *E. coli* the hydrogenase gene of *Citrobacter freundii*. Several recombinant *E. coli* strains were tested, and some exhibited high hydrogenase activity in vitro and a somewhat elevated rate from various organic substrates in vivo. This group also demonstrated hydrogen production from glucose by the genetically engineered *E. coli* in an immobilized bioreactor, achieving rates of hydrogen production about twice those observed with the wild-type strain. However, at a maximum, only about 0.5 mol of H_2 were produced per glucose utilized. This demonstrates that simply transferring a hydrogenase gene (with an "almost unknown" mechanism of hydrogen evolution, to quote these authors) is not sufficient to achieve the goals of this project. A concerted effort is required to supply *E. coli* not only with a suitable hydrogenase gene expression level, but also with the other proteins required to reduce hydrogenase (e.g., the electron transport pathway). Equally important will be to reduce the enzymatic pathways that lead to alternative, energetically more favorable (to the bacteria) electon sinks, such as lactate and ethanol. This will require application of the tools of metabolic engineering (Bailey, 1991; Cameron and Tong, 1993).

3.1. Low-Copy Plasmids

Engineering a cell to produce a desirable product and remain viable for long periods with little or no growth requires expression vectors different from those commonly used to overproduce a heterologous protein, the focus of much genetic engineering research and development for the past decade. First, the balance of macromolecular synthesis and energy production must be carefully manipulated and accurately controlled so that it is possible to separate, possibly through a nutrient limitation trigger, the growth (cell replication) and product (hydrogen) formation stages of the bacterial culture. Second, cloning entire metabolic pathways into a new host organism may involve the introduction of large amounts of DNA, so the genetic vector must be capable of carrying and replicating large pieces of DNA. Third, the genetic vector must be maintained in all cells of a culture to maximize the usefulness of this biocatalyst and to decrease the possibility of plasmid-free cells overtaking the culture. Finally, this genetic vector must have a narrow copy number distribution in all cells of a culture, since cells that have higher-than-average plasmid content will exhibit a greater metabolic burden than cells that have lower-than-average plasmid content. In short, this new cloning vehicle should have many of the characteristics of the chromosome.

The first tool that we have developed for metabolic engineering is a low-copy, segregationally stable plasmid for the expression of one or more genes (Jones and Keasling, 1998). The low-copy expression vector was constructed from a 9-Kb region of the *Escherichia coli* F plasmid that contains the *oriV* and *oriS* origins of replication. This plasmid carries the β-lactamase gene to confer resistance to ampicillin, the *ara*BAD promoter and the *ara*C gene for arabinose-inducible gene expression, and transcription terminators. A derivative of this plasmid that carries a *lacZ* gene positioned in the multi-cloning site behind the arabinose promoter was also constructed and was found to be stably maintained for at least 150 generations, both uninduced and under induced expression of *lacZ* (Table 1). In contrast, multicopy plasmids constructed with the ColE1-type origin of replication were stably maintained when uninduced, but no plasmid-bearing segregants re-

Table 1. Comparison of low- and high-copy plasmids. The low-copy plasmid is the mini-F plasmid containing the *lacZ* gene under control of the arabinose-inducible promoter (P_{BAD}). The high-copy plasmid is a ColE1-based vector containing the same reporter system and antibiotic resistance

	Low-copy plasmid		High-copy plasmid	
	Uninduced	Induced	Uninduced	Induced
Growth rate of host (hr^{-1})	0.69	0.69	0.69	0.73
β-gal production (units)	15	4000	150-200	12500
Segregational stability (generations)	>150	>150	>150	25

The segregational stability studies were performed in the absence of selection pressure (antibiotic).
Data taken from Jones and Keasling (1998).

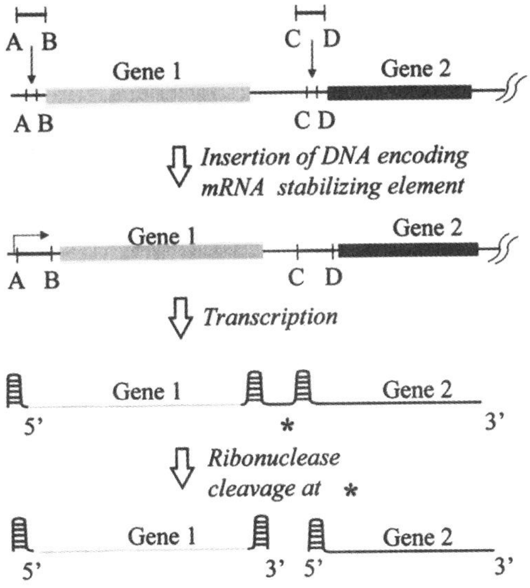

Figure 1. Insertion of hairpins between two genes in an operon. The hairpin technology can be extended to operons by placing between the genes cassettes that will cause the resulting mRNA to form hairpins. By incorporating a ribonuclease site at the 5′ base of the hairpin, one can differentially and independently control the stability of the two transcripts. Restriction endonuclease sites are indicated by A and B. The DNA cassette placed between the A and B sites forms a hairpin in the mRNA. The 5′ hairpin acts as a stabilizing element for the mRNA.

mained after 60 generations when induced with 0.001% arabinose. Besides their extreme segregational stability, the low-copy plasmids imposed no observable metabolic burden on the host cell (Table 1). In contrast, the presence of the high-copy plasmid reduced the host-cell growth rate under induced and uninduced conditions.

These results strongly support the viability of low-copy plasmids as excellent alternatives to multi-copy plasmids for expression of heterologous genes in *E. coli* when extreme stability and low metabolic burden imposed on the cells are desired. This is a particularly important requirement for hydrogen production, where little excess metabolic energy would be available.

3.2. Operon Technologies

Metabolic engineering of hydrogen production may require the introduction of a large number of genes. As the number of inducible promoters available is limited, it is necessary to place as many genes as possible under control of a single promoter, much like native operons. Even though one may want to induce the expression of multiple genes from a single promoter, it may be necessary to express those genes at different levels. For example, multi-subunit enzymes may require the subunits in different stoichiometries. Also, multi-enzyme pathways may require enzymes at different levels to prevent bottlenecks at a particular enzyme in the pathway. In prokaryotes, differential expression of genes in an operon is accomplished through differential stability of the coding regions in the multicistronic mRNA, which is affected by secondary structures, hairpins, at the 3′ and 5′ ends of the transcript and between coding regions. Thus, it should be possible to introduce synthetic hairpins between the coding regions of the multi-cistronic mRNA and engineer new operons that will produce enzyme subunits at stoichiometric levels sufficient for a functioning enzyme complex or multiple enzymes in a metabolic pathway at sufficient levels to prevent bottlenecks at one particular enzyme (Figure 1).

The second tool we have developed is a method to control the stability of mRNA by the introduction of DNA cassettes into the region between the transcription and translation

Table 2. Stability of, and protein synthesis from, transcripts with and without hairpins

	mRNA half-life (min)	Protein production (units)
No hairpin	2.6	48
Hairpin	8.2	195

The protein production levels are units of β-galactosidase as defined by Miller (1972).
The data were taken from Carrier and Keasling (1997b).

start sites of a gene of interest (Carrier and Keasling, 1997a; Carrier and Keasling, 1997b). This cassette system was used to engineer mRNA stability through the introduction of hairpins at the 5' end. The hairpin-containing mRNA exhibited a half-life three times that of the mRNA with no hairpin, resulting in increases in both mRNA and protein levels (Table 2). These results indicate that it is possible to engineer mRNA stability as an additional means of controlling gene expression. These hairpins may be particularly important for metabolic engineering to coordinate regulation of the many genes necessary for hydrogen fermentations.

3.3. Improved Control of Gene Expression

Once the genes encoding a metabolic pathway have been placed under the control of an inducible promoter, it is important that the promoter of interest be well-controlled. The phosphate-starvation promoter of *E. coli* has been widely used for the expression of heterologous genes. During starvation for external P_i, the cell induces the phosphate-starvation response, in which alkaline phosphatase (AP) is secreted into the periplasmic space to liberate P_i from organophosphates (Wanner, 1996). Since there is no known internal sensor for phosphate and only an external sensor, the cell cannot sense the amount of phosphate or polyphosphate stored intracellularly. By placing a gene of interest under control of the phosphate-starvation promoter, one can use phosphate starvation to control the expression of heterologous genes during the non-growth phase.

Work in our laboratory indicates that even though there is no internal sensor for phosphate, internal stores of polyphosphate modulate the phosphate-starvation response (Sharfstein et al., 1996). Polyphosphate is synthesized by polyphosphate kinase (PPK) from ATP and hydrolyzed by polyphosphatase (PPX) to inorganic phosphate (P_i) (Kornberg, 1995). To examine the effect of polyphosphate on the phosphate-starvation response, two plasmids were used: one plasmid carried the polyphosphate kinase gene (*ppk*) and the polyphosphatase gene (*ppx*) under control of their native promoter; the other plasmid carried only the *ppk* gene under control of the native promoter (Akiyama et al., 1993; Akiyama et al., 1992). These plasmids were transformed into a strain in which the *ppk* and *ppx* genes on the chromosome had been inactivated by insertion of a kanamycin resistance gene into the *ppk* gene.

In strains that are able to make large quantities of polyphosphate during growth under surplus phosphate conditions but not degrade it during phosphate starvation (*ppk* overexpressed, no *ppx*), alkaline phosphatase was expressed at very high levels indicating the overexpression of the phosphate-starvation response (Table 3). In strains that are able to make large quantities of polyphosphate and then rapidly degrade it, the polyphosphate level was low during starvation and high during surplus. The phosphate-starvation response was induced to only a very low level since polyphosphate could be degraded to P_i

Table 3. Polyphosphate levels and alkaline phosphatase production in wild-type and genetically engineered *E. coli* strains

Genotype		polyP levels (relative)	AP levels (relative)
ppk	*ppx*		
chrom.	chrom.	1.0	1.0
none	none	ND	1.67
plasmid	plasmid	10	0.167
plasmid	none	90	2.33

The polyphosphate and alkaline phosphatase levels are relative to those in the wild-type strain.
The data were taken from Sharfstein et al. (1996). Abbreviations: chrom., chromosomal; ND, not detectable.

by PPX and secreted from the cell (Van Dien et al., 1997). In a strain that could neither make nor degrade polyphosphate, there was no internal polyphosphate, and alkaline phosphatase was induced to nearly the same level as in the strain that was able to make polyphosphate but not degrade it. Finally, a wild-type *E. coli* strain that could make small quantities of polyphosphate and degrade it had no polyphosphate during starvation and low levels during surplus; alkaline phosphatase was induced to intermediate levels. Thus, for maximal and reproducible expression of genes under control of the phosphate-starvation response, it is important to eliminate synthesis of PPX. Elimination of polyphosphate by inactivating *ppk* would accomplish a similar goal. The new strain, in which *ppk* and *ppx* have been inactivated, has higher expression levels from, and more predictable control of, the phosphate starvation response than does the wild-type strain.

The production of hydrogen could be maximized by limiting growth, so that substrate would not be used for growth. A very effective method for controlling gene expression and growth is phosphate starvation. By placing genes necessary for enhanced hydrogen production under control of the phosphate-starvation promoter, one could control growth and hydrogen formation simultaneously and inexpensively. The new strain described above will allow one to reproducibly control expression of the phosphate-starvation response.

3.4. A Theoretical Framework to Engineer Metabolic Fluxes

The fourth tool is a theoretical framework to predict how the introduction of a new pathway or the enhancement of existing pathways will affect cell growth and metabolism, and to predict the appropriate expression levels for the genes of a new pathway (Pramanik and Keasling, 1998; Pramanik and Keasling, 1997). A steady-state mathematical model of the fluxes through the various metabolic pathways in *Escherichia coli* was formulated from the known stoichiometry:

$$d\mathbf{x}/dt = \mathbf{S} \cdot \mathbf{v} - \mathbf{b}$$

where $\mathbf{x}$ is the vector of metabolites, $\mathbf{S}$ is the stoichiometric matrix, $\mathbf{v}$ is the vector of reaction rates, and $\mathbf{b}$ is the demand vector. The demand vector is essentially the requirements to build a cell at a given growth rate (i.e., the necessary amino acids, nucleotides, fatty acids, cell wall components, etc.). As the composition of the cell and the products secreted into the medium vary with growth rate, the demand vector is a function of the growth rate (μ) and the amount and composition of the desired product (p):

$$b = fn(\mu, p)$$

Since many metabolites participate in multiple metabolic pathways and since there are cyclic pathways in the cell (i.e., TCA cycle), there are more reactions than metabolites (more equations than unknowns). At steady-state, one can solve for **v** using linear programming by maximizing (or minimizing) a combination of the variables and by placing constraints on the variables. Since the model predicts how the fluxes through the metabolic pathways must be distributed in order to achieve a specific growth rate and product, one can compare the fluxes through the metabolic pathways under various growth conditions and predict which metabolic pathways will be regulated.

As an example of the power of such a model, we have compared the model predictions with the experimentally determined values (Walsh and Koshland, 1984) for a cell growing on glucose and acetate. The model predictions are, on average, within 15% of the experimentally determined values. In addition, we have compared the metabolic fluxes for a cell growing in glucose minimal medium to a cell growing on acetate. The model predicted a significant flux through the glyoxylate bypass during growth on acetate, in agreement with experimental data. In contrast, the glyoxylate shunt remains closed during growth on glucose or glucose plus acetate. Thus, the model is able to predict which metabolic pathways must be regulated for optimal growth or product formation.

The model predictions and experimental data for the rates of metabolite production during fermentation of glucose are compared in Table 4. The prediction for the hydrogen flux differs from the experimental data by approximately 40%, whereas the predictions for the other metabolites differ by only 6%. The fluxes through the major metabolic pathways (glycolysis and TCA cycle) are shown in Figure 2. Note that the model predicts that the TCA cycle branches during fermentation and the glyoxylate shunt is not used. This regulation is found experimentally (Neidhardt et al., 1990).

This model should find important application in predicting how changes in cellular metabolism will affect the production of hydrogen and the appropriate expression levels for the genes encoding the enzymes involved in hydrogen production.

4. CONCLUSIONS

As discussed in the introduction, two arguments have been raised against high-yield hydrogen fermentations:

Table 4. End-products of anaerobic fermentation of glucose by *E. coli*

Products	Moles per 100 moles of glucose fermented	
	Model	Experiment
Ethanol	49.80	46.70
Formic acid	2.43	2.28
Acetic acid	36.50	34.26
Lactic acid	79.50	74.59
Succinic acid	10.70	10.03
Carbon dioxide	88.00	56.06
Hydrogen	75.00	40.78

Model predictions are compared to experimental data taken from Wood (1962).

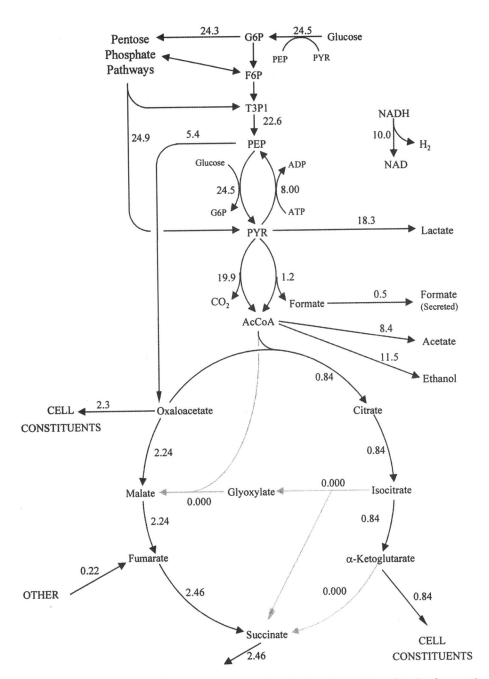

Figure 2. Model predictions for the fluxes through the central metabolic pathways of *E. coli* during fermentation of glucose. The units for the fluxes are mmol/(g dry cell weight hr). The fluxes at all metabolites are not balanced as shown, as metabolites are drained from the pathways shown for synthesis of biomass. The abbreviations are as follows: G6P, glucose-6-phosphate; F6P, fructose-6-phosphate, T3P1, glyceraldehyde-3-phosphate; PYR, pyruvate; AcCoA, acetyl-CoA; PEP, phosphoenol pyruvate.

1. hydrogen fermentations are thermodynamically unfavorable and do not allow for sufficient metabolic energy for the growth of the organisms; and
2. there are no metabolic pathways for hydrogen fermentations, as evidenced by the lack of high-yield hydrogen fermentations found in nature.

Both arguments have been countered above. Although the metabolic machinery is available for such a process, the major limitation to hydrogen fermentations is the evolutionary selection against a high-yield process in wild-type bacteria. However, as presented above, we now have the practical and theoretical tools at our disposal to alter, control, and redirect fundamental metabolic machinery, without the need to undergo the labor-intensive and uncertain step-wise strain improvement methods used to develop traditional industrial fermentations. Of the biological hydrogen production systems currently being explored, hydrogen fermentations rank highest in their potential for both the likely economics of such a process and environmental sustainability (Benemann, 1997).

ACKNOWLEDGMENT

We thank the National Science Foundation (BES-9502495) for funding this research.

REFERENCES

Akiyama, M., Crooke, E., and Kornberg, A., 1993, An exopolyphosphatase of *Escherichia coli:* the enzyme and its *ppx* gene in a polyphosphate operon, *J. Biol. Chem.*, 268:633–639.
Akiyama, M., Crooke, E., and Kornberg, A., 1992, The polyphosphate kinase gene of *Escherichia coli*: isolation and sequence of the *ppk* gene and membrane location of the protein, *J. Biol. Chem.*, 267:22556–22561.
Archer, D.B., and Thompson, L.A., 1987, Energy production through the treatment of wastes by micro-organisms, *J. Appl. Bact. Symp.*, Supp:59s-70s.
Bailey, J.E., 1991, Towards a science of metabolic engineering, *Science*, 252:1668–1675.
Benemann, J.R., 1977, Hydrogen and methane production through microbial photosynthesis, in *Living Systems as Energy Converters*, Buvet, R. (ed.), Elsevier/North-Holland Biomedical Press, Amsterdam, pp. 285–298.
Benemann, J.R., 1997. *Photobioreactors.*
Benemann, J.R., Berenson, J.A., Kaplan, N O., and Kamen, M.D., 1973, Hydrogen evolution by a chloroplast-ferredoxin-hydrogenase system, in *Proceedings of the National Academy of Sciences (USA)*, 70:2317–2320.
Cameron, D.C., and Tong, I.T., 1993, Cellular and metabolic engineering, *Applied Biochemistry and Biotechnology*, 38:105–140.
Carrier, T.A., and Keasling, J.D., 1997a, Controlling messenger RNA stability in bacteria: strategies for engineering gene expression, *Biotechnol. Prog.*, 13:699–708.
Carrier, T.A., and Keasling, J.D., 1997b, Engineering mRNA stability in *E. coli* by the addition of synthetic hairpins using a 5' cassette system, *Biotechnol. Bioeng.*, 55:577–580.
Dolfing, J., 1992, The energetic consequences of hydrogen gradients in methanogenic ecosystems, *FEMS Microb. Ecol.*, 101:183–187.
Ferris, G., (ed.), 1993, *Methanogenesis*, Academic Press, San Diego, California.
Jones, K.L., and Keasling, J.D., 1998,. Construction and characterization of F plasmid-based expression vectors, *Biotechnol. Bioeng.*, in press.
Kanayama, H., Sode, K., and Karube, I., 1988, Continuous hydrogen evolution by immobilized recombinant *E. coli* using a bioreactor, *Biotechnol. Bioeng.*, 32:396–399.
Kanayama, H., Sode, K., Yamamoto, T., and Karube, I., 1988, Molecular breeding of hydrogen-evolving bacteria, *Biotech. Genetic Eng. Rev.*, 6:379–401.
Kornberg, A., 1995, Inorganic polyphosphate: toward making a forgotten polymer unforgettable, *J. Bacteriol.*, 177:491–496.
Miller, J.H., 1972, *Experiments in Molecular Genetics*, Cold Spring Harbor Laboratory, Cold Spring Harbor, New York.

Neidhardt, F.C., Ingraham, J.L., and Schaechter, M., 1990, *Physiology of the Bacterial Cell: A Molecular Approach*, Sinauer Associates, Sunderland, Massachusetts.

Pramanik, J., and Keasling, J.D., 1998, Effect of carbon source and growth rate on biomass composition and metabolic flux predictions of a stoichiometric model, *Biotechnol. Bioeng.*, in press.

Pramanik, J., and Keasling, J.D., 1997, A stoichiometric model of *Escherichia coli* metabolism: incorporation of growth-rate dependent biomass composition and mechanistic energy requirements, *Biotechnol. Bioeng.*, 56:398–421.

Schink, B., 1997, Energetics of syntrophic cooperation in methanogenic degradation, *Microbiol. Molec. Biol. Rev.*, 61:262–280.

Sharfstein, S.T., Van Dien, S.J., and Keasling, J.D., 1996, Modulation of the phosphate-starvation response by genetic manipulation of the polyphosphate pathways, *Biotechnol. Bioeng.*, 51:434–438.

Tanisho, S., 1995, presented at the 2nd International Conference of New Energy Systems and Conversions.

Thauer, R., 1976, Limitation of microbial hydrogen formation via fermentation, in *Microbial Energy Conversion*, Schlegel, H.G., and Barnea, J. (eds.), Erich Goltze, Gottingen, Germany, pp. 201–294.

Thauer, R., Jungerman, K., and Decker, K., 1977, Energy conservation in chemotrophic anaerobic bacteria, *Bacteriol. Rev.*, 41:100–180.

Van Dien, S.J., Keyhani, S., Yang, C., and Keasling, J.D., 1997, Manipulation of independent synthesis and degradation of polyphosphate in *Escherichia coli* for investigation of phosphate secretion from the cell, *Appl. Environ. Microbiol.*, 63:1689–1695.

Walsh, K., and Koshland. D.E., 1984, Determination of flux through the branch point of two metabolic cycles, *J. Biol. Chem.*, 259:9646–9654.

Wanner, B.L., 1996, Phosphorus assimilation and control of the phosphate regulon, in *Escherichia coli and Salmonella; Cellular and Molecular Biology*, Neidhardt, F.C., Curtiss, R., Ingraham, J.L., Lin, E.C.C., Low, K.B., Magasanik, B., Reznikoff, W.S., Riley, M., Schaechter, M., and Umbarger, H.E., (eds.), ASM Press, Washington, D.C., pp. 1357–1381.

Wood, W.A., 1962, Fermentation of carbohydrates and related compounds, in *The Bacteria*, vol. 2, Gunsalus, I.C., and Stanier, R.Y., (eds.), Academic Press, New York, New York.

ELECTRON TRANSPORT AS A LIMITING FACTOR IN BIOLOGICAL HYDROGEN PRODUCTION

Patrick C. Hallenbeck, Alexander F. Yakunin, and Giuseppa Gennaro

Département de microbiologie et immunologie
Université de Montréal
CP 6128 Centre-ville
Montréal, Québec H3C 3J7, Canada

Key Words

nitrogenase, electron transport, ferredoxin, flavodoxin

1. SUMMARY

Increasing the capability of biological systems for hydrogen production requires an understanding of the fundamental level of potential limiting factors. Degradation of various wastes with the concomitant evolution of hydrogen by photosynthetic bacteria has long been proposed as a possible biohydrogen source. Here we present evidence, using growth studies with different nitrogen sources, which indicates that the maximal in vivo activity of the nitrogenase (N_2ase) system of the photosynthetic bacterium, *Rhodobacter capsulatus*, is probably restricted at the level of electron flow. Thus, an increase in catalytic efficiency of at least threefold is possible. We are presently investigating the physiological mechanisms responsible for this metabolic gating. Although very little is presently known in general about the determinants for reaction specificity and efficiency between low-potential redox carriers and enzymes functioning in biological hydrogen production, these results suggest that a consideration of these factors may be extremely important. As a model system, we are undertaking studies of the role of electron carriers in controlling reductant flux through the pyruvate oxidoreductase (POR), N_2ase system. The methods we are using and results obtained should be applicable to other hydrogen-producing systems, for example, POR and hydrogenase.

BioHydrogen, edited by Zaborsky *et al.*
Plenum Press, New York, 1998

2. INTRODUCTION

Biological hydrogen production is based on the coupling of the generation of reducing power, either by the cellular machinery (Benemann and Weare, 1974) or in vitro systems (Benemann et al., 1973), to a terminal enzyme capable of reducing protons to hydrogen (either hydrogenase or the hydrogenase reaction of nitrogenase). These terminal enzymes are in general not reduced by specific reductases and require an electron transport factor to be able to draw off the low-potential electrons that have been generated. In many organisms, the suitable electron transport factors are low molecular weight soluble proteins, flavodoxin (e.g., NifF) or ferredoxin (e.g., FdI). Despite the isolation and extensive characterization of many of these electron carriers and their respective electron acceptors, relatively little is known about what factors control their efficiency and reaction specificity. These factors may play important roles in regulating electron flux through the in vivo systems and certainly need to be considered when genetic modifications of pre-existing systems are contemplated with a view towards increasing hydrogen production capability. We present results that indicate that under some circumstances, electron flux to the nitrogenase system limits in vivo nitrogenase activity. The results of stopped-flow spectrophotometric kinetic analysis show that NifF forms a tight complex with Fe-protein whereas FdI does not, suggesting that electron flow is from FdI (E_m =-510 mV) → NifF (E_m =-474mV) → N_2ase (E_m =-470mV, [Hallenbeck, 1983]). We have initiated studies on the interaction of NifF and POR (both from *K. pneumoniae*) and devised schemes for selection of NifF mutants impaired in their interaction with either POR or N_2ase.

3. EXPERIMENTAL

3.1. Measurement of in Vivo Nitrogenase Activity and Content

For experiments in which in vivo nitrogenase activity and Fe-protein modification state were analyzed, batch cultures of *R. capsulatus* SB1003 were grown with -N RCV medium in 1.6 x 20.5 cm tubes sealed with rubber stopppers with needles inserted for gas sparging (argon or N_2; 5–10 mL/min). Seven mM glutamate, N_2, or limiting amounts of NH_4^+ were used as nitrogen sources. In vivo nitrogenase activity was determined by the acetylene reduction method and the amounts of modified and unmodified N_2ase proteins were determined by chemiluminescent detection of immunoblots (Yakunin and Hallenbeck, 1997).

3.2. Stopped-Flow Kinetic Analysis

NifF was purified from an over-producing strain of *E. coli* (Hallenbeck and Gennaro, 1998). FdI and nitrogenase were purified as previously described by Hallenbeck et al. (1982a, b). Anaerobic conditions were maintained by using either an anaerobic chamber or purified nitrogen lines. Carriers were reduced with sodium dithionite. Oxidized species were exchanged into 25 mM HEPES, 10 mM $MgCl_2$ (pH 7.4) by gel filtration (PDG-8) in the anaerobic chamber. $NifF_{SQ}$ (NifF-semiquinone) was conveniently generated by mixing equimolar amounts of $NifF_{ox}$ and $NifF_{HQ}$ (NifF-hydroquinone). All kinetic experiments were carried out at 25 ± 0.1 °C in 25mM HEPES, 10 mM $MgCl_2$ in the anaerobic chamber using a Hi-Tech SF-61 Stopped-Flow Spectrophotometer, and data were analyzed using the associated software. Kinetic analysis followed standard methods for treating scheme 1 under pre-equilibrium and non-steady state conditions.

$$A_{ox} + B_{red} \underset{k-1}{\overset{k+1}{\longleftrightarrow}} AB \overset{k+2}{\longrightarrow} A_{red} + B_{ox}$$

4. RESULTS

4.1. In Vivo Nitrogenase Activity Is Not Limited by the Amount of Nitrogenase Components

As part of a larger study on the control of in vivo N_2ase activity, we have examined the influence of the degree of nitrogen limitation on in vivo activity and Fe-protein component forms (unpublished). In spite of the large differences in in vivo nitrogenase activity observed between the *R. capsulatus* cultures grown with different nitrogen sources, immunoblot analysis demonstrates that the amounts of MoFe-protein and total Fe-protein were similar in these cells. Depending on the nitrogen source used, the concentration of MoFe-protein varied between 10 and 30 µM and that of Fe-protein between 18 and 65 µM (Table 1). The molar ratio of Fe-protein/MoFe-protein in *R. capsulatus* cells (1.8–2.6) is close to that previously reported for *A. vinelandii*. It can be seen from the data presented in Table 1 that the concentration of nitrogenase proteins in N_2-grown cells was two times higher than that in NH_4^+-limited, high-density (17.1 initial ammonium) cultures, although they had the same maximal in vivo nitrogenase activity (87–99 nmol $C_2H_4 \cdot min^{-1} \cdot mg^{-1}$). Glutamate-grown and NH_4^+-limited, low-density (5.7 mM initial ammonium) cultures had similar levels of nitrogenase protein, while in vivo nitrogenase activity was higher in the glutamate-grown cells. Similarly, the calculated catalytic activity of either the MoFe-protein or the Fe-protein, obtained by using the measured in vivo activity and the concentration of the respective proteins determined by quantitative immunoblotting, shows a wide variation with culture conditions (Table 1). Therefore, in *R. capsulatus*, the level of in vivo nitrogenase activity is not quantitatively related with the content of nitrogenase proteins or the modification state of the Fe-protein, and may therefore reflect a modulation at the level of electron transport.

4.2. Redox Properties of FdI and NifF and Interaction of NifF and FdI with Nitrogenase Fe-Protein

FdI, which has two 4Fe-4S clusters, demonstrates a single mid-point potential of −490 mV (EPR titration) or -510 mV (cyclic voltametry). We have determined the mid-

Table 1. Maximal concentrations of nitrogenase proteins and their catalytic activity in *R. capsulatus* cells grown with different nitrogen sources

Nitrogen source	Concentration[a] (µM)		Ratio Fe protein/ MoFe protein)	Catalytic activity[b] (nmol $C_2H_4 \cdot min^{-1} \cdot mg^{-1}$ N_2ase component)	
	MoFe protein	Fe-protein		MoFe-protein	Fe-protein
NH_4^+ (5.7mM)	27.0–30.4	50.5–51.1	1.77	5200	10700
NH_4^+ (17.1 mM)	10.1–10.2	18.7–21.9	2.02	5900	10000 (14300)
N_2	17.7–19.7	41.3–48.9	2.40	2800	4200 (32300)
Glutamate	23.7–24.0	56.2–65.4	2.56	10950	15600 (26000)

[a]The concentration of nitrogenase proteins was calculated from the amount of protein (in µg/mg total protein) by using an internal volume for *R. capsulatus* cells of 5 µL/mg total protein (8) and an Mr 230000 for MoFe-protein and 63000 for Fe-protein (10). The amounts of the nitrogenase proteins in each culture type were determined by Western blotting (Yakunin and Hallenbeck, 1998).
[b]In vivo nitrogenase activity was measured by acetylene reduction. The activities obtained were calculated on the basis of the concentration of each nitrogenase protein component as measured by Western blotting. The numbers in parentheses represent the values correcting for the content of unmodified Fe-protein.

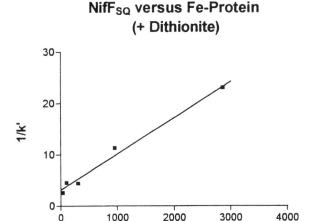

Figure 1. Reduction of FdI_{OX} by reduced Fe-Protein. Final FdI_{OX} concentration was 15.7 µM. [k_2 (apparent second-order rate constant) $7.79 \cdot 10^2$ M^{-1} s^{-1}]. [K (pre-equilibrium binding constant) ≥ 325 µM].

point potential of the SQ,HQ couple of NifF from *R. capsulatus* to be -474 mV (Hallenbeck and Gennaro, 1995), intermediate between the corresponding proteins from *A. vinelandii* and *K. pneumoniae*. We examined the interaction of NifF and FdI with the Fe-protein of nitrogenase by examining their reduction by Fe-protein. For the reduction of FdI, a plot of k' versus [Fe-Protein] was linear (Figure 1), indicating that the reaction was not saturated. The slope therefore gives an apparent bimolecular rate Kk_{et}. A lower limit to K can be estimated as K $\geq$ 325 µM, if one assumes that curvature would have been observed at the highest [Fe-protein] if it were 20% of K. Thus, FdI binds only weakly to reduced Fe-Protein. On the other hand, the reduction of $NifF_{SQ}$ by reduced Fe-Protein showed saturation, and the pre-equilibrium binding constant K could be directly determined from a plot of 1/k' vs. 1/[Fe-Protein] (Figure 2). The determined K was 0.44 µM; thus, binding of NifF to Fe-Protein is three orders of magnitude greater than the binding of FdI. We examined the potential role of electrostatic interactions in the reaction between NifF and Fe-Protein by examining the variation in k_{obs} with ionic strength (varied by addi-

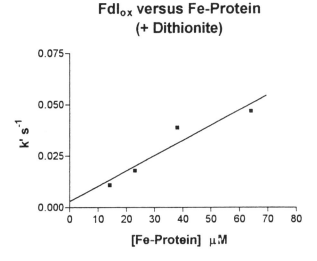

Figure 2. Reduction of $NifF_{SQ}$ by reduced Fe-Protein. Final $NifF_{SQ}$ concentration was 10.5 µM. [k_2 (apparent second-order rate constant) $7.05 \cdot 10^6$ M^{-1} s^{-1}]. [K (pre-equilibrium binding constant) 0.44 µM].

tion of NaCl up to a final concentration of 0.81 M). The linear portion of a plot of ln k_{obs} vs. $[NaCl]^{1/2}$ gave a slope of -0.485 ± 0.084 (results not shown), indicating that electrostatic forces, either positive or negative, do not play a significant role in the interaction of NifF and Fe-Protein. These results suggest that NifF is the proximal electron donor to N_2ase and that electron flow to N_2ase is: FdI (E_m = -510 mV) → NifF (E_m = -474 mV) → N_2ase (E_m = -470 mV).

4.3. Interaction of NifF$_{Kp}$ and NifJ$_{Kp}$

We have also initiated studies on the nif-specific electron transport system of *K. pneumoniae*, since this is the only organism for which all the players are known for certain. In addition, it is interesting in its own right because it is a soluble, well-defined, electron-generating transport system operating at or well below the potential of the hydrogen electrode. We have examined, using stopped-flow spectrophotometric analysis, the interaction of NifF and NifJ (pyruvate flavodoxin oxido-reductase). Our results (unpublished) indicate that electron transfer involves the formation of a tight complex between the two redox partners (K = 18.6 μM). Previously, NifF was found to form a tight complex with the cognate nitrogenase (K = 13 μM; Thorneley and Deistung, 1988). Even though tight binding is observed, electron transfer is slow, 0.88 s^{-1}, perhaps explaining why high levels of NifJ are necessary in vivo.

We are interested in the future in determining the NifF structural elements that determine binding specificity to the electron donor (NifJ) and the electron acceptor (N_2ase). We are creating mutant NifFs using doped PCR. Mutants affected in electron transport are screened on the basis of their inability to complement a NifF$^-$ strain of *K. pneumoniae*. We will be able to further classify the obtained mutants using controlled expression (mutants with weakened interactions) and metronidazole sensitivity (mutant NifFs affected in interaction with NifJ). The results we hope to obtain will be the first dissection of the specificity elements of a low-potential electron transport system. These results and methodology should be directly applicable to the interaction of ferredoxins with H$_2$ase and the corresponding electron-generating systems.

ACKNOWLEDGMENTS

This work was supported by grants from the Natural Sciences and Engineering Research Council of Canada (OGP0036584) and the Fonds de Recherche en Santé du Québec (Bourse de perfectionnement pour chercheur autonome (941776)).

REFERENCES

Benemann, J.R., Berenson, J.A., Kaplan, N.O., and Kamen, M.D., 1973, Hydrogen evolution by a chloroplast-ferredoxin-hydrogenase system, in *Proceedings of the National Academy of Sciences (USA)*, 70:2317–2320.

Benemann, J.R., and Weare, N.M., 1974, Hydrogen evolution by nitrogen-fixing *Anabaena cylindrica* cultures, *Science*, 184:174–175.

Hallenbeck, P.C., 1983, Nitrogenase reduction by electron carriers: influence of redox potential on activity and the ATP/2e- ratio, *Arch. Biochem. Biophys.*, 220:657–660.

Hallenbeck, P.C., and Gennaro, G., 1998, Stopped-flow kinetic studies of low potential electron carriers of the photosynthetic bacterium, *Rhodobacter capsulatus:* Ferredoxin I and NifF, *Biochim. Biophys. Acta*, in press.

Hallenbeck, P.C., Jouanneau, Y., and Vignais, P.M., 1982a, Purification and molecular properties of a soluble ferredoxin from Rhodopseudomonas capsulata, *Biochim. Biophys. Acta*, 681:168–176.

Hallenbeck, P.C., Meyer, C.M., and Vignais, P.M., 1982b, Nitrogenase from the photosynthetic bacterium Rhodopseudomonas capsulata: purification and properties, *J. Bacteriol.*, 149:708–717.

Thorneley, R.N.F., and Deistung, J., 1988, Electron-transfer studies involving flavodoxin and a natural redox partner, the iron protein of nitrogenase, *Biochem. J.*, 253:587–595.

Yakunin, A.F., and Hallenbeck, P.C., 1997, Regulation of synthesis of pyruvate carboxylase in the photosynthetic bacterium Rhodobacter capsulatus, *J. Bacteriol.*, 179:1460–1468.

Yakunin, A.F., and Hallenbeck, P.C., 1998, A luminol/iodophenol chemiluminescent detection system for Western immunoblots, *Analyt. Biochem.*, in press.

13

RECONSTITUTION OF AN IRON-ONLY HYDROGENASE

Hugh McTavish

University of Minnesota
Department of Biochemistry
1479 Gortner Avenue
St. Paul, Minnesota 55108

Key Words

Clostridium pasteurianum hydrogenase I, reconstitution

1. SUMMARY

Extracts *from Clostridium pasteurianum* were denatured in 7 M urea anaerobically to eliminate hydrogenase activity. Hydrogenase from the denatured extracts could be renatured by incubation anaerobically with sulfide, reduced thiol reagents, and Fe^{2+}. Reconstituted extracts recovered 2% of the original hydrogen evolution activity. Reconstitution was also successful with purified hydrogenase I that was denatured anaerobically. This recovered 3.6% of its original hydrogen evolution activity. These results suggest that accessory proteins are not necessary for assembly of iron-only hydrogenases as they are for nickel-iron hydrogenases.

2. INTRODUCTION

The best-characterized hydrogenases fall into two groups: nickel-iron and iron-only hydrogenases. The nickel-iron hydrogenases are known to require several accessory gene products for processing the enzyme to an active form (Vignais and Toussaint, 1994). Some of these are involved in assembling the unique nickel-containing active site. The iron-only hydrogenases also have an active site metal cluster whose structure appears to be unique in biology—the H cluster, containing six Fe atoms (Adams, 1990). It has not yet been determined whether this active site also requires accessory proteins for its assembly.

For industrial use of hydrogenases and industrial "improvement" of hydrogenases, it will be desirable to express hydrogenases from their cloned genes in heterologous hosts. To do that, if accessory genes are necessary, the accessory genes will need to be expressed along with the structural genes. Expression of an iron-only hydrogenase—the hydrogenase of *Desulfovibrio vulgaris* (Hildenborough)—was previously attempted from the enzyme's cloned structural genes in *E. coli*, but that expression was unsuccessful, which was taken as evidence that accessory proteins might be necessary for the enzyme's assembly.

In the nickel-iron hydrogenases, some accessory genes are always located adjacent to and in the same operon as the structural genes (Vignais and Toussaint, 1994). In that respect, circumstantial evidence from DNA sequencing suggests that at least one iron-only hydrogenase—*Clostridium pasteurianum* hydrogenase I—does not require accessory gene products. The regions flanking that enzyme's structural gene have been sequenced, and the hydrogenase structural gene appears to belong to a monocistronic operon. Also, some of the flanking gene products have been assigned functions, and the functions are not connected with hydrogenase (Meyer, 1995).

In this paper, the question of whether an iron-only hydrogenase requires accessory proteins for assembly of an active enzyme has been addressed by attempting to reconstitute an active enzyme in vitro from denatured *C. pasteurianum* hydrogenase I. It has previously been shown that the iron-sulfur clusters of a bacterial ferredoxin could be reconstituted by incubating apoferredoxin with ionic iron, inorganic sulfide, and a mercaptide reducing agent (Malkin and Rabinowitz, 1966). Similar procedures were used here to attempt to reconstitute *C. pasteurianum* hydrogenase I.

3. EXPERIMENTAL

C. pasteurianum Winogradsky 5 ATCC 6013 was grown under N_2 at 35 °C in YEM medium (Palosaari and Rogers, 1988). All steps in *C. pasteurianum* extract and hydrogenase purification were done anaerobically with 2 mM sodium dithionite added to all buffers. To prepare the extract, cell paste was resuspended in 100 mM Tris-HCl, pH 8.3, and cells were broken by lysozyme digestion and one freeze-thaw cycle and then treated with DNAse and RNAse. Debris was pelleted by centrifugation for 15 min in a JA-10 rotor at 8000 rpm. Hydrogenase I was purified according to Chen and Mortenson (1974). Protein was quantified according to Bradford (1976).

C. pasteurianum extract and purified hydrogenase I were denatured by adding 0.54 g solid urea per 1 mL solution volume anaerobically under N_2 gas. This increased the volume by 29% and gave a final 7 M urea. For renaturation, the denatured solutions were diluted anaerobically under N_2 into 40 mM sodium phosphate, pH 7.0, 5 mM dithiothreitol (DTT), 2 mM cysteine, 0.3 mM ferrous ammonium sulfate, and 2 mM sodium dithionite. Sodium sulfide (0.3 mM) was sometimes also included in the renaturation mix.

Hydrogen evolution was measured amperometrically (Sweet et al., 1980) in 40 mM sodium phosphate, pH 7.0, 2 mM methyl viologen, and 5 mM sodium dithionite, at 23 °C.

4. RESULTS AND CONCLUSIONS

Proteins in soluble fraction anaerobic extracts of *C. pasteurianum* were denatured by adding and dissolving solid urea as described in section 3 (Experimental) and heating to 50 °C for 5 min. The original extracts had hydrogen evolution activity of 144 nmol

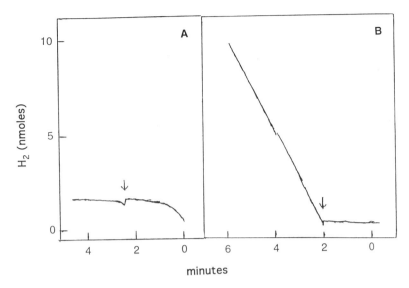

Figure 1. Hydrogen evolution at 23 °C in 1.5 mL of 40 mM sodium phosphate, pH 7.0, 5 mM dithionite. At the arrows, methyl viologen was added to 2 mM. (A) Catalyzed by urea-denatured *C. pasteurianum* extract (0.65 mg protein). (B) Catalyzed by urea-denatured *C. pasteurianum* extract reconstituted 70 min as described in section 4 (Results) (0.65 mg protein).

H_2/min/mg protein. After urea denaturation, there was no detectable hydrogen evolution activity (Figure 1A).

The denatured extract was diluted 1:19 into a renaturation mix including 0.3 mM Na_2S and incubated anaerobically under nitrogen gas at 23 °C for 70 min. After this time, it had regained hydrogen evolution activity of 2.77 nmol H_2/min/mg (Figure 1B). This is 2.0% of the original hydrogen evolution activity of the extract. The activity did not increase with further incubation in the reconstitution mix overnight.

As a control, the renaturation mix was incubated as above without denatured extract added and was tested for hydrogen evolution activity. There was no hydrogen evolution, demonstrating that the observed activity was not a chemical artifact of the renaturation mix.

Denaturation and reconstitution was also performed with purified hydrogenase I. Hydrogenase I (9.1 μmol H_2-evolved/min/mg protein) in 25 mM Tris-HCl, pH 8.1, 0.2 M NaCl, and 2 mM dithionite, was denatured by the addition of solid urea to 7.5 M and incubation at 36 °C for 15 min. This treatment caused the sample to lose all hydrogen evolution activity. The sample was diluted 1:14 into the reconstitution buffer containing 0.3 mM Na_2S and incubated 2.5 h. It regained 17 nmol/min/mg hydrogen evolution activity—0.19% of the starting activity.

This was a poorer recovery of activity than was accomplished with denatured crude extracts. One possible reason is that crude extract contained sulfide and sulfhydryl compounds, while the purified protein did not. Sulfide and sulfhydryls were present in the reconstitution mix to which purified enzyme was added, but they were not present when the enzyme was denatured. To test if the presence of these compounds during denaturation affected the subsequent reconstitution, 5 mM DTT and 1 mM Na_2S were added to the purified enzyme before adding urea to 7.5 M and treating at 36 °C for 15 min. This treatment again totally abolished enzyme activity (Figure 2A). When the denatured enzyme was diluted 1:14 into the reconstitution mix lacking sulfide and incubated at 23 °C 2.5 h, it re-

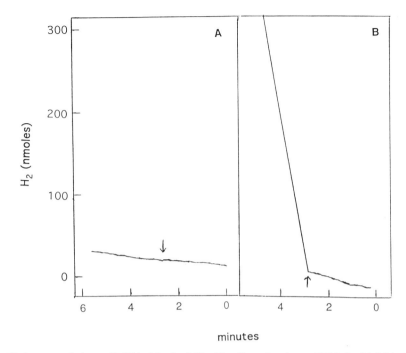

Figure 2. Hydrogen evolution at 23 °C in 1.5 mL of 40 mM sodium phosphate, pH 7.0, 5 mM dithionite. At the arrows, methyl viologen was added to 2 mM. Note that more hydrogenase was used in A than in B. (A) Catalyzed by *C. pasteurianum* hydrogenase I (500 µg) denatured by urea in the presence of 5 mM DTT, 1 mM Na_2S, as described in section 4 (Results). (B) Catalyzed by *C. pasteurianum* hydrogenase I (89 µg) denatured by urea in 5 mM DTT, 1 mM Na_2S, and reconstituted 2.5 h as described in section 4 (Results).

covered hydrogen evolution activity of 327 nmol/min/mg—3.6% of the original activity. This is 18 times better recovery of activity than with enzyme denatured in the absence of sulfide and DTT.

As a control, the reconstitution mix lacking sulfide was tested for hydrogen evolution without the addition of denatured hydrogenase I. No H_2 evolution was detected, again showing that the observed hydrogen evlution is not a chemical artifact.

The presence of DTT and sulfide during denaturation may protect protein sulfhydryls against some type of irreversible alteration. Malkin and Rabinowitz (1966) found that ferredoxin could only be reconstituted if its sulfhydryl groups had been protected by the mercurial sodium mersalyl.

Denatured *C. pasteurianum* hydrogenase I can be reconstituted in vitro in the presence of sulfide, reduced thiol reagents, and ferrous iron to 3.6% of the original enzyme activity. This indicates that accessory proteins are not necassary for assembly of active enzyme. This suggests that it should be possible to express recombinant *C. pasteurianum* hydrogenase I in heterologous hosts from the structural gene alone.

The previous failure to express the *D. vulgaris* (Hildenborough) iron-only hydrogenase might have been due to that enzyme being a two-polypeptide periplasmic enzyme that uses an unusual signal sequence for transport across the membrane. The transmembrane transport and processing were not carried out well in *E. coli* (van Dongen et al.,

1988). *C. pasteurianum* hydrogenase I is a one-polypeptide cytoplasmic protein (Adams, 1990), so it should not suffer from the same problems.

ACKNOWLEDGMENTS

I thank Professor Larry Wackett for the use of his laboratory space and for helpful discussions. This work was supported by U.S. Department of Agriculture grant number 94-37500-1073.

REFERENCES

Adams, M.W.W., 1990, The structure and mechanism of iron-hydrogenases, *Biochim. Biophys. Acta*, 1020:115–145.
Bradford, M., 1976, A rapid and sensitive method for the quantitation of microgram quantities of protein utilizing the principle of protein-dye binding, *Anal. Biochem.*, 72:248–254.
Chen, J.S., and Mortenson, L.E., 1974, Purification and properties of hydrogenase from *Clostridium pasteurianum* W5, *Biochim. Biophys. Acta*, 371:283–298.
Malkin, R., and Rabinowitz, J.C., 1966, The reconstitution of Clostridial ferredoxin, *Biochim. Biophys. Res. Comm.*, 23:822–827.
Meyer, J., 1995, Sequence of a 10.5-kbp fragment of *Clostridium pasteurianum* genomic DNA encompassing the hydrogenase I gene and two spore germination genes, *Anaerobe*, 1:169–174.
Palosaari, N.R., and Rogers, P., 1988, Purification and properties of the inducible coenzyme A-linked butyraldehyde dehydrogenase from *Clostridium acetobutylicum*, *J. Bacteriol.*, 170:2971–2976.
Sweet, W.J., Houchins, J.P., Rosen, P.R., and Arp, D.J., 1980, Polarographic measurement of H_2 in aqueous solutions, *Anal. Biochem.*, 107:337–340.
van Dongen, W., Hagen, W., van den Berg, W., and Veeger, C., 1988, Evidence for an unusual mechanism of membrane translocation of the periplasmic hydrogenase of *Desulfovibrio vulgaris* (Hildenborough), as derived from expression in *Escherichia coli*, *FEMS Microbiol. Lett.*, 50:5–9.
Vignais, P.M., and Toussaint, B., 1994, Molecular biology of membrane-bound H_2 uptake hydrogenases, *Arch. Microbiol.*, 161:1–10.
Voordouw, G., Hagen, W.R., Kruse-Wolters, M., van Berkel-Arts, A., and Veeger, C., 1987, Purification and characterization of *Desulfovibrio vulgaris* (Hildenborough) hydrogenase expressed in *Escherichia coli*, *Eur. J. Biochem.*, 162:31–36.

14

ATTEMPT AT HETEROLOGOUS EXPRESSION OF CLOSTRIDIAL HYDROGENASE IN CYANOBACTERIA

Yoji Koike,[1] Katsuhiro Aoyama,[4] Masato Miyake,[2] Junko Yamada,[3] Ieaki Uemura,[4] Jun Miyake,[5] and Yasuo Asada[2]

[1]Research Institute of Innovative Technology for the Earth
[2]National Institute of Bioscience and Human-Technology
Higashi 1-1, Tsukuba-shi, Ibaraki, 305
[3]Applied Biochemistry
University of Tsukuba
Tennohdai, Tsukuba-shi, Ibaraki, 305
[4]Frontier Technology Research Institute
Tokyo Gas Co., Ltd.
Suehiro-chou 1-7-7, Tsurumi-ku, Yokohama-shi, 230
[5]National Institute for Advanced Interdisciplinary Research
Japan

Key Words

photohydrogen production, cyanobacteria, *Synechococcus* sp. PCC7942, *Clostridium*, hydrogenase

1. SUMMARY

A shuttle vector (pKE4-9) system for a cyanobacterium, *Synechococcus* sp.PCC7942 and *Escherichia coli*, was examined for expression of clostridial hydrogenase. pKE4-9 carries the chloramphenicol-resistant (*cat*) gene and a strong promoter we have isolated from PCC7942. The clostridial hydrogenase gene was cloned upstream of the *cat* gene in pKE4-9, which was transformed into PCC7942. Although the *cat* protein was expressed, the clostridial hydrogenase was not detected by Western blot analysis. The hydrogenase mRNA was detected in PCC7942; it was co-transcribed with *cat*. The problems in the expression of foreign proteins in cyanobacteria are discussed.

2. INTRODUCTION

Since Benemann and Weare (1974) demonstrated hydrogen evolution by nitrogen-fixing *Anabaena cylindrica*, hydrogen evolution by cyanobacteria has been extensively studied with the goal of developing a new energy production method (Miyamoto, 1994).

Although nitrogen-fixing cyanobacteria are much more sustainable than hydrogenese-containing green algae under light conditions, nitrogenase-dependent hydrogen production is theoretically less efficient than hydrogenase-dependent production because of ATP requirements. We are aiming to apply hydrogenase enzymes to hydrogen evolution. Coupling of the photochemical reaction with hydrogenase via ferredoxin has been demonstrated by Benemann et al. (1973). Fry et al. (1977) reported that hydrogenase of *Clostridium pasteurianum* produced hydrogen gas in vitro with the ferredoxin of *Spirulina maxima* as the electron mediator. Recently, Miyake and Asada (1997) demonstrated photo-hydrogen evolution by cyanobacteria that had clostridial hydrogenase protein introduced by electroporation.

In order to permanently incorporate clostridial hydrogenase in cyanobacterial cells, we have worked to develop a genetic engineering system for the non-nitrogen-fixing cyanobacterium, *Synechococcus* sp. PCC7942, and obtained a gene fragment showing strong promoter activity in PCC7942 (Miyake et al., 1996). Using the plasmid pKE4-9 carrying the promoter, we have examined the expression of clostridial hydrogenase in PCC7942.

3. MATERIALS AND METHODS

Synechococcus sp. PCC7942 was cultured with air-bubbling under continuous illumination of 1-3 klux (fluorescent lamps) at 30 °C in culture medium BG-11 (Allen 1968). Transformants of *Synechococcus* sp. PCC7942 are as follows: PCC4 is harboring plasmid pKE4; PCC4-9 is harboring plasmid pKE4-9; and PCC4-9H2 is harboring plasmid pKE4-9H2 (Figure 1).

Total DNA was extracted from PCC7942 as described by Mak and Ho (1991). The clostridial hydrogenase gene that was sequenced by Meyer and Gagnon (1991) was amplified by PCR with chromosomal DNA of *Clostridium pasteurianum* ATCC6013.

The clostridial hydrogenase gene was inserted between the strong promoter fragment and *cat* gene in pKE4-9.

The cell-free extracts, prepared by ultra-sonicator, were subjected to SDS-polyacrylamide gel electrophoresis (SDS-PAGE) (Laemmli, 1970). Separated proteins were electro-transferred to nitrocellulose membranes. The membranes were probed with rabbit antiserum against the synthesized partial polypeptide fragment of hydrogenase (16 amino acid residues from cysteine 356 to aspartate 371, based on the sequence by Meyer and Gagnon, 1991).

Total mRNA was isolated by using a procedure previously described by Reith et al. (1986). Northern analysis used an ECL detection kit (Amersham) with *cat* or hydrogenase DNA as a probe.

4. RESULTS AND DISCUSSION

Western blot analysis of soluble and membrane fractions from *E. coli* transformants harboring plasmid pKE4-9H2 with specific hydrogenase antibody revealed positive bands

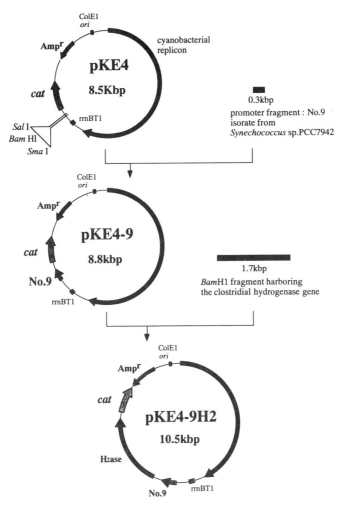

Figure 1. Construction strategy of plasmid, pKE4-9H2. PCR-amplified DNA fragment containing the clostridial hydrogenase structure gene was restricted by *Bam*HI and inserted in the pKE4-9 (see also Miyake et al., 1996).

of approximately 64 kDa, which corresponded to the size of hydrogenase (Meyer and Gagnon, 1991) (data not shown). On the other hand, in both soluble and membrane fractions from *Synechococcus* transformant, positive signals could not be detected (Figure 2).

Although we did not observe expression of hydrogenase in PCC4-9H2, *cat* activity was detected in both PCC4-9 and PCC4-9H2 (Aoyama et al., 1996), which suggested the occurrence of co-transcription of hydrogenase and *cat* because the hydrogenase gene did not retain a terminator sequence. We therefore conducted a northern blot analysis with fragments from hydrogenase and *cat* as a probe (Figure 3). In PCC4-9H2, only one positive band was detected at 2.2 kbp (Figure 3A, B, lane 3) with each probe. As the *cat* and hydrogenase genes were 0.7 kbp (Figure 3A, lane 2) and 1.7 kbp (Figure 3B, lane 4), respectively, the 2.2-kbp band in PCC4-9H2 was estimated to be the *cat* and hydrogenase co-transcript.

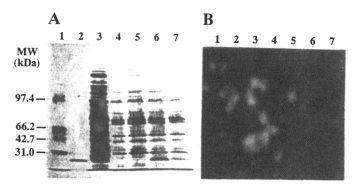

Figure 2. Western blot analysis of cell-free extracts from *Synechococcus* sp. PCC7942. *Synechococcus* extracts were subjected to SDS-PAGE on 5–20% gradient gel (A) and western blot analysis (B). Western blot was carried out with the probe, which was rabbit antiserum against the hydrogenase polypeptides. Lane 1, molecular marker; lane 2, *cat* protein; lane 3, hydrogenase fraction from *Clostridium pasteurianum*; lanes 4, 5, 6, and 7, total protein fractions from wild type, PCC4, PCC4-9, and PCC4-9H2, respectively.

Even though the hydrogenase mRNA was co-transcribed with the *cat* gene, the hydrogenase protein was not detected by Western analysis. From these results, we concluded that the obstacles encountered in the expression of hydrogenase might be in the translational phase, such as translation initiation, elongation, termination or differences in codon usage. One probable translation problem could be an unsuitable ribosomal binding site (Shine-Dalgarno: SD sequence). Takeshima et al. (1994) reported that the translation efficiency depends on SD sequence and the distance between SD and the initiation codon. They succeeded in controlling the foreign gene expression in cyanobacteria by changing the SD sequence as well as the distance between the SD and ATG start codon. In our case, the SD sequence of clostridial hydrogenase gene (-AGGAGG-) is different from the *cat* SD (-AAGGAA-) (Brosius, 1984), which is equal to that of cyanobacteria, that is, D1 proteins of PCC7942 (Bustos et al., 1990). Our efforts will continue to be directed at overcoming genetic obstacles in expression of the clostridial hydrogenase. We are now constructing a new expression vector where the SD sequence of the hydrogenase gene is changed to the same as the *cat* SD.

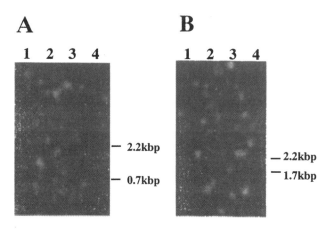

Figure 3. Northern blot analysis of total mRNA from *Synechococcus* sp. PCC7942. *cat* DNA as a probe (A) and hydrogenase DNA as a probe (B). Lane 1, mRNA from PCC7942 wild type; lane 2, PCC49; lane 3, PCC49H2; lane 4, *Clostridium pasteurianum*.

ACKNOWLEDGMENTS

This work was performed under the management of the Research Institute of Innovative Technology for the Earth as a part of the Project for Biological Production of Hydrogen supported by the New Energy and Industrial Technology Development Organization. The authors are grateful to Dr. G. Grandi and his co-workers, Eniricerche, Milan, for the generous gift of the hydrogenase gene from *C. pasteurianum*.

REFERENCES

Allen, M.M., 1968, Simple conditions for the growth of unicellular blue-green algae on plates, *J. Phycol.*, 4:1–4.

Aoyama, K., Miyake, M., Yamada, J., Miyake, J., Uemura, I., Hoshino, T., Asada, Y., 1996, Application of vector pKE4–9 carrying a strong promoter to the expression of foreign protein in *Synechococcus* PCC7942, *J. Marine Biotech.*, 4:64–67.

Benemann, J.R., Berenson, J.A., Kaplan, N.O., and Kamen, M.D., 1973, Hydrogen evolution by a chloroplast-ferredoxin-hydrogenase system, in *Proceedings of the National Academy of Sciences (USA)*, 70:2317–2320.

Benemann, J.R., and Weare, N.M., 1974, Hydrogen evolution by nitrogen-fixing *Anabaena cylindrica* culture, *Science*, 184:174–175.

Brosius, J., 1984, Plasmid vectors for selection of promoters, *Gene*, 27:151–60.

Bustos, S.A., Schaefer, M.R., and Golden, S.S., 1990, Different and rapid responses of four cyanobacterial psbA transcripts to changes in light intensity, *J. Bacteriol.*, 172:1998–2004.

Laemmli, U.K., 1970, Cleavage of structural proteins during the assembly of the head of bacteriophage T4, *Nature*, 227:680–685.

Mak, Y.M., and Ho, K.K., 1992, An improved method for the isolation of chromosomal DNA from various bacteria and cyanobacteria, *Nucleic Acids Res.*, 20:4101–4102.

Meyer, J., and Gagnon, J., 1991, Primary structure of hydrogenase I from *Clostridium pasteurianum*, *Biochem.*, 30:9697–9704.

Miyake, M., and Asada, Y., 1997, Direct electroporation of clostridial hydrogenase into cyanobacterial cells, *Biotech. Techniques*, 11:787–790.

Miyake, M., Yamada, J., Aoyama, K., Uemura, I., Hoshino, T., Miyake, J., and Asada, Y., 1996, Strong expression of foreign protein in *Synechococcus* sp. PCC7942, *J. Marine Biotech.*, 4:61–63.

Miyamoto, K., 1994, Hydrogen production by photosynthetic bacteria and microalgae, in *Recombinant Microbes for Industrial and Agricultural Applications*, Murooka, Y., and Imanaka, T., (eds.), Marcel Dekker, Inc., pp. 771–786.

Reith, M.E., Laudenbach, D.E., and Straus, N.A., 1986, Isolation and nucleotide sequence analysis of the ferredoxin I gene from the cyanobacterium *Anacystis nidulans* R2, *J. Bacteriol.*, 168(3):1319–1324.

Takeshima, Y., Takatsugu, N., Sugiura, M., and Hagiwara, H., 1994, High level expression of human superoxide dismutase in the cyanobacterium *Anacystis nidulans* 6301, in *Proceedings of the National Academy of Sciences (USA)*, 91:9685–9689.

15

STUDY ON THE BEHAVIOR OF PRODUCTION AND UPTAKE OF PHOTOBIOHYDROGEN BY PHOTOSYNTHETIC BACTERIUM *Rhodobacter sphaeroides* RV

Reda M. A. El-Shishtawy, Yoji Kitajima, Seiji Otsuka, Shozo Kawasaki, and Masayoshi Morimoto

Kajima Laboratory
Project for Biological Production of Hydrogen
Research Institute of Innovative Technology for the Earth (RITE)
2-19-1 Tobitakyu
Chofu, Tokyo 182, Japan

Key Words

photosynthetic bacteria, *Rhodobacter sphaeroides* RV, H_2 production fermentation, CO_2 fixation

1. SUMMARY

In this study, using a standard GL system and Roux flask of 1 cm width, the growth rate was saturated at 100 W/m^2 and the production rate of H_2 increased steadily with light energy from 50–150 W/m^2. The production of H_2 under halogen lamp illumination (155 W/m^2) using a Roux flask of 3 cm width was independent of the initial cell concentration in terms of OD_{660} range 0.18–1.04. The photostationary state of growth was reached at 42 h. Simultaneous consumption of lactate and excretion of organic acids were revealed during the course of H_2 production under standard conditions. Fermentation and CO_2 fixation by *Rhodobacter sphaeroides* RV was possibly revealed by HPLC analysis of the excreted organic acids during the course of H_2 production and the increase of cell concentration as a result of bacterial growth on CO_2/H_2 (16/84) gas mixture under light (300 W/m^2) and dark conditions. These results suggest that the RV strain has an inherent property to perform fermentation while under light conditions, a phenomenon that might lead the way to further investigations aimed at developing an effective strain for wastewater H_2 production process.

BioHydrogen, edited by Zaborsky *et al.*
Plenum Press, New York, 1998

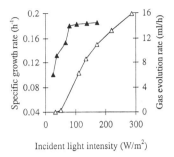

Figure 1. Rate of H_2 evolution and growth of *Rodobacter sphaeroids* RV at different light intensities. Conditions: halogen lamp, Roux flask of 1 cm width (84 mL, 84 cm^2), no stirring, 30 °C. ▲: specific growth rate; △: production rate.

2. INTRODUCTION

Photobiohydrogen is a promising clean energy source. Photosynthetic bacteria produce molecular H_2 with solar energy using suitable electron donors and ammonia-starved conditions (Sasikala et al., 1993). The production of H_2 by photosynthetic bacteria (Weaver et al., 1980) is limited by several factors. The most significant limitation is the low light intensity at which the photosynthetic process saturates, meaning that the very high light intensity of sunlight cannot be efficiently used and the possibility that photosynthetic bacteria would have to perform dark fermentation and/or CO_2 fixation. Since light, dark, and cloudy conditions are inevitable, it is indispensable to have an accurate investigation into the behavior of bacterial functions under such conditions so as to enhance H_2 productivity by engineering and/or genetic manipulation.

In this work, the photosynthetic bacterium *Rhodobacter sphaeroides* RV, as a highly active strain for H_2 production (Mao et al., 1986; Miyake and Kawamura, 1987), was used to study the effect of light intensity on cell growth and H_2 production, as well as to study the behavior of acid consumption and/or excretion during the course of H_2 production. Furthermore, the behavior of cell growth in light and dark using a 16/84 CO_2/H_2 gas mixture is described.

3. EXPERIMENTAL

3.1. Microorganism and Culture Conditions

Rhodobacter sphaeroides RV from the National Institute of Bioscience and Human Technology in Tsukuba, Japan, was pre-cultured in a basal medium (inorganic salts and vitamins) containing 0.1% (w/v) yeast extract, 10 mM ammonium sulfate, and 60 mM sodium succinate (Mao et al., 1986; Miyake et al., 1984). The strain was grown anaerobically in a completely full Roux flask of 6 cm width with a 272-cm^2 area under illumination (43 W/m^2 from a tungsten lamp) for about three days at 30 °C. Generally, the H_2 production medium (HPM) consisted of a 1/3 (v/v) ratio of the preculture and a freshly prepared GL medium of basal medium containing 72 mM sodium D,L-lactate and 10 mM sodium glutamate (Nakada et al., 1995), respectively. At a constant initial lactate concentration (67 mM lactate, Figure 3) in the HPM, different initial bacterial concentrations were obtained by mixing different ratios of the preculture, GL medium, and basal medium. HPM was transferred to Roux flasks of 1 cm (Figure 1) and 3 cm (Figures 2 and 3) widths and 84 and 60 cm^2 illuminated area, respectively.

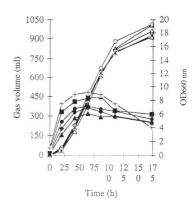

Figure 2. Effect of initial cell concentration on the production of H_2. Conditions: halogen lamp (155 W/m^2), Roux flask of 3 cm width (180 mL, 60 cm^2), no stirring, 30 °C. △: Gas evolution using OD1; ◇: Gas evolution using OD2; ○: Gas evolution using OD3; □: Gas evolution using OD4; ✻: Gas evolution using OD5; ▲: OD1 660 nm = 0.18; ◆: OD2 660 nm = 0.33; ●: OD3 660 nm = 0.475; ■: OD4 660 nm = 0.72; +: OD5 660 nm = 1.04.

HPM that contained 3.88 mM lactate was transferred to two cylindrical plane reactors of 1 cm depth (490 cm^2) to make cultures A and B. Cultures A and B were continuously illuminated using halogen lamps (300 W/m^2) for about two days, when there was no more H_2 production and starving conditions had been attained, at 30 °C with shaking. After reaching the starving conditions, that is, no detection of organic substrate, illumination was halted and the gas in A and B was replaced with a 16/84 CO_2/H_2 gas mixture to make culture A' for light icubation and culture B' for dark incubation (Figure 4).

Cell growth was determined either photometrically by measuring the optical density of the culture at 660 nm or gravimetrically by weighing the amount of dry cell after overnight desiccation at 120 °C. A linear relationship between OD_{660} and cell dry weight values was observed and the cell dry weight was calculated from the graph (1 OD_{660} = 0.34 mg 9 (dry weight/mL).

The specific growth rate (μ) was calculated using the following equation (Pirt, 1975):

$$\mu = (\ln x - \ln x_0)/t$$

where x_0 is bacterial cell concentration (g/L) that increases exponentially to x in time t.

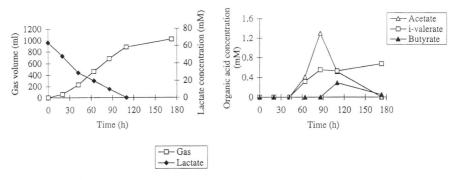

Figure 3. (A) Organic acid consumption and/or excretion during the course of H_2 production. Conditions: halogen lamp (155 W/m^2), Roux flask of 3 cm width (180 mL, 60 cm^2), no stirring, 30 °C. (B) Excretion and consumption of new organic acids, other than lactate, during the course of H_2 production (0-180 hr).

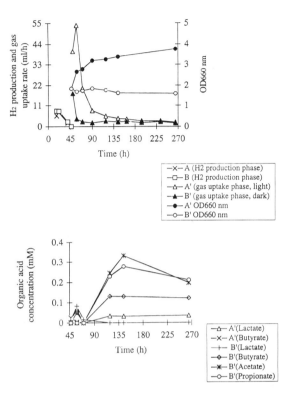

Figure 4. H_2 production and CO_2/H_2 (16/84) gas mixture uptake vs. time by RV strain under dark and light conditions (300 W/m^2) at 30 °C and with piston shaking (160 min^{-1}) using a 1-cm deep plane reactor (490 cm^2, 500 mL). Excretion of organic acids as a result of gas mixture uptake in dark and light conditions

3.2. Analytical Methods

The light intensity at the surface of the reactor was measured with a photometric sensor (LI-COR Inc., Model LI-189). The volume of gas evolved was measured by displacement of acidic water at room temperature and the H_2 content was analyzed by Sensidyne detector tubes (Gastec Corporation) and/or gas chromatography. Organic acid excretion and lactate consumption in the medium were measured by HPLC using a Shim-Pack SCR-102H column (8 × 300 mm, Shimadzu) and a conductivity detector (CCD-6A, Shimadzu).

4. RESULTS AND DISCUSSION

Rates of growth and H_2 production as a function of light intensity are depicted in Figure 1. The growth rate saturated at 100 W/m^2 and the rate of H_2 increased steadily with light energy from 50–150 W/m^2 and then started to increase in a curved manner. Likewise, the case of *Rhodopseudomonas capsulata* (Hillmer and Gest, 1977) might reflect that at the saturation point of growth, the maximum rate of photophosphorylation is attained when the production rate of H_2 is no longer proportional with light intensity. In other words, it is known that nitrogenase is the H_2-evolving enzyme in photosynthetic bacteria and since nitrogenase synthesis is related to the bacterial photophosphorylation activity (Boïchenko and Hoffmann, 1994), the stationary state of growth presumably reflects the maximal rate of photophosphorylation.

As shown in Figure 2, the effect of initial cell concentration on the production of H_2 revealed that the photostationary state of growth was reached after 42 h and was independent of the initial cell concentrations (OD_{660} range 0.18–1.04). Also, the production of H_2 was independent of the initial cell concentration. Such a good correlation between the photostationary state of growth and H_2 production confirms that light is the rate-limiting factor.

Investigating acid consumption and/or excretion during the course of H_2 production (Figure 3) has revealed that excretion of organic acids could be detected by HPLC at the stationary state of growth, that is, after 42 h. At this stage, the evolution of H_2 started to increase steadily over a certain time (about 60 h). It is noteworthy that the excretion of organic acids started while lactate was present in more than one-third of the started concentration. This phenomenon might indicate that bacteria would need to make additional ATP synthesis by performing autotrophic CO_2 fixation using H_2 as the electron donor and/or dark fermentation of the endogenous sugars at the dark zone of the reactor, the case that has been reported for many facultative bacteria (Thauer et al., 1977; Anderson and Fuller, 1967).

This finding was further clarified by conducting two similar experiments (cultures A and B) using low concentrations of lactate (38.8 mM) to produce H_2 for two days. After that, illumination (300 W/m^2) was halted and the culture was starving, that is, no detection of organic acid substrate was observed. Then, a mixture of CO_2/H_2 (16/84) was fed to the starved cultures (A´ and B´) to conduct A´ under light (300 W/m^2) and B´ under dark conditions. It should be mentioned that 300 W/m^2 was selected based on our experience of its suitability for the RV strain and that it is more or less the average value of the simulated solar energy by sine curve mode of light 0–1000 W/m^2:Light$_{12h}$-Dark$_{12h}$. CO_2/H_2 (16/84) was also selected to simulate the real case of H_2 production using the RV strain. As shown in Figure 4, as the culture age increased, the capability of *Rhodobacter sphaeroides* RV to perform gas mixture uptake decreased. Also, gas mixture uptake under light conditions was much higher than that under dark conditions, which nicely correlated with cell growth. Furthermore, the excretion of organic acids was clearly detected by HPLC to show more excretion under dark conditions than light. The fact that the RV strain grows on a CO_2/H_2 mixture in the absence of an organic substrate and with the excretion of organic acids indicates that this process should be coupled with phosphorylation. This behavior might indicate the inherent property of *Rhodobacter sphaeroides* RV to perform autotrophic CO_2 fixation. This paper has presented the possibility of autotrophic CO_2 fixation and dark fermentation; however, biochemical analysis of the metabolism in RV strain remains to be investigated.

ACKNOWLEDGMENTS

R.M.A. El-Shishtawy, on leave from the National Research Centre, Cairo, Egypt, was supported by a New Energy and Industrial Technology Development Organization Fellowship Award to work under the management of the Research Institute of Innovative Technology for the Earth. The authors thank Miss Sakamoto for her analytical work.

REFERENCES

Anderson, L., and Fuller, R.C., 1967, *Plant Physiol.*, 42:487–490.
Boïchenko, V.A., and Hoffmann, P., 1994, *Photosynthetica*, 30:527–552.
Hillmer, P., and Gest, H., 1977, *J. Bacteriol.*, 129:724–731.
Mao, X.Y., Miyake, J., and Kawamura, S., 1986, *J. Ferment. Technol.*, 64:245–249.
Miyake, J., and Kawamura, S., 1987, *Int. J. Hydrogen Energy*, 12:147–149.
Miyake, J., Mao, X.Y., and Kawamura, S., 1984, *J. Ferment. Technol.*, 62:531–535.
Nakada, E., Asada, Y., Arai, T., and Miyake, J., 1995, *J. Ferment. Bioeng.*, 80:53–57.
Pirt, S.J., 1975, *Principles of Microbe and Cell Cultivation*, 1st ed., Oxford, Blackwell.
Sasikala, K., Ramana, C.V., and Raghuveer Rao, P., 1993, in *Advances in Applied Microbiology*, vol. 38, Academic Press, Inc., pp. 211–295.
Thauer, R.K., Jungermann, K., and Decker, K., 1977, *Bacteriol. Rev.*, 41:100–180.
Weaver, P.F., Lien, S., and Seibert, M., 1980, *Sol. Energy*, 24:3–45.

16

CHARACTERIZATION OF A NOVEL LIGHT-HARVESTING MUTANT OF *Rhodobacter sphaeroides* WITH RELATION TO PHOTOHYDROGEN PRODUCTION

Lyudmila Vasilyeva,[1,2,3] Masato Miyake,[1] Masayuki Hara,[1,4] Eiju Nakada,[3] Satoshi Nishikata,[3] Yasuo Asada,[1] and Jun Miyake[1,4]

[1]National Institute of Bioscience and Human Technology
1-1-3 Higashi, Tsukuba, Ibaraki 305
[2]Research Institute of Innovative Technology for the Earth
[3]Fuji Electric Corporate R&D, Ltd.
2-2-1 Nagasaka, Yokosuka City, 240-01
[4]National Institute of Advanced Interdisciplinary Research
1-1-4 Higashi, Tsukuba, Ibaraki 305, Japan

Key Words

Rhodobacter sphaeroides, photosynthetic light-harvesting system, reaction center, *puf* operon

1. SUMMARY

We have isolated and characterized a UV mutant of *Rb. sphaeroides* that has considerable alterations in its light-harvesting system, but shows a normal growth rate over a wide range of light intensities (5–150 W/m^2). To examine possible rearrangements in the photosynthetic system of the P3 mutant, quantitative analysis of pigment-protein complexes solubilized from the membranes was accomplished. After sucrose density gradient centrifugation, the amounts of core complexes and RC:LH1:LH2 complexes in corresponding fractions of P3 were two times lower compared to those in corresponding fractions of the wild-type strain. At the same time, the content of both the reaction center and LH2 in the upper fraction of P3 was considerably higher (threefold and 1.4-fold, respectively) compared to that in the corresponding fraction of the wild-type strain. A model of structural rearrangements in the photosynthetic light-harvesting system of P3 was proposed. We have suggested that the direct interaction of unassociated reaction centers with LH2 complexes could retain a high efficiency of light-energy transfer. P3 provided an interesting model for studying the contribution of each antenna complex to hydrogen production.

2. INTRODUCTION

Purple nonsulfur bacteria have extremely versatile metabolisms, which make them attractive models for basic and applied studies. *Rhodobacter sphaeroides* strain RV is a prospective producer of photohydrogen (Miyake et al., 1984). This process is catalyzed by nitrogenase and requires the energy of ATP provided by photosynthesis. However, the efficiency of photohydrogen production is still not sufficient for practical application, and further increases are necessary. Among the relevant factors, the size and organization of the photosynthetic unit are considered to be important in determining the efficiency of light-energy transfer (Pearlstein, 1982). However, the relationship between the composition of the photosynthetic system and the hydrogen production capacity has not been clarified. In this respect, mutants somewhat defective in the synthesis of light-harvesting (LH) complexes would be useful in understanding the regulatory mechanisms that can control the efficiency of light-energy conversion to molecular hydrogen.

The photosynthetic apparatus of *Rb. sphaeroides* is localized in the intracytoplasmic membrane system. It comprises the reaction center (RC) and two types of LH complexes, designated B_{875} (LH1, core antenna) and $B_{800-850}$ (LH2, peripheral antenna) on the basis of their absorption maxima in the near-infrared (Sistrom, 1962). The number of all antenna Bchl per one RC is defined as a photosynthetic unit, but in vivo, RC-LH1 core complexes appear to be associated in functional units of up to 150 Bchl a, in combination with 2, 3, or 4 RC (Hunter, 1995). All existing mutants of *Rb. sphaeroides,* whether they lack the intact LH1 complex (Meinhardt et al., 1985), contain low levels of LH1 and RC complexes (DeHoff et al., 1988), or have the LH1 genes deleted (Jones et al., 1992), grow slowly at low light intensity compared to the wild-type. It was assumed that in LH1-deficient mutants, much of the light energy absorbed by LH2 antenna complexes could not be passed to the RC but was instead emitted as fluorescence (Meinhardt et al., 1985). However, according to more recent data, LH1 is not a unique donor of energy to the reaction center because direct and efficient energy transfer from LH2 to RC has been demonstrated in an LH1-deficient mutant (Hess et al., 1993). Currently, it is still not clear whether RC can directly interact with LH2 in the presence of the core antenna and how it might affect the efficiency of light-energy transfer.

In this study, we described the novel mutant of *Rb. sphaeroides*, named P3, with alterations in the LH system. Despite the considerable reduction of the LH1 complex content, P3 grows at the same rate as the wild-type strain even at low light intensity, demonstrating high efficiency of light-energy transfer. According to recent results, under certain conditions of illumination the photosynthetic membranes of P3 can convert light energy to hydrogen more efficiently than the wild-type strain (Miyake et al., 1998). To examine possible structural rearrangements in the LH system of P3, we isolated and analyzed the pigment protein complexes from the photosynthetic membranes of both the mutant and wild-type strains. Based on the data obtained, we suggest that unassociated reaction centers could interact with LH2 complexes directly, retaining high efficiency of light-energy transfer in photosynthetic membranes of P3.

3. EXPERIMENTAL

3.1. Bacterial Strains and Growth Conditions

The P3 mutant was isolated in the process of screening for mutants in the photosynthetic apparatus after UV irradiation of plates with the wild-type strain *Rb. sphaeroides* RV (Miyake, et al., 1984) (30 sec, UV transilluminator TM-20, Funakoshi, Japan). Both the wild-type and P3 mutant strains were cultivated at 30 °C, 60 W/m^2, in ASY medium as previously described by Miyake et al. (1984). *E. coli* strains JM109 and S17-1 were routinely cultured at 37 °C in LB medium (Sambrook, et al., 1989).

3.2. DNA Manipulations and Analysis

Standard cloning procedures were performed as described by Sambrook et al. (1989). Oligonucleotide primers for PCR were produced based on published sequences from *Rb. sphaeroides* strain NC1B (Hunter et al., 1991) using a Millipore DNA synthesizer. The oligonucleotides were used to amplify the 1.1-kb DNA fragment from *Rb.sphaeroides* genomic DNA: 5'-CGG ATC CCA ACC CTC TTT CAT CGC TGC CTC-3' and 5'-GCG ACG GAG AAG CTT CCC GAT TAC TC-3'. The fragment, flanked by the engineered BamHI and HindIII restriction sites (in boldface type), was introduced in trans in pRKM-415. The plasmid pRKM-415 is a derivative of pRK-415 with the Tc^R gene disrupted by Km^R gene insertion (Sode, unpublished). Plasmid constructs were introduced into *E. coli* S17-1 and then transferred into *Rb. sphaeroides* by conjugation (Simon et al., 1983). Genomic DNA purified from the wild-type strain as described by Mak and Ho (1991), was used as a template for PCR.

3.3. Preparation of Membranes and Solubilized RC Complexes

Intracellular membranes and crude extracts were prepared as described by Hara, et al. (1994). The RC complexes were extracted from intracellular membranes using a mixture of 15 mM n-octyl-β-D-glucopyranoside (OG) and 15 mM deoxycholate (DOC), as described by Molenaar, et al. (1988), and subsequently separated on 10–40% (w/v) sucrose gradient. Equal amounts of chromatophores, isolated from the RV and P3 mutant strains (600 μmol of bacteriochlorophyll), were layered on the gradient. Gradients were centrifugated in a Beckman SW 40 Ti rotor (22 h, 35000 rpm, 6 °C).

3.4. Spectral and Pigment-Protein Analysis

Room-temperature absorption spectra of chromatophores or sucrose-gradient fractions were obtained using a Shimadzu spectrophotometer. Analyses of RC and LH complexes by SDS-PAGE electrophoresis were performed with 15–25% linear gradient gel (Daiichi Pure Chemicals Co., Ltd.).

Protein concentration was determined with the Biorad standard protein assay using bovine serum albumin as a standard.

The total bacteriochlorophyll (Bchl) was determined after acetone-methanol (7:2) extraction as described by Clayton (1966). The levels of B$_{875}$ complexes were calculated from the spectral data by using A$_{878-820}$ with an extinction coefficient of ε = 73 mM^{-1}·cm^{-1}. The B$_{800-850}$ complex amounts were measured from A$_{849-900}$ with an extinction coefficient of ε = 96 mM^{-1}·cm^{-1} (Meinhardt, et al., 1985). For thin–layer chromatography (TLC) analysis, carotenoids were extracted as described previously (Yokoyama et al., 1995) using hexane:acetone (7:3) as a solvent.

RC content in sucrose-gradient fractions was measured by flash-induced (200 mS) oxidation of cyt c as described previously by Hara, et al. (1997), with the exception that in the near-infrared region an ITF-50S-83RT filter (Toshiba, Japan) was used to minimize stray light. The concentration of RC was calculated using a cyt c molar extinction coefficient of $\varepsilon = 27.6$ mM^{-1}·cm^{-1} at 550 nm (Margoliash and Frohwirt, 1959). The amounts of the RC in the membranes of the wild-type strain and P3 mutant were compared using ELISA as previously described by Douillard and Hoffman (1983), with antibodies raised against the RC-M subunit as a primary antibody and anti-rabbit IgG antibody-peroxidase conjugate as a secondary antibody.

4. RESULTS AND DISCUSSIONS

4.1. Characterization of the Novel Mutant of *Rb. sphaeroides*

A stable mutant, P3, was obtained after UV irradiation of the wild-type strain and screening of mutants in the photosynthetic apparatus. This mutant was greenish-brown, whereas the wild-type strain was brown. Photoheterotrophic growth rates of both strains were very similar at light intensities ranging from 5–150 W/m^2. Membranes from the P3 exhibited reduced absorbance at 875 nm, characteristic of the LH1 antenna complex (Figure 1A). According to the results of SDS-PAGE, the amounts of LH1-α polypeptide in P3 were decreased compared to the wild-type strain (7.8 and 17% of total protein, respectively; Figure 1B). The RC content observed in SDS PAGE (Figure 1B) or measured by ELISA (data not shown) in the two strains appeared to be similar (Figure 1B). Consistent with the other results, spectral analysis revealed that in the photosynthetic membranes of the P3 mutant, the LH1 complex content was approximately 2.5-fold lower and the LH2 complex content was 1.4-fold higher compared to those in the wild-type strain (Figure 1C). Carotenoid analysis by TLC showed no qualitative differences in carotenoid content between the two strains (data not shown).

Phenotypical P3 mutation appeared as a decrease in LH1 complex content. It was suggested that the mutation might be localized in the *puf* operon because *puf*BA genes are encoding the LH1 α and β polypeptides. For complementation analysis, two DNA fragments, 0.8 kb and 1.1 kb, comprising the *puf* promoter and coding sequences of the *puf*Q (or *puf*QBA structural genes, respectively), were amplified by PCR from a wild-type chromosome template and cloned in trans into pRKM-415. The resulting plasmids pQ-1 and pQBA-1, as well as the control plasmid pRKM-415, were transformed to a P3 mutant. The wild-type absorption spectrum was restored in the P3 (pPQBA-1) strain, but not in the P3 (pQ-1) or P3 (pRKM-415) strains (data not shown), suggesting that *puf*BA genes might bear the P3 mutation. However, sequencing analysis of the 0.6-kb PCR *puf* BA DNA fragment, amplified from the P3 mutant chromosome, revealed that it was identical to the corresponding fragment amplified from the wild-type chromosome (data not shown). Therefore, it is likely the wild-type absorption spectrum was restored to the mutant not by complementation, but by the increase in the LH1 structural gene copy number in the P3 (pQBA-1) strain. Transformant strains P3 (pPQBA-1) and P3 (pRKM-415) were used for further hydrogen production experiments (Miyake et al., 1998).

Characterization of a Novel Light-Harvesting Mutant of *Rhodobacter sphaeroides*

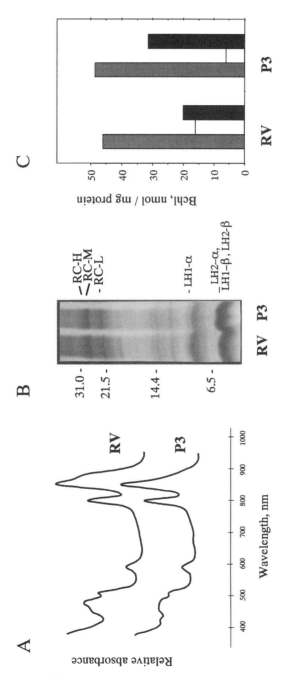

Figure 1. (A) Absorbtion spectra, (B) SDS-PAGE gel electrophoresis, and (C) Bchl contents in membranes from the wild-type strain (RV) and P3 mutant of *Rb. sphaeroides*. Shaded bars = total Bchl, open bars = LH1 Bchl, black bars = LH2 Bchl.

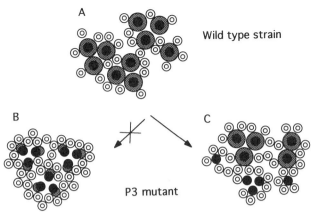

Figure 2. (A) Schematic model of organization of photosynthetic complexes in the membrane of *Rb. sphaeroides* wild-type strain and (B, C) possible structural rearrangements in the membrane of P3 mutant. Reaction centers are represented by black circles; LH1 complex by dark rings; and LH2 complex by white rings.

4.2. Analysis of Pigment-Protein Complexes

The ability of P3 to grow efficiently even at low light intensity despite the decreased level of LH1 complexes indicated that all RC were correctly assembled and functional. Accelerated hydrogen production of the mutant at 800 and 850 nm showed that under certain conditions the photosynthetic membrane of P3 mutant can convert light energy to biohydrogen even more efficiently than the wild-type strain. Thus, the question arises as to what sort of rearrangements could occur inside the photosynthetic unit of the P3 mutant. We have suggested that the decrease in the overall level of LH1 observed in membranes from the P3 mutant could reflect a uniform reduction in the core antenna size (Figure 2B), as demonstrated for *Rb. sphaeroides* LH2$^-$, *pufA* deletion mutants (McGlynn et al., 1996). Alternatively, the membranes of the P3 mutant could contain a mixture of RC:LH1:LH2 complexes similar in size to those in the wild-type strain, and an extra RC directly associated with peripheral antenna complexes (Figure 2C). The possibility of direct RC-LH2 interaction has been demonstrated previously in a LH1$^-$ mutant of *Rb. sphaeroides* (Hess et al., 1993).

To investigate possible rearrangements in the photosynthetic system of P3 mutant, membranes from the wild-type strain and mutant were solubilized with a mixture of detergents OG and DOC. When the solubilized pigment-protein complexes were separated by sucrose density gradient centrifugation, four pigmented bands were seen, labeled 1 to 4 in Figure 3. Distribution of total Bchl in these bands was found to be different for the wild-type strain and P3 mutant (Figure 3).

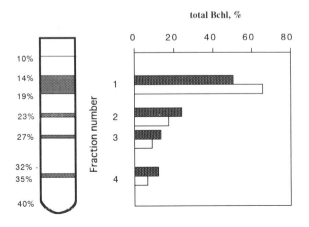

Figure 3. Distribution of total Bchl after 10–40% sucrose density gradient centrifugation of solubilized membranes from *Rb. sphaeroides* wild-type strain (shaded bars) and P3 mutant (open bars).

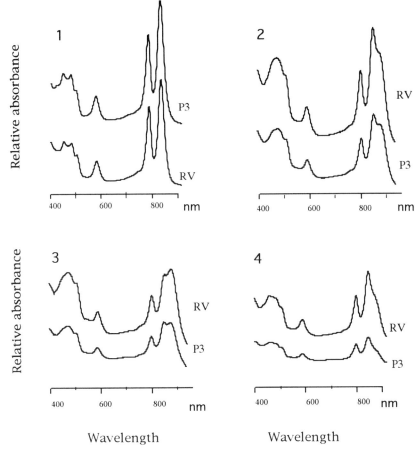

Figure 4. Spectra of 10–40% sucrose density gradient fractions 1, 2, 3, and 4, respectively.

According to the spectral analysis (Figure 4-1, Table 1) and SDS PAGE (data not shown), fraction-1 contained RC and LH2 complexes. The presence of RC:LH1:LH2 complexes was observed in fraction-2 (Figure 4-2). Fraction-3 had a major peak at 874 nm, indicating the presence of RC:LH1 core complexes together with some amounts of LH2 complex (Figure 4-3). The spectra of fraction-4 were indistinguishable from correspondent spectra of intact membranes and were assumed to be undissociated RC:LH1:LH2 complexes (Figure 4-4).

The amounts of RC and individual antenna Bchl were estimated in each fraction (Table 1). It was shown that the amounts of RC-complexes in fractions P3-2, P3-3, and P3-4 were more than two times lower compared to those in corresponding RV fractions (Table 1). However, the RC:LH1 ratio calculated for fractions RV-2, RV-3, P3-2, and P3-3 were similar. Our data are consistent with the previous observations that the core complex is present in the photosynthetic membrane in a fixed stoichiometry at approximately 29-32 LH1 Bchls per RC (Francke and Amesz, 1995; Karrasch et al., 1995) and indicate that no considerable reduction of core complex size occurred in photosynthetic membranes of the P3 mutant (Figure 2B).

Table 1. Amounts of RC and LH complexes in sucrose gradient fractions[a]

Individual complexes	Fractions							
	Wild-type strain				P3 mutant			
	RV-1	RV-2	RV-3	RV-4	P3-1	P3-2	P3-3	P3-4
LH2 Bchl[b]	371 ± 6	93 ± 2	23 ± 0.5	51 ± 1	511 ± 8	49 ± 1	14 ± 0.2	18 ± 0.5
LH1 Bchl[b]	0	77 ± 2	43 ± 1	21 ± 0.5	0	47 ± 1	25 ± 0.5	2 ± 0.25
RC[c]	0.8 ± 0.1	2.4 ± 0.25	1.45 ± 0.1	0.85 ± 0.1	2.5 ± 0.2	1.8 ± 0.1	0.9 ± 0.1	n.d.
LH1:RC	—	32	30	n.d.	—	26	28	n.d.

[a] Shown as nmol/mg protein.
[b] Estimated using spectral data and corresponding extinction coefficients (Meinhardt et al., 1985).
[c] Measured by flash-induced oxidation of cyt c.
n.d. -not determined.

In contrast to the considerable decrease in pigment-protein complex yield in all other P3 fractions, in the P3-1 fraction the amounts of RC and LH2 complexes were increased (threefold and 1.4-fold, respectively), compared to those in the RV-1 fraction (Table 1). Moreover, the RC content in the P3-1 fraction was higher than in any other P3 fraction. The data suggest considerable structural changes in the light-harvesting system of P3. We have proposed that in the P3 membrane, those RC that were not included in RC:LH1:LH2 complexes due to the decrease of core antenna amount could associate with LH2 directly, forming RC-LH2 complexes (Figure 2C).

In the LH1$^-$ mutant of *Rb. sphaeroides,* the efficiency of excitation energy transfer from LH2 to the reaction center was comparable to that of native RC-LH1 core complexes (Hess et al., 1993). Thus, we assume that the high growth efficiency of P3 at low light intensity could be explained by the rearrangements in the LH system. Recently in our laboratory, accelerated photohydrogen production by P3 under illumination by 800 nm and 850 nm monochromatic light was demonstrated (Miyake et al., 1998). The data can be considered as indirect evidence of the coexistence of RC-LH2 and RC-LH1-LH2 complexes in the photosynthetic system of P3. The P3 mutant appears to be a useful strain for investigation of the structure and function of a photosynthetic unit. It also provides an interesting model for studying the contribution of each antenna complex to hydrogen production.

ACKNOWLEDGMENTS

We thank Dr. Koji Sode of the Tokyo University of Agriculture and Technology for providing the pRKM-415, Takao Ueno and Yasuhiro Morimoto for the preparation of antibodies, and Kinuko Koike and Kiminori Kataoka for their technical assistance. This work was performed under the management of the Research Institute of Innovative Technology for the Earth (Japan) as part of the R&D Project on Environmentally Friendly Production Technology supported by the New Energy and Industrial Technology Development Organization, Japan.

REFERENCES

Clayton, R.K., 1966, Spectroscopic analysis of bacteriochlorophylls in vitro and in vivo, *Photochem. Photobiol.,* 5:669–677.

DeHoff, B.S., Lee, J.K., Donohue, T.J., Gumport, R.I., and Kaplan, S., 1988, In vivo analysis of *puf* operon expression in *Rhodobacter sphaeroides* after deletion of a putative intercystronic transcription terminator, *J. Bacteriol.*, 170:4681–4692.

Douillard, J.Y., and Hoffman, T., 1983, Enzyme-linked immunosorbent assay for screening monoclonal antibody production using anzyme-labeled second antibody, *Methods Enzymol.*, 92:168–174.

Francke, C., and Amesz, J., 1995, The size of the photosynthetic unit in the purple bacteria, *Photosynth. Res.*, 46:347–352.

Hara, M., Shigeno, M., Asada,Y., and Miyake, J., 1994, Sheet formation of membrane proteins from photosynthetic bacteria, *Materials Science and Engineering*, C2:13–28.

Hara, M., Ueno, T., Fujii, T., Yang, Q., Asada, Y. and Miyake, J., 1997, Orientation of photosynthetic reaction center reconstituted in neutral and charged liposomes, *Biosci. Biotech. Biochem.*, 61:1577–1579.

Hess, S., Visscher, K., Ulander, J., Pullerits, T., Jones, M.R., Hunter, C.N., and Sundström, V., 1993, Direct energy transfer from the peripheral LH2 antenna to the reaction center in a mutant of *Rhodobacter sphaeroides* that lacks the core LH1 antenna, *Biochemistry*, 32:10314–10322.

Hunter, C.N., 1995, Genetic manipulation of the antenna complexes of purple bacteria, in *Anoxygenic Photosynthetic Bacteria*, Blankenship, R.E., et al. (eds.), Kluwer Academic Publisher, Dordrecht, The Netherlands, pp. 473–501.

Hunter, C.N., McGlynn, P., Ashby, M.K., Burgess, J.G., and Olsen, J.D., 1991, DNA sequencing and complementation/deletion analysis of the *bchA-puf* operon region of *Rhodobacter sphaeroides*: in vivo mapping of the oxygen-regulated *puf* promoter, *Mol. Microbiol.*, 5:2649–2661.

Jones, M.R., Fowler, G.J.S., Gibson, L.C.D., Grief, G.G., Olsen, J.D., Crielaard, W., and Hunter, C.N., 1992, Mutants of *Rhodobacter sphaeroides* lacking one or more pigment-protein complexes and complementation with reaction-center, LH1, and LH2 genes, *Mol. Microbiol.*, 6:1173–1184.

Karrasch, S., Bullough, P.A., and Ghosh, R., 1995, The 8.5 Å projection map of the light-harvesting complex I from *Rhodospirillum rubrum* reveals a ring composed of 16 subunits, *EMBO J.*, 14:631–638.

Mak, Y.M., and Ho, K.K., 1991, An improved method of the isolation of chromosomal DNA from various bacteria and cyanobacteria, *Nucleic Acids Res.*, 20:4101–4102.

Margoliash, F., and Frohwirt, N., 1959, Spectrum of horse heart cytochrom, *Biochem. J.*, 71:570–572.

McGlynn, P., Westerhuis, W.H.J., Jones, M.R., and Hunter, C.N., 1996, Consequences for the organization of reaction center-light harvesting antenna 1 (LH1) core complexes of *Rhodobacter sphaeroides* arising from deletion of amino acid residues from the C terminus of the LH1 α polypeptide, *J. Biol. Chem.*, 271:3885–3892.

Meinhardt, S.W., Kiley, P.J., Kaplan, S., Crofts, A.R., and Harayama, S., 1985, Characterization of light-harvesting mutants of *Rhodopseudomonas sphaeroides*; Measurement of the efficiency of energy transfer from light-harvesting complexes to the reaction center, *Arch. Biochem. Biophys.*, 236:130–139.

Miyake, J., Mao, X.Y., and Kawamura, S., 1984, Photoproduction of hydrogen from glucose by a co-culture of a photosynthetic bacterium and *Clostridium butyricum*, *J. Ferment. Technol.*, 62:531–535.

Miyake, M., Sekine, M., Vasilieva, L., Nakada, E., Wakayama, T., Asada, Y., and Miyake, J., 1998, Improvement of bacterial light-dependent hydrogen production by altering photosynthetic pigment ratio, in *Proceedings of BioHydrogen '97*, Hawaii, Kona, Plenum Press, in press.

Molenaar, D., Crielaard, W., and Hellingwerf, K.J., 1988, Characterization of protonmotive force generation in liposomes reconstituted from phosphatidylethanolamine, reaction centers with light-harvesting complexes isolated from *Rhodopseudomonas palustris*, *Biochemistry*, 27:2014–2023.

Pearlstein, R.M., 1982, Exciton migration and trapping in photosynthesis, *Photochem. Photobiol.*, 35:835–844.

Sambrook, J., Fritsch, E.F., and Maniatis, T., 1989, *Molecular Cloning: A Laboratory Manual*, New York, Cold Spring Harbor Lab Press.

Simon, R., Priefer, U., and Puhler, A., 1983, A broad host range mobilization system for in vivo genetic engineering: transposon mutagenesis in Gram-negative bacteria, *Bio/Technology*, 1:784–791.

Sistrom, W.R., 1962, The kinetics of the synthesis of the photopigments in *Rhodopseudomonas sphaeroides*, *J. Gen. Microbiol.*, 28:607–616.

Yokoyama, A., Sandmann, G., Hoshino, T., Adachi, K., Sakai, M., and Shizuri, Y., 1995, Thermozeaxanthins, new carotenoid-glycoside-esters from thermophilic eubacterium *Thermus termophilus*, *Tetrahedron Lett.*, 36:4901–4904.

17

HYDROGEN AND 5-AMINOLEVULINIC ACID PRODUCTION BY PHOTOSYNTHETIC BACTERIA

Ken Sasaki

Materials Science and Engineering
Hiroshima-Denki Institute of Technology
6-20-1, Nakano
Akiku, Hiroshima, 739-03, Japan

1. ABSTRACT

Hydrogen production by *Rhodobacter sphaeroides* from organic acids was investigated. The maximum hydrogen production rate, 80.3 mL/g cell/h, was obtained when L-malic acid was used as a substrate with glutamate-Na as a nitrogen source. In addition, 30.0 mL/g cell/h of hydrogen production was observed with a mixed substrate containing acetic and propionic acids (VFA). Small amounts of NH_4 did not inhibit nitrogenase and hydrogen production in this bacterium.

5-aminolevulinic acid (ALA) was produced extracellularly from a VFA medium prepared from a post-anaerobic digestion liquor of swine waste and sewage sludge using the cells of *R. sphaeroides* with a dense resting cell system. About 9 mM ALA was produced after six days of incubation. A culture broth containing 4 mM ALA has a relatively strong herbicide effect in clover. In addition, other applications of ALA as a growth-promoting factor in plants, a means of improving salt tolerance in plants, and cancer treatment for humans were reported.

Thus, the combination of hydrogen production and ALA production might be promising for the bioconversion of organic wastes.

2. INTRODUCTION

Hydrogen production from agro-industrial wastes using photosynthetic bacteria is attractive because energy can be recovered from wastes derived from renewable resources (Mitsui, 1985). Photosynthetic bacteria can produce hydrogen from organic acids (Hillmer and Gest, 1977; Takahashi, 1984), as well as generate various kinds of valuable materials including high amounts of protein with balanced essential amino acids, carotenoids, ubiquinone, and vitamin B_{12} (Sasaki et al., 1991). In addition, we reported that photosynthetic bacteria can produce 5-aminolevulinic acid (ALA), which is attractive as a biodegradable herbicide and growth-promoting factor for plants (Sasaki et al., 1997; Tanaka et al., 1992). Photosynthetic bacteria are able to utilize various kinds of organic substrates, such as sugars, organic acids, and volatile fatty acids (VFA). Therefore, they may be useful for hydrogen production from organic wastes.

Miyake et al. (1984) and Mao et al. (1986) observed a relatively high amount of hydrogen production from the newly isolated photosynthetic bacteria *Rhodopseudomonas* sp. RV using butyric and lactic acids. Takahashi et al. (1981) reported hydrogen production by photosynthetic bacteria utilizing lactic acid as a carbon source and electron donor. However, hydrogen production from VFA, which is easily obtained from anaerobic treatment of agro-industrial wastes, was not investigated.

In this project, we investigated the characteristics of hydrogen and ALA production from VFA using the photosynthetic bacterium *Rhodobacter sphaeroides*. Furthermore, applications of ALA for agricultural and medical fields were examined.

3. MATERIALS AND METHODS

Rhodobacter sphaeroides S (Sasaki, et al., 1991) and *R. sphaeroides* IFO12203 (Sasaki, et al., 1987) were used for hydrogen and ALA production, respectively.

A glutamate-malate medium (Sasaki et al., 1987) was used for preculture. For hydrogen production, the carbon source was changed to L-malic acid and acetic and propionic acids from DL-malic acid. For ALA production, the post-anaerobic digestion liquor of swine and sewage sludge waste was used as the VFA medium. The VFA medium was prepared by collecting the supernatant of anaerobic digestion liquor of the wastes (Sasaki et al., 1990: Sasaki et al., 1991).

Preculture was carried out with a 300 mL conical flask (100 mL medium) under static light conditions with 5 klux (17.5 W/m^2) light illumination (tungsten bulbs) on the surface of the vessel at 30 °C for 2–3 days.

Measurements of hydrogen production were carried out using the set-up shown in Figure 1. A 1.5-L Roux bottle (1 L medium) was used as the culture vessel. Hydrogen produced during culture (5–7 days) was trapped in a gas holder filled with saturated NaCl solution (Sasaki et al., 1994). Illumination was provided at 5–30 klux (17.5–105 W/m^2) as shown in Figure 1. Temperature was controlled at 30 ± 0.2 °C, and the pH was controlled at 7.0 ± 0.1 by adding 4 N HCl or 4 N NaOH solutions.

Measurements for ALA production were carried out using a 1.5-L Roux bottle (1 L VFA medium), the same culture vessel as shown in Figure 1, except that the hydrogen gas ports were eliminated. Levulinic acid (LA, 30 mM), a competitive inhibitor of ALA dehydratase, and glycine (30–60 mM), a precursor of ALA, were added as indicated. During ALA production, any gas produced was released from the vessel though a porous silicon stopper. Illumination, temperature, and pH control were carried out as described above for hydrogen production.

Cell mass, hydrogen, CO_2, residual L-malic acid, acetic and propionic acids, levulinic acid, and ALA were measured as described previously (Sasaki et al., 1996). Nitrogenase, ALA synthetase, and ALA dehydratase activities were also measured as described previously (Sasaki et al., 1991).

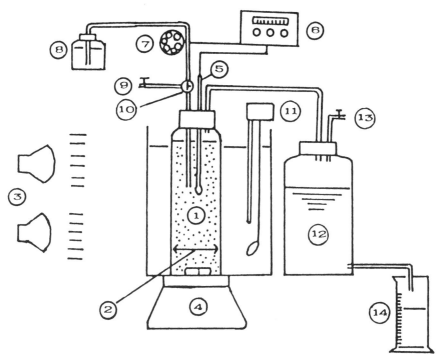

Figure 1. Schematic diagram of the experimental set-up for photohydrogen production by *R. sphaeroides* S.: (1) culture broth (1.5-L Roux bottle); (2) light path ca. 5 cm; (3) light source (two tungsten bulbs); (4) magnetic stirrer; (5) pH electrode; (6) pH controller; (7) pelistalic pump; (8) 4N HCl solution (pH control); (9) sampling port; (10) three-direction cock; (11) water bath with temperature control unit (30 ± 0.2 °C); (12) saturated NaCl solution; (13) port; (14) graduated cylinder.

4. RESULTS AND DISCUSSION

4.1. Effect of Nitrogen Source on Hydrogen Production

Photosynthetic bacteria produce hydrogen mainly through the nitrogenase enzyme in the cells. However, the induction of nitrogenase activity is affected by the nitrogen source used (Hillmer and Gest, 1977; Takahashi, 1984). Effects of different nitrogen sources on hydrogen production by *R. sphaeroides* S are shown in Figure 2. When glutamate-Na was used, large amounts of hydrogen were produced with a maximum production rate of 80.3 mL/g cell/h. Maximum hydrogen production observed here was comparable to that of *Rhodopseudomonas* sp. RV (85.1 mL/g cell/h) (Miyake et al., 1984) and *Rhodopseudomonas* strain TN3 (80 mL/g cell/h) (Kim et al., 1981). The gas components were 90% hydrogen and 10% CO_2. In addition, relatively high amounts of hydrogen were produced with $(NH_4)_2HPO_4$ as the nitrogen source (Figure 2b). In this case, hydrogen was produced under the presence of residual NH_4 along with the induction of nitrogenase activity. This result is advantageous considering that practical hydrogen production must use agro-industrial wastes and these wastes sometimes contain NH_4. However, large amounts of residual NH_4 suppressed the nitrogenase activity and consequently hydrogen production (Figure 2b).

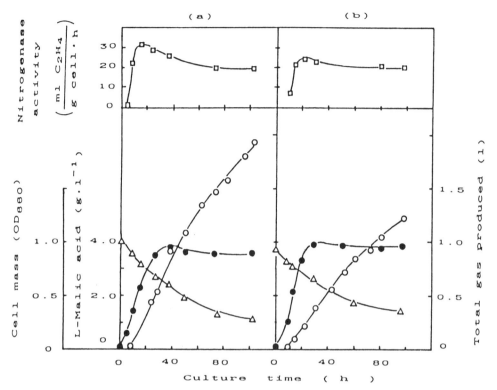

Figure 2. Effect of nitrogen source on photohydrogen production from malic acid as a substrate by *R. sphaeroides* S (30 °C, 10 klux): (a) nitrogen source, glutamate-Na (0.17 g/L); (b) nitrogen source, $(NH_4)_2HPO_4$ (0.13 g/L); ○ = total gas evolved; ● = cell mass; △ = residual L-malic acid; □ = nitrogenase activity.

4.2. Effect of Light Intensity and pH

The effect of light intensity in the range of 5–30 klux on hydrogen production was investigated using glutamate-Na when L-malic acid was used as the electron donor. The result (not shown) was that 10 klux induced the maximum hydrogen production rate. Higher light intensity (20 and 30 klux) suppressed hydrogen production (Sasaki et al., 1996). The effects of pH on hydrogen production at pH of 5.5, 6.0, 7.0, 8.0, and 9.0 were also investigated by means of pH-controlled cultures. The maximum production rate was observed at pH 7.0 (Sasaki et al., 1996).

4.3. Hydrogen Production from VFA

As shown in Figure 3, hydrogen production with acetic acid (a) and a mixture of acetic and propionic acids (b) was investigated using glutamate-Na as a nitrogen source. Hydrogen was produced in both cases, although the maximum production rate (33 mL/g cell/h) was low compared to L-malic acid (Figure 2a). However, it was confirmed that hydrogen can be produced from VFA, suggesting that the VFA medium from the agro-industrial wastes might be applicable for hydrogen production using *R. sphaeroides* S. Few reports have appeared on hydrogen production from VFA (Mao et al., 1986).

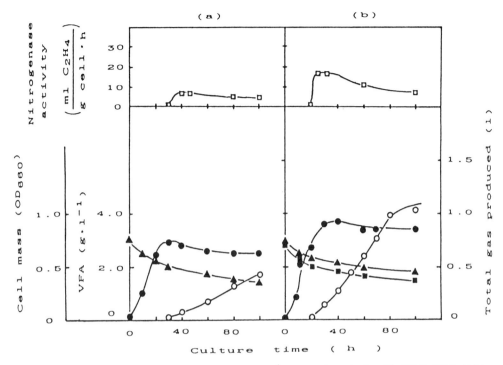

Figure 3. Photohydrogen production by *R. sphaeroides* S from acetic and propionic acids (VFA) (30°C, 10 klux, pH 7.0 ± 0.1): (a) acetic acid 3 g/L + glutamate-Na 0.17 g/L; (b) acetic acid 3 g/L + propionic acid 3 g/L + glutamate-Na 0.17 g/L; ● = cell mass; ○ = total gas evolved; ▲ = residual acetic acid; ■ = residual propionic acid; □ = nitrogenase activity.

4.4. ALA Production from VFA

ALA is a valuable physiological active substance for use as a biodegradable herbicide, insecticide (Rebeiz et al., 1984), and growth-promoting factor for plants (Sasaki et al., 1995). Therefore, the possibility of using the cells obtained after hydrogen production for ALA production was investigated.

Our strategy for ALA production by *R. sphaeroides*, in which ALA is synthesized via the Shemin pathway (Lascelles, 1978) from glycine and succinate, is summarized in Figure 4. By adding levulinic acid (LA) to the medium, ALA dehydratase is partially inhib-

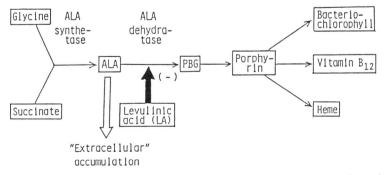

Figure 4. ALA extracellular accumulation induced by levulinic acid (LA) addition in *R. sphaeroides*. ALA: 5-aminolevulinic acid; PBG: porphobilinogen.

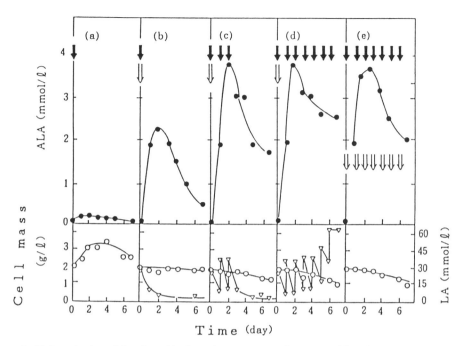

Figure 5. ALA production of *R. sphaeroides* from the supernatant of post-anaerobic swine waste digest. Culture conditions: anaerobic light illumination (5 klux); initial pH = 6.5; temperature 30°C; dense inoculation of preculture cells (ca. 2 g/L). LA (solid arrow, 30 mM, one time (a,b), 3 times repeatedly (c), and seven times repeatedly (d,e)) and glycine (open arrow, 60 mM, one time (b) and seven times repeatedly (d,e)) were added. ● = ALA, ○ = cell mass, ▽ = LA.

ited and ALA is excreted. The amount of LA addition is important because excess LA inhibits growth completely. Therefore, modes of LA additions for extracellular ALA production were investigated using a VFA medium that was prepared from the supernatant of post-anaerobic swine waste digest (Sasaki et al., 1990).

In Figure 5, only LA addition (a) did not produce a considerable amount of ALA; however, with LA and glycine (b), ALA excretion increased dramatically from the cells. Repeated addition of LA three times (c) accelerated ALA excretion up to 3.9 mM. This 3.9 mM of ALA accumulation was almost twofold compared with that of another ALA producer, *Chlorella vulgaris* (Beale, 1970). It was suggested that the repeated addition of LA might maintain ALA dehydratase activity at a lower level in the cells. However, when LA was added seven times (d), ALA accumulation was not enhanced compared with (c). Repeated addition of LA together with glycine seven times (e) also did not enhance ALA compared with (c). Accumulated ALA was re-utilized when LA disappeared.

Profiles of ALA, cell mass, LA, and glycine during dense culture of *R. sphaeroides* were investigated when a VFA medium from post-anaerobic digestion of sewage waste was used. As shown in Figure 6, extra-cellular ALA was produced at up to 9 mM with repeated additions of LA and a one-time addition of glycine. Propionic acid was mainly utilized during ALA production. This phenomenon was also observed in VFA medium from swine waste. This suggests that propionic acid might play an important role for ALA formation as a source of succinate supply via the methylmalonyl-CoA pathway (Sasaki et al., 1990).

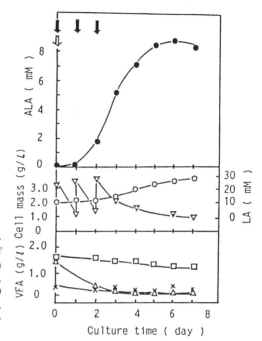

Figure 6. Profiles of ALA, cell mass, LA, and VFA after dense inoculation of *R. sphaeroides* (ca. 2 g/L of cells, 5 klux, 30 °C, pH 7.0 ± 0.1) on VFA medium prepared from sewage sludge. LA (solid arrow, 30 mM, three times repeatedly) and glycine (open arrow, 60 mM, one time) were added. ● = ALA, ○ = cell mass, ▽ = LA, □ = acetic acid, △= propionic acid, × = *n*-butyric acid.

4.5. Applications of ALA

4.5.1. Herbicide. As a practical use of ALA-containing culture broth (Figure 5c), ca. 4 mM ALA was tested as a herbicide in the same manner reported by Rebeiz et al. (1984). The herbicide effect is caused by the singlet oxygen produced by light irradiation with enhanced tetrapyrrol synthesis derived from ALA (Rebeiz et al., 1984). The culture broth indicated a relatively strong herbicide activity within three days after spraying the culture broth directly onto the leaves and stems of *Trifolum repense* (a clover). In addition, almost the same herbicide effect could be observed for this culture broth in worm wood, day flower, and creeping woodsorre. On the other hand, little effect was observed for monocot weeds such as crabgrass and goosegrass (Sasaki et al., 1991, 1995).

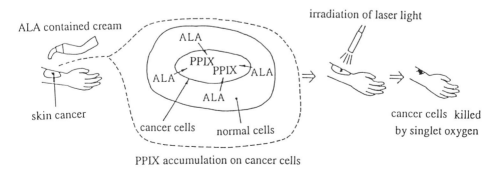

Figure 7. Photodynamic therapy on skin cancer, based on information from Kennedy (1991).

Table 1. Photodynamic therapy on several cancers

Cancer	Study
Skin cancer	Practical experiment level
Oral cancer	Clinical experiment level
Esophagus cancer	Clinical experiment level
Colon cancer	Clinical experiment level
Duodenal cancer	Clinical experiment level
Pancreatic cancer	Animal experiment level
Bladder cancer	Animal experiment level

In addition, ALA is also applicable as a herbicide accelerator. For example, 100 mg/L of ALA addition accelerates the herbicide effect of 0.3 g/L NIP (2,4-dichlorophenyl-4-nitrophenyl ether) up to ca. 40% (an acceleration of 1.4 times). It suggests that the amount of chemical herbicides can be reduced, which is attractive from an environmental perspective (Tanaka, 1993).

4.5.2. Medical Treatments for Cancer. Kennedy et al.(1991) proposed the medical application of ALA for photodynamic therapy (PDT) on cancer in humans. He applied a cream containing ALA on lesions of skin cancer and irradiated them with laser light after a few hours. The cancer cells were killed by singlet oxygen produced by PPIX (protoporphyrin IX) as shown in Figure 6 because ALA is accumulated preferentially in cancer cells compared with normal cells. The mode of action of ALA is similar to that of the herbicide proposed by Rebeiz et al. (1984). Other applications for the treatment of several cancers are summarized in Table 1.

ALA can also be applied for diagnosis of heavy-metal poisoning appearing in patient urine (Fujita and Harada, 1994). ALA inhibits peptidase activity. It is suggested that ALA can be applicable for nasal drag derivative systems (medicines for colds and influenza).

4.5.3. Growth-Promoting Factor for Plants. In plants, ALA plays an important biological role as an intermediary in the biosynthesis of chlorophyll. ALA biosynthesis is the rate-limiting step in chlorophyll biosynthesis. Recently, Tanaka et al. (1992) reported that low concentrations of ALA increased chlorophyll content and accelerated growth of plant tissue and rice seedlings. Folia, soil, and root treatment with ALA were considered to be effective in increasing the yield of crops. Yield increase tests for various plants are summarized in Table 2. The proper concentration of ALA was in the range of 10 to 100 mg/L

Table 2. Effect of ALA on yield of several crops

Crop	Yield (%)	ALA conc.
Spinach (*Spinacia oleracea* L.)	140	100 ppm
Rape (*Brassica campestris* L. subsp. *napus*)	140	100 ppm
Garlic (*Allium satiyum* L.)	139	100 ppm
Potato (*Solanum tuberosum* L.)	163	100 ppm
Radish (*Raphanus sativas*, var. *radicula* DC.)	145	100 ppm
Barley (*Hordium vulgar* L.)	141	30 ppm
Wheat (*Triticum aestivum* L.)	108	30 ppm
Rice (*Oriza sativa* L.)	121	100 g/10 a

Foliar application (rice: soil application)

Table 3. Increasing salt tolerance of cotton by ALA

Salt (%)	ALA (ppm)	Root (wt %)					Aerial part (wt %)				
		Na	K	Mg	Ca	Si	Na	K	Mg	Ca	Si
0.0	0	0.49	4.68	0.42	0.39	—	0.07	2.49	0.54	2.03	—
	3	0.52	2.30	0.39	0.37	0.30	0.09	1.96	0.49	2.15	0.02
	10	0.51	2.70	0.42	0.41	—	0.08	1.91	0.46	2.16	—
	30	0.52	2.70	0.42	0.37	0.27	0.10	2.06	0.48	2.14	0.03
	100	0.47	1.80	0.40	0.64	0.23	0.09	2.09	0.50	2.19	0.03
1.0	0	1.03	2.47	0.34	0.38	—	1.54	2.10	0.50	2.16	—
	3	0.84	2.30	0.36	0.38	0.34	1.27	2.27	0.49	2.20	0.07
	10	0.68	1.77	0.35	0.47	—	0.91	2.04	0.50	2.16	—
	30	0.58	1.90	0.31	0.53	0.23	1.70	2.10	0.47	2.26	0.02
	100	0.71	2.20	0.38	0.46	0.16	1.31	2.00	0.50	2.25	0.03
1.5	0	1.01	2.05	0.33	0.50	—	3.26	1.34	0.43	1.68	—
	3	0.56	2.70	0.35	0.51	0.23	2.26	1.60	0.47	2.10	0.08
	10	0.61	1.95	0.29	0.45	—	1.34	2.07	0.46	2.10	—
	30	0.52	1.20	0.33	0.62	0.20	1.66	1.52	0.49	2.11	0.07
	100	0.47	1.90	0.33	0.45	0.20	0.72	1.92	0.47	2.07	0.06

(ppm) when it was applied to roots (Tanaka et al., 1994). This growth-promoting effect might be caused by the enhancement of the CO_2 gas uptake rate and the photosynthetic activity (Tanaka et al., 1994; Sasaki et al., 1997).

4.5.4. Increasing Salt Tolerance. ALA was found to increase salt tolerance in cotton (Kuramoto et al.,1996). Cotton seedlings treated with ALA could grow in soil containing 1.5% (W/W) NaCl. Usually, cotton cannot grow at such high NaCl concentrations. The Na ion concentration in the roots treated with ALA was suppressed at low ALA concentrations. However, the concentration of Ca and Si in the root did not change by ALA application as shown in Table 3. These results suggest that ALA may raise the osmotic pressure and suppress the absorption of Na into the roots.

This observation is quite significant if humanity is to expand farmlands to near desert areas. Furthermore, increased cold tolerance and fuructan content of plants by ALA application has been observed (Hotta et al., 1995).

5. CONCLUSIONS

Hydrogen production by the photosynthetic bacterium *Rhodobacter sphaeroides* using VFA from agro-industrial wastes may be possible. In addition, *R. sphaeroides* cells have the ability to produce up to 4–9 mM ALA extracellularly under anaerobic light conditions from a VFA medium prepared from post-anaerobic digest of agro-industrial wastes.

ALA can be applied to agricultural fields as a herbicide, insecticide, herbicide accelerator, growth-promoting factor, and enhancer of salt tolerance for plants. In addition, ALA also has application in the medical field as a cancer treatment for diagnosis of heavy-metal poisoning and as a medication.

Thus, combinations of hydrogen production and ALA production from agro-industrial wastes appear to be promising, considering the need for recycling of organic wastes.

REFERENCES

Beale, S.I., 1970, The biosynthesis of 5-aminolevulinic acid in *Chlorella, Plant. Physiol.*, 45:504–506.
Fujuta, H., and Harada, K., 1994, The lead, in *Toxicology Today*, Sato, Y. (ed.), Kinhoudo, Kyoto, pp. 129–147.
Hillmer, P., and Gest, H., 1977, H_2 metabolism in the photosynthetic bacterium *Rhodopseudomonas capsulata*: production and utilization of H_2 by resting cells, *J. Bacteriol.*, 129:732–739.
Hotta, Y., Tanaka, T., Yoshida, R., Takeuchi, Y., and Konnai, M., 1995, Promotive effects of 5-aminolevulinic acid on rice plants, in *Proceedings of the Second Asian Crop Science Conference*, Fukui, Japan, pp. 524–525.
Kennedy, J., 1991, Method of detection and treatment of malignant and non-malignant lesion by photochemotherapy, International patent WO91/01727.
Kim, J.S., Ito, K., and Takahashi, H., 1981, Production of molecular hydrogen by *Rhodopseudomonas* sp., *J. Ferment. Technol.*, 59:185–190.
Kuramoti, H., Watanabe, K., Tanaka, T., Hotta, Y., Takeuchi, Y., and Konnai, M., 1996, Increasing salt tolerance of cotton by 5-aminolevulinic acid, in *Proceedings of the 31st Plant Growth Regulator Society of Japan*, Kyoto, Japan , pp. 88–89.
Lascelles, J., 1978, Regulation of pyrrol synthesis, in *The Photosynthetic Bacteria*, Clayton, R.K., and Sistrom, W.R., (eds.), Plenum Press, New York and London, pp. 795–808.
Mao, X.Y., Miyake, J., and Kawamura, S., 1986, Screening photosynthetic bacteria for hydrogen production from organic acids, *J. Ferment. Technol.*, 64:45–249.
Miyake, J., Mao, X.Y., and Kawamura, S., 1984, Photoproduction of hydrogen from glucose by a co-culture of a photosynthetic bacterium and *Clostridium butyricum*, *J. Ferment. Technol.*, 62:531.
Mitsui, A., Matsunaga, H., Ikemoto, H., and Renuka, B.R., 1985, Organic and inorganic waste treatment and simultaneous photoproduction of hydrogen by immobilized photosynthetic bacteria, in *Development in Industrial Microbiology*, Society of Industrial Microbiology, pp. 209–222.
Rebeiz, C.A., Montazer, Z.A., Hopen, H.J., and Wu, S.M., 1984, Photodynamic herbicides; concept and phenomenology, *Enzyme Microb. Technol.*, 6:390–401.
Sasaki, K., Ikeda, S., Nishizawa, Y., and Hayashi, M., 1987, Production of 5-aminolevulinic acid by photosynthetic bacteria, *J. Ferment. Technol.*, 65:511–515.
Sasaki, K., Tanaka, T., Nishizawa, Y., and Hayashi, M., 1990, Production of herbicide, 5-aminolevulinic acid by *Rhodobacter sphaeroides* using the effluent of swine waste from an anaerobic digestor, *Appl. Microbiol. Biotechnol.*, 32:727–731.
Sasaki, K., Noparatnaraporn, N., and Nagai, S., 1991, Use of photosynthetic bacteria on the production of SCP and chemicals from agroindustrial wastes, in *Bioconversion of Waste Materials to Industrial Products,* Martin, A.M. (ed.), Elsevier Science Publishers, New York and London, pp. 223–265.
Sasaki, K., Tanaka, T., and Hotta, Y., 1995, Production of 5-aminolevulinic acid from organic wastes using photosynthetic bacteria and applications for herbicide and growth promoting factor for plants, *Mizushyori-Gijutu*, Japan, 6:135–145.
Sasaki, K., Takeno, K., and Emoto, Y., 1996, Utilization of organic acids and volatile fatty acids and hydrogen production by photosynthetic bacteria, *J. Soc. Water Environ.*, 19:63–70.
Sasaki, K., Tanaka, T., and Nagai, S., 1997, Use of photosynthetic bacteria for the production of SCP and chemicals from organic wastes, in *Bioconversion of Waste Materials to Industrial Products*, 2nd ed., Martin, A.M. (ed.), Chapman and Hall, pp. 247–292.
Tanaka, T., Takahasi, K., Hotta, Y., Takeuchi, Y., and Konnai, M., 1992, Promotive effects of 5-aminolevulinic acid on the yields of several crops, in *Proceedings of the 19th Annual Meeting of Plant Growth Regulator Society of America*, San Francisco, pp. 237–241.
Tanaka, T., 1993, Herbicide and weeding methods, *Japan Patent Koukai*, 5–117110.
Takahashi, H., 1984, Hydrogen production by nitrogenase of photosynthetic bacteria, in *Kougousei-Saikin* (Photosynthetic Bacteria), Kitamura, H. et al. (eds.), Gakkai-Shyppan Center, Tokyo, pp. 196–204.

18

CONTINUOUS HYDROGEN PRODUCTION BY *Rhodobacter sphaeroides* O.U.001

İnci Eroğlu,[1] Kadir Aslan,[1] Ufuk Gündüz,[2] Meral Yücel,[2] and Lemi Türker[3]

[1]Department of Chemical Engineering
[2]Department of Biology
[3]Department of Chemistry
Middle East Technical University
06531, Ankara, Turkey

Key Words

hydrogen production, photobioreactor, *Rhodobacter sphaeroides*

1. SUMMARY

This paper describes hydrogen gas production by *Rhodobacter sphaeroides O.U.001* using a column photobioreactor in batch and continuous operation. The effect of substrates on the hydrogen production rate was investigated in batch-type photobioreactor experiments. Substrate concentrations (L-malic acid and sodium glutamate) were measured by using high-pressure liquid chromatography. The gas produced was analyzed by gas chromatography.

In order to achieve prolonged and stable hydrogen production, pH, cell, and substrate concentrations (L-malic acid and sodium glutamate) were maintained within a certain range by periodically diluting the culture with fresh feed. Optimum conditions obtained in a previous study (150-mL volume) were applied to a photobioreactor having an inner volume of 400 mL. Light and dark period cycles were applied to simulate outdoor conditions. After every 100 h, the feed containing 100 mL of fresh medium replaced the same volume of culture removed. In some experiments, after every 100 h, 100 mL of growth medium and 30 mL culture grown overnight (fresh inoculation) replaced the same volume of culture removed. In operations without fresh inoculation to the medium, hydrogen evolution ceased after 25 days; however, by adding fresh medium, 3.6 L of gas was evolved in 70 days.

In conclusion, the L-malic acid to sodium glutamate concentration ratio and cell concentration are important factors affecting the hydrogen production rate. In this regard, the amount of feed and time intervals of the dilution must be optimized for a continuous system.

2. INTRODUCTION

It is known that using sunlight some photosynthetic microorganisms produce hydrogen, which is considered to be a promising, ideal, and renewable form of energy. Hydrogen is also used industrially in the synthesis of ammonia, hydrogenation reactions, and many other important applications. In different countries, this subject has been investigated in laboratories or in institutions. In India, Sasikala et al. (1991, 1992, 1995) produced hydrogen with *Rhodobacter sphaeroides O.U.001*. Miyake et al. in Japan are working on hydrogen production with *Rhodobacter spheroides RV* (Tsygankov et al., 1993).

In Turkey, at the Middle East Technical University, we have been working on three different bio-systems aimed at producing high-purity hydrogen. These bio-systems are for hydrogen production by *Rhodobacter sphaeroides O.U.001*, coupled systems of *Halobacterium halobium* and *E. coli*, and photoelectrochemical hydrogen production by *Halobacterium halobium*.

In previous work (Arik et al., 1996), a 150-mL glass column photobioreactor was constructed for the production of hydrogen in light under anaerobic conditions by *Rhodobacter sphaeroides O.U001*. The optimum hydrogen production conditions were obtained (pH between 7.3 and 7.8, temperature 31 to 36 °C, light intensity 200 W/m^2, and cell concentration 1.6 to 1.8 mg dry wt/mL) and the maximum hydrogen production rate obtained was 0.047 L H_2/h/L culture with 99% purity. The objective of this study is to scale-up the continuous photobioreactor to 400 mL and determine the effect of substrate concentrations on the hydrogen production rate in batch systems.

3. EXPERIMENTAL

3.1. Growth Conditions and Experimental Set-Up

At 32 °C, with illumination using a tungsten lamp having a light intensity of 200 W/m^2, *Rhodobacter sphaeroides O.U.001* (DSM 5648) was grown anaerobically under sterile conditions in a minimal medium of Biebl and Pfennig (1981), supplied with L-malic acid and sodium glutamate as carbon and nitrogen sources, respectively. Argon gas was used to flush out air for anaerobic conditions.

The experimental set-up is shown in Figure 1. The column photobioreactor (PBR) is made of a glass cylinder that has an inner volume of 400 mL and is surrounded by a water jacket. At the top of the reactor are an inlet for the medium and outlets for the argon and produced gas that is collected in a gas-measuring burette. Fresh medium is added from a reservoir placed above the photobioreactor. Microorganisms are inoculated through the septum.

At the bottom of the column photobioreactor are an outlet for the culture and an inlet for argon gas. Bacterial cell concentrations were measured as an increase in absorbance at 660 nm (Hitachi Spectrometer). L-malic acid and sodium glutamate concentrations

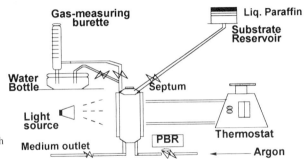

Figure 1. Experimental set-up for batch and continuous biohydrogen production.

were determined by high-pressure liquid chromatography (Shimadzu HPLC, BIORAD Aminex Ion Exclusion Column). The gas produced was analyzed by gas chromatography (Hewlett-Packard 5890, Series II).

3.2. Hydrogen Production Studies

The hydrogen production was studied in two ways: batch or continuous operation. In batch operation, samples were taken out at certain time intervals while flushing the column with argon. Cell concentration and pH were measured. The samples were centrifuged and the supernatant was analyzed by HPLC to measure L-malic acid and sodium glutamate concentrations.

In continuous operation, cyclic 12-h long light and dark periods were applied to the photobioreactor, simulating outdoor conditions. The substrate concentrations were 7.5 mM L-malic acid and 10 mM sodium glutamate, which were taken from a previous study (Arik et al., 1996) After every 100 h, 100 mL of fresh medium replaced the same volume of culture removed. In some experiments, after every 100 h, 100 mL of growth medium and 30 mL of culture grown overnight replaced the same volume of culture removed in order to investigate the effect of a new inoculation. Samples of the effluent culture were analyzed.

4. RESULTS AND CONCLUSIONS

4.1. Batch Operation

The experimental data obtained in hydrogen production experiments using the batch system are summarized in Table 1. The details as a function of time are shown in Figure 2.

Table 1. Results of the batch-type hydrogen production experiments

	Initial substrate conc. (mM)		H_2 production			Maximum cell conc. (mg dry wt/mL)
Run no.	L-malic acid	Na Glu	Volume (mL)	Duration (h)	Rate (L H_2/h/Lc)	
1	7.5	10	52	55	0.0020	8.5
2	7.5	2	117	73	0.0040	4.5
3	30	10	101	87	0.0029	8.5
4	30	2	477	287	0.0042	5.4

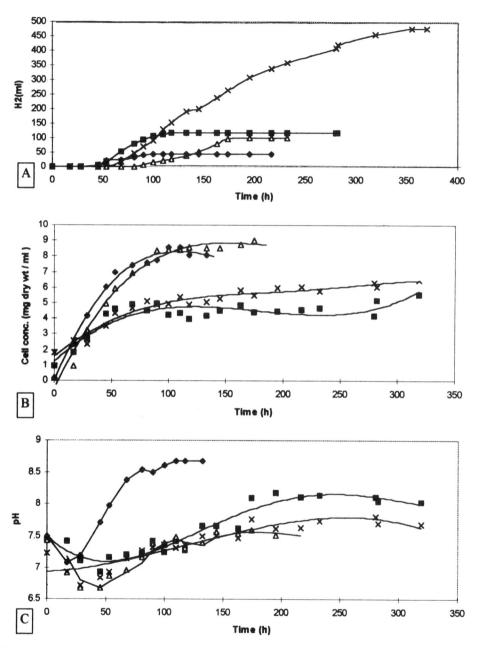

Figure 2. Comparison of experimental data obtained in four runs: (A) total hydrogen production; (B) growth curves of the bacteria; (C) change of pH with respect to time. ◆ = Run 1, ■ = Run 2, ∆ = Run 3, x = Run 4.

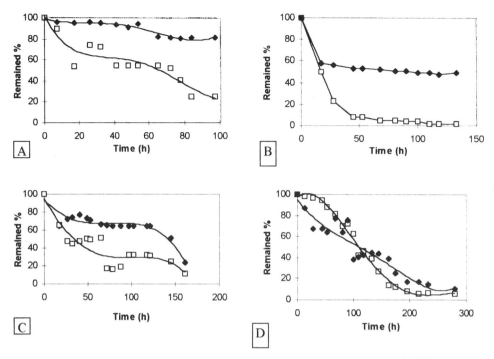

Figure 3. Percent consumption of L-malic acid and sodium glutamate with respect to time in different runs, (A) Run 1, (B) Run 2, (C) Run 3, (D) Run 4, (♦) sodium glutamate, (□) L-malic acid.

Figure 2A illustrates the total hydrogen production obtained in four different runs. In Figure 2B, the growth curves of *Rhodobacter sphaeroides O.U.001* are shown for the same set of runs. In Figure 2C, the change of pH with respect to time is shown where the pH vs. time curve for run 1 is quite different than the others. Most probably at a low L-malic acid to sodium glutamate concentration, the metabolism of the bacteria is different than the other cases in which the above-mentioned concentration ratios are much higher.

Figure 3 displays the percent consumption of L-malic acid and sodium glutamate with respect to time on different runs. The maximum cell concentration obtained in run 1 was almost two times greater than the maximum cell concentration obtained in run 2. In both runs, the same amounts of initial L-malic acid concentrations were used (7.5 mM). However, the initial amount of sodium glutamate in run 1 was five times greater than that in run 2. This result implies that excess sodium glutamate enhances cell growth. Similar results were obtained in run 3 compared to run 4. It is interesting to note that cell growth reached a steady state value after all the L-malic acid had been consumed, even if some sodium glutamate remained in the solution. The total amount of hydrogen produced was also found to be related to L-malic acid utilization. Hydrogen production stopped in runs 1 and 2 when the cell concentration reached a steady value, whereas L-malic acid had been utilized completely. It should be emphasized that in run 2, the consumption rate of L-malic acid was much greater than it was in run 1, in parallel with the increase in hydrogen production rate. Similar results were obtained in runs 3 and 4. However, the maximum hydrogen production rate was obtained in run 4. As can be seen from Figure 3D, in this run both the L-malic acid and sodium glutamate were utilized in a similar manner.

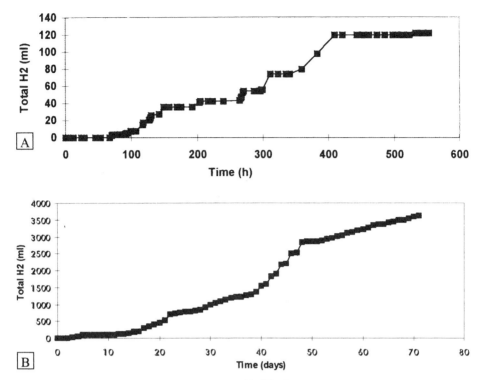

Figure 4. Continuous hydrogen production (A) dilution with 100 mL fresh medium, (B) dilution with 100 mL fresh medium and 30 mL overnight grown cells (7.5 mM L-malic acid, 10 mM sodium glutamate, 2.2 - 3.6 mg dry wt/mL).

4.2. Continuous Hydrogen Production

Figure 4A gives the total hydrogen production while the system was diluted with 100 mL of fresh medium every 100 h. A total of 120 mL of hydrogen was produced during 500 h. Figure 4B shows the total hydrogen production while 100 mL of fresh medium and 30 mL of culture grown overnight were added to the medium every 100 h. In this experiment, 3.6 L of hydrogen was produced in 70 days. Produced gas contained 99% hydrogen. It has been observed that the addition of overnight-grown culture enhanced hydrogen production. Under these conditions, the maximum hydrogen production rate obtained was 0.02 L H_2/h/L culture. Jee et al. (1987) showed hydrogen production at a rate of 0.014 L H_2/h/L culture using a batch system. Sasikala et al. (1991) and Kim et al. (1987) reported hydrogen production at a rate of 0.006 L H_2/h/L culture and 0.027 L H_2/h/L culture, respectively, using a continuous system. It should be noted that the maximum hydrogen production rate obtained in the present study was lower (about one-half) than that obtained in previous studies (Arik et al., 1996). The reason might be due to the reactor geometry. The diameter-to-length ratio of the reactor in the present study was increased about four times, which may have caused a change in the efficiency of the light energy utilization.

The following conclusions can be drawn from this study:

1. Both L-malic acid and sodium glutamate are necessary for cell growth. As sodium glutamate concentration increases, the maximum cell concentration increases.

2. There is a relationship between cell concentration and the hydrogen production rate. However, the hydrogen production rate basically depends on the L-malic acid to sodium glutamate ratio. As this ratio increases, the hydrogen production rate increases. The optimum L-malic acid to sodium glutamate concentration ratio is 15.
3. A slight decrease is observed in the pH of the medium during the cell growth period; however, pH increases as hydrogen is produced.
4. Hydrogen production starts after a lag period (40–80 h) and continues even after the cell concentration has leveled out. This is an interesting finding compared to conclusions by Arik et al. (1996) and Sasikala et al. (1992); they found that hydrogen is mainly produced in the exponential phase.

5. RECOMMENDATIONS

It has been recommended that continuous hydrogen production must be optimized with the experience gained in batch operations; in particular, new experiments must be carried out with optimum L-malic acid to sodium glutamate concentration ratios. The suggested value is 15. It is believed that the hydrogen production rate will be positively affected by this factor.

ACKNOWLEDGMENT

This research has been supported by the Turkish Scientific Research Council (TUBITAK) project number TBAG 1535, and the Middle East Technical University (METU) Research Fund, project number AFP-96-07-02-02.

REFERENCES

Arik, T., Gunduz, U., Yucel, M., Turker, L., Sediroglu, V., and Eroglu, I., 1996, Photoproduction of hydrogen by *Rhodobacter sphaeroides O.U.001*, in *Proceedings of the 11th World Hydrogen Energy Conference*, Stuttgart, Germany, 3:2417–2424.
Biebl, H., and Pfenning, N., 1981, Isolation of member of the family *Rhodosprillacaea*, *The Prokaryotes*, Springer-Verlag, New York, pp. 267–273.
Jee, H.S., and Ohashi, T., 1987, Limiting factor of nitrogenase system mediating hydrogen production of *Rhodobacter sphaeroides O.U.001*, *J. Fermentation Technol.*, 65:153–158.
Kim, J.S., and Ito, K., 1987, Production of molecular hydrogen by semi-continuous outdoor culture of *Rhodopseudomanas sphaeroides*, *Agric. Biol. Chem.*, 51(4):1173.
Sasikala, K., Ramana, C.H.V., and Rao, P.R., 1991, Environmental regulation for optimal biomass yield and photoproduction of hydrogen by *Rhodobacter sphaeroides O.U.001*, *Int. J. Hydrogen Energy*, 16:597–601.
Sasikala, K., Ramana, C.H.V., and Rao, P.R., 1992, Photoproduction of hydrogen from the waste water of a distillery by *Rhodobacter sphaeroides O.U.001*, *Int. J. Hydrogen Energy*, 17:23–27.
Sasikala, K., Ramana, C.H.V., and Rao, P.R., 1995, Regulation of simultaneous hydrogen photoproduction during growth by pH and glutamate in *Rhodobacter sphaeroides O.U.001*, *Int. J. Hydrogen Energy*, 20:123–126.
Tsygankov, A.A., Hirata, Y., Miyake, M., Asada, Y., and Miyake J., 1993, Hydrogen evolution photosynthetic bacterium *Rhodobacter sphaeroides RV* immobilized on porous glass, *New Energy Systems and Conversions*, Universal Academy Press, Inc., pp. 229–233.

19

PHOTOBIOLOGICAL HYDROGEN PRODUCTION BY *Rhodobacter sphaeroides* O.U.001 BY UTILIZATION OF WASTE WATER FROM MILK INDUSTRY

Serdar Türkarslan,[1] Deniz Özgür Yigit,[1] Kadir Aslan,[2] Inci Eroglu,[2] and Ufuk Gündüz[1]

[1]Department of Biology
[2]Department of Chemical Engineering
Middle East Technical University
06531, Ankara, Turkey

Key Words

hydrogen production, photobioreactor, *Rhodobacter sphaeroides*, waste water, milk industry

1. SUMMARY

Hydrogen production with photosynthetic microorganisms contributes to the protection of the environment, not only in producing a clean fuel but also in waste treatment processes. In this study, *Rhodobacter sphaeroides*, a photosynthetic bacteria, is used in photobiological hydrogen production by using waste water from the milk industry.

Rhodobacter sphaeroides was grown in a cylindrical glass-column bioreactor (150 mL). A tungsten lamp was placed 40 cm from the column (200 W/m^2 intensity). The waste material was diluted threefold and added as a carbon source by replacing the malic acid in the standard growth medium. The growth of bacteria was supported by the waste, even with shorter lag and longer log periods, compared to the case where malic acid was the carbon source. Under anaerobic conditions, the hydrogen was continuously produced in this batch bioreactor. Eighty-five milliliters of H$_2$ gas with 99% purity was collected for 90 h continuously. Thus, waste water from the milk industry seems to be a very promising source for the inexpensive generation of hydrogen using *Rhodobacter sphaeroides*.

2. INTRODUCTION

Hydrogen is an ideal, clean fuel that does not create environmental pollution. The production of hydrogen by utilization of solar light seems to be one of the most promising

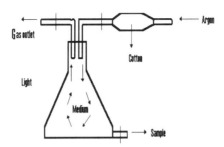

Figure 1. Apparatus for anaerobic cultivation of *Rhodobacter sphaeroides*.

technological processes. Photosynthetic bacteria that can utilize solar light as an energy source and generate hydrogen by decomposition of water and organic matter have many excellent features. Under anaerobic conditions, they can give high hydrogen yields. There is no need for further treatment because the gas evolved is mainly hydrogen with low CO_2 content. Photosynthetic bacteria have a broad range of light intensity that they can use under different environmental conditions. These bacteria can utilize many organic wastes and contribute to waste treatment processes as they produce clean fuel. In addition to photosynthetic microorganisms, hydrogen photoproduction by means of cyanobacteria has also been shown to be effective (Hall et al., 1995; Markov et al., 1995).

There are several groups in the world who work on hydrogen production by using photosynthetic microorganisms. Some of these groups have utilized organic wastes from the sugar industry (Vincenzini et al., 1982; Tanisho and Ishiwata, 1994; Bolliger et al., 1985) or distilleries (Sasikala, 1992). There are also studies on using glucose (Miyake et al., 1984) and organic acids (Miyake et al., 1986). Sasikala et al. (1991, 1992) in India and Tsygankov et al. (1993) in Japan obtained promising results with *Rhodobacter sphaeroides*. Miyake and Asada (1993) have developed a process for hydrogen production that is suitable for large-scale operation.

Rhodobacter sphaeroides O.U.001 is used in this study for photobiological hydrogen production. This photosynthetic bacterium evolves hydrogen gas under anaerobic conditions. The organic waste from the milk industry was added to the growth medium by substituting the carbon source.

3. MATERIALS AND METHODS

Rhodobacter sphaeroides O.U.001 (DSM 5684) was grown on a standard growth medium containing KH_2PO_4 (0.5 g/L), $MgSO_4 \cdot 7H_2O$ (0.2 g/L), $CaCl_2 \cdot 2H_2O$ (0.05 g/L), NaCl (0.4 g/L), sodium glutamate (1.8 g/L), malic acid (1.0 g/L), vitamins, and minerals. Argon gas was passed through the medium to provide anaerobic conditions.

In order to study the growth of *Rhodobacter sphaeroides*, an apparatus shown in Figure 1 was used. In the growth medium, waste from the milk industry was substituted for malic acid. The waste material was diluted threefold, 1.3 mL of actively growing culture of *Rhodobacter sphaeroides* was inoculated, and at suitable time intervals, 5-mL samples were taken from the 500 mL medium. Continuous argon circulation provided anaerobic conditions. The culture was incubated at 31 °C under illumination (200 W/m^2). Growth was determined by OD measurements at 660 nm.

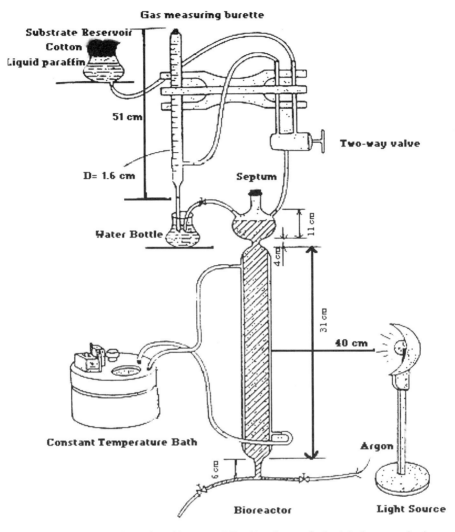

Figure 2. Cylindrical glass-column bioreactor (150 mL) and set-up for batch hydrogen production.

Waste from a milk factory (a transparent solution; chemical composition to be determined) was immediately filtered by coarse filtration and then sterilized with 0.2 μ filters. Sterilization can also be done by autoclaving, which does not seem to harm the quality of the medium. The waste material was diluted threefold and added as a carbon source by substituting the malic acid in the standard growth medium.

Rhodobacter sphaeroides was grown in a cylindrical, glass-column bioreactor (1000 mL), which was sterilized by autoclaving. The set-up is shown in Figure 2.

The sterile medium was diluted 1/10 times with the milk waste and the reactor was filled with the medium. The temperature was kept constant at 31 °C by water circulation in the column jacket. Twenty milliliters of *Rhodobacter sphaeroides* cell suspension (0.36 mg dry wt/mL) from a log phase culture was inoculated into the medium. The reactor was illuminated with a 150-W tungsten lamp placed 40 cm from the column (200 W/m^2). Hy-

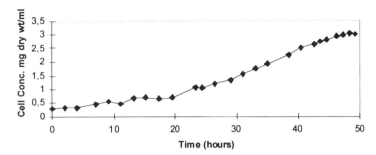

Figure 3. Growth curve (1/3 dilution waste water from milk industry added to malic acid).

drogen was collected in a burette and the volume of the collected gas was measured. Gas chromatography was used to determine the composition of the collected gas.

4. RESULTS AND DISCUSSION

When milk waste was used as the only nutrient source for *Rhodobacter sphaeroides*, no growth was observed. The nutrient in waste from the milk factory is not sufficient to support growth. Most probably, this waste is lacking some vitamins or essential minerals. (However, this study has shown that milk waste is a good source for the enrichment of the growth medium). When malic acid was added to the milk waste, shorter lag periods and long log phases were observed, compared to the case where malic acid was the carbon source. It took 50 h to reach the stationary phase where the cell concentration in the reactor was 3.03 mg dry wt/mL (Figure 3). This corresponds to the maximum cell concentration. Since the accumulation of the dead microorganisms and depletion of the medium may slow growth, the process may be improved by the addition of fresh medium to the reactor, either continuously or periodically (semi-batch operation). The suitable dilution rate needs to be determined.

The effect of the milk factory waste on H_2 production was also studied. The initial pH of the medium was found to be critical and should be maintained at around pH 6.8. In a batch operation, under anaerobic conditions, hydrogen gas was produced. The gas evolution began immediately after a short lag period (10–15 h) and continued for several hours at a high rate (0.94 mL/h) as seen in Figure 4a. About 85 mL of gas was collected in 90 h. At the end of 90 h, gas production seemed to stop, probably due to a critical nutrient depletion in the medium. The pH was increased (up to pH 9.2) during the process (Figure 4c). The increase in cell concentration with respect to time is indicated in Figure 4b, and the change of pH in the batch process is shown in Figure 4c. It was observed that the cell density and pH were very critical parameters for hydrogen production. In semi-batch operations, where new medium was added and old medium was removed to keep the pH around 6.8, and where the cell density was maintained at 1.6–1.8 mg dry wt/mL, it was possible to maintain hydrogen gas production for extended periods (Arik et al., 1996).

It was interesting to observe that hydrogen production was affected positively by replacing malic acid with waste. High amounts of H_2 gas were produced at a higher rate with a shorter lag period. It seemed that the milk industry waste had no inhibitory effect; in fact, certain components were even stimulatory, enhancing the gas production.

a)

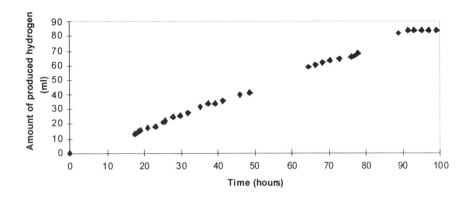

b)

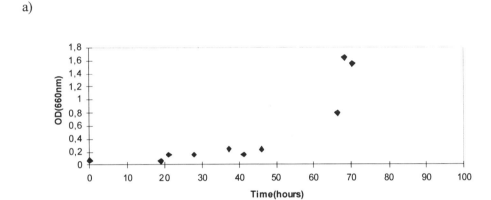

c)

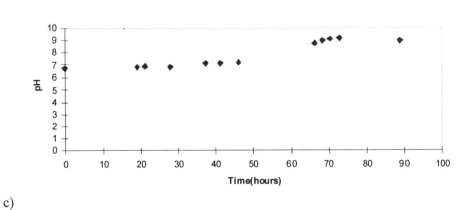

Figure 4. (a) Amount of hydrogen produced in column bioreactor, (b) *Rhodobacter sphaeroides* cell density in column bioreactor, and (c) pH of the medium in column bioreactor.

ACKNOWLEDGMENT

This research is being supported by the Turkish Scientific Research Council (TUBITAK) and Middle East Technical University (METU).

REFERENCES

Arik, T., Gündüz, U., Yücel, M., Türker, L., Sediroglu, V., and Eroglu, I., 1996, Photoproduction of hydrogen by *Rhodobacter sphaeroides* O.U.001, *Proceedings of the 11th World Hydrogen Energy Conference*, Stuttgart, Germany, 3:2417–2424.
Bolliger, R., Zürrer, H., and Bachofen, R., 1985, Photoproduction of molecular hydrogen from waste water of a sugar refinery by photosynthetic bacteria, *Appl. Microbiol. Biotechnol.*, 23:147–151.
Hall, D.O., Markov, S.A., Watanabe, Y., and Rao, K.K., 1995, The potential applications of cyanobacterial photosynthesis for clean technologies, *Photosynthesis Research*, 46:159–167.
Markov, S.A., Bazin, M.J., and Hall, D.O., 1995, The potential of using cyanobacteria in photobioreactors for hydrogen production, *Advances in Biochem. Eng. Biotech.*, 52:60–81.
Miyake, J., Mao, X.Y., and Kawamura, S.,1984, Photoproduction of hydrogen from glucose by a co-culture of a photosynthetic bacterium and *Clostridium butyricum*, *J. Ferment. Technol.*, 62:531–535.
Miyake, J., Mao, X.Y., and Kawamura, S., 1986, Screening photosynthetic bacteria for hydrogen production from organic acids, *J. Ferment. Technol.*, 64:245–249.
Sasikala, K., Ramana, C.V., and Rao, P.R., 1991, Environmental regulation for optimal biomass yield and photoproduction of hydrogen by *Rhodobacter sphaeroides* O.U.001, *Int. J. Hydrogen Energy*, 16:597.
Sasikala, K., Ramana, C.V., and Rao, P.R., 1992, Photoproduction of hydrogen from the waste water of a distillery by *Rhodobacter sphaeroides* O.U.001, *Int. J. Hydrogen Energy*, 17:23.
Tanisho, S., and Ishiwata, Y., 1994, Continuous hydrogen production from molasses by the bacterium *Enterobacter aerogenes*, *Int. J. Hydrogen Energy*, 19:807–812.
Tsygankov, A.A., Hirata, Y., Miyake, M., Asada, Y., and Miyake, J., 1993, Hydrogen evolution by photosynthetic bacterium *Rhodobacter sphaeroides* immobilized on porous glass, *New Energy Systems and Conversions*, pg. 229.
Vincenzini, M., Florenzano, G., Materassi, R., and Tredici, M.R., 1982, Hydrogen production by immobilized cells; H_2 photoevolution and wastewater-treatment by agar-entrapped cells of *Rhodopseudomonas palustris* and *Rhodosprillum molischianum*, *Int. J. Hydrogen Energy*, 7(9):725–728.

POLYHYDROXYBUTYRATE ACCUMULATION AND HYDROGEN EVOLUTION BY *Rhodobacter sphaeroides* AS A FUNCTION OF NITROGEN AVAILABILITY

Emir Khatipov,[1,2] Masato Miyake,[1] Jun Miyake,[3] and Yasuo Asada[1]

[1]Research Institute of Innovative Technology for the Earth (RITE)
[2]National Institute of Bioscience and Human Technology
Higashi 1-1
[3]National Institute of Advanced Interdisciplinary Research, AIST/MITI
Higashi 1-1-4, Tsukuba, Ibaraki, 305, Japan

Key Words

polyhydroxybutyrate accumulation, hydrogen evolution, purple nonsulfur bacteria, nitrogen availability

1. SUMMARY

Polyhydroxybutyrate (PHB) is a useful by-product of hydrogen production using photosynthetic bacteria. This current study was aimed at understanding the effect of different carbon substrates and nitrogen availability on the levels of PHB accumulation and H_2 evolution by *Rhodobacter sphaeroides* strain RV. Media containing acetate, lactate, and pyruvate as carbon sources and ammonium or glutamate as nitrogen sources were used, as well as nitrogen-free media. The highest levels of PHB were observed in the cells grown on ammonium and acetate (35–38% of dry cells), while those in the cells grown on other carbon substrates were low and/or stable. PHB accumulation under nitrogen-deprived conditions was observed on acetate (>38–41%), lactate (~12–25%), and to a lesser extent, on pyruvate (~15%). Hydrogen evolution by the culture was observed on lactate and, to a lesser extent, on pyruvate under nitrogen-deprived conditions and when cells were grown on glutamate. PHB levels, as well as H_2 evolution, were lower when cells were grown on carbon substrates with glutamate as the sole nitrogen source. The data presented indicate the possibility of controlling the bacterial cultivation process, whichever final product, PHB or H_2, is preferred.

BioHydrogen, edited by Zaborsky *et al.*
Plenum Press, New York, 1998

2. INTRODUCTION

Photosynthetic bacteria are promising organisms for biotechnological production of polyhydroxybutyrate (PHB), a source of biodegradable plastics. Production of PHB by phototrophic bacteria would provide a major advantage as an environmentally acceptable and friendly technology, where sunlight energy and CO_2 or organic acids of waste water could be converted directly to the polymer.

The purple nonsulfur bacterium, *Rhodobacter sphaeroides* strain RV, has been studied in our laboratory for its possibility for biotechnological application in H_2-gas-producing photobioreactors (Miyake et al., 1984; Mao et al., 1986; Tsygankov et al., 1994; Nakada et al., 1996; Nagamine et al., 1996). Recent work in our laboratory has shown that ammonium- and acetate-fed continuous pH-stat cultures of this strain accumulate higher amounts of polyhydroxybutyrate (PHB) at higher pHs (Suzuki et al., 1995). However, the PHB content in cells grown under pH-uncontrolled conditions was shown to be even higher. We are currently selecting cultivation conditions for *R. sphaeroides* RV that lead to both H_2 production and final accumulation of considerable amounts of PHB as a by-product. For this purpose, the relationship of the two processes has to be studied, as well as regulation of PHB accumulation.

The goal of the present investigation was to study the interrelationship of H_2 evolution and PHB accumulation processes during growth of *R. sphaeroides* RV on different carbon substrates under nitrogen-sufficient (with NH_4^+ or glutamate as a source of nitrogen) and nitrogen-deprived growth conditions.

3. EXPERIMENTAL

Rhodobacter sphaeroides strain RV (Miyake et al., 1984) was cultivated anaerobically in 15-mL, screw-cap glass vials on medium with sodium succinate (40 mM, if not otherwise stated), ammonium sulfate (1.25 g/L), and yeast extract (1.0 g/L), at 30 °C, 2000 lux (~20 W/m^3) and initial pH 6.8. Cultures for further use in experiments with nitrogen-sufficient growth conditions were pre-grown 12–15 h in 300 mL polystyrene flat (3 cm) culture flasks (Iwaki, Japan) at 5000 lux (~45 W/m^3) and 30 °C on the same medium (6–8% inoculate). Cultures for experiments with nitrogen-limited conditions or with glutamate as a sole nitrogen source were pre-grown at the same time on the media with succinate substituted for acetate, lactate, or pyruvate (40 mM). Cells were harvested by centrifugation (6000 g, 10', 18 °C); resuspended in the corresponding fresh medium (OD_{660} ~1.2); transferred from 45-mL to 60-mL polystyrene, see-through, double-seal test tubes (Iwaki, Japan); flushed with argon; and grown further under standard conditions.

Culture tubes were illuminated from two sides by two 100-W tubular tungsten lamps with multiple incandescence points that provided light intensity of 5000 lux and energy of ~45 W/m^2 on the front surface of the tubes from both sides. In the case of nitrogen-deprived conditions, or when glutamate was used as a sole nitrogen source, the polystyrene tubes were closed with a silicone stopper with a gas outlet. The hydrogen gas evolved during the culture growth was collected inside plastic graduated cylinders connected to each test tube. The initial cell concentration in the latter case was adjusted to OD_{660} ~2.0. Cells in experiments were grown 14 h on media with ammonium, and 24 h on media with glutamate, or under nitrogen-deprived conditions. Test tubes were taken out sequentially at fixed time intervals to determine medium pH, cell dry weight, and PHB content of the cells. Determination of PHB was performed by gas chromatography (Braunegg et al.,

1978) with a Shimadzu GC-14A gas chromatograph using a 3 mm × 3.1 m glass column filled with 60/80 mesh Chromosorb GAW.

4. RESULTS AND CONCLUSIONS

PHB accumulation has been studied in various photosynthetic bacteria (Brandl et al., 1991; Liebergessell et al., 1991; Dephilippis et al., 1992; Fuller, 1995; Mas and Van Gemerden, 1995). As in heterotrophic bacteria (Anderson and Dawes, 1990), PHB accumulation in photosynthetic bacteria is normally reversibly associated with nitrogen availability in the growth medium (Brandl et al., 1990). Most of the data published on the PHB accumulation by *Rhodobacter sphaeroides* are for the cultures grown on acetate, mixed carbonic acids, or on different aliphatic organic acids (Brandl et al., 1991; Liebergessell et al., 1991; Hustede et al., 1993). There have been no data published so far on the comparison of H_2 evolution and PHB accumulation levels in *R. sphaeroides* with different carbon substrates, except in a recent paper describing the acquisition of H_2-evolving activity by the PHB⁻ mutants (Hustede at al., 1993).

Various carbon substrates for growing *R. sphaeroides* RV (acetate, lactate, and pyruvate) were used to study the interrelationship between H_2 evolution and PHB accumulation under different conditions of nitrogen availability. The following nitrogen availability conditions were employed: nitrogen sufficiency, with either ammonium or glutamate used as nitrogen substrates; or nitrogen deprivation, when no source of nitrogen was added to the growth medium. The highest level of PHB accumulation was observed on acetate in cells grown under nitrogen sufficient conditions (Table 1). PHB accumulation began at the inception of the growth cycle and reached 35–38% of cell dry weight in the early stationary phase. With cells grown on lactate and pyruvate, the final PHB content was 2% lower until the early stationary phase. No hydrogen evolution was observed under these conditions on any carbon substrate.

PHB accumulation exhibited rather particular patterns in cells grown on different carbon substrates under nitrogen deprivation or in the presence of glutamate (Table 1). Glutamate appeared to be a good substrate for cell growth on lactate and pyruvate. The cell yield with these carbonic acids with glutamate was 60 and 50% higher, respectively. However, no growth induction was achieved with glutamate on acetate. Nitrogen deprivation of acetate-grown cells led to an increase in PHB content, while supplying the cells with glutamate led to preservation of the unchanged PHB level.

The PHB content and H_2 evolution differed after cell growth on glutamate and under nitrogen-deprived conditions with lactate or pyruvate as carbon substrates. The PHB content in lactate- and pyruvate-fed nitrogen-deprived cells was correspondingly 33 and 40% higher than in the cells grown with glutamate. It is noteworthy that in both cases, the specific hydrogen evolution rate was also higher under nitrogen-deprived conditions. The ratio of specific H_2 evolution rate by the cells grown under nitrogen-deprived conditions to that in the cells grown on glutamate reached 1.6 and 3.8 with lactate and pyruvate, respectively. However, the latter substrate supported comparatively low H_2-evolving activity of the culture. No hydrogen evolution by the cultures was observed on acetate under nitrogen-deprived conditions and on glutamate.

Another important aspect of the study is the relationship of PHB accumulation and H_2 evolution to the pH of the growth media. It is common knowledge that pH-uncontrolled batch cultures normally grown on media with carbonic acids buffered with weak phosphate buffer increase their pH by the end of the cultivation process. The degree of this increase is

Table 1. PHB accumulation and H_2 evolution by *R. sphaeroides* RV grown with various carbon substrates under different conditions of nitrogen availability

Carbon substrate	Nitrogen substrate	Final cell dry wt. (mg/mL)	PHB initial (% dry cells)	PHB final (% dry cells)	H_2 (mL/mg dry cells)	Final pH
Acetate	None	1.4	27.5	40.9	0	10.9
	NH_4^+	2.2	3.9	38.4	0	10.3
	Glutamate	1.6	25.4	25.7	0	10.0
Lactate	None	1.8	3.0	11.4	1.42	7.1
	NH_4^+	2.9	4.1	1.4	0	9.4
	Glutamate	2.9	3.5	7.6	0.90	7.3
Pyruvate	None	2.0	1.9	12.7	0.23	7.3
	NH_4^+	2.2	4.1	0.4	0	7.1
	Glutamate	3.0	2.2	5.1	0.06	7.9

a function of the carbon and nitrogen source and the rate of their utilization. We found that in nitrogen-deprived, lactate-grown cells, the initial medium pH 7.5 increased to a pH of ~10. At the same time, PHB accumulation increased to 28% (data not presented), that is, more than twice that compared to cells grown from an initial pH of 6.8 (see Table 1). H_2 evolution in this case decreased more than eightfold. The simultaneous decrease in H_2 evolution and increase in PHB level indicate the balance of both processes depending on the cell energy status. Increasing the pH of the medium leads to the dissipation of the cell membrane potential, thereby inhibiting cell growth, and this may force cells to accumulate PHB. The pH-dependent accumulation of PHB may also reflect the existence of a regulatory mechanism leading to accelerated synthesis of the polymer in response to sub-optimal pHs.

Further work will be aimed at the study of the interrelation of both processes in *R. sphaeroides* RV grown continuously. It is also planned to monitor changes in activities of enzymes involved in PHB metabolism: acetate kinase, PHB synthase, and phosphotransacetylase, under different cultivation conditions in order to discover the regulatory mechanisms involved in the PHB synthesis. Understanding PHB synthesis regulation would provide the opportunity to control the cultivation process in order to achieve a desired product, whether PHB or H_2 is preferred.

ACKNOWLEDGMENTS

This work was performed under the management of RITE as a part of the R&D Project on Environmentally Friendly Technology for the Production of Hydrogen supported by New Energy Development Organization of Japan (NEDO).

REFERENCES

Anderson, A.J., and Dawes, E.A., 1990, Occurrence, metabolism, metabolic role, and industrial uses of bacterial polyhydroxyalcanoates, *Microbiol. Rev.*, 54(4):450–472.

Brandl, H., Gross, R.A., Lenz, R.W., and Fuller, R.C., 1989, Ability of the photosynthetic bacterium *Rhodospirillum rubrum* to produce various poly (3-hydroxyalcanoates): potential sources for biodegradable polyesters, *Int. J. Biol. Macromol.*, 11:49–55.

Brandl, H., Gross, R.A., Lenz, R.W., Lloyd, R., and Fuller, R.C., 1991, The accumulation of poly (3-hydroxyalcanoates) in *Rhodobacter sphaeroides*, *Arch. Microbiol.*, 155:337 340.

Braunegg, G., Sonnenleitner, B., and Lafferty, R.M., 1978, A rapid gas chromatographic method for the determination of poly-ß-hydroxybutyric acid in microbial biomass, *Eur. J. Appl. Microbiol. Biotechnol.*, 6:29–37.

Dephilippis, R., Ena, A., Guastini, M., Sili, C., and Vincenzini, M., Factors affecting poly-ß-hydroxybutyrate accumulation in cyanobacteria, *FEMS Microbiol. Rev.*, 103(2–4):187–194.

Fuller, R.C., 1995, Polyesters and photosynthetic bacteria; from lipid cellular inclusions to microbial thermoplastics, in *Anoxygenic Photosynthetic Bacteria*, Blankenship, R., et al. (eds.), Kluwer, pp. 973–990.

Hustede, E., Steinbüchel, A., and Schlegel, H.G., 1993, Relationship between the photoproduction of hydrogen and the accumulation of PHB in non-sulphur purple bacteria, *Appl. Microbiol. Biotechnol.*, 39:87–93.

Liebergessell, M., Hustede, E., Timm, A., Steinbüchel, A., Fuller, R.C., Lenz, R.W., and Schlegel, H.G., 1991, Formation of poly (3-hydroxyalcanoates) by phototrophic and chemolithotrophic bacteria, *Arch. Microbiol.*, 155:415–421.

Mao, X., Miyake J., and Kawamura, S., 1986, Screening photosynthetic bacteria for hydrogen production, *J. Ferment. Technol.*, 64:245–249.

Mas, J., and Van Gemerden, H., 1995, Storage products in purple and green sulfur bacteria, in *Anoxygenic Photosynthetic Bacteria*, Blankenship, R., et al. (eds.), Kluwer, pp. 973–990.

Miyake J., Mao, X.Y., and Kawamura, S., 1984, Photoproduction of hydrogen from glucose by a co-culture of a photosynthetic bacterium and *Clostridium butyricum*, *J. Ferment. Technol.*, 62(6):531–535.

Nagamine Y., Kawasugi, T., Miyake, M., Asada, Y., and Miyake, J., 1996, *J. Mar. Biotechnol.*, 4:34–37.

Nakada, E., Nishikata, S., Asada, Y., and Miyake, J., 1996, *J. Mar. Biotechnol.*, 4:38–42.

Suzuki, T., Tsygankov, A., Miyake, J., Tokiwa, Y., and Asada, Y., 1995, *Biotechnol. Lett.*, 17(4):395–400.

Tsygankov, A., Hirata, Y., Miyake, M., Asada, Y., and Miyake, J., 1994, *J. Ferment. Bioenerg.*, 77(5):575–578.

21

PHOTOSYNTHETIC BACTERIA OF HAWAII

Potential for Hydrogen Production

Mitsufumi Matsumoto,[1,2] Brandon Yoza,[1] JoAnn C. Radway,[1] and Oskar R. Zaborsky[1]

[1]University of Hawaii at Manoa
Hawaii Natural Energy Institute
Marine Biotechnology Center
School of Ocean and Earth Science and Technology
2540 Dole Street, Holmes Hall 246
Honolulu, Hawaii 96822
[2]Tokyo University of Agriculture and Technology

Key Words

photosynthetic bacteria, marine biodiversity, tropical marine microbes, hydrogen

1. SUMMARY

Photosynthetic bacteria isolated in Hawaii were evaluated for hydrogen evolution. A number of hydrogen-producing strains, provisionally identified as *Rhodospirillaceae*, were obtained from a canal at Ala Moana Beach Park in Honolulu. One isolate, designated M0006, generated up to 55 µL/mg dry weight/h hydrogen under laboratory conditions. In comparison, *Rhodobacter sphaeroides* RV produced 31 µL/mg dry weight/h under similar experimental conditions. Our study points to the untapped potential of microbes found in marine environments.

2. INTRODUCTION

The production of hydrogen by biological means, that is, biohydrogen, is of worldwide interest. Much attention has focused on hydrogen generation using photosynthetic bacteria. The highest rates of hydrogen production by these organisms have been achieved with strains belonging to the Rhodospirillaceae, of which the best studied is *R. sphaeroides* RV, isolated by Miyake (Mao et al., 1986). These findings suggest the possibility that isolation and screening efforts could yield additional photosynthetic bacteria that are both capable of high rates of hydrogen production and are robust microorganisms able to operate sustainably under real-world conditions.

Over the years, many researchers, in particular Mitsui at the University of Miami (Kumazawa and Mitsui, 1981), have examined various environments for unusual hydrogen-producing microorganisms (Ohta et al., 1991; Mao et al., 1986). Many of the Mitsui strains are now contained in the Hawaii Culture Collection. However, the vast biodiversity offered by tropical marine environments has hardly been tapped. A primary interest of the Marine Biotechnology Center is to examine the production of hydrogen from organic acids and wastewater by photosynthetic bacteria found in marine environments. Here we report on the isolation and initial screening of several promising strains.

3. EXPERIMENTAL

3.1. Isolation of Photosynthetic Bacteria

Sediment and liquid samples were collected from a canal at Ala Moana Beach Park, Honolulu, Hawaii, using sterile 70-mL plastic centrifuge tubes. Canal water was brackish at the time of collection due to seawater influx. Black, silty sediment, 3–120 cm deep, covered the bottom of major portions of the canal. Sediment pH was ca. 7.5, while water pH was 8.2. Samples were inoculated into 30-mL vials filled with basal medium containing 0.1% yeast extract and a mix of trace elements and vitamins (Whittenbury, 1971). The inoculated vials were incubated at 32 °C under 3000 lux illumination from a 300-W halogen light source. After ca. one week, the red, green, or brown liquid cultures were spread on ASY agar (Miyake, 1984) and incubated anaerobically for 5–10 days (32 °C, 3000 lux). Individual colonies were picked and inoculated into liquid ASY medium. The process was repeated several times to assure purity. Isolates were maintained in liquid and solid ASY medium.

3.2. Growth Measurements

Halotolerance of isolates was examined by inoculating cultures into 30-mL test tubes filled with ASY medium amended with 0–5% NaCl. Growth was measured on a dry weight basis.

3.3. Hydrogen Production

Hydrogen production experiments were performed with exponential phase cultures. Isolates were precultured for three days at 32 °C under 3000 lux unilateral illumination, using 60-mL BOD bottles completely filled with ASY medium. Cells were harvested by centrifugation (4000 rpm, 20 min, 25 °C) and resuspended in a 72-mL flat tissue culture flask containing 70 mL GL medium (Miyake and Kawamura, 1987). Key components of GL medium were 1.34 g/L D,L-lactic acid, 1.25 g/L glutamic acid, and 1.5 g/L sodium bicarbonate. Flask headspaces were flushed with argon gas, after which the bacterial suspensions were incubated in a 32 °C water bath with 10 klux unilateral illumination. Evolved gas was then collected over water in an inverted graduated cylinder via a Tygon tube, and hydrogen content was quantified by gas chromatography.

Table 1. Hydrogen production rates for Hawaiian photosynthetic bacterial isolates

Strain	Rate (µL/mg dw/h)
M0001	13
M0002	40
M0003	46
M0004	46
M0005	52
M0006	55
M0007	39
M0008	33
M0009	31
M0010	26
R. sphaeroides RV	31

4. RESULTS AND CONCLUSIONS

Thirteen strains, provisionally identified as members of the Rhodospirillaceae, were isolated from the Ala Moana Canal. All showed a wide salinity tolerance, growing best without added NaCl, but exhibiting stable growth at NaCl levels as high as 4% (data not shown). Hydrogen production rates of the isolates are shown in Table 1. Cultures of strains M0005 and M0006, containing 2.5–3.0 mg dry wt /mL, showed the highest rates of total gas production (11.6 and 11.2 mL/h from the 70-mL suspensions; data not shown) as well as the highest rates of hydrogen production on a dry weight basis (Table 1). Figure 1 illustrates the time course of gas evolution by strain M0006. Similar gas production curves were seen with other isolates.

R. sphaeroides RV (used for comparison purposes) evolved only 7.3 mL/h total gas (data not shown) and 31 µL H_2/mg/h (Table 1). These rates are lower than those reported by other researchers. We have not yet been able to determine the reason for this discrepancy. Obvious factors that have been examined include light source, composition of the medium, and temperature. We have also examined two different *R. sphaeroides* samples supplied to us by the Miyake laboratory. Although this matter needs to be resolved and is being further investigated, the data suggest that Strain M0006, in particular, is capable of high rates of hydrogen production and may hold potential for field applications.

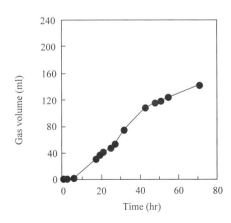

Figure 1. Gas evolution profile of isolate M0006 under 6000 lux illumination.

ACKNOWLEDGMENTS

This material is based upon work supported by the Department of Energy Hydrogen Program, under award no. DE-FC36-97GO10202 to O. R. Zaborsky, principal investigator. Any opinions, findings and conclusions, or recommendations expressed in this publication are those of the authors and do not necessarily reflect the views of the Department of Energy. The Ministry of Education, Japan, is acknowledged for financial assistance to Mr. Matsumoto for studies at the University of Hawaii. We thank Dr. Jun Miyake, National Institute for Advanced Interdisciplinary Research, Japan, for his generosity in providing *R. sphaeroides* RV and for advice on analytical protocols.

REFERENCES

Kumazawa, S., and Mitsui, A., 1981, Characterization and optimization of hydrogen photoproduction by a saltwater blue-green alga, *Oscillatoria* sp. Miami BG-7, I; enhancement through limiting the supply of nitrogen nutrients, *Int. J. Hydrogen Energy*, 6(4):339–348.

Mao, X.Y., Miyake, J., and Kawamura, S., 1986, Screening photosynthetic bacteria for hydrogen production from organic acids, *J. Ferment. Technol.*, 64(3):245–249.

Miyake, J., and Kawamura, S., 1987, Efficiency of light energy conversion to hydrogen by the photosynthetic bacteria *Rhodobacter sphaeroides*, *Int. J. Hydrogen Energy*, 12(13):147–149.

Miyake, J., Mao, X.Y., and Kawamura, S., 1984, Photoproduction of hydrogen from glucose by co-culture of photosynthetic bacterium and *Clostridium butyricum*, *J. Ferment. Technol.*, 62(6):531–535.

Ohta, Y., Frank, J., and Mitsui, A., 1981, Hydrogen production by marine photosynthetic bacteria: effect of environmental factors and substrate specificity on the growth of a hydrogen-producing marine photosynthetic bacterium, *Chromatium* sp. Miami PBS 1071, *Int. J. Hydrogen Energy*, 6(5):451–460.

Whittenbury, R., 1971, Isolation of anaerobes, in *Society for Applied Bacteriology Technical Series No. 5*, Shipton, D.A., and Board, R.G. (eds.), Academic Press, New York, pg. 241.

22

CONVERSION EFFICIENCIES OF LIGHT ENERGY TO HYDROGEN BY A NOVEL *Rhodovulum* sp. AND ITS UPTAKE-HYDROGENASE MUTANT

Akiyo Yamada, Tomoyuki Hatano, and Tadashi Matsunaga

Department of Biotechnology
Tokyo University of Agriculture and Technology
2-24-16 Naka-cho
Koganei, Tokyo 184-8588, Japan

Key Words

marine, *Rhodovulum*, uptake-hydrogenase mutant, hydrogen production, conversion efficiency

1. SUMMARY

A novel marine photosynthetic bacterium, *Rhodovulum* sp. NKPB160471R, that produces H_2 at a high rate was isolated from marine mud in Micronesia. The maximum H_2 production rate of this strain was 8.8 µmol/mg dry weight/h at a Xenon light intensity of 1800 W/m^2. An uptake-hydrogenase mutant of this strain, H-1, was obtained using transposon mutagenesis. H-1 showed approximately 25% less uptake hydrogenase activity than NKPB160471R even under H_2-saturated conditions. As a result, the conversion efficiency of light energy to hydrogen was improved in H-1. Maximum efficiencies of 26 and 35% were obtained at a light intensity of 13 W/m^2 in NKPB160471R and H-1, respectively. Therefore, utilization of uptake-hydrogenase-deficient mutants can be an efficient method of obtaining higher conversion efficiencies.

2. INTRODUCTION

Biological production of H_2 by marine photosynthetic bacteria is potentially useful because the marine environment has a large surface area to absorb solar energy, and sea water contains nutrients essential for the growth of marine photosynthetic bacteria. Improvements in the conversion efficiency of light energy to H_2 are required because they will lower the cost of the H_2 produced. Most H_2 production studies have been done on fresh water strains (Miyake and Kawamura, 1987; Warthmann et al., 1993). They have included the inactivation of uptake hydrogenase (Takakuwa et al., 1983; Kern et al., 1994) and the utilization of a PHB negative mutant (Hustede et al., 1993). There have been few attempts to increase the H_2 production efficiency in marine photosythetic bacteria (Matsunaga and Mitsui, 1982, Yamada et al., 1995), and uptake-hydrogenase-deficient mutants have not been employed for efficient hydrogen production in these organisms. On the other hand, much progress has been made in the development of recombinant DNA technology for marine photosynthetic bacteria (Matsunaga et al., 1986, 1990, 1991).

In this study, isolation of marine photosynthetic bacteria capable of producing hydrogen at high rates was accomplished, and conversion efficiencies of light energy to H_2 were improved by utilizing an uptake-hydrogenase-deficient mutant.

3. EXPERIMENTAL

3.1. Strains and Culture Conditions

Strain NKPB160471R is one of 300 isolates of marine photosynthetic bacteria collected from coastal waters and sediments around Japan, the Philippines, and Micronesia. This culture collection is maintained at the Department of Biotechnology of the Tokyo University of Agriculture and Technology. These marine photosynthetic bacteria were grown in an RCVBN medium (Beatty and Gest, 1981). The cultures were maintained anaerobically at 30 °C under continuous illumination with cool white fluorescent lamps at a light intensity of about 10 W/m^2.

3.2. Construction of an Uptake-Hydrogenase Mutant

Transposon (Tn5) mutagenesis of strain NKPB160471R was carried out (Takeyama et al., 1996) and strains that could not grow with H_2 under photoautotrophic conditions were isolated as uptake-hydrogenase mutants (Takakuwa et al., 1983). The uptake hydrogenase activity was measured using methylene blue reduction under H_2-saturated conditions, and the time course of the methylene blue reduction was measured anaerobically using a colorimeter (Model 21150, ICM, United States).

3.3. H_2 Production Rates and Conversion Efficiency

Hydrogen production rates were determined using a dissolved H_2 electrode (Type B-2, Able, Japan). A schematic diagram of the experimental apparatus is shown in Figure 1. A Xenon lamp with a UV cutoff filter (C.S. 3-74, Corning, United States) and IR cutoff filter (C.S. 1-75, Corning) was used as a light source. Light intensity was measured with a photometer (LI-189, LI-COR, United States). The medium for the H_2 production experiment was RCVBN (Beatty and Gest, 1981) without NH_4Cl; DL-malate was used as an electron donor.

The conversion efficiency of light energy to H_2 was defined by the following equation:

Efficiency (%) = (combustion energy of hydrogen produced/energy of irradiated light)
$\times 100^*$

---

[*] (Miyake and Kawamura, 1986)

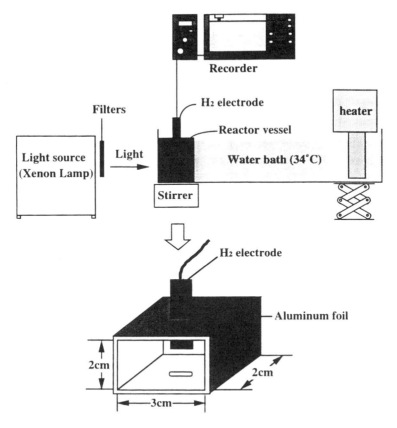

Figure 1. Schematic diagram of the experimental apparatus.

In this equation, consumption of the substrate for H_2 production was not considered.

4. RESULTS AND DISCUSSION

4.1. Isolation and Identification

A novel marine photosynthetic bacterium, strain NKPB160471R, which produces hydrogen at a high rate, was isolated from marine mud in Micronesia. This strain was identified as a *Rhodovulum* species based on 16S rDNA analysis. The highest similarity (96.7%) was observed with *Rhodovulum sulfidophilum*.

4.2. Uptake Hydrogenase Activity

An uptake hydrogenase mutant of this strain, H-1, was obtained by transposon mutagenesis. Southern hybridization analysis showed that Tn5 was successfully transferred into the genomic DNA of H-1 (data not shown).

Uptake hydrogenase activity of strain NKPB160471R and H-1 is shown in Figure 2. The uptake hydrogenase activity of mutant H-1 had 25% less uptake hydrogenase activity than the NKPB160471R strain under H_2-saturated conditions.

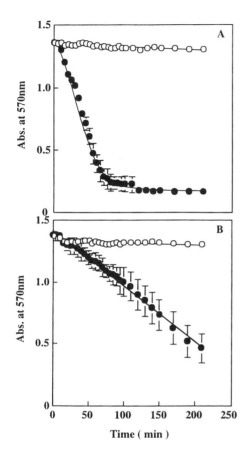

Figure 2. Time courses of methylene blue reduction by uptake hydrogenase of *Rhodovulum* sp. NKPB160471R (A) and H-1 (B). Open and closed circles indicate argon and H_2 gas-saturated conditions, respectively. Cell concentration in this reaction was adjusted to 0.28 mg (dry weight)/mL.

4.3. Hydrogen Production Rates and Conversion Efficiencies of Light Energy to H_2

At a light intensity of 1800 W/m^2, maximum hydrogen production rates of 8.8 and 11.2 μmol/mg dry weight/h were obtained by NKPB160471R and H-1, respectively (Figure 3). On the other hand, at a light intensity of 13 W/m^2, maximum conversion efficiencies of 26 and 35% were obtained in NKPB160471R and H-1, respectively (Figure 3). These rates were higher than reported in fresh water photosythetic bacteria capable of producing hydrogen efficiently (Miyake and Kawamura, 1987). This difference may be due to differences in the light quality and intensity.

In order to obtain the higher conversion efficiencies using marine photosynthetic bacteria, utilization of uptake hydrogenase mutants will be a great benefit.

ACKNOWLEDGMENT

This project is partially supported by the New Energy Industrial Technology Development Organization.

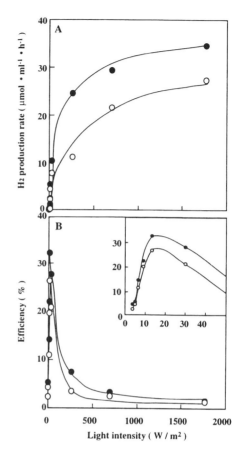

Figure 3. Hydrogen production rates (A) and conversion efficiencies (B) of light energy to H_2 in *Rhodovulum* sp. NKPB160471R and H-1 under various light intensities. Open and closed circles indicate NKPB060471R and H-1, respectively. Cell concentration was adjusted to 3.1 mg (dry weight)/mL.

REFERENCES

Beatty, J.T., and Gest, H., 1981, Generation of succinyl coenzyme A in photosynthetic bacteria, *Arch. Microbiol.*, 129:335–340.

Hustede, E., Steinbuchel, A., and Schlegel, H.G., 1993, Relationship between the photoproduction of hydrogen and the accumulation of PHB in non-sulphur purple bacteria, *Appl. Microbiol. Biotechnol.*, 39:87–93.

Kern, M., Klipp, W., and Klemme, J.H., 1994, Increased nitrogenase-dependent H_2 photoproduction by hup mutant of *Rhodospirillum rubrum*, *Appl. Environ. Microbiol.*, 60:1768–1774.

Matsunaga, T., and Mitsui, A., 1982, Seawater-based hydrogen production by immobilized marine photosynthetic bacteria, *Biotechnol. Bioeng. Symp.*, 12:441–450.

Miyake, J., and Kawamura, S., 1987, Efficiency of light energy conversion to hydrogen by the photosynthetic bacterium *Rhodobacter sphaeroides*, *Int. J. Hydrogen Energy*, 12:147–149.

Takakuwa, S., Odom, J.M., and Wall, J.D., 1983, Hydrogen uptake deficient mutants of *Rhodopseudomonas capsulata*, *Arch. Microbiol.*, 136:20–25.

Takeyama, H., Sunarjo, J., Yamada, A., and Matsunaga, T., 1996, ß-carotene production in a novel hydrogen-producing bacterium *Rhodovulum sulfidophilum* by cloning the *Erythrobacter longus* Och101 *crt*I and *crt*Y genes, *J. Mar. Biotechnol.*, 4:224–229.

Warthmann, R., Pfennig, N., and Cypionka, H., 1993, The quantum requirement for H_2 production by anoxygenic phototrophic bacteria, *Appl. Microbiol. Biotechnol.*, 39:358–362.

Yamada, A., Takano, H., Burgess, J.G., and Matsunaga, T., 1996, Enhanced hydrogen photoproduction by a marine photosynthetic bacterium, *Rhodobacter marinus*, immobilized onto light-diffusing optical fibers, *J. Mar. Biotechnol.*, 4:23–27.

23

HYDROGENASE-MEDIATED HYDROGEN METABOLISM IN A NON-NITROGEN-FIXING CYANOBACTERIUM, *Microcystis aeruginosa*

Yasuo Asada,[1] Masato Miyake,[1] Youji Koike,[2] Katsuhiro Aoyama,[3] Ieaki Uemura,[3] and Jun Miyake[4]

[1]National Institute of Bioscience and Human-Technology
Agency of Industrial Science and Technology
Ministry of International Trade and Technology
[2]Research Institute of Innovative Technology for the Earth
[3]Frontier Technology Research Institute
Tokyo Gas Co., Ltd.
[4]National Institute for Advanced Interdisciplinary Research
Agency of Industrial Science and Technology
Ministry of International Trade and Industry, Japan

Key Words

hydrogen production, hydrogenase, cyanobacteria, fermentation

1. SUMMARY

Hydrogen metabolism in a non-nitrogen-fixing cyanobacterium, *Microcystis aeruginosa*, was studied. The cyanobacterium evolved hydrogen gas under dark and anaerobic conditions. The rate of hydrogen evolution was higher in cells grown under nitrogen-limited conditions than under nitrogen-sufficient conditions. Hydrogen uptake occurred immediately when light was irradiated to the hydrogen-evolving cells in darkness. The light-dependent uptake of hydrogen was not sensitive to the inhibitor for photosystem II, DCMU [3-(3,4-dichloro-phenyl)-1,1-dimethyl urea], but sensitive to the antagonist of plastoquinone, DBMIB (2,5-dibromo-3-methyl-6-isopropyl-*p*-benzoquinone). These results suggest that hydrogen may be produced with degradation of endogenous storage materials, such as glycogen, and be taken up by the light-dependent reaction of photosystem I via plastoquinone.

BioHydrogen, edited by Zaborsky *et al.*
Plenum Press, New York, 1998

2. INTRODUCTION

There have been many studies on hydrogenase-mediated hydrogen production by green algae under light and dark conditions (Miyamoto, 1993) ever since Gaffron and Rubin's (1942) discovery of hydrogen production by *Scenedesmus*. On the other hand, most hydrogen production by cyanobacteria is nitrogenase-mediated (Houchins, 1984; Smith et al., 1992; Miyamoto, 1993; Markov et al., 1995). Hydrogenases in cyanobacteria were first investigated in the context of the recovery of hydrogen gas produced by a nitrogenase-dependent mechanism (Eisbrenner et al., 1981; Stewart, 1980). Subsequent studies, however, have demonstrated that some hydrogenases can be active for the evolution of molecular hydrogen, particularly in dark and fermentative metabolism (Asada and Kawamura, 1984,1986; Howarth and Codd, 1985; Almon and Boeger, 1988; Oost and Cox, 1989; Aoyama et al., 1997).

We have demonstrated hydrogen evolution by a non-nitrogen-fixing and uni-cellular cyanobacterium, *Microcystis aeruginosa* NIES 44, under dark and anaerobic conditions (Asada and Kawamura, 1984). Here we will describe some characteristics of hydrogen evolution and uptake by *M. aeruginosa*.

3. EXPERIMENTAL

An axenic culture of the cyanobacterium *Microcystis aeruginosa* NIES-44 was obtained from the National Institute of Environmental Studies of the Environment Agency of Japan. The cyanobacterium was cultivated at 30 °C photoautotrophically in light/dark cycles, as reported previously by Asada and Kawamura (1984).

The glycogen content in *M. aeruginosa* cells was determined according to Ernst et al. (1984).

The cyanobacterial cells harvested in the late logarithmic growth phase were washed and suspended in 50 mM Tris-HCl, pH 7.5, and then passed through a French press (Ohtake, Tokyo) at 800 kg/cm^2. After centrifugation at 2400 x g for 10 min, the supernatants were used as cell-free extracts.

The protein concentrations of the cell-free extracts were determined according to Peterson (1977).

Evolution and uptake of hydrogen gas were measured by continuous monitoring of dissolved hydrogen gas with a hydrogen-oxygen electrode system (Asada and Kawamura, 1984) at a temperature of 25 °C. The experiments were conducted under anaerobic conditions by previously sparging the reaction mixture with argon gas.

Purification of cytochrome c_{550} was carried out as described by Asada et al. (1993).

Redox titration of cytochrome c_{550} was carried out according to Takamiya (1981), as originally based on Dutton and Wilson (1974).

4. RESULTS AND CONCLUSIONS

4.1. Hydrogen Evolution under Dark and Anaerobic Conditions

Figure 1 indicates the time course of hydrogen evolution of *M. aeruginosa* NIES-44 using the hydrogen-oxygen electrode system. Hydrogen evolution in darkness began after the dissolved oxygen reached nearly zero and continued for at least an hour. A reduced hy-

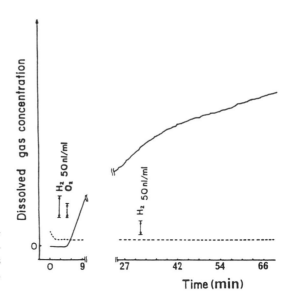

Figure 1. Time-course of hydrogen evolution by *M. aeruginosa* under dark and anaerobic conditions. Concentration of cyanobacterial cells was 1.86 mg/mL of 10 mM sodium phosphate buffer, pH 8.0. Electrode chamber was closed at time 0.

drogen-evolving rate was observed, which was due to the decrease of evolving activity or the increase of reverse reaction.

The initial hydrogen-evolving rates of intact cells with and without reduced methyl viologen and glycogen content were compared with the cells grown with sufficient and limited concentrations of nitrate (Table 1). The initial hydrogen-evolving rates and glycogen content were much higher in the cells grown with limited nitrate than with sufficient nitrate. The hydrogen-evolving rate with reduced methyl viologen is responsible for hydrogenase activity. As described previously for *Spirulina platensis* (Aoyama et al., 1997), the hydrogen evolution is probably due to anaerobic degradation of intracellular glycogen. The increase in the hydrogen-evolving rate by the intact cells grown in limited nitrate could result from increased hydrogenase activity and enhanced supply of reducing power because of higher glycogen content.

4.2. Hydrogen Evolution by Cell-Free Extracts

The hydrogen-evolving activity by cell-free extracts of *M. aeruginosa* in the presence of various electron donors was assayed (Table 2). The maximal rate of hydrogen pro-

Table 1. Hydrogen-evolving activity of *M. aeruginosa* grown with sufficient or limited amounts of nitrate

Nitrate salts	Glycogen (% glucose/ dry wt)	H_2 evolution (nmol/h/mg dry wt)	
		By intact cells	From MV_{red}
Sufficient	4.5	6.70	104.9
Limited	32.5	15.89	320.9

100 mL of preculture (about 0.5 mg dry cell wt/mL) of *M. aeruginosa* was inoculated into 1000 mL of fresh culture media with and without nitrate salts. Therefore, "limited" indicates less than 1/11 of ordinary concentration of nitrate salts. Cultivation was for 7.7 days.

Table 2. Hydrogen evolution by cell-free extracts of *M. aeruginosa*

Additive	Relative value of H_2 production (%)*
None	n.d.**
Pyruvate 20 mM	n.d.
Pyruvate 20 mM + Coenzyme A 0.4 mM	100
α-Ketoglutarate 20 mM + Coenzyme A 0.4 mM	n.d.
Formate 10 mM	n.d.
NADH 5 mM	n.d.
NADH 5 mM + ATP 5 mM	n.d.
NADPH 5 mM	n.d.
NADPH 5mM + ATP 5 mM	n.d.
Hyrdosulfite 5 mM	118
Methyl viologen 1 mM + hydrosulfite 5 mM	10540

The reaction mixture contained 6.2 mg/mL of protein and the additive.
*100% H_2 production was equivalent to 38.4 nmoles/h/mL of the reaction mixture.
**Not detected (less than about 0.4 %).

duction was observed in the presence of an artificial electron mediator, methyl viologen reduced by an excess amount of hydrosulfite. Among the physiological electron donors, only pyruvate plus CoA was effective, giving a rate near that in the presence of hydrosulfite alone, which could fully reduce an electron donor in the extracts.

The results in Table 2 suggest that the supply of reducing power for hydrogenase may involve CoA-dependent cleavage of pyruvate by the reaction of pyruvate-ferredoxin oxidoreductase, which has been shown to be the case in *Anabaena cylindrica* (Bothe et al., 1974):

$$\text{pyruvate} + Fd_{ox} + CoA \rightarrow \text{acetyl-CoA} + CO_2 + Fd_{red}$$

Therefore, the electron donation for hydrogenase is thought to involve ferredoxin. However, soluble ferredoxin purified from *M. aeruginosa* NIES-44 did not work as an electron mediator for the partially purified hydrogenase (Asada et al., 1987). This point is discussed below.

4.3. Hydrogen Uptake by Light Irradiation

Green algae that have been incubated under dark and anaerobic conditions produce hydrogen gas when light is irradiated (Greenbaum, 1980). Contrarily, light irradiation to the hydrogen-evolving cyanobacterial cells resulted in immediate hydrogen uptake (Figure 2A). The light-dependent uptake was not sensitive to the inhibitor for photosystem II in the reducing side, DCMU [3-(3,4-dichloro-phenyl)-1,1-dimethyl urea] (Figure 2B), but completely inhibited by the antagonist of plastoquinone, DBMIB (2,5-dibromo-3-methyl-6-isopropyl-*p*-benzoquinone) (Figure 2C). The results suggest that hydrogen may be taken up by the light-dependent reaction of photosystem I via plastoquinone.

5. CONCLUSIONS

The hydrogenase system in cyanobacteria is active for evolution only under dark and anaerobic conditions. From the results above, it is suggested that hydrogen may be produced with degradation of endogenous glycogen via pyruvate. More detailed studies

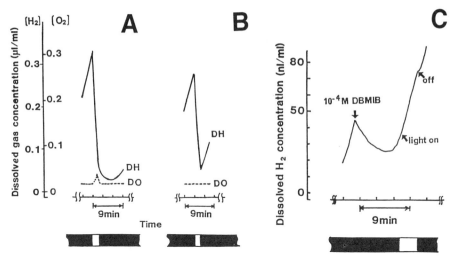

Figure 2. Effects of DCMU and DBMIB on light-dependent hydrogen uptake by *M. aeruginosa*. Cyanobacterial cell concentration was 2.13 mg/mL in A and B, 2.50 mg/mL in C. Cells were suspended in the same buffer as in Figure 1. Black and white belt indicates dark and light (2 klux, an iodine lamp) periods, respectively. (A) Light-dependent hydrogen uptake without additives. (B) Light-dependent hydrogen uptake in the presence of 10^{-5} M DCMU [3-(3,4-dichloro-phenyl)-1,1-dimethyl urea]. (C) Light-dependent hydrogen uptake in the simultaneous presence of 10^{-5} M DCMU and 10^{-4} M DBMIB (2,5-dibromo-3-methyl-6-isopropyl-*p*-benzoquinone).

are required to clarify the reason light irradiation results not in evolution, but in uptake of hydrogen gas. The hypothesis by Krogmann (1991) that cytochrome c_{550} may be the direct electron carrier for cyanobacterial hydrogenase and mediate electrons from ferredoxin needs to be proven with purified proteins. However, if the hypothesis is true, the phenomena described here may be understandable. The redox titration of purified cytochrome c_{550} from *M. aeruginosa* NIES-44 was carried out (Figure 3). The mid-point is calculated to be as relatively high as -220 mV for hydrogen evolution. The involvement of cytochrome c_{550}

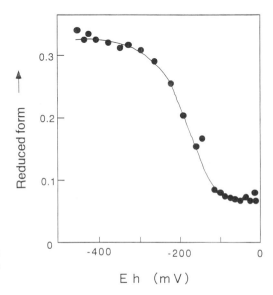

Figure 3. Redox titration of cytochrome c_{550} from *M. aeruginosa*. "Reduced form" equivalent to the difference of absorption at 550 nm and 600 nm.

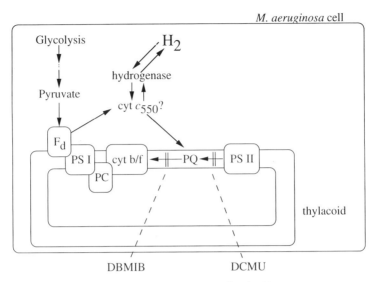

Figure 4. Hypothetical hydrogen metabolism by *M. aeruginosa*.

as an electron mediator appears to make the system favorable for hydrogen uptake. Hydrogen evolution occurs only in limited cases when the intracellular redox level is as low as about -400 mV, at which cytochrome c_{550} is mostly reduced to evolve hydrogen. The hydrogen metabolism in cyanobacteria is roughly hypothesized to be as shown in Figure 4 at this time.

ACKNOWLEDGMENTS

This work was partially performed under the management of the Research Institute of Innovative Technology for the Earth as a part of the Project for Biological Production of Hydrogen supported by the New Energy and Industrial Technology Development Organization.

REFERENCES

Almon, H., and Boeger, P., 1988, Hydrogen metabolism of the unicellular cyanobacterium *Chroococcidiopsis thermalis* ATCC29380, *FEMS Microbiol. Lett.*, 49:445–449.

Aoyama, K., Uemura, I., Miyake, J., and Asada, Y., 1997, Fermentative metabolism to produce hydrogen gas and organic compounds in a cyanobacterium, *Spirulina platensis*, *J. Ferment. Bioeng.*, 83:17–20.

Asada, Y., and Kawamura, S., 1984, Hydrogen evolution by *Microcystis aeruginosa* in darkness, *Agric. Biol. Chem.*, 48:2595–2596.

Asada, Y., and Kawamura, S., 1986, Screening for cyanobacteria that evolve molecular hydrogen under dark and anaerobic conditions, *J. Ferment. Technol.*, 64:553–556.

Asada, Y., Kawamura, S., and Ho, K.K., 1987, Hydrogenase from the unicellular cyanobacterium, *Microcystis aeruginosa*, *Phytochemistry*, 26:637–640.

Asada, Y., Kurashima, H., Shirai, M., and Satoh, A., 1993, Purification of cytochrome c_{550} from a cyanobacterium, *Microcystis aeruginosa* NIES-44, *Report of the National Institute of Biosciences and Human-Technology*, 1:89–92.

Bothe, H., Falkenberg, B., and Nolteernsting, U., 1974, Properties and function of the pyruvate: ferredoxin oxidoreductase and from the blue-green alga *Anabaena cylindrica*, *Arch. Microbiol.*, 96:291–304.
Dutton, P.L., and Wilson, D.F., 1974, Redox potentiometry in mitochondrial and photosynthetic bioenergetics, *Biochim. Biophys. Acta*, 346:165–212.
Eisbrenner, G., Roos, P., and Bothe, H., 1981, The number of hydrogenases in cyanobacteria, *J. Gen. Microbiol.*, 125:383–390.
Ernst, A., Kirschenlohr, H., Diez, J., and Boeger, P., 1984, Glycogen content and nitrogenase activity in *Anabaena variabilis*, *Arch. Microbiol.*, 140:120–125.
Gaffron, H., and Rubin, J., 1942, Fermentative and photochemical production of hydrogen in algae, *J. Gen. Physiol.*, 26:219–240.
Greenbaum, E., 1977, Simultaneous photoproduction of hydrogen and oxygen by photosynthesis, *Biotech. Bioeng. Symp.*, 10:1–13.
Houchins, J. P., 1984, The physiology and biochemistry of hydrogen metabolism in cyanobacteria, *Biochim. Biophys. Acta*, 768:227–255.
Howarth, D.C., and Codd, G.A., 1985, The uptake and production of molecular hydrogen by unicellular cyanobacteria, *J. Gen. Microbiol.*, 131:1561–1569.
Krogmann, D.W., 1991, The low-potential cytochromes in cyanobacteria and algae, *Biochim. Biophys. Acta*, 1058:35–37.
Markov, S.A., Bazin, M.J., and Hall, D.O., 1995, The potential of using cyanobacteria in photobioreactors for hydrogen production, *Adv. Biochem. Eng. Biotech.*, 52:60–86.
Miyamoto, K. 1993: Hydrogen production by photosynthetic bacteria and microalgae, *Recombinant Microbes for Industrial and Agricultural Applications*, Murooka, Y., and Imanaka, T. (eds.), Marcel Dekker, New York, pp. 771–786.
Oost, J., and Cox, R.P., 1989, Hydrogenase activity in nitrate-grown cells of the unicellular cyanobacterium *Cyanothece* PCC 7822, *Arch. Microbiol.*, 151:40–43.
Peterson, G.L., 1977, A simplification of the protein assay method of Lowry et al., which is more generally applicable, *Analyt. Biochem.*, 83:346–356.
Schulz, R., 1996, Hydrogenase and hydrogen production in eukaryotic organisms and cyanobacteria, *J. Marine Biotech.*, 4:16–22.
Smith, G.D., Ewart, G.D., and Tucker, W., 1992, Hydrogen production by cyanobacteria, *Int. J. Hydrogen Energy*, 17:695–698.
Stewart, W.D.P., 1980, Some aspects of structure and function in N_2-fixing cyanobacteria, *Ann. Rev. Microbiol.*, 34:497–536.
Takamiya, K., 1981, Analysis of electron transport by redox posing method, *Kougousei Kenkyuhou* (Methods in Photosynthesis Research, in Japanese), Katoh, S., Miyachi, S., and Murata, Y. (eds.), Kyouritu Shuppan, Tokyo, pp. 396–402.

24

IDENTIFICATION OF AN UPTAKE HYDROGENASE GENE CLUSTER FROM *Anabaena* sp. STRAIN PCC7120

Jyothirmai Gubili[1,2] and Dulal Borthakur[1]

[1]Department of Plant Molecular Physiology
[2]Department of Microbiology
University of Hawaii
Honolulu, Hawaii 96822

Key Words

PCR cloning, *hup* genes, *Anabaena*, cyanobacteria, uptake hydrogenase

1. ABSTRACT

Nitrogen-fixing filamentous cyanobacterium *Anabaena* possesses an uptake hydrogenase capable of recycling nitrogenase-evolved H_2 via an energy-conserving respiratory electron transport chain. A cluster of *hup* genes of *Anabaena* sp. strain PCC7120 required for the synthesis of uptake hydrogenase was isolated from a genomic DNA library using a cloning and screening strategy that involved the polymerase chain reaction (PCR). Two highly conserved stretches of amino acids were identified within the HupB sequences of various organisms and two degenerate PCR primers were synthesized based on these sequences. By using these primers, a part of the *hupB* gene was amplified from the *Anabaena* genomic DNA. This PCR fragment was cloned into the pGEM cloning vector and sequenced. Based on this sequence, two high-stringency PCR primers were synthesized that were used to isolate two lambda clones containing the *hupB* gene by PCR-screening of an *Anabaena* genomic library. From one of these clones, a 15-kb DNA fragment containing a cluster of *hup* genes was isolated. A restriction map of this fragment showing the locations of six *hup* genes was developed. These genes are located upstream of the *hupL* gene in the adjacent fragment in the *Anabaena* chromosome.

2. INTRODUCTION

Anabaena sp. strain PCC7120 is a filamentous photosynthetic cyanobacterium capable of aerobic nitrogen fixation. In the absence of combined nitrogen, such as ammonia, *An-*

abaena filaments develop certain specialized cells for nitrogen fixation, called heterocysts, at regular intervals of 10–12 vegetative cells (Haselkorn, 1978). Heterocysts are formed at regular intervals along the filament (approximately every tenth cell) and provide an anaerobic environment for nitrogen fixation as the enzyme nitrogenase is inactivated by oxygen. The product of nitrogen fixation, ammonia, is assimilated into glutamine that is exported to neighboring vegetative cells and, at the same time, carbohydrate produced through photosynthetic carbon dioxide fixation is imported from vegetative cells to provide reductant for nitrogen fixation (Haselkorn et al., 1985). During nitrogen fixation in the heterocysts, hydrogen is produced as an obligate by-product that consumes at least 25% of the energy (adenosine triphosphate or ATP) input into the nitrogenase-catalyzed reaction (Eisbrenner and Evans, 1983). Fortunately, most nitrogen-fixing cyanobacteria and eubacteria possess an uptake hydrogenase (Hup) capable of recycling nitrogenase-evolved H_2 via an energy-conserving respiratory electron transport chain. Electrons from H_2 can replenish the electron transport chain, resulting in ATP synthesis. The transfer of a pair of electrons from H_2 to an electropositive acceptor such as O_2 releases sufficient energy to support the synthesis of several molecules of ATP. The overall effect of uptake hydrogenase, therefore, is to increase the metabolic efficiency of the nitrogen fixation process.

The uptake hydrogenase has been characterized in several nitrogen-fixing bacteria such as *Bradyrhizobium japonicum, rhizobium leguminosarum, Azotobacter chroococcum, Azotobacter vinelandii*, and *Rhodobacter capsulatus* (Vignais and Toussaint, 1994). The genes involved in hydrogen metabolism (*hyp* and *hup*) have been cloned and sequenced in these organisms. Twenty genes have been identified in the *Rhodobacter capsulatus hup* gene cluster (Colveau et al., 1993). In *Anabaena*, only the gene encoding the large subunit of uptake hydrogenase (*hupL*) so far has been cloned and sequenced (Carrasco et al., 1995). The aim of the present study is to isolate and characterize other *hup* genes from *Anabaena*.

3. MATERIALS AND METHODS

3.1. DNA Preparation and Southern Hybridization

Growth of *Anabaena* cultures, genomic DNA isolation, and Southern hybridizations were done as previously described (Borthakur and Haselkorn, 1989), except that DNA probes were labeled by random priming using the dioxigenin labeling and detection kit from Boehringer Mannheim (Indianapolis, United States).

3.2. Restriction Analysis

Restriction enzymes and lygase were purchased from Promega Corporation (Madison, Wisconsin). DNA fragments were separated by electrophoresis on 1% agarose gels. Restriction maps of various cloned DNA fragments were constructed using *Hin*dIII, *Hin*cII, and *Sac*I. The map of the 15-kb region was constructed by a combination of several methods, such as Southern hybridization, partial digestion, and deletion cloning.

3.3. DNA Sequence Analysis

Selected clones were partially sequenced using the pUC18 forward and reverse primers with an automated DNA sequencer (Model 373A, Applied Biosystems Corporation, California) at the Biotechnology and Molecular Biology Instrumentation Facilities, University of

Hawaii. The sequencing data was analyzed by GCG program (Genetics Computer Group, 1994) and the Experimental BLAST Network Service (Altschul et al., 1990).

3.4. PCR Conditions

The PCR reactions were done using Gene Amp PCR System 2400 (Perkin Elmer). Standard PCR reactions were done in a final volume of 50 µL, which included 100 ng of sample DNA; 200 µM each of dATP, dCTP, dGTP, and dTTP; 0.2 µM of each specific primer; 2.5 mM $MgCl_2$; and 1.5 U of *Taq* polymerase (Promega, Madison, Wisconsin). Thermal cycling was done as follows for 30 cycles: denaturation for 1 min at 94 °C, primer annealing at 58 °C, and extension at 72 °C for 3 min. For screening the lambda library, a 2-µL aliquot from the lambda mixture eluted in phage buffer was used as the template and the PCR conditions were as follows: The template was denatured at 94 °C for 15 min, followed by heating at 80 °C for 45 min. Reactions were then done under the same conditions as mentioned above after adding all the reagents. PCR products were electrophoresed on 1.5% agarose gels and visualized by staining with ethidium bromide.

3.5. Construction of PCR Primers

The *hupB* genes of *A. chroococcum* (Tibelius et al., 1993) and the *hypB* genes of *E. coli* (Lutz et al., 1991), *R. leguminosarum* (Rey et al., 1993), *B. japonicum* (Fu et al., 1995), *R. capsulatus* (Colveau et al., 1993), and *A. vinelandii* (Chen and Mortensen, 1992) have homology at the deduced amino acid level. By aligning the amino acid sequences of these genes, four highly conserved regions were identified between amino acid numbers 128 and 221. Two stretches of 10 amino acids from two conserved regions were selected and converted into DNA sequences using the most probable codon usage for *Anabaena* (Figure 1a). The sequence of 29 nucleotides of the forward strand from region 1 was selected for primer 1 and another sequence of 29 nucleotides from the reverse strand corresponding to region 2 was selected for the reverse primer. These two primers had similar GC content and melting temperatures and were not homologous to each other. Primers were synthesized in the Biotechnology and Molecular Biology Instrumentation Facilities.

3.6. Construction of the Lambda Genomic Library

A genomic library of *Anabaena* sp. strain PCC7120 was constructed by inserting 10–16 kb partially digested *Sau*3AI fragments into the *Bam*HI site of the lambda Gem11 vector (Promega, Madison, Wisconsin). The library of approximately 15000 clones was amplified in *E. coli* strain LE392 (Gubili and Borthakur, 1996).

3.7. Plating and Screening of the Lambda Library

The library was plated in ten 85-mm plates, each containing approximately 3000 plaques and screened by using a PCR screening method (Gubili and Borthakur, 1996). The phages were soaked with 2.5 mL of phage buffer (50 mM Tris-HCl, pH 7.5, 100 mM NaCl, and 8 mM $MgSO_4$) by incubating for 2 h at 37°C. One milliliter of phage buffer containing the phages was collected from each plate and centrifuged in a microfuge to remove the debris and stored at 4°C. A 2-µL aliquot from each phage stock was used for PCR reactions. One positive aliquot was selected and further plated in 10 plates to obtain approximately 500 plaques per plate. These plates were screened as above and a positive

(a) Construction of *hupB* specific PCR primers

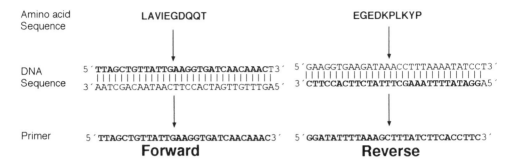

(b) DNA sequence of a 220-bp region of *Anabaena hupB*

```
    1
5' AATTAACGATGCCGAAAAAATTAAAGAAACTGGCTGTAAAGTAGTCCAAATTAACACGGGAACAGGTTGTCATT
   TGGATGCTTCCATGATTGAAAGGGGTTTACAACAATTAAACCCACCGATAAACTCTGTTTTGATGATTGAAAAT
   GTCGGTAATTTAGTTTGTCCGGCTTTATTTGATTTGGGTGAACAAGCGAAAGTCGTGATTCTTTCGGTGACA 3'
                                                                          220
```

(c) Homology of HupB of *Anabaena* and *Rhodobacter capsulatus*

```
HupB(An):    5 NDAEKIKEIGCKVVQINTGTGCHLDASMIERGLQQLNPPINSVLMIENVGNLVCPALFDL 184
               NDA++I+  G   +Q+NTG GCHLD +M+E+ L   L P  ++L IENVGNLVCPA FDL
HupB(Rc):  187 NDADRIRATGAPAIQVNTGKGCHLDGAMVEQALAHLPLPAGALLFIENVGNLVCPAAFDL 246

HupB(An):  185 GEQAKVVILSVT 220
               GE AKV ILSVT
HupB(Rc):  247 GEDAKVAILSVT 258
```

Figure 1. (a) Construction of two degenerate PCR primers for amplification of a section of the *Anabaena hupB* gene. Two highly conserved stretches were identified among the HupB sequences of several organisms. The forward and the reverse degenerate primers were synthesized using the most probable codon preferences for *Anabaena*. (b) DNA sequence of the 220-bp PCR fragment showing homology with the *hupB* sequences of several organisms. The sequences from which the two high-stringency primers were derived are shown as underlined. (c) Homology of the deduced amino acid sequence of the 220-bp *Anabaena* PCR fragment with the HupB of *Rhodobacter capsulatus* (Colbeau et al., 1993).

aliquot was selected. This aliquot was plated in 100 plates, each containing approximately 100 plaques. The phages were eluted with phage buffer as described above and transferred separately to 100 1.5-mL tubes. Ten-µL aliquots from these 100 samples were combined in groups of five in 20 tubes, mixed and screened by PCR. A positive mixture was selected and the five original tubes, from which this mixture was made, were individually screened by PCR. In this way, a positive sample was identified which originated from a plate containing 95 plaques. The sample that appeared positive with a strong band was identified and its five component plaques were further screened to identify a positive single plaque sample. Several dilutions of this sample were plated to obtain isolated individual plaques.

Ten plaques from one of these plates were individually picked and screened by PCR to identify the positive clone.

4. RESULTS

4.1. Identification and Cloning of a Cluster of *hup* Genes from *Anabaena* sp. Strain PCC7120

We amplified a region of *Anabaena hupB* by PCR using two degenerate primers. This fragment was cloned into the pGEM-T vector and sequenced. The nucleotide sequence of this fragment, excluding the two 29-bp regions covered by the primers on both sides, is shown in Figure 1b. The deduced amino acid sequence shows 61% identity and 75% similarity with *Rhodobacter capsulatus* HupB sequences (Figure 1c).

The two degenerate primers also amplified a faint PCR fragment with the *E. coli* genomic DNA as the template. Therefore, it was not possible to use them for PCR screening of lambda clones containing the *Anabaena hupB* gene. Based on the 220-bp *Anabaena hupB* sequence, two 23-nucleotide primers were synthesized for high-stringency PCR specific to the *Anabaena hupB* DNA. These primers have similar GC content and melting temperature and did not have homology to each other. When they were used against *Anabaena* genomic DNA in a PCR reaction, a 166-bp fragment was observed as expected (data not shown). This fragment was purified from the gel and sequenced in both directions using the same two primers. The sequence showed that this fragment was the inner segment of the previous fragment generated by using the two degenerate primers.

The lambda library of *Anabaena* genomic DNA was screened using PCR as described in section 3 (Materials and Methods). In this way, two lambda clones were isolated that generated the 166-bp hupB fragment in PCR reactions (Figure 2a). When the 166-bp PCR fragment generated by the two high-stringency primers were used as the probe, it hybridized with a 3.5-kb *Hin*dIII fragment of these clones and a fragment of the same size in the *Hin*dIII-digested genomic DNA of *Anabaena* 7120 (Figures 2b and c).

4.2. Restriction Map of the 15-kb *hup* Region in *Anabaena* 7120

The 15-kb *Sac*I fragment from one of the lambda clones was cloned in pUC18 in both orientations to generate the plasmids pUHA2 and pUHA3. These two plasmids and their subcloned fragments were used to develop a restriction map of this 15-kb region (Figure 3). This fragment comprised three *Hin*dIII fragments of sizes 3.3, 3.5, and 1.8 kb and two terminal *Hin*dIII-*Sac*I fragments of sizes 2.5 and 3.9 kb. These fragments were subcloned in pUC18 and each fragment was partially sequenced from both ends using M13 forward and reverse primers. Sequence analysis showed that the 2.5-kb *Sac*I-*Hin*dIII fragment contained at least 108 bp from the 5′ end of the *hupS* gene, which extends beyond the 15-kb *Sac*I fragment. Similarly, the sequence from one end of the 1.8-kb *Hin*dIII fragment showed high homology at the amino acid level with the N-terminus segment of HypF of *R. capsulatus*, *E. coli*, and several other organisms, suggesting that this fragment contains at least 78 bp the *Anabaena hupF* gene. The remaining segment of the *Anabaena hupF* gene is located in the 3.3-kb *Hin*dIII fragment since the sequence obtained from one end of this fragment showed high homology at the amino acid level with the HypF of *E. coli*, and several other bacteria (Figure 3). Similar sequence comparison showed that the other end of the 3.3-kb *Hin*dIII fragment contained at least 400 bp from the 5′ end of the

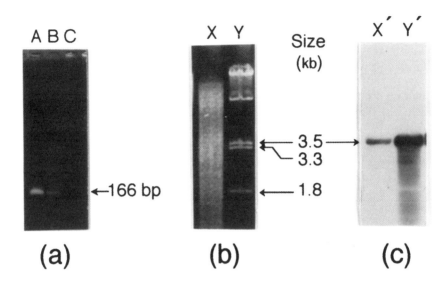

Figure 2. (a) Identification of a lambda clone from the library of *Anabaena* genomic DNA by PCR screening using two high-stringency primers. Individual plaques from a plate that was found positive in the previous round of PCR screening using the plate lysate were each suspended in 100 μL phage buffer. One plaque was found positive when 2 μL lysate from individual plaques were screened by PCR (lane A); 2 μL of the lysate from the library was used as the positive control (lane B) and lambda DNA (lane C) was used in the negative control. (b) HindIII digests of *Anabaena* 7120 genomic DNA (lane X) and purified DNA of lambda clone λJG1 (lane Y). (c) Identification of *hupB* in the 3.5-kb fragment of λJG1 by Southern hybridization. The HindIII-digested DNA in panel (b) above were transferred into a nylon membrane and the 166-bp *hupB* fragment obtained by PCR amplification of the *Anabaena* 7120 genomic DNA was used as a probe in Southern hybridization. Autoradiograph shows a 3.5-kb fragment in both *Anabaena* genomic DNA (lane X′) and λJG1 (lane Y′).

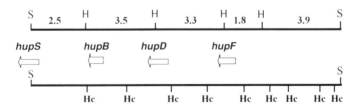

Figure 3. Restriction map of the 15-kb region containing *hup* genes of *Anabaena* sp. strain PCC7120 indicating the HindIII (H) and HincII (Hc) sites. The positions of *hupF*, *hupD*, *hupB* and *hupS* in this fragment are shown with open arrows. The numbers indicate the size of the fragments in kilobase pairs. S indicates SacI restriction site that originates from the lambda cloning vector.

Anabaena hupD gene. The 3′ end of the *hupD* gene is located at one end of the 3.5-kb *Hin*dIII fragment, which also contains the *hupB* gene at the other end. *Anabaena hup* genes are oriented in one direction as the *hup* genes of other organisms. There may be additional *hup* genes between *hupS* and *hupB*, *hupB* and *hupD*, and also *hupD* and *hupF*.

5. DISCUSSION

We have cloned a 15-kb region of *Anabaena* DNA containing a number of *hup* genes. This fragment, however, does not contain all the genes in the *hup* gene cluster of *Anabaena* 7120. Previously, Carrasco et al. (1995) cloned and sequenced *hupL* and the 3′ end of the *hupS* gene. The small and large subunits of the membrane-bound uptake hydrogenase are encoded by the *hupS* and *hupL* genes, respectively. Sequence analysis of the 2.5-kb *Sac*I-*Hin*dIII fragment shows that the *Sac*I end of this fragment has homology with both *hupS* and *hupU* genes of several organisms. The deduced amino acid sequence of this segment showed 82% identity and 88% similarity with the N-terminus of the small subunit of *Thiobacillus ferrooxidans* cytochrome-C3 hydrogenase (Fischer et al., 1996). Based on sequence identity, we concluded that the 5′ end of the *hupS* gene identified in this study is a part of the *hupS* that is located upstream to the *hupL* gene and has been identified and sequenced recently (J. Golden, personal communication). The 15-kb *Sac*I fragment we have cloned is therefore located upstream of the *hupL* and *hupS* in the *Anabaena* 7120 chromosome.

The filamentous cyanobacterium *Anabaena* contains at least two forms of hydrogenase that catalyze either hydrogen formation or uptake (Houchins and Burris, 1981). An *Anabaena* hydrogenase having both hydrogen formation and uptake functions, known as bidirectional hydrogenase, has been characterized and four *hox* genes involved in its synthesis have been described recently (Schmitz et al., 1995). *Anabaena* also contains another form of hydrogenase enzyme that catalyzes only the utilization of H_2 in the presence of an electron acceptor. Such an uptake hydrogenase occurs also in other cyanobacteria and eubacteria (for a review, see Vignais and Toussaint, 1994). Bacterial or cyanobacterial uptake hydrogenases are heterodimeric enzymes containing a large (about 65 kDa) and a small (about 34 kDa) subunit, as well as nickel and iron-sulfur (NiFe) clusters (Przybyla et al., 1992; Wu and Mandrand, 1993). The 15-kb DNA fragment cloned in this study contains four or more *hup* genes. Sequencing of this fragment may show the presence of additional *hup* genes in this fragment. At least 20 *hup* or *hyp* genes have been identified in several Ni-Fe hydrogenase-containing bacteria such as *E. coli* (Lutz et al., 1991), *R. capsulatus* (Colbeau et al., 1993), *A. vinelandii* (Chen and Mortenson, 1992), *A. chroococcum* (Tibelius et al., 1993), *Rhizobium leguminosarum* (Rey et al., 1993), and *Bradyrhizobium japonicum* (Fu and Maier, 1994).

Uptake hydrogenases are present in both N_2-fixing and non-N_2-fixing bacteria and cyanobacteria. *Anabaena* is different in that the O_2-evolving photosynthesis and H_2-evolving N_2 fixation in this filamentous cyanobacterium are physically separated in vegetative cells and heterocysts, respectively. Previously, Peterson and Wolk (1978) showed that uptake hydrogenase in *Anabaena* was present in the heterocysts only. Recently, Carrasco et al. (1995) also demonstrated that that *hupL* transcript was present only after induction of the heterocysts.

ACKNOWLEDGMENTS

This work was supported by a grant (DE-FG04-94AL85804) to the Hawaii Natural Energy Institute from the U.S. Department of Energy via the National Renewable Energy Laboratory.

REFERENCES

Altschul, S.F, Gish, W., Miller, W., Myers, E.W., and Lipman, D.J., 1990, Basic local alignment search tool, *J. Mol. Biol.*, 215:403–10.
Borthakur, D., and Haselkorn, R., 1989, Tn5 mutagenesis of *Anabaena* sp. strain PCC7120: isolation of a new mutant unable to grow without combined nitrogen, *J. Bacteriol.*, 10:5759–5761.
Carrasco, C.D., Buettner, J.A., and Golden, J.W., 1995, Programmed DNA rearrangement of a cyanobacterial *hupL* gene in heterocysts, in *Proceedings of the National Academy of Sciences (USA)*, 92:791–795.
Chen, J.C., and Mortenson, L.E., 1992, Identification of six open reading frames from a region of the *Azotobacter vinelandii* genome likely involved in dihydrogen metabolism, *Biochem. Biophys. Acta*, 1131:199–202.
Colbeau, A., Richaud, P., Toussaint, B., Caballero, F.J., Elster, C., Delphin, C., Smith, R.L., Chabert, J., and Vignais, P.M., 1993, Organization of the genes necessary for hydrogenase expression in *Rhodobacter capsulatus*: sequence analysis and identification of two *hyp* regulatory mutants, *Mol. Microbiol.*, 8:15–29.
Eisbrenner, G., and Evans, H.J., 1983, Aspects of hydrogen metabolism in nitrogen-fixing legumes and other plant-microbe interactions, *Annu. Rev. Plant Physiol.*, 34:105–136.
Fischer, J., Quentmeier, A., Kostka, S., Kraft, R., and Friedrich, C.G., 1996, Purification and characterization of the hydrogenase from *Thiobacillus ferrooxidans*, *Arch. Microbiol.*, 165:289–296.
Fu, C., and Maier, R.J., 1994, Sequence and characterization of three genes within the hydrogenase gene cluster of *Bradyrhizobium japonicum*, *Gene*, 14:47–52.
Fu, C., Olson, J.W., and Maier, R.J., 1995, HypB protein of *Bradyrhizobium japonicum* is a metal-binding GTPase capable of binding 18 divalent nickel ions per dimer, in *Proceedings of the National Academy of Sciences (USA)*, 92:2333–2337.
Genetics Computer Group, 1994, *Program Manual for the Wisconsin Package 8*, Madison, Wisconsin, United States.
Gubili, J., and Borthakur, D., 1996, The use of a PCR cloning and screening strategy to identify lambda clones containing the hupB gene of *Anabaena* sp. strain PCC7120, *J. Microbiol. Methods*, 27:175–182.
Haselkorn, R., 1978, Heterocysts, *Ann. Rev. Plant Physiol.*, 29:319–344.
Haselkorn, R., Curtis, S.E., Golden, J.W., Lammers, P.J., Nierzwicki-Bauer, Robinson, S.J., and Turner, N.E., 1985, Organization and transcription of *Anabaena* genes regulated during heterocyst differentiation, in *Molecular Biology of Microbial Differentiation*, Hoch, J.A., and Setlow, P. (eds.), American Society of Microbiology, Washington, D.C., pp 261–266.
Houchins, J.P., and Burris, R.H., 1981, Comparative characterization of two distinct hydrogenases from *Anabaena* sp. strain 7120, *J. Bacteriol.*, 146:215–221.
Lutz, S., Jacobi, A., Schlensog, V., Böhm, R., Sawers, G., and Böck, A., 1991, Molecular characterization of an operon (*hyp*) necessary for the activity of the three hydrogenase isoenzymes in *Escherichia coli*, *Mol. Microbiol.*, 5:123–135.
Przybyla, A.E., Robbins, J., Menon, N., and Peck, H.D., Jr., 1992, Structure-function relationships among the nickel-containing hydrogenases, *FEMS Microbiol. Rev.*, 88:109–136.
Peterson, R.B., and Wolk, C.P., 1978, Localization of an uptake hydrogenase in *Anabaena*, *Plant Physiol.*, 61:688–691.
Rey, L., Murillo, J., Hernando, Y., Hidalgo, E., Cabrera, E., Imperial, J., and Ruiz-Argüeso, T., 1993, Molecular analysis of a microaerobically induced operon required for hydrogenase synthesis in *Rhizobium leguminosarum* biovar viciae, *Mol. Microbiol.*, 8:471–481.
Schmitz, O., Boison, G., Hilscher, R., Hundeshagen, B., Zimmer, W., Lottspeich, F., and Bothe, H., 1995, Molecular biological analysis of a bidirectional hydrogenase from cyanobacteria, *Eur. J. Biochem.*, 233:266–276.
Tibelius, K.H. Du, L., Tito, D., and Stejskal, F., 1993, The *Azotobacter chroococcum* hydrogenase gene cluster: sequences and genetic analysis of four accessory genes, *hupA*, *hupB*, *hupY* and *hupC*, *Gene*, 127:53–61.
Vignais, P.M., and Toussaint, B., 1994, Molecular biology of membrane-bound H_2 uptake hydrogenases, *Arch., Microbiol.*, 161:1–10.
Wu, L.F., and Mandrand, M.A., 1993, Microbial hydrogenases: primary structure, classification, signatures and phylogeny, *FEMS Microbiol. Rev.*, 104: 243–270.

25

HYDROGENASE(S) IN *Synechocystis*

Tools for Photohydrogen Production?

Jens Appel, Saranya Phunpruch, and Rüdiger Schulz

FB Biology/Botany
Philipps-University
Karl-von-Frisch-Str.
D-35032 Marburg, Germany

Key Words

reversible NAD(P)$^+$-reducing hydrogenase, complex I, respiration, photosynthesis, mutational analysis, cyanobacteria

1. SUMMARY

From the unicellular non-N_2-fixing cyanobacterium *Synechocystis* sp. PCC 6803, we isolated and characterized a *hox* gene cluster encoding the large and small subunit of a NAD(P)$^+$-reducing NiFe hydrogenase (reversible hydrogenase) and at least three additional polypeptides. Until now, only 11 of the 14 polypeptides of the NADH-dehydrogenase were found in *Synechocystis*. By sequence homologies, we suggested that the missing subunits of the peripheral part of the dehydrogenase, containing most of the FeS clusters, are encoded by three ORFs of this gene cluster. Since there are no other homologs of these subunits in the entire sequence of *Synechocystis* (which is now available on the Internet), joint use of these polypeptides as the diaphorase moiety of the complex I and the reversible hydrogenase seems obvious. Complex I of cyanobacteria is located at the interface of the photosynthetic and respiratory electron transport in the thylakoid membrane. Our working hypothesis is that the reversible hydrogenase could be an important controlling device of the redox conditions in the different membranes. It could, for example, function as an electron valve under high light intensities at the photosynthetic membrane. To prove this hypothesis, deletion mutants of the corresponding *hox* genes were generated. Besides confirming the suggested function of HoxH as the large subunit of the reversible hydrogenase, we confirmed parts of the presented idea tested so far about the physiological role of the reversible hydrogenase in *Synechocystis* in relation to fermentation, respiration, and photosynthesis.

BioHydrogen, edited by Zaborsky *et al.*
Plenum Press, New York, 1998

2. INTRODUCTION

Hydrogenases are enzymes that catalyze the oxidation of hydrogen and the reduction of protons to hydrogen. They contain several iron-sulfur clusters, and the active site mostly consists either of a nickel and an iron ion in the case of NiFe hydrogenases (Fontecilla-Camps, 1996), or a special iron-sulfur cluster (H-cluster) in "Fe-only" hydrogenases (Adams, 1990).

The occurrence of hydrogenases in algae (Gaffron, 1939; Gaffron and Rubin, 1942) and cyanobacteria (Frenkel et al., 1950; Houchins, 1984) has been known for decades. The existence of at least two different enzymes in cyanobacteria, called uptake and reversible hydrogenases, has been shown by biochemical, physiological, and immunological methods (Houchins, 1984).

Recently, several hydrogenase genes were cloned from different cyanobacterial species: Through the study of a genomic rearrangement in heterocysts of *Anabaena* sp. PCC 7120, the sequence for the large subunit of a membrane-bound uptake hydrogenase was found (Carrasco et al., 1995). In *Anabaena variabilis* (Schmitz et al., 1995), *Synechocystis* sp. PCC 6803 (Appel and Schulz, 1996), and *Anacystis nidulans* (Boison et al., 1996), sequences for reversible NAD(P)$^+$-reducing hydrogenases were reported. Such a type of reversible NiFe hydrogenases is best characterized in *Alcaligenes eutrophus* (Tran-Betcke et al., 1990).

3. EXPERIMENTAL

All molecular biological methods were performed according to standard lab manual protocols (for example, Sambrook et al., 1989). For PCR (polymerase chain reaction) degenerated primers (A A/G CG T/G AT T/C TG T/C GG T/C A/G T T/G TG upstream, A/G CA A/G GC A/C A A/G A/G CA A/G GG A/G TC downstream) and for genomic screening, a λ-ZAP genomic library of *Synechocystis* sp. PCC 6803 (kindly provided by Dr. McIntosh, East Lansing, Michigan, United States) was used as described in Appel and Schulz (1996). Isolated clones were subjected to deletion subcloning with the Erase-a-Base kit from Promega. Sequencing was performed by the chain-termination method of Sanger et al. (1977). The first strand was sequenced with the deletion subclones and the second by primer sequencing. Sequence analysis was performed by the blast server for sequence databanks (Altschul et al., 1990) and with the DNASIS program (Hitachi). To construct *hox* deletion mutants, parts of the *hox* genes were deleted and replaced by a kanamycin cassette (Taylor and Rose, 1988; Williams, 1988).

4. RESULTS, DISCUSSION, AND CONCLUSIONS

4.1. Isolation and Characterization of a Hydrogenase Gene Cluster from *Synechocystis*

By application of the PCR technique with primers derived from the conserved Ni-binding motifs of NiFe hydrogenases (Przybyla et al., 1992) and subsequent screening of a genomic library, genes for a NAD(P)$^+$-reducing NiFe hydrogenase were isolated from the unicellular non-N$_2$-fixing cyanobacterium *Synechocystis* sp. PCC 6803 (Appel and Schulz,

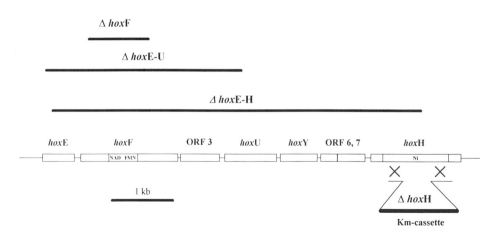

Figure 1. Structure of the *hox* gene cluster from *Synechocystis* sp. PCC 6803. As an example, construction of deletion mutant Δ*hox*H is shown. This was performed by active homologous recombination in *Synechocystis* after transformation of a plasmid-construct, in which part of the *hox*H gene is replaced by a kanamycin resistant cassette. In addition, localizations of deletion mutants Δ*hox*F, Δ*hox*E-U, and Δ*hox*E-H, generated by the same method, are shown.

1996). The overall order of the characterized gene cluster is the same as in the *hox*S operon of *Alcaligenes eutrophus* (Tran-Betcke et al., 1990). Besides the genes of the small and large hydrogenase subunits (*hox*Y and *hox*H), six additional ORFs (*hox* E, F, U, and three unidentified ORFs) were found (Figure 1).

To prove the identification of *hox*H as a functional hydrogenase gene, a deletion mutant was made in *Synechocystis* by replacing part of *hox*H with a kanamycin-cassette (Taylor and Rose, 1988; Williams, 1988) (Figure 1). Mutant cells completely lack the methylviologen-dependent hydrogen evolution activity of the wild-type.

4.2. Similarities of Isolated *hox*-Genes with Complex I Genes

For many years, the homology of the NAD^+-reducing hydrogenase subunits HoxF and HoxU with proteins of the NADH-dehydrogenase (complex I) was emphasized (Pilkington et al., 1990; Preis et al., 1991; Walker, 1992; Albracht, 1993). Interestingly, we found in *Synechocystis* the same arrangement of the genes *hox*E, *hox*F, and *hox*U as in the *nuo* operon (genes *nuo*E, F, and G) for the NADH-dehydrogenase of *E. coli* with a similar pattern of putative FeS-cluster binding motifs (Appel and Schulz, 1996). In spite of intensive work on this field (Friedrich et al., 1995), these subunits were still not described for the complex I of *Synechocystis*. Other homologs of these ORFs were not found in the entire sequence of *Synechocystis*, which is now available in the genebank or Internet (http://www.kazusa.or.jp/cyano/cyano.html).

4.3. Hypothesis for Direct Coupling of Hydrogenase to Complex I

The described homologies (Appel and Schulz, 1996) led us to the conclusion that the peripheral part of the complex I with most of the FeS-clusters is built by HoxE, F, and U (Figure 2).

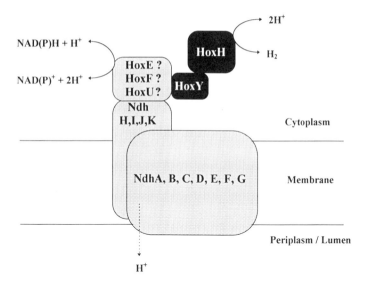

Figure 2. Proposed model for the interaction of NAD(P)$^+$-reducing hydrogenase and NAD(P)H-dehydrogenase (complex I) of *Synechocystis* sp. PCC 6803 (adapted from Berger et al., 1993).

Since HoxF and U are part of the soluble hydrogenase in *A. eutrophus* (Tran-Betcke et al., 1990), it seems plausible that the small and large subunits of the reversible hydrogenase are also coupled to the NAD(P)H-dehydrogenase at the cytoplasmic face of the cytoplasmic membrane as well as the thylakoid membrane (Figure 3). Other investigators have found a membrane association of this type of hydrogenase in two different cyanobacteria (Kentemich et al., 1989; Serebryakova et al., 1994). Interestingly, Schmitz and Bothe (1996) independently reached a similar conclusion for *Anabaena variabilis* by mainly comparing different HoxU sequences and suggesting this subunit as the connection between diaphorase, hydrogenase, and complex I.

4.4. Interaction of the Reversible Hydrogenase with Complex I and the Photosynthetic Apparatus

Our hypothesis about the coupling of the reversible hydrogenase to complex I in *Synechocystis* is supported by the long-known oxyhydrogen (knallgas) reaction that could be maintained by cyanobacteria in the dark (Houchins, 1984) and implies some new exciting pathways of the electron transport especially during photosynthesis.

The complex I of cyanobacteria is located at the interface of the photosynthetic and respiratory electron transport in the thylakoid membrane (Figure 3). It can participate in the cyclic electron flow around photosystem I (PS I) (Mi et al., 1992); it could be reduced by reversed electron flow from plastoquinone (PQ) driven by the electrochemical potential (Mulkidjanian and Junge, 1996); or it could just oxidize NAD(P)H and reduce PQ during respiration. Our hypothesis is that in transition states where light reactions are faster than dark reactions or where photosystem II (PS II) is working at a higher pace than PS I (PQ-pool mainly reduced), the reversible hydrogenase, if coupled to complex I, could function as an electron valve. This fine-tuning of the electron transport may be important since autotrophic cultures of *Synechocystis* permanently show low but detectable hydrogenase

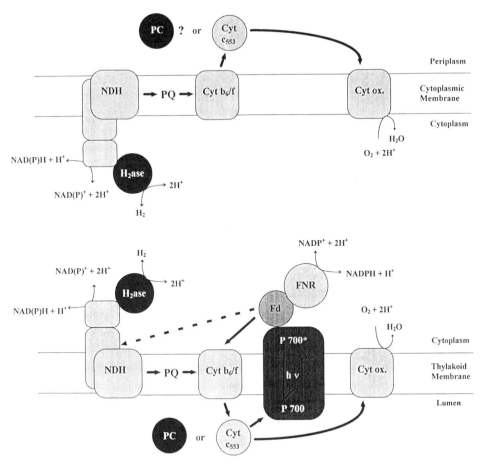

Figure 3. Proposed model for the interaction of reversible $NAD(P)^+$-reducing hydrogenase, photosynthesis, and respiration in the cytoplasmic and thylakoid membrane of the cyanobacterium *Synechocystis* sp. PCC 6803 (adapted from Friedrich et al., 1995 and Bryant, 1994).

activity with artificial electron donors like methylviologen. Other protective mechanisms like state transition might be too slow under rapidly changing light conditions.

On the other hand, after an anaerobic adaptation, *Synechocystis* cells do not show evolution of hydrogen in the light as do some green algae (Schulz, 1996). This apparent contradiction could easily be explained by the same arrangement of hydrogenase and complex I localized at the cytoplasmic membrane, recapturing hydrogen at this site and driving the oxyhydrogen reaction (Figure 3).

One possible method of verifying this model could be an analysis of a mutant that does not have a complex I in the cytoplasmic membrane, but only in the thylakoid membrane. Through immunogold studies, it could be shown that such a *Synechocystis* mutant (ΔndhD1) is available already. It was constructed by Dr. Steinmüller (Düsseldorf, Germany) by deletion of gene *ndh*D1. In comparison with the wild-type, this mutant indeed shows photohydrogen evolution without any artificial electron donor. Thus, the cyanobacterium *Synechocystis* was genetically engineered to produce hydrogen by a hydrogenase coupled to oxygenic photosynthesis.

4.5. Mutational Analysis of *hox*-Genes in *Synechocystis*

To investigate the function of the structural genes of the isolated gene-cluster, we generated several deletion mutants (ΔhoxH, ΔhoxF, ΔhoxEFU, and ΔhoxEFUYH) (Figure 1). All mutant strains were tested by PCR and Southern blotting to show correct generation and the genetic purity of the mutants.

No significant differences in the growth rate of the mutants in comparison with the wild-type under autotrophic conditions were obtained by recording the paramenter's optical density at 730 nm and chlorophyll content. But by counting the cells during growth under autotrophic conditions, it was observed that in mutant cultures, cell numbers are less than those of the wild-type, while the cells are relatively larger. This explains the missing differences in optical density and chlorophyll content.

All mutants missing parts of *hox*H had absolutely no hydrogen-producing activity by using methylviologen as an electron donor.

4.6. Future Work

The main goal of our future research is the application of hydrogenases in photohydrogen production on a technical scale (Schulz and Borzner, 1996). This goal absolutely needs to answer the question why there are hydrogenases in photosynthetic organisms. We believe that *Synechocystis* should be a good model to solve this enigma for at least some cyanobacteria. The presented results show the role of the hydrogenase during photosynthesis under special conditions. Investigations are in progress to determine if the hydrogenase is permanently necessary in cyanobacterial photosynthesis. In addition, effects of the deletion mutations on the photosynthetic apparatus will be studied.

Since green algae are able to produce high amounts of hydrogen in the light, we are planning to use *Synechocystis* as a genetic tool to study the algal hydrogenases (see paper of Schulz et al. in this issue) and to look for their properties in a cyanobacterial context.

ACKNOWLEDGMENTS

We are very grateful to Prof. Dr. Horst Senger (Marburg, Germany) for his personal encouragement, support and helpful discussions. Support by Prof. Dr. Lee McIntosh (East Lansing, Michigan, United States) in the initial phase of this project is gratefully acknowledged. This work was supported financially by BMFT (Bundesministerium für Forschung und Technologie der Bundesrepublik Deutschland, "Biologische Wasserstoffgewinnung"), DFG (Deutsche Forschungsgemeinschaft, "Hydrogenasen oxygener Mikroorganismen"), COST Action 8.18 (European Co-Operation in the Field of Scientific and Technical Research, "Hydrogenases and their Biotechnical Applications"), and DAAD (Deutscher Akademischer Austauschdienst).

REFERENCES

Adams, M.W.W., 1990, The structure and mechanism of iron-hydrogenases, *Biochim. Biophys. Acta*, 1020:115–145.
Albracht, S.P.J., 1993, Intimate relationship of the large and the small subunits of all nickel hydrogenases with two nuclear-encoded subunits of the mitochondrial NADH:ubiquinone oxidoreductase, *Biochim. Biophys. Acta*, 1144:221–224.

Altschul, S.F., Warren, G., Miller, W., Myers, E.W., and Lipman, D.J., 1990, Basic local alignment search tool, *J. Mol. Biol.*, 215:403–410.

Appel, J., and Schulz, R., 1996, Sequence analysis of an operon of a NAD(P)-reducing nickel hydrogenase from the cyanobacterium *Synechocystis* sp. PCC 6803 gives additional evidence for direct coupling of the enzyme to NAD(P)H-dehydrogenase (complex I), *Biochim. Biophys. Acta*, 1298:141–147.

Berger, S., Ellersiek, U., Kinzelt, D., and Steinmüller, K., 1993, Immunopurification of a subcomplex of the NAD(P)H-plastoquinone oxidoreductase from the cyanobacerium *Synechocystis* sp. PCC 6803, *FEBS Lett.*, 326:246–250.

Boison, G., Schmitz, O., Mikheeva, L., Shestakov, S., and Bothe, H., 1996, Cloning, molecular analysis, and insertional mutagenesis of the bidirectional hydrogenase genes from the cyanobacterium *Anacystis nidulans*, *FEBS Lett.*, 329:153–158.

Bryant, D.A., 1994, *The Molecular Biology of Cyanobacteria*, Kluwer Academic Publishers, Dordrecht, Netherlands.

Carrasco, C.D., Buettner, J.A., and Golden, J.W., 1995, Programmed DNA rearrangement of a cyanobacterial *hup*L gene in heterocysts, in *Proceedings of the National Academy of Sciences (USA)*, 92:791–795.

Fontecilla-Camps, J.C., 1996, The active site of Ni-Fe hydrogenases: model chemistry and crystallographic results, *J. Bioinorg. Chem.*, 1:91–98.

Frenkel, A., Gaffron, H., and Battley, E.H., 1950, Photosynthesis and photoreduction by the blue-green alga *Synechococcus elongatus* Näg, *Biol. Bull.*, 99:157–162.

Friedrich, T., Steinmüller, K. and Weiss, H., 1995, The proton-pumping respiratory complex I of bacteria and mitochondria and its homologue in chloroplasts, *FEBS Lett.*, 367:107–111.

Gaffron, H., 1939, Reduction of carbon dioxide with molecular hydrogen in green algae, *Nature*, 143:204–205.

Gaffron, H., and Rubin, J., 1942, Fermentative and photochemical production of hydrogen in algae, *J. Gen. Physiol.*, 26:219–240.

Houchins, J.P., 1984, The physiology and biochemistry of hydrogen metabolism in cyanobacteria, *Biochim. Biophys. Acta*, 768:227–255.

Kentemich, T., Bahnweg, M., Mayer, F., and Bothe, H., 1989, Localization of the reversible hydrogenase in cyanobacteria, *Z. Naturforsch*, 44c:384–391.

Mi, H., Endo, T., Schreiber, U., Ogawa, T., and Asada, K., 1992, Electron donation from cyclic electron transport and respiratory flows to the photosynthtic intersystem chain is mediated by pyridine nucleotide dehydrogenase in the cyanobacterium *Synechocystis* PCC 6803, *Plant Cell Physiol.*, 33:1233–237.

Mulkidjanian, A.Y., and Junge, W., 1996, New photosynthesis or old? *Nature*, 379:304–305.

Pilkington, S.J., Skehel, M., Gennis, R.B., and Walker, J.E., 1990, Relationship between mitochondrial NADH-ubiquinone reductase and a bacterial NAD-reducing hydrogenase, *Biochemistry*, 30:2166–2175.

Preis, D., Weidner, U., Conzen, C., Azevedo, J.E., Nehls, U., Röhlen, D., van der Pas, J., Sackmann, U., Schneider, R., Werner, S., and Weiss, H., 1991, Primary structure of two subunits of NADH:ubiquinone reductase from *N. crassa* concerned with NADH-oxidation; relationship to a soluble hydrogenase of *A. eutrophus*, *Biochim. Biophys. Acta*, 1090:133–138.

Przybyla, A.E., Robbins, J., Menon, N., and Peck, H.D., 1992, Structure-function relationship among the nickel-containing hydrogenases, *FEMS Microbiol. Rev.*, 88:109–136.

Sambrook, J., Fritsch, E.F., and Maniatis, T., 1989, *Molecular Cloning: A Laboratory Manual*, Cold Spring Harbor Laboratory, Cold Spring Habor, New York.

Sanger, F., Nickelsen, S., and Coulson, A.R., 1977, DNA sequencing with chain-termination inhibitors, in *Proceedings of the National Academy of Sciences (USA)*, 74:5463–5467.

Schmitz, O., Boison, G., Hilscher, R., Hundeshagen, B., Zimmer, W., Lottspeich, F., and Bothe, H., 1995, Molecular biological analysis of a bidirectional hydrogenase from cyanobacteria, *Eur. J. Biochem.*, 233:266–276.

Schmitz, O., and Bothe, H., 1996, The diaphorase subunit HoxU of the bidirectional hydrogenase as electron transferring protein in cyanobacterial respiration? *Naturwissenschaften*, 83:525–527.

Schulz, R., 1996, Hydrogenases and hydrogen production in eukaryotic organisms and cyanobacteria, *J. Mar. Biotechnol.*, 4:16–22.

Schulz, R., and Borzner, S., 1996, Wasserstoffproduzierende Hydrogenasen - Schlüsselenzyme zur Lösung des Energieproblems? *Chemie in Labor und Technik*, 47:496–500.

Serebryakova, L.T., Zorin, N.A., and Lindblad, P., 1994, Reversible hydrogenase in *A. variabilis* ATCC 29413, *Arch. Microbiol*, 161:140–144.

Taylor, L.A., and Rose, R.E., 1988, A correction in the nucleotide sequence of the Tn903 kanamycin resistance determinant in pUC4K, *Nucleic Acids Res.*, 16:358.

Tran-Betcke, A., Warnecke, U., Böcker, C., Zaborosch, C., and Friedrich, B., 1990, Cloning and nucleotide sequence of the genes for the subunits of NAD-reducing hydrogenase of *A. eutrophus* H16, *J. Bacteriol.*, 172:2920–2929.

Walker, E.J., 1992, The NADH-ubiquinone oxidoreductase (complex I) of respiratory chains, *Q. Rev. Biophys.*, 25:253–350.

Williams, J.G.K., 1988, Construction of specific mutations in photosystem II photosynthetic reaction center by genetic engineering methods in *Synechocystis* PCC 6803, *Methods Enzymol.*, 167:766–778.

26

DETECTION OF MARINE NITROGEN-FIXING CYANOBACTERIA CAPABLE OF PRODUCING HYDROGEN BY USING DIRECT NESTED PCR ON SINGLE CELLS

Haruko Takeyama and Tadashi Matsunaga

Department of Biotechnology
Tokyo University of Agriculture and Technology
Koganei, Tokyo 184-8588, Japan

Key Words

direct nested PCR, nitrogen-fixing cyanobacteria, hydrogen production, *nif H* gene, single-cell PCR

1. ABSTRACT

In order to detect nitrogen-fixing cyanobacteria capable of producing hydrogen at the single-cell level in natural seawater samples, direct PCR targeting the *nifH* gene was employed. Several primers were designed from the *nifH* gene sequence of *Anabaena* PCC7120. The optimum amplification condition for the *nifH* gene was investigated in *Anabaena* PCC7120, and effective direct amplification was observed at a 55 °C annealing temperature. Direct nested PCR was applied to a single cell of several marine nitrogen-fixing cyanobacteria, which were isolated under an epifluorescence microscope. Primers designed to amplify the 446-bp fragment in the *nifH* gene were used in the first PCR, followed by a second PCR using internal primers designed to amplify a 317-bp fragment. Stable and reliable amplification of *nifH* gene fragments was confirmed by using this direct nested PCR method.

2. INTRODUCTION

Nitrogen-fixing cyanobacteria have been observed in a wide variety of environments, where they have an important role in the nitrogen cycle (Young, 1992). Furthermore, they are advantageous in culture systems producing hydrogen or useful substances without the addition of a nitrogen source (Krishna et al., 1996; Suda et al., 1992). Screening exclusively depends on the primary culture followed by single-cell isolation, and therefore many unculturable strains may remain cryptic (Thornhill et al., 1995; Ward et al., 1992). In order to detect unculturable nitrogen-fixing cyanobacteria and to elucidate their phylogenetic position, the *nifH* gene encoding the subunits of the Fe protein of nitrogenase has been used as a molecular marker (Zehr et al., 1997). The *nifH* genes amplified from several nitrogen-fixing cyanobacteria have high degrees of similarity (Ben-Porath et al., 1994). The extracted DNAs from natural samples have been subjected to PCR amplification of the *nifH* genes. However, subsequent nucleotide sequence or hybridization analysis is required for identification as the cyanobacterial *nifH* gene because the *nifH* gene is present in all nitrogen-fixing eubacteria and archaeobacteria. Combining microscopic observation and direct PCR gene amplification may be useful in overcoming this problem, and accurate evaluation of the cyanobacterial *nifH* gene may be possible. Identification at the single-cell level will provide more information about unculturable strains of nitrogen-fixing cyanobacteria in natural environments.

Conventional PCR involving template DNA purification is time-consuming, and there is a potential risk of losing small amounts of DNA molecules during purification. For more rapid and effective detection, direct PCR without extraction and purification of DNA is required. In a small number of available targets, nested PCR is known to be effective for highly sensitive detection (Arias et al., 1995; Herman et al., 1995; Ibeas et al., 1996; Puig et al., 1994). In this study, a method combining microscopic observation and direct nested PCR targeting a *nifH* gene has been performed for the stable detection of nitrogen-fixing cyanobacteria at the single-cell level.

3. MATERIAL AND METHODS

3.1. Strains and Culture Conditions

Several nitrogen-fixing cyanobacterial strains (freshwater strain, *Anabaena* sp. PCC7120; marine strains, *Nostoc* sp. MiamiBG 039501 and 104401, *Anabaena* sp. MiamiBG 060002 and 094001, N_2-fixing unicell MiamiBG 039331 and 044012) were used in this study. They were cultured at room temperature in a BG11 medium (ATCC catalogue, medium no. 617, supplemented with 3% of NaCl) or a nitrogen-deficient A-N medium (Kumazawa et al., 1981) at a light intensity of 50 µE m^{-2} s^{-1} under aerobic conditions. Natural water samples were collected from coastal areas of Okinawa and Kochi and a hot spring at Izu in Japan and used for experiments within a week after collection.

3.2. Direct PCR Amplifications

Four PCR primers (Figure 1) were designed from the *nifH* sequence of *Anabaena* PCC 7120 (GenBank accession number, V00001) (Mevarech et al., 1980). Cyanobacterial cell filaments were sheared using sonication (Biorupter UCD-200T, Cosmo Bio, Tokyo, Japan), resulting in short filaments consisting of a few or single cells. Cell suspensions were then washed and resuspended in sterile water. Ten µL of serially diluted cell suspensions were used for direct PCR amplification. The PCR reaction mixture (50 µL) contained 10 µL of cell suspension, 1.0 µM of each oligonucleotide primer, 5 µL of 10 x PCR buffer, 0.2 mM of each dNTP, and 0.25 unit of *Taq* DNA polymerase (Perkin Elmer) and

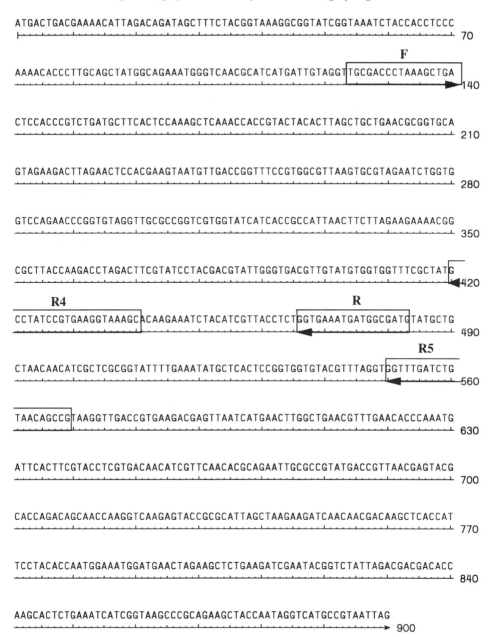

Figure 1. Primer positions in the *nifH* gene of *Anabaena* sp. PCC7120.

sterilized water. The reaction mixture was preheated for 3 min at 95 °C, followed by 30 cycles of amplification which consisted of denaturing at 95 °C for 1 min, annealing at several temperatures for 2 min, and extension at 72 °C for 2 min using a GeneAmp PCR 9600 (Perkin Elmer). After the last cycle, extension was continued for an additional 15 min to allow for completion of the reaction. PCR products were separated on a 1% agarose gel. The gel was stained in ethidium bromide for 10 min and visualized using a fluorescence image analyzer (Fluor Imager 575, Molecular Dynamics, California, United States).

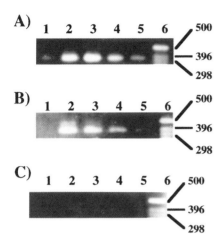

Figure 2. Gel electrophoresis of the *nifH* genes amplified from whole *Anabaena* PCC7120 cells using various cell numbers and at various annealing temperatures. Annealing temperature for PCR, (A) 55 °C, (B) 62 °C, (C) 65 °C; Lane 1, 10^5 cells/mL; Lane 2, 10^4 cells/mL; Lane 3, 10^3 cells/mL; Lane 4, 10^2 cells/mL; Lane 5, 10^1 cells/mL; Lane 6, molecular size marker.

3.3. Direct Nested PCR Amplifications at the One-Cell Level and Microscopic Observation

A single cell was isolated using a three-dimensional micromanipulator (MN151, Narishige, Tokyo, Japan), and subjected to direct nested PCR. A 1-µL sample from the first PCR using a single cell was applied to the second PCR. PCR conditions were the same as described above.

Cells were observed with an epifluorescence microscope (Olympus Optical Co., Ltd., IMT-2, Tokyo, Japan) under green light excitation (wavelength, 450–550 nm). Cells containing chlorophyll and/or phycocyanin were distinct as red fluorescent cells.

4. RESULTS AND DISCUSSION

4.1. Direct Amplification of the *nifH* Gene of *Anabaena* PCC7120 at Various Annealing Temperatures

After shearing the filaments of *Anabaena* PCC7120 by sonication, a few or single cells were obtained. They were used for the direct amplification of the *nifH* gene at various annealing temperatures, using an F-R primer set that was designed to amplify a partial 359-bp fragment of the *nifH* gene (Zehr et al., 1995). Clear bands (ca. 360 bp) were observed when fewer than 10^4 cells were used at a 55 °C annealing temperature (Figure 2). On the other hand, at a 62 °C annealing temperature less amplification was observed at low cell concentrations. No amplification was observed at 65 °C. These results indicate that the *nifH* gene amplification in cyanobacteria is possible by direct PCR methods without extraction of DNA and optimum conditions were obtained at a 55 °C annealing temperature. Less amplification was observed when more than 10^5 cells were used, suggesting that this may be caused by an overdose of genomic DNA and/or other cellular components.

4.2. Detection of the *nifH* Gene from Water Samples and Isolation of Nitrogen-Fixing Cyanobacteria

The diluted algal mat samples from the hot spring and 20 times concentrated seawater samples were used for direct amplification of the *nifH* genes. Figure 3 shows the gel

Detection of Marine Nitrogen-Fixing Cyanobacteria Capable of Producing Hydrogen

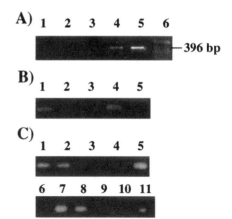

Figure 3. Gel electrophoresis of the *nifH* genes amplified from several natural water samples. (A) Amplified *nifH* gene from the algal mat from Izu hot spring: Lane 1, undiluted; Lane 2, 10^{-1} dilution; Lane 3, 10^{-2} dilution; Lane 4, 10^{-3} dilution; Lane 5, 10^{-4} dilution; Lane 6, molecular size marker. (B) Okinawa samples. (C) Kouchi samples; lane number is sample number.

electrophoresis of PCR products from these natural samples at a 55 °C annealing temperature. The *nifH* gene was well amplified at 10^{-4} dilutions from an algal mat. On the other hand, amplification of the *nifH* gene was observed in two samples from Okinawa and six samples from Kochi seawater samples. These samples that were confirmed to have nitrogen-fixing microorganisms were subjected to further isolation procedures. However, only three cyanobacteria were successfully isolated under nitrogen-deficient conditions (data not shown). This may show that the *nifH* gene amplified from samples may have originated from other nitrogen-fixing bacteria. Since the *nifH* gene appears to be highly conserved at the amino acid level in diazotrophs, the primer sets used in this experiment could be used to amplify the *nifH* gene of the other diazotrophs.

4.3. Direct Nested PCR Amplification of the *nifH* Gene from Single Cells of Nitrogen-Fixing Cyanobacteria

In order to overcome the above problem, characterization using direct nested PCR, combined with epifluorecence microscopy at the single-cell level, was carried out. The primers for the nested PCR were designed from the sequence of *Anabaena* PCC7120 *nifH* gene (Figure1). Using two primer sets (F-R4 and F-R5), the *nifH* genes were amplified from all tested nitrogen-fixing cyanobacteria except for the non-nitrogen-fixing microalga, *Chlorella* sp. (data not shown).

The nested PCR for amplifying the *nifH* gene from a single cell was carried out using two primer sets where the F-R5 primer set was employed for the first PCR step, followed by the F-R4 primer set for the second PCR step. A single cell was isolated under fluorescence microscopic observation and subjected to the direct nested PCR. Several fragments became visible after the second PCR step (Figure 4). The same-sized fragments (ca. 317 bp) were amplified in all samples, which were the authentic PCR products from the *nifH* gene using F-R4 primers.

Several red fluorescent cells in natural seawater samples were isolated under the epifluorescence microscope and subjected to the nested PCR under the same conditions described above. The *nifH* gene amplification was observed in two of four unicells tested.

Successful detection of the *nifH* gene at the single-cell level was demonstrated using direct nested PCR. This procedure may be useful for the identification of unknown strains in seawater samples. Furthermore, this method can be used to identify samples containing

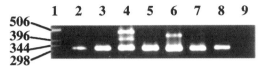

Figure 4. Gel electrophoresis of *nifH* gene sequences amplified from a single cell of several nitrogen-fixing cyanobacteria using direct nested PCR. Lane 1, molecular size marker; Lane 2, *Nostoc* sp. MiamiBG 039501; Lane 3, *Nostoc* sp. MiamiBG 104401; Lane 4, *Anabaena* sp. MiamiBG 060002; Lane 5, *Anabaena* sp. MiamiBG 094001; Lane 6, nitrogen-fixing unicell MiamiBG 039331; Lane 7, nitrogen-fixing unicell MiamiBG 044012; Lane 8, *Anabaena* PCC7120; Lane 9, negative control (PCR reaction buffer without cell).

nitrogen-fixing cyanobacteria from many aquatic samples at the first screening, thus making the isolation procedures easy and reliable.

REFERENCES

Arias, C.R., and Garay, E,R.A., 1995, Nested PCR method for rapid and sensitive detection of *Vibrio vulnificus* in fish, sediments, and water, *Appl. Environ. Microbiol.*, 61:3476–3478.

Ben-Porath, J., and Zehr, J.P., 1994, Detection and characterization of cyanobacterial *nifH* genes, *Appl. Environ. Microbiol.*, 60:880–887.

Herman, L.M., De Block, J.H.G.E., and Moermans, R.J.B., 1995, Direct detection of *Listeria monocytogenes* in 25 milliliters of raw milk by a two-step PCR with nested primers, *Appl. Environ. Microbiol.*, 61:817–819.

Ibeas J.I., Lozano, I., Perdigones, F., and Jimenez, J., 1996, Detection of Dekkera-Brettanomyces strains in sherry by a nested PCR method, *Appl. Environ. Microbiol.*, 62:998–1003.

Krishna, K., and Hall, D.O., 1996, Hydrogen production by cyanobacteria: potential, problems, and prospects, *J. Mar. Biotechnol.*, 4:10–15.

Kumazawa, S., and Mitsui, A., 1981, Characterization and optimization of hydrogen photoproduction by a saltwater blue-green alga, *Oscillatoria* sp. Miami BG7. I enhancement through limiting the supply of nitrogen nutrients, *Int. J. Hydrogen Energy*, 6:339–348.

Mevarech, M., Rice, D., and Haselkorn, R., 1980, Nucleotide sequence of a cyanobacterial *nifH* gene coding for nitrogenase reductase, in *Proceedings of the National Academy of Sciences (USA)*, 77:6476–6480.

Porter, D.R., 1988, DNA transformation, in *Methods in Enzymology*, Paker, L., and Glazer, A.N. (eds.), Academic Press, pp. 703–712.

Puig, M., Jofre, J., Lucena, F., Allard, A., Wadell, G., and Girones, R., 1994, Detection of adenoviruses and enteroviruses on polluted water by nested PCR amplification, *Appl. Environ. Microbiol.*, 60:2963.

Suda, S., Kumazawa, S., and Mitsui, A., 1992, Change in the H_2 photoproduction capability in a synchronously grown aerobic nitrogen-fixing cyanobacterium, *Synechococcus* sp. Miami BG043511, *Arch. Microbiol.*, 158:1–4.

Thornhill, R.H., Burgess, J.G., and Matsunaga, T., 1995, PCR for direct detection of indigenous uncultured magnetic cocci in sediment and phylogenetic analysis of amplified 16S ribosomal DNA, *Appl. Environ. Microbiol.*, 61:495–500.

Ward, D.M., Weller, R., and Bateson, M.M., 1990, 16S rRNA sequences reveal numerous uncultured microorganisms in a natural community, *Nature*, 345:63–5.

Young, J.P.W., 1992, Phylogenetic classification of nitrogen-fixing organisms, in *Biological Nitrogen Fixation*, Stacey, G. et al. (eds.), Chapman and Hall, New York, pp. 43–86.

Zehr, J.P., Mellon, M.T., Braun, S., Litaker, W., Steppe, T., and Paerl, H.W., 1995, Diversity of heterotrophic nitrogen fixation genes in a marine cyanobacterial mat, *Appl. Environ. Microbiol.*, 61:2527–2532.

Zehr, J.P., Mellon, M.T., and Hiorns, W.D., 1997, Phylogeny of cyanobacterial *nifH* genes: evolutionary implications and potential applications to natural assemblages, *Microbiology*, 143:1443–1450.

27

PROGRAMMED DNA REARRANGEMENT OF A HYDROGENASE GENE DURING *Anabaena* HETEROCYST DEVELOPMENT

Claudio D. Carrasco, Joleen S. Garcia, and James W. Golden

Department of Biology
Texas A&M University
College Station, Texas 77843-3258

Key Words

cyanobacteria, site-specific recombination, integrase, transcription, nitrogen fixation, uptake hydrogenase

1. SUMMARY

Anabaena PCC 7120 heterocysts undergo at least three programmed DNA rearrangements that affect the nitrogen-fixation genes *nifD* and *fdxN*, and the uptake hydrogenase gene *hupL* RT-PCR showed that *hupL* is expressed only after heterocysts have formed. An open reading frame that encodes a product similar to *hupS* or *hupU* is upstream of *hupL*. A 10.5-kb DNA element interrupts the *hupL* open reading frame and is excised during heterocyst differentiation. Correct expression of the presumed *hupSL* operon requires both transcriptional regulation and excision of the element. The *hupL* element encodes its own site-specific recombinase, XisC, which is 46% identical to the predicted XisA recombinase present on the PCC 7120 *nifD* element. The XisA and XisC proteins show sequence similarity to the catalytic domains of the integrase family of site-specific recombinases and contain the absolutely conserved catalytic Tyr residue. However, the *Anabaena* recombinases represent a distinct subset of the integrase family.

2. INTRODUCTION

In *Anabaena* sp. strain PCC 7120, DNA elements are excised from within the *nifD* and *fdxN* nitrogen-fixation genes during heterocyst development by site-specific recombination (Golden et al., 1987; Golden et al., 1985; Mulligan and Haselkorn, 1989). Recently, a third developmentally regulated DNA rearrangement was found that results in the excision of an element from a hydrogenase gene (Carrasco et al., 1995). *Anabaena* PCC 7120 is a filamentous photosynthetic cyanobacterial strain that is able to fix atmospheric nitrogen. Under nitrogen-limiting conditions, *Anabaena* PCC 7120 grows as a simple multicellular organism composed of two interdependent cell types: vegetative cells and heterocysts. Heterocysts are highly specialized, terminally differentiated cells that supply vegetative cells with fixed nitrogen, probably in the form of glutamine (Wolk et al., 1994). Heterocysts provide the microaerobic environment required for nitrogen fixation and uptake hydrogenase. The heterocyst envelope consists of a polysaccharide layer that surrounds an inner glycolipid layer (Wolk, 1996). Both layers are external to the outer cell membrane and provide a barrier that limits the oxygen rate of entry (Wolk, 1996). Heterocysts are connected to vegetative cells via narrow pores. These pores are not obstructed by the glycolipid layer and allow for the movement of reductant and fixed nitrogen between the two cell types (Wolk, 1996).

3. EXPERIMENTAL

Anabaena sp. strain PCC 7120 was grown and induced to form heterocysts as described by Golden et al. (1991). Heterocysts were isolated as described by Carrasco et al. (1994).

DNA manipulations and recombinant DNA techniques were performed by standard procedures (Ausubel et al., 1994). *Anabaena* PCC 7120 DNA was prepared from vegetative cells, purified heterocysts, and induced samples as described by Golden et al. (1991). Restriction endonucleases and other DNA-modifying enzymes were used according to the supplier's recommendations or standard protocols. For Southern analysis, DNA was transferred from agarose gels to MagnaCharge membrane (MSI) with 50 mM NaOH–1 M NaCl. DNA fragments were labeled with a random primer kit (Boehringer Mannheim), and Southern hybridization was performed as previously described by Golden et al. (1991).

RNA for Northern blot analysis and RT-PCR were prepared from frozen samples of vegetative cells, heterocysts, and heterocyst-induction samples as described by Golden et al. (1991).

4. RESULTS AND CONCLUSIONS

A 10.5-kb DNA element interrupts an open reading frame in *Anabaena* PCC 7120 vegetative cells that shows homology to *hupL* genes (Carrasco et al., 1995). *hupL* genes encode the large subunit of [NiFe] hydrogenases (Przybyla et al., 1992; Wu and Mandrand, 1993). The predicted *Anabaena* PCC 7120 HupL polypeptide is homologous to a large number of H_2-uptake [NiFe] hydrogenases; for example, it is 55% similar and 31% identical to the membrane-bound HupL of *Rhodobacter capsulatus* (Leclerc et al., 1988), and 53% similar and 31% identical to the periplasmic HydB from *Desulfovibrio fructosovorans* (Rousset et al., 1990). The conserved nickel-binding motifs found in the N- and C-terminal domains of [NiFe] hydrogenase large subunits are present in the *Anabaena* PCC 7120 protein. However, the *Anabaena* PCC 7120 C-terminal motif contains a serine in a position occupied by a proline in other hydrogenases (Wu and Mandrand, 1993).

The *Anabaena* PCC 7120 *hupL* gene shows transcriptional regulation during heterocyst induction similar to that of the *nifH* gene (Carrasco et al., 1995). The *nifHDK* operon is transcribed late during heterocyst induction (Golden et al., 1991) in differentiating heterocysts (Elhai and Wolk, 1990). Northern blot analysis of *hupL* expression with a random primer-labeled DNA probe or a high-specific-activity RNA probe showed very weak hybridization signals to degraded transcripts only in total RNA samples isolated from late stages of heterocyst induction and from purified heterocysts. These results suggested that the *hupL* gene is developmentally regulated, and showed that *hupL* is only weakly expressed or that its message has a relatively short half-life compared to the *nifH* message. RT-PCR was used to detect *hupL* message during heterocyst development to confirm the Northern analysis results. PCR-amplified products from reverse transcribed cDNA for *nifH* and *hupL* showed similar patterns of expression during heterocyst induction. For both messages, RT-PCR product was produced from total RNA isolated from heterocysts and from filaments 30 h after nitrogen step-down. This shows that *hupL* message is present only under heterocyst-inducing conditions and suggests that its expression is developmentally regulated similarly to the *nif* genes.

Sequence analysis of the regions near the *hupL* element recombination sites identified the *xisC* gene, whose predicted amino acid sequence shows similarity to XisA, the PCC 7120 *nifD*-element recombinase (Carrasco et al., 1995). The carboxy-terminal (C-terminal) region of XisC showed limited similarity to the integrase family of site-specific recombinases. Comparison of the C-terminal regions of XisC, and analysis of the *xisA* sequence with a codon usage program, indicated that the *xisA* sequence had errors at its 3' end. Correction of these errors resulted in extension of the *xisA* open reading frame from 399 to 472 amino acids. Comparison of the predicted XisC amino acid sequence to the corrected XisA sequence showed that they can be aligned throughout their length and are 64% similar and 46% identical. The corrected XisA protein now shows sequence similarity to conserved regions in the C-terminal catalytic domain of the integrase family of site-specific recombinases.

Representative members of the Integrase family of site-specific recombinases include the Int protein of phage Lambda, the Cre protein of phage P1, and Flp from the 2 μ circle plasmid of *Saccharomyces cerevisiae* (Argos et al., 1986; Kwon et al., 1997). This family of recombinases contains the invariant residues Arg, His, Arg, and Tyr in their C-terminal region (Argos et al., 1986; Kwon et al., 1997). These four residues have been shown to form part of the catalytic domain of Int and Flp. The Tyr becomes covalently linked to DNA during recombination and the Arg-His-Arg triad is strictly conserved and may play a role in DNA cleavage or ligation (Chen et al., 1992; Han et al., 1994; Kwon et al., 1997; Pargellis et al., 1988; Parsons et al., 1988). The XisC and XisA proteins contain the conserved Arg from region I and the conserved Arg and Tyr from region II. However, unlike all other members of the Int family (Argos et al., 1986; Kwon et al., 1997), the conserved His residue in region II is replaced by a Tyr in both XisC and XisA. Therefore, XisA and XisC represent a distinct subclass of the integrase family.

The 3' end of a partial open reading frame that showed similarity to *hupS*, which encodes the small subunit of [NiFe] hydrogenases, was identified upstream of the *hupL* gene (Carrasco et al., 1995). Additional DNA sequence was obtained upstream of *hupL*. An open reading frame was identified whose predicted product showed similarity to the small subunit of several uptake hydrogenases including HydA from *Desulfovibrio gigas* (55% similarity and 39% identity) and HupS from *R. capsulatus* (49% similarity and 30% identity) (Leclerc et al., 1988; Li et al., 1987). This predicted *Anabaena* PCC 7120 HupS protein also showed 62% similarity and 38% identity to HupU, a 339 amino acid protein, from

Bradyrhizobium japonicum (Black et al., 1994). *hupU* forms part of an operon with *hupV* in *B. japonicum* and of the *hupTUV* operon in *R. capsulatus* (Black et al., 1994; Elsen et al., 1996). HupU, along with HupV, may be involved in the regulation of uptake hydrogenase and could possibly function by sensing the concentration of hydrogen, oxygen, nickel, or all three ions (Black et al., 1994; Elsen et al., 1996). The small subunit of uptake hydrogenases, including HydA, contain several Cys residues involved in the formation of Fe-S clusters and a leader peptide that is needed for translocation to or across the membrane. HupU proteins contain the Cys arrangement needed to form the Fe-S clusters, but lack the leader peptide. The *Anabaena* PCC 7120 HupS protein is 321 amino acids, contains the Cys arrangement, and lacks the leader peptide. However, it is possible that the uptake hydrogenase in *Anabaena* PCC 7120 is located in the cytoplasm instead of the periplasmic membrane. There is evidence that suggests that uptake hydrogenase is located in the cytoplasm of the cyanobacterial strain *Anabaena variabilis* (Serebriakova et al., 1994).

The difficulty in clearly establishing the identity of the *Anabaena* PCC 7120 *hupS* gene prompted us to reexamine the *hupL* sequence. The predicted HupL peptide shows 56% similarity and 34% identity to *hupV* from *B. japonicum* (Black et al., 1994), which is comparable to the percent similarity and identity that it showed to hydrogenase large subunit proteins. The *Anabaena* PCC 7120 HupL contains the Ni-binding sites and the C-terminal sequence that are characteristic of the large subunit of uptake hydrogenases. The C-terminal sequence is cleaved during enzyme maturation (Wu and Mandrand, 1993). However, the C-terminal, Ni-binding motifs contain a serine in a position occupied by an invariant proline in other hydrogenases (Carrasco et al., 1995; Wu and Mandrand, 1993). The *Anabaena* PCC 7120 HupL appears to be more closely related to the large subunit of uptake hydrogenase than to HupV because of its C-terminal sequence.

We cannot unambiguously identify the *Anabaena* PCC 7120 genes as either *hupSL* or *hupUV*–this assignment awaits further studies. Analysis of an *xisC* knockout mutant strain, AMC414 (unpublished data), could help determine the function and identity of the *Anabaena* PCC 7120 *hup* genes that are affected by the DNA rearrangement. Measuring the levels of uptake hydrogenase protein and activity in AMC414 could determine whether these genes produce, repress, or activate uptake hydrogenase in *Anabaena* PCC 7120 heterocysts.

ACKNOWLEDGMENTS

This work was supported by Public Health Service grant GM36890 from the National Institutes of Health.

REFERENCES

Argos, P., Landy, A., Abremski, K., Egan, J.B., Haggard-Ljungquist, E., Hoess, R.H., Kahn, M.L., Kalionis, B., Narayana, S.V.L., Pierson III, L.S., Sternberg, N., and Leong, J.M., 1986, The integrase family of site-specific recombinases: regional similarities and global diversity, *EMBO J.*, 5:433–440.

Ausubel, F.M., Brent, R., Kingston, R.E., Moore, D.D., Seidman, J.G., Smith, J. A., and Struhl, K., 1994, *Current protocols in molecular biology*, Greene Publishing Associates and Wiley-Interscience, New York.

Black, L.K., Fu, C., and Maier, R.J., 1994, Sequences and characterization of *hupU* and *hupV* genes of *Bradyrhizobium japonicum* encoding a possible nickel-sensing complex involved in hydrogenase expression, *J. Bacteriol.*, 176:7102–7106.

Carrasco, C.D., Buettner, J.A., and Golden, J.W., 1995, Programmed DNA rearrangement of a cyanobacterial *hupL* gene in heterocysts, in *Proceedings of the National Academy of Sciences (USA)*, 92:791–795.

Carrasco, C.D., Ramaswamy, K.S., Ramasubramanian, T.S., and Golden, J.W., 1994, *Anabaena xisF* gene encodes a developmentally regulated site-specific recombinase, *Genes Dev.*, 8:74–83.
Chen, J.W., Lee, J., and Jayaram, M., 1992, DNA cleavage in trans by the active site tyrosine during Flp recombination: switching protein partners before exchanging strands, *Cell*, 69:647–658.
Elhai, J., and Wolk, C.P., 1990, Developmental regulation and spatial pattern of expression of the structural genes for nitrogenase in the cyanobacterium *Anabaena*, *EMBO J.*, 9:3379–3388.
Elsen, S., Colbeau, A., Chabert, J., and Vignais, P.M., 1996, The *hupTUV* operon is involved in negative control of hydrogenase synthesis in *Rhodobacter capsulatus*, *J. Bacteriol.*, 178:5174–5181.
Golden, J.W., Mulligan, M.E., and Haselkorn, R., 1987, Different recombination site specificity of two developmentally regulated genome rearrangements, *Nature* (London), 327:526–529.
Golden, J.W., Robinson, S.J., and Haselkorn, R., 1985, Rearrangement of nitrogen fixation genes during heterocyst differentiation in the cyanobacterium *Anabaena*, *Nature* (London), 314:419–423.
Golden, J.W., Whorff, L.L., and Wiest, D.R., 1991, Independent regulation of *nifHDK* operon transcription and DNA rearrangement during heterocyst differentiation in the cyanobacterium *Anabaena* sp. strain PCC 7120, *J. Bacteriol.*, 173:7098–7105.
Han, Y.W., Gumport, R.I., and Gardner, J.F., 1994, Mapping the functional domains of bacteriophage lambda integrase protein, *J. Mol. Biol.*, 235:908–25.
Kwon, H.J., Tirumalai, R., Landy, A., and Ellenberger, T., 1997, Flexibility in DNA recombination—structure of the lambda integrase catalytic core, *Science*, 276:126–131.
Leclerc, M., Colbeau, A., Cauvin, B., and Vignais, P.M., 1988, Cloning and sequencing of the genes encoding the large and the small subunits of the H_2 uptake hydrogenase (*hup*) of *Rhodobacter capsulatus*, *Molecular & General Genetics*, 214:97–107.
Li, C., Peck Jr., H.D., LeGall, J., and Przybyla, A.E., 1987, Cloning, characterization, and sequencing of the genes encoding the large and small subunits of the periplasmic [NiFe]hydrogenase of *Desulfovibrio gigas*, *DNA*, 6:539–551.
Mulligan, M.E., and Haselkorn, R., 1989, Nitrogen-fixation (*nif*) genes of the cyanobacterium *Anabaena* sp. strain PCC 7120: the *nifB-fdxN-nifS-nifU* operon, *J. Biol. Chem.*, 264:19200–19207.
Pargellis, C.A., Nunes-Duby, S.E., de Vargas, L.M., and Landy, A., 1988, Suicide recombination substrates yield covalent lambda integrase-DNA complexes and lead to identification of the active site tyrosine, *J. Biol. Chem.*, 263:7678–7685.
Parsons, R.L., Prasad, P.V., Harshey, R.M., and Jayaram, M., 1988, Step-arrest mutants of FLP recombinase: implications for the catalytic mechanism of DNA recombination, *Mol. Cell. Biol.*, 8:3303–3310.
Przybyla, A.E., Robbins, J., Menon, N., and Peck Jr., H.D., 1992, Structure-function relationships among the nickel-containing hydrogenases, *FEMS Microbiol. Rev.*, 8:109–135.
Rousset, M., Dermoun, Z., Hatchikian, C.E., and Belaich, J.P.P., 1990, Cloning and sequencing of the locus encoding the large and small subunit genes of the periplasmic [NiFe] hydrogenase from *Desulfovibrio fructosovorans*, *Gene*, 94:95–101.
Serebriakova, L., Zorin, N.A., and Lindblad, P., 1994, Reversible hydrogenase in *Anabaena variabilis* ATCC 29413: presence and localization in non-N_2 fixing cells, *Arch. Microbiol.*, 161:140–144.
Wolk, C.P., 1996, Heterocyst formation, *Annu. Rev. Genet.*, 30:59–79.
Wolk, C.P., Ernst, A., and Elhai, J., 1994, Heterocyst metabolism and development, in *The Molecular Biology of Cyanobacteria*, Bryant, D.A. (ed.), Kluwer Academic Publishers, Dordrecht, pp. 769–823.
Wu, L.F., and Mandrand, M.A., 1993, Microbial hydrogenases: primary structure, classification, signatures, and phylogeny, *FEMS Microbiol. Rev.*, 10:243–269.

28

CONSTRUCTION OF TRANSCONJUGABLE PLASMIDS FOR USE IN THE INSERTION MUTAGENESIS OF *Nostoc* PCC 73102 UPTAKE HYDROGENASE

Alfred Hansel, Fredrik Oxelfelt, and Peter Lindblad

Department of Physiological Botany
Uppsala University
Villavägen 6, S-752 36 Uppsala, Sweden

Key Words

uptake hydrogenase, *hupSL* genes, *hupL*-inactivation mutant, conjugation, triparental mating, cyanobacteria

1. SUMMARY

The N_2-fixing, filamentous, heterocystous cyanobacterium *Nostoc* PCC 73102 is capable of forming symbiotic associations with a variety of organisms, but can also grow independently from its hosts photoautotrophically as well as heterotrophically. During the N_2-fixation process, molecular hydrogen is produced, which is metabolized by an uptake hydrogenase. In contrast to several other cyanobacteria, *Nostoc* PCC 73102 seems to possess no bidirectional (reversible) hydrogenase. On the other hand, *Nostoc* cells grown under N_2-fixing conditions show hydrogen uptake activity. Recently, genes homologous to *hupSL* in *Anabaena* PCC 7120, as well as genes encoding uptake hydrogenases in several other bacteria, were cloned in our laboratory. They might encode the large and small subunit of a *Nostoc* PCC 73102 uptake hydrogenase, respectively. Investigation of a mutant lacking a functional uptake hydrogenase in comparison with the wild type could provide insight into the role of this enzyme in *Nostoc*. We present here the construction of vectors for use in horizontal gene transfer experiments. They might allow the insertional inactivation of the uptake hydrogenase by single- or double-crossover in this cyanobacterial strain.

2. INTRODUCTION

Cyanobacteria are a morphologically diverse but phylogenetically cohesive group of prokaryotes that are phenotypically united by their ability to perform oxygen-evolving, plant-type photosynthesis (Cohen et al., 1994; Summers et al., 1995). Many cyanobacteria fix N_2 in order to deal with the lack of combined nitrogen in their environment. Under such conditions, the formation of a small fraction of specialized cells, the heterocysts, may be induced in several filamentous strains (Wolk, 1982). The process of N_2 fixation leads to a by-product, molecular hydrogen. Therefore, filamentous cyanobacteria able to fix nitrogen have been cultivated in bioreactors for the biological conversion of water to hydrogen (Rao and Hall, 1996). However, the conversion efficiencies achieved are low, as the net H_2 production is the result of H_2 evolution by nitrogenase(s) and its recycling by mainly uptake hydrogenase(s) (Rao and Hall, 1996; Schulz, 1996). Such uptake hydrogenases are found in most N_2-fixing bacteria (for a review, see Maier and Triplett, 1996). Additionally, cyanobacteria contain bidirectional hydrogenases that might catalyze both, the reduction as well as oxidation of H_2 (Smith, 1990).

Nostoc PCC 73102 has the ability to grow heterotrophically in complete darkness (Rippka et al., 1979). This strain was originally isolated from symbiotic associations with the cycad *Macrozamia sp.* (Rippka et al., 1979). It can, by hormogonia infection, reconstitute symbiotic associations with the bryophyte *Anthoceros punctatus* (Campbell and Meeks, 1989) and the angiosperm *Gunnera manicata* (Johansson and Bergman, 1994). As *Nostoc* PCC 73102 is able to fix N_2, the in vivo hydrogen-uptake activity in this strain was examined by measurements with an H_2 electrode. The influence on the H_2 uptake of factors such as the substrate itself, the putative cofactor nickel, as well as carbon and nitrogen, could be demonstrated (Oxelfelt et al., 1995). Subsequently, it was shown by in vivo and in vitro physiological methods as well as molecular techniques that *Nostoc* PCC 73102, in contrast to the other strains used in that study (*Anabaena* PCC 7120, *Anabaena variabilis*, and *Nostoc muscorum*), possesses no bidirectional hydrogenase activity (Tamagnini et al., 1997). Recently, genes encoding cyanobacterial hydrogenases have been cloned in some unicellular as well as filamentous strains: bidirectional hydrogenase genes in *Synechococcus* PCC 6301 (Boison et al., 1996), *Anabaena variabilis* (Schmitz et al., 1995), and *Synechocystis* PCC 6803 (Appel and Schulz, 1996); and uptake-hydrogenase gene *hupL* in *Anabaena* PCC 7120 (Carrasco et al., 1995). Parts of these genes were utilized as probes in Southern hybridizations with *Nostoc* PCC 73102 DNA. No DNA regions homologous to bidirectional hydrogenase genes could be detected. On the other hand, the same studies indicated the presence of DNA regions homologous to genes encoding uptake hydrogenases (Tamagnini et al., 1997). Subsequently, the two genes *hupS* and *hupL*, putatively encoding an uptake-hydrogenase in *Nostoc* PCC 73102, were cloned in our laboratory (Oxelfelt et al., 1998).

The physiological versatility of *Nostoc* PCC 73102, together with the fact that this strain seems to possess no bidirectional hydrogenase activity/genes, makes it an interesting candidate for studying the effects of a deletion mutant lacking the uptake-type hydrogenase encoded by the cloned *hupSL* genes. With the development of protocols making use of horizontal DNA transfer, gene inactivations are also possible in bacteria that are not taking up DNA from their environment naturally. Protocols for conjugation of *Escherichia coli* with several cyanobacteria were developed and utilized with good success (Elhai and Wolk, 1988). *Nostoc* PCC 73102 belongs to this group of organisms (Flores and Wolk, 1985; Cohen et al., 1994; Summers et al., 1995). Physiological studies of a deletion mutant will show if *Nostoc* PCC 73102 possesses alternative biochemical pathways to recycle the H_2 generated by nitrogenase activity, or if these mutants then evolve molecular hydrogen.

3. EXPERIMENTAL

3.1. Bacterial Strains, Plasmids, and Growth Media

The bacterial strains and plasmids used in this study are listed in Table 1. *E. coli* strains were grown in liquid or on agar-solidified (1.5% w/v Bacto agar, Difco) Luria broth (LB; Sambrook et al., 1989). Cell concentrations were determined by measurement of the optical density of cultures at 600 nm (A_{600}). Concentrations of antibiotics in the medium, if required for plasmid selection, were as follows: Ampicillin (Ap) 100 mg/L; Kanamycin (Km) 50 mg/L; Streptomycin (Sm) 25 mg/L; or Spectinomycin (Sp) 100 mg/L. All antibiotics were purchased from Sigma. *E. coli* XL1-Blue was utilized as a host for transformations during plasmid constructions.

Nostoc sp. PCC 73102 (type strain of *Nostoc punctiforme*; Rippka and Herdman, 1992) was grown in buffered (10 mM HEPES) liquid or on agar-solidified (1% w/v) BG11$_0$ medium (Rippka et al., 1979), supplied with 5 mM NH$_4$Cl ($\rightarrow$ BG11 medium). To select for resistance in *Nostoc* PCC 73102, Streptomycin (2 mg/L), Spectinomycin (10

Table 1. Bacterial strains and plasmids

Bacterial strain or plasmid	Characteristics	Source or reference
A. Bacterial strain		
Nostoc sp. PCC 73102	Type strain of *Nostoc punctiforme*	Rippka and Herdman, 1992
E. coli HB101	F$^-$ *recA13 bsdS20 supE44* (r$_B^-$ m$_B^-$)	Sambrook et al., 1989
E. coli XL1-Blue	*RecA1 endA1* Tn*10* (Tc^R)	Stratagene
E. coli AM1460	HB101 containing the conjugal plasmid pRK2013	N. Tsinoremas, United States
B. Plasmid		
pBluescriptII SK	pUC19-derived vector (Ap^R)	Stratagene
pGEM3Z	pUC18-derived vector (Ap^R)	Promega
pBR322	Cloning vector (Ap^R, Tc^R)	Bolivar et al., 1977
pBR322K	pBR322 with KmR casette from pUC4K in *Pst*I-site (Km^R, Tc^R)	This study
pBR322S	pBR322 with SmR casette from pDW9 in *Pst*I-site (Sm^R, Tc^R)	This study
pRK2013	RK2 derivative with ColE1 *oriV* (Km^R)	Figurski and Helinski, 1979
pDS4101	Conjugal helper plasmid encoding *mob* gene	Twigg and Sherratt, 1980
pDW9	pUC18 derivative allowing isolation of Sm^R Ω fragment with enzymes cutting pUC18/19 MCS (Sm^R)	Golden and Weist, 1988
pUC4K	Carries Tn*903* Km^R GenBlock (Km^R)	Pharmacia
pFO1	pBluescriptII SK with 2.8-kB fragment encoding *hupL* and 3´half of *hupS* in *Eco*RI/*Hind*III (Ap^R)	Oxelfelt, unpublished
pAL500	pBR322 carrying 0.7-kB *Mun*I *hupL* fragment in EcoRI site (Ap^R, Tc^R)	This study
pAL555	pGEM3 derivative carrying 2.8-kB *hupSL* fragment from pFO1 in *Pst*I/*Sal*I	This study
pAL726	pAL500 derivative with Sm^R from pDW9 in *Pst*I site	This study
pAL850	pBR322K carrying EcoRI *hupSL* fragment from pAL555	This study
pAL990	*Eco*RI restricted Sm^R Ω casette ligated into 7.8-kB *Mun*I fragment from pAL850	This study

mg/L), or Neomycin (Nm; 10 mg/L) was added to the medium (Cohen et al., 1994; Golden and Wiest, 1988).

3.2. DNA Isolations and Manipulations

Small-scale preparations of plasmid DNA from *E. coli* were performed using standard protocols (Sambrook et al., 1989). Large-scale plasmid purifications from *E. coli* were carried out using a commercial kit, following the instructions of the supplier (Qiagen). All enzymes were purchased from Pharmacia, except for restriction enzyme *Mun*I (Gibco BRL).

3.3. DNA Transfer Methods

Plasmids were transformed into *E. coli* cells using standard protocols (Sambrook et al., 1989). DNA was mobilized into *Nostoc* by horizontal transfer from *E. coli* (Cohen et al., 1994). Tri-parental matings were conducted on sterile conjugation filters (HATF 082; Millipore) placed on BG11-agar plates containing 0.5% (v/v) LB medium. The bacteria involved in the matings were prepared as follows:

a. The *E. coli* strains, either containing the conjugal plasmid pRK2013 (conjugal strain AM1460; obtained from N. Tsinoremas, United States) or the helper plasmid pDS4101 (obtained from J. Elhai, United States; Twigg and Sheratt, 1980) in combination with the plasmid bearing the interrupted *hupL* gene (donor strain), were grown to an $A_{600} < 1$ in LB supplemented with the antibiotics required for selection. Equal volumes of the cells were mixed, pelleted by centrifugation (2000 x g for 10 min), and resuspended to a final concentration of 9–10 A_{600}.
b. *Nostoc* filaments were fragmented to lengths of 4–6 cells/filament by sonic cavitation. For regeneration, the fragmented cells were cultivated for 6 h in BG11 medium before being harvested (2000 x g, 10 min) and resuspended to obtain a 10x concentrated cell suspension in fresh BG11 medium.

For each conjugation filter, 0.5 mL each of *Nostoc* and *E. coli* cells were mixed prior to centrifugation at 4000 x g for 30 s. The pelleted cells were resuspended in ca. 200 µL of the supernatant and the suspension immediately spread onto the filters. The plates were incubated at 25 °C at low light intensity (ca. 1.5 W/m^2).

3.4. Screening for Recombinants

To allow the expression of resistances, the filters were transferred to fresh BG11 plates, but without LB, and incubated at low light intensity for another 2 days. The plates were then moved to high light intensity (8 W/m^2) for an additional two-day incubation. The filters were transferred to selective plates containing either neomycin or spectinomycin. At approximately one-week intervals, the filters were transferred to new plates. Mutants are expected to appear as colonies after 10–30 days (Cohen et al., 1994).

4. RESULTS AND CONCLUSION

4.1. General Considerations

During our studies concerning the hydrogen metabolism of the filamentous heterocyst-forming cyanobacterium *Nostoc* PCC 73102, we discovered that this strain exhibits

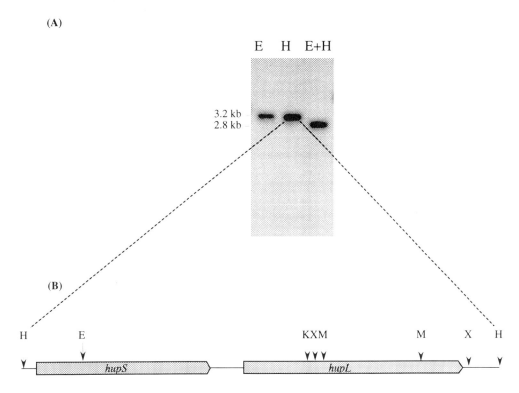

Figure 1. (A) Southern blot hybridization of *Nostoc* PCC 73102 genomic DNA with a probe derived from the *Anabaena* PCC 7120 *hupL* gene (Carrasco et al., 1995). Total genomic DNA isolated from *Nostoc* was digested with *Eco*RI (E), *Hin*dIII (H), or double-digested using both enzymes (E+H). The sizes of the hybridizing fragments are indicated. (See also Tamagnini et al., 1997.) (B) Restriction map of the *Nostoc* sp. PCC 73102 DNA region corresponding to the hybridization signals in Figure 1A. The 3.2-kB *Hin*dIII hybridizing fragment was subcloned from a cosmid cloned from a genomic library consisting of *E. coli* clones harboring SuperCos cosmids with partially digested *Sau*3AI *Nostoc*-DNA fragments ranging in sizes from 10 to 40 kb. Restriction and sequence analysis of this fragment led to the establishment of the physical map. The sequence of the DNA fragment is deposited in GenBank (Accession number AF030525).

H_2 uptake activity in vivo (Oxelfelt et al., 1995). However, no bidirectional hydrogenase activity could be detected (Tamagnini et al., 1997). In Southern hybridizations using parts of the *Anabaena variabilis* ATCC 29413 *hoxXY* genes (Schmitz et al., 1995) as probes, no signals could be detected with *Nostoc* DNA, in contrast to the other strains tested (Tamagnini et al., 1997). On the other hand, hybridization signals could be observed with a probe deduced from the *hupL* gene from *Anabaena* PCC 7120 (Carrasco et al., 1995) using DNA isolated from this strain (Figure 1a; Tamagnini et al., 1997).

Our further molecular studies led to the cloning of *hupSL* genes putatively encoding the small and large subunit of an uptake hydrogenase in this strain (Oxelfelt et al., 1998). Restriction and sequence analysis of a cloned 3.3-kB *Hin*dIII fragment encoding the complete *hupS* and *hupL* genes resulted in the establishment of its physical map (Figure 1b). Using this information, a plasmid containing the *hupL* open reading frame interrupted by a casette conferring resistance to Sm/Sp was constructed, applicable in double-crossover experiments (Cohen et al., 1994; Summers et al., 1995). An alternative to using double-crossover plasmids for gene inactivation is the construction of single-crossover plasmids,

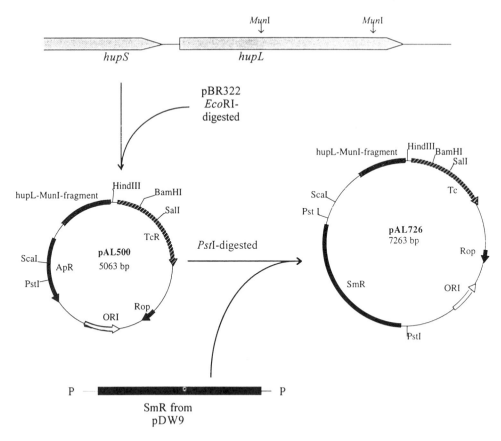

Figure 2A. Construction of the single-crossover plasmid pAL720: the 0.7-kB *Mun*I fragment from *hupL* (see also restriction map, Figure 1B) was ligated into the *Eco*RI site of pBR322S, which carried the Sm^R/Sp^R casette from pDW9 in its *Pst*I site.

which contain a fragment of the gene of interest lacking the 5' and 3' end, flanked by a resistance gene (Tsinoremas et al., 1994; Thiel, 1994). As selective markers, we used either the Km^R/Nm^R gene from the plasmid pUC4K (Pharmacia), or the Sm^R/Sp^R gene from the Ω fragment of pHP45Ω. This fragment was cloned into a vector allowing its isolation by using all of the restriction sites present in the multiple cloning site of pUC18/19 except *Sph*I (plasmid pDW9, obtained from J. Golden; Golden and Wiest, 1988).

The plasmid vector pBR322, which contains an *oriT*, was chosen as the carrier of the casette-insertion construct into *Nostoc*. The presence of the *oriT*, or *bom* site, a region that can be nicked by the product of the *mob* gene on helper plasmids, is a prerequisite for mobilization (Thiel, 1994). Besides, the plasmids should not be able to replicate in the host. The vector pBR322 (Bolivar et al., 1977) fulfills these requirements, as it contains an *oriT*, but possesses no replication origin functioning in cyanobacteria (Thiel, 1994).

4.2. Single-Crossover Experiments

The plasmid pAL726 for use in single-crossover experiments was constructed by cloning the *Mun*I fragment from pFO1 into the *Eco*RI site of pBR322S, which contained the Ω fragment from pDW9 cloned into the *Pst*I site (Figure 2A). The resulting plasmid

Construction of Transconjugable Plasmids

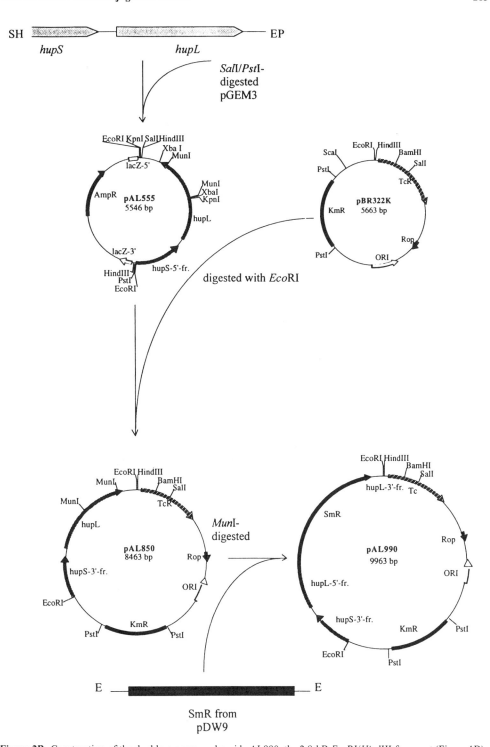

Figure 2B. Construction of the double-crossover plasmid pAL990: the 2.8-kB EcoRI/HindIII fragment (Figure 1B) was removed from the plasmid pFO1 by restriction with PstI (P) and SalI (S), and cloned into pGEM3 digested with the same enzymes. The fragment was removed from the resulting plasmid pAL555 by restriction with EcoRI and then placed in the EcoRI site of pBR322K, which contained a Km^R cassette in the PstI site. The resulting plasmid was digested with MunI and the Sm^R/Sp^R casette, removed from pDW9 with EcoRI (E), and ligated into the MunI sites.

was transformed into the *E. coli* strain XL1Blue containing the helper plasmid pDS4101. This plasmid contains a *mob* gene, the product of which is required for mobilization of pBR322 derivatives with *oriT* (Thiel, 1994). The resulting strain was used in triparental matings with the strain AM1460, an *E. coli* HB101 strain containing the conjugal plasmid pRK2013 (Figurski and Helinski, 1979), and with *Nostoc* PCC 73102. In this case, recombination of the homologous parts on the cargo plasmid with the chromosome should lead to inactivation of the gene by producing a merodiploid. Such a mutant has two incomplete copies of the gene, interrupted by the plasmid DNA (Tsinoremas et al., 1994). The plasmid DNA then confers a selective resistance. It can be detected by Southern hybridization. Until now, no positive conjugants could be obtained by mobilization of pAL726 into *Nostoc* PCC 73102.

4.3. Double-Crossover Experiments

The plasmids for double-crossover experiments were constructed as follows (Figure 2B): a 2.8-kb *Eco*RI/*Hin*dIII subfragment containing the 3'-half of *hupS* and the complete *hupL* gene (see Figure 1a), was ligated into pBluescript, predigested with the same enzymes and dephosphorylated. The insert was restricted from the resulting plasmid with the enzymes *Pst*I and *Sal*I, and ligated into pGEM3 digested with the same enzymes. The fragment can be recovered from the resulting plasmid pAL555 by digestion with either *Eco*RI or *Hin*dIII instead of using both enzymes. Obtained fragments were then ligated into plasmid pBR322K, which contained the Km^R casette from pUC4K ligated into the *Pst*I site of pBR322, or plasmid pBR322S containing the Sm^R/Sp^R gene cloned into the same restriction site. The presence of one of the resistance casettes in the vector provides a basis for distinguishing between double-crossover mutants and merodiploids. Merodiploids are the results of single-crossover events and contain a complete as well as an incomplete copy of the gene, interrupted by the vector DNA (Thiel, 1994).

Cloning of the 2.8-kB *Hin*dIII fragment from pAL555 into pBR322K led to construction of plasmid pAL850. The Sm^R/Sp^R gene from pDW9 was then recovered by restriction with *Eco*RI. pAL850 was digested with the enzyme *Mun*I, which cuts twice within the *hupL* ORF (see Figure 1B). The recognition hexamer for this enzyme is CAATTG, and the enzyme leaves a 5' overhang AATT, compatible with the overhangs created by *Eco*RI. Ligation of *Mun*I fragments into *Eco*RI sites or vice versa leads to loss of the restriction sites. This feature is usable as a negative control for successful experiments. The resulting plasmid pAL990 is used in conjugation experiments. An alternative possibility is the restriction of pAL850 with *Xba*I. This enzyme cuts once within the *hupL* ORF, once downstream from its stop codon (see Figure 1B). The Sm^R/Sp^R gene from pDW9 in this case is recovered by restriction with the same enzyme. Ligation of the casette into the 7.5-kB fragment obtained from pAL850 leads to construction of pAL970, which also can be used in matings.

Triparental matings using plasmid pAL990 are presently underway. The creation and characterization of a mutant will allow a better understanding of the role of the proteins encoded by the cloned *hupSL* genes. As *Nostoc* PCC 73102 does not contain any bidirectional hydrogenase (Tamagnini et al., 1997), physiological characterization of the mutant will provide an insight into the role of the *hupSL* gene products in the recycling of H_2 produced during N_2 fixation of this strain.

ACKNOWLEDGMENTS

This work was supported by the Swedish Natural Science Research Council, Swedish National Board for Industrial and Technical Development, and Carl Tryggers and OK Environment Foundations.

REFERENCES

Appel, J., and Schulz, R., 1996, Sequence analysis of an operon of a NADP-reducing nickel hydrogenase from the cyanobacterium *Synechocystis* sp. PCC 6803 gives additional evidence for direct coupling of the enzyme to NAD(P)H-dehydrogenase (complex I), *Biochim. Biophys. Acta*, 1298:141–147.

Boison, G., Schmitz, O., Mikheeva, L., Shestakov, S., and Bothe, H., 1996, Cloning, molecular analysis, and insertional mutagenesis of the bidirectional hydrogenase genes from the cyanobacterium *Anacystis nidulans*, *FEBS Lett.*, 394:153–158.

Bolivar, F., Rodriguez, R., Greene, P., Betlach, M., Heynecker, H., Boyer, H., Crosa, J., and Falkow, S., 1977, Construction and characterization of new cloning vehicles. II. Multipurpose cloning system, *Gene*, 2:95–113.

Campbell, E.L., and Meeks, J.C., 1989, Characteristics of hormogonia formation by symbiotic *Nostoc* spp. in response to the presence of *Anthoceros punctatus* or its extracellular products, *Appl. Environ. Microbiol.*, 55:125–131.

Carrasco, C.D., Buettner, J.A., and Golden, J.W., 1995, Programmed DNA rearrangement of a cyanobacterial *hupL* gene in heterocysts, in *Proceedings of the National Academy of Sciences (USA)*, 92:791–795.

Cohen, M.F., Wallis, J.G., Campbell, E.L., and Meeks, J.C., 1994, Transposon mutagenesis of *Nostoc* sp. strain ATCC 29133, a filamentous cyanobacterium with multiple cellular differentiation alternatives, *Microbiology*, 140:3233–3240.

Elhai, J., and Wolk, C.P., 1988, Conjugal transfer of DNA to cyanobacteria, *Meth. Enzymol.*, 167:47–754.

Figurski, D.H., and Helinski. D.R., 1979, Replication of an origin containing derivative of plasmid RK2 dependent on a plasmid function provided in trans, in *Proceedings of the National Academy of Sciences (USA)*, 76: 1648–1652.

Golden, J.W., and Wiest, D., 1988, Genome rearrangement and nitrogen fixation in *Anabaena* blocked by inactivation of *xisA* gene, *Science*, 242:1421–1423.

Johansson, C., and Bergman, B., 1994, Reconstitution of the symbiosis of *Gunnera manicata* Linden: cyanobacterial specificity, *New Phytologist*, 126:643–652.

Maier, R.J., and Triplett, E.W., 1996, Toward more productive, efficient, and competitive nitrogen-fixing symbiotic bacteria, *Crit. Rev. Plant Sci.*, 15:191–234.

Oxelfelt, F., Tamagnini, P., Salema, R., and Lindblad, P., 1995, Hydrogen uptake in *Nostoc* strain PCC 73102: effects of nickel, hydrogen, carbon, and nitrogen, *Plant Physiol. Biochem.*, 33:617–623.

Oxelfelt, F., Tamagnini, P., and Lindblad, P., 1998, Hydrogen uptake in *Nostoc* strain PCC 73102: cloning and characterization of a *hupSL* homologue, *Arch. Microbiol.*, 169:267–274.

Rao, K.K., and Hall, D.O., 1996, Hydrogen production by cyanobacteria: potential, problems, and prospects, *J. Mar. Biotechnol.*, 4:10–15.

Rippka, R., and Herdman. M., 1992, Pasteur culture collection of cyanobacteria in pure axenic culture, Institute Pasteur, Paris, pp. 44–57.

Rippka, R., Deruelles, J.B., Waterbury, J.B., Herdman, M., and Stanier, R.Y., 1979, Generic assignments, strain histories, and properties of pure cultures of cyanobacteria, *J. Gen. Microbiol.*, 143:1–61.

Sambrook, J., Fritsch, E.F., and Maniatis, T., 1989, *Molecular Cloning: A Laboratory Manual*, 2nd ed., Cold Spring Harbor Laboratory Press, Cold Spring Harbor, United States.

Schmitz, O., Boison, G., Hilscher, R., Hundeshagen, B., Zimmer, W., Lottspeich, F., and Bothe, H., 1995, Molecular biological analysis of a bidirectional hydrogenase from cyanobacteria, *Eur. J. Biochem.*, 233:266–276.

Schulz, R., 1996, Hydrogenases and hydrogen production in eukaryotic organisms and cyanobacteria, *J. Mar. Biotechnol.*, 4:16–22.

Smith, G.D., 1990, Hydrogen metabolism in cyanobacteria, in *Phycotalk 2*, Kumar, H.O. (ed.), Rastogi and Company, Meerut, India, pp. 131–143.

Summers, M.L., Wallis, J.G., Campbell, E.L., and Meeks, J.C., 1995, Genetic evidence of a major role for glucose-6-phosphate dehydrogenase in nitrogen-fixation and dark growth of the cyanobacterium *Nostoc* sp. strain ATCC 29133, *J. Bacteriol.*, 177:6184–6194.

Tamagnini, P., Troshina, O., Oxelfelt, F., Salema, R., and Lindblad, P., 1997, Hydrogenases in *Nostoc* sp. strain PCC 73102, a strain lacking a bidirectional enzyme, *Appl. Env. Microbiol.*, 63:1801–1807.

Thiel, T., 1994, Genetic analysis of cyanobacteria, in *The Molecular Biology of Cyanobacteria*, Bryant, D.E. (ed.), Kluyver Academic Publishers, Dordrecht, Netherlands, pp. 581–611.

Tsinoremas, N.F., Kutach, A.K., Strayer, C.A., and Golden, S.S., 1994, Efficient gene transfer in *Synechococcus* sp. strain PCC 7942 and PCC 6301 by interspecies conjugation and chromosomal recombination, *J. Bacteriol.*, 176:6764–6768.

Twigg, A.J., and Sheratt, D., 1980, Trans-complementable copy-number mutants of plasmid ColE1, *Nature*, 283:216–218.

Wolk, C.P., 1982, Heterocysts, in *The Biology of Cyanobacteria*, Carr, N.G., and Whitton, B.A. (eds.), Blackwell Scientific Publications Ltd., London, pp. 359–386.

29

EFFECT OF EXOGENOUS SUBSTRATES ON HYDROGEN PHOTOPRODUCTION BY A MARINE CYANOBACTERIUM, *Synechococcus* sp. MIAMI BG 043511

Yao-Hua Luo,[1] Shuzo Kumazawa,[2] and Larry E. Brand[1]

[1]Marine Biology and Fisheries
Rosenstiel School of Marine and Atmospheric Science
University of Miami
4600 Rickenbacker Causeway
Miami, Florida 33149
[2]Department of Marine Science
School of Marine Sciences and Technology
Tokai University
3-20-1 Orido
Shimizu 424, Japan

Key Words

cyanobacteria, glucose, hydrogen production, nitrogenase activity, pyruvate, *Synechococcus* sp. 043511

1. SUMMARY

In this study, the effects of exogenous substrates on hydrogen production were examined using synchronously grown cells of *Synechococcus* sp. strain Miami 043511. When synchronously grown cells were used, enhancement of hydrogen production was observed by the addition of pyruvate only after about a 20-h incubation period. This enhancement, however, was not so large as observed in batch-grown, late-log phase cells. The period when the enhancement began coincided with the period when substantial cessation in hydrogen production in control cells was observed. This can be interpreted to mean that the utilization of exogenous substrate occurs only when an endogenously available carbohydrate is depleted. Although glucose and pyruvate were good substrates for hydrogen production in late-log to stationary-phase, batch-grown cells, substantial enhancement was not observed in synchronously grown cells. Pyruvate but not glucose was effective to some extent in prolonging hydrogen production in synchronously grown cells. This enhancement of hydrogen production was ascribed to prolonged activity of nitrogenase due to the addition of pyruvate.

2. INTRODUCTION

Synechococcus sp. Strain Miami 043511, a marine unicellular cyanobacterium, was shown to be one of the Miami strains with the highest rate of H_2 production (Kumazawa and Mitsui, 1989). The strain is characterized by certain properties that others may lack: high oxygen-resistance ability (Nemoto, 1995; Luo and Brand, 1997) and the absence of uptake-hydrogenase activity (Luo, unpublished data). When synchronously grown cells were used, nitrogenase activity and cellular carbohydrate content alternatively oscillated during the light and dark cycle (Mitsui et al., 1986). The oscillation can be explained by the data of Kumazawa and Mitsui (1994), who demonstrated that nitrogenase-catalyzed hydrogen production correlates with the rate of carbon dioxide evolution from the incubated cells. Our recent experiments with batch cultures of the strain have shown that the content of cellular carbohydrates changed during growth and that H_2 photoproduction capability decreased with reduction of cellular carbohydrate content. The peak of hydrogen production activity was observed when early-log phase cells were used, but it decreased substantially after the late-log phase. However, the reduced ability for H_2 production could be reactivated by the addition of exogenous substrates such as sugars, pyruvate, and glycerol (Luo and Mitsui, 1994). In this study, we examined the effect of exogenous substrates on hydrogen production by synchronously grown cells at the phase with high carbohydrate content.

3. EXPERIMENTAL

Synechococcus sp. strain Miami BG 043511 (Mitsui et al., 1986) was grown in a combined nitrogen-free and sodium bicarbonate-enriched (2.5 g/L) artificial seawater medium, A-N, at 30 °C and aerated with 4% carbon dioxide-enriched air, as described previously (Leon et al., 1986; Mitsui and Cao, 1988). After inoculation, cells were grown under continuous illumination with fluorescent lamps (an intensity of 150 µmol · m^{-2} · s^{-1} for 3 days). The cultures were then treated to promote synchronous growth for a period of 16-h dark,16-h light, 16-h dark. The composition of incubation medium was the same as that of the synchronous culture medium except with HEPES buffer (10 mM) added and sodium bicarbonate omitted. Aliquot of 5-mL cell suspensions with and without added substrate were placed into 25-mL micro-Fernbach flasks and sealed with butyl rubber stoppers for H_2 assays. The flasks were incubated in an illuminated shaker water bath at 30 °C. The amount of hydrogen evolved was determined by gas chromatography using the method described previously (Kumazawa and Mitsui, 1981). Nitrogenase activity was determined by the acetylene reduction method as described previously (Leon et al., 1986).

4. RESULTS AND DISCUSSION

The data in Figure 1 demonstrate that hydrogen production capability changes depending on the phase of cells used during batch growth (Luo and Mitsui, 1994). As al-

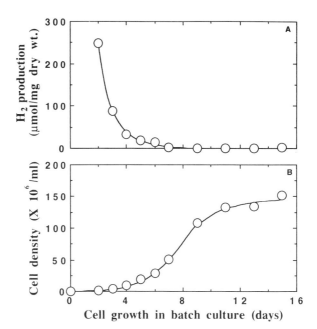

Figure 1. Changes in maximum H_2 accumulation (A) and cell density (B) during batch-culture growth of *Synechococcus* sp. strain Miami BG 043511. The data indicates a relationship between the hydrogen production capabilities in cells and cell growth. This figure is a modified version of previously published data by Luo and Mitsui (1994).

ready reported, exogenous substrates such as glucose (Figure 2A) and pyruvate did not enhance hydrogen production when early-log phase cells were used, but did enhance it substantially when late-log phase or stationary phase cells were used (Figure 2). When late log phase cells were used, enhancement of hydrogen production by glucose (Figure 2B) or pyruvate (data not shown) was about ten times higher (Luo and Mitsui, 1994). This observation was interpreted to mean that utilization of endogenous carbohydrate precedes the utilization of exogenously added substrates.

The effect of glucose and pyruvate addition on hydrogen production by synchronously grown cells is shown in Figure 3. The effect was examined at two different cell densities. Under both cell-density conditions, the active phase of hydrogen production in control cells ceased after about one day. However, the duration of hydrogen production differed under the two different cell densities. Under high cell-density conditions (Figure 3B), the rate of hydrogen production in control cells decreased substantially after one day's incubation and thereafter continued at much lower rates. On the other hand, little increase in hydrogen production in control cells was observed under low cell-density conditions (Figure 3A). The difference in the duration of hydrogen production in control cells at different cell densities could be due to the self-shading effect. More light was available under low cell-density conditions; hence, more active hydrogen production occurred during the initial phase and more carbohydrates were degraded. On the other hand, limited light under high cell-density conditions reduced the rates of hydrogen production and carbohydrate degradation; hence, hydrogen production continued for a longer period, though at a much reduced rate.

As shown in Figure 3, no enhancement of hydrogen production by the exogenous substrates was observed during the initial active hydrogen production period. Some enhancement of hydrogen production by the addition of pyruvate is visible after active hydrogen production in control cells ceased. This is consistent with previous observations using batch-grown cells (Luo and Mitsui, 1994). Using batch-grown cells, it was observed that the enhancement occurred only when endogenous carbohydrates were exhausted. At the end of the incubation shown in Figure 3, more hydrogen accumulated in vessels with

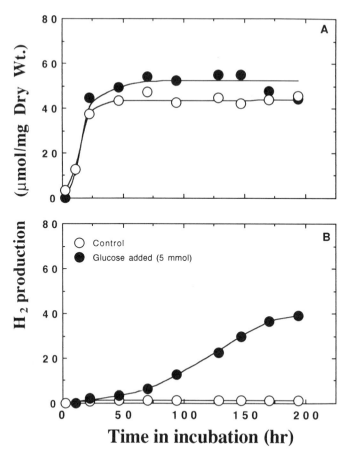

Figure 2. Effect of glucose on the hydrogen production by early-log phase (A) and late-log phase (B) cells. This figure is a modified version of previously published data by Luo and Mitsui (1994).

added substrate. Pyruvate but not glucose was to some extent effective in the sustained production of hydrogen.

The pattern of hydrogen production and the change in nitrogenase activity during incubation are shown in Figure 4. The cells for nitrogenase activity assays were harvested once at the end of synchronous culture, then the cell suspension was equally (5 mL) separated into different incubation vessels that were incubated under the same conditions. The vessels were individually removed from the incubation bank at the periods shown in Figure 4 for C_2H_2 reduction measurements. Each vessel was incubated another 30 min on the shaker after injection of C_2H_2. On the other hand, the assay for hydrogen production was based on the amount of hydrogen accumulated in the vessels from the beginning of the incubation. Thus, Figures 4A and 4B may not be tightly correlated. The maximum nitrogenase activity occurred about 8 h after incubation commenced (Figure 4B), which is consistent with the previous observations (Suda et al., 1992). As shown in the inset of Figure 4B, nitrogenase activity in control cells was somewhat higher than cells with exogenous substrates added during the initial period. This may indicate that added substrates may have some inhibitory effect on the nitrogenase activity. Higher concentrations of added substrate reduced the extent of the enhancement of hydrogen production (Luo and Mitsui, 1994). However, vessels with added pyruvate maintained higher nitrogenase activ-

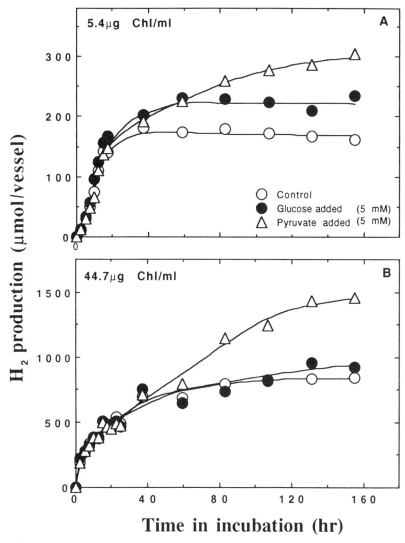

Figure 3. Effect of exogenous substrates on hydrogen production by synchronously grown cells of *Synechococcus* sp. strain Miami BG 043511. Glucose and pyruvate (final concentration, 5 mM) were added, respectively, to the cell suspensions immediately after the 48-h synchronization treatment. The effect was examined under low (5.4 µg Chl/mL, A) and high cell density (44.7 µg Chl/mL, B) conditions. ○ = control; ● = glucose-added samples; △ = pyruvate-added samples.

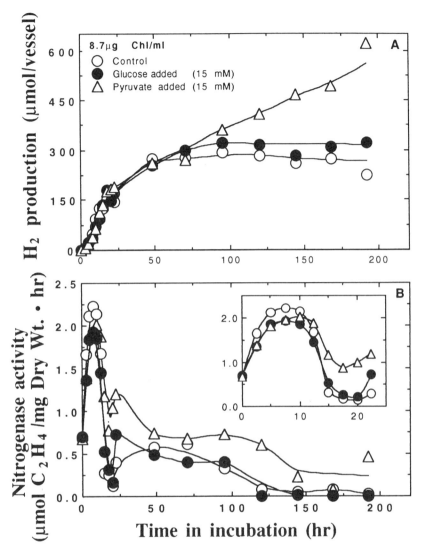

Figure 4. Pattern of hydrogen accumulation (A) and changes in nitrogenase activity (B) during incubation under hydrogen production conditions. For the measurement of nitrogenase activity (acetylene reduction activity), cells incubated under hydrogen production conditions for periods shown in the figure were removed and incubated an additional 30 min with acetylene added.

ity for longer periods. Therefore, enhancement of hydrogen production by pyruvate suggests that pyruvate serves as a substrate for nitrogenase rather than hydrogenase and, hence, nitrogenase-catalyzed hydrogen production.

ACKNOWLEDGMENTS

This work was supported partly by grants to the late professor A. Mitsui from the Research Institute of Innovative Technology for the Earth, the New Energy and Industrial

Technology Development Organization in Japan, and the U.S. Department of Energy, National Renewable Energy Laboratory.

REFERENCES

Kumazawa, S., and Mitsui, A., 1981, Characterization and optimization of hydrogen photoproduction by a saltwater blue-green alga, *Oscillatoria* sp. Miami BG 7. I. Enhancement through limiting the supply of nitrogen nutrients, *Int. J. Hydrogen Energy*, 6:339–348.
Kumazawa, S., and Mitsui, A., 1989, Algae for hydrogen generation, in *Biomass Handbook*, Kitani, O., and Hall, C.W. (eds.), Gordon and Research Science Publishers, New York, pp. 219–228.
Kumazawa, S., and Mitsui, A., 1994, Efficient hydrogen photoproduction by synchronous grown cells of a marine cyanobacterium, *Synechococcus* sp. Miami BG 043511, under high cell density conditions, *Biotechnol. and Bioeng.*, 44:854–858.
Leon, C., Kumazawa, S., and Mitsui, A., 1986, Cyclic appearance of aerobic nitrogenase activity during synchronous growth of unicellular cyanobacterium, *Synechococcus*, *Current Microbiol.*, 13:149–153.
Luo, Y.H. and Brand, L., 1997, High O_2-resistant characterization of a unicellular cyanobacterium during nitrogenase activity stage, *Hypothesis II*, Grimstad, Norway.
Luo, Y.H., and Mitsui, A., 1994, Hydrogen production from organic substrates in an aerobic nitrogen-fixing marine unicellular cyanobacterium *Synechococcus* sp. strain Miami BG 043511, *J. Biotech. Bioeng.*, 44:1255–1260.
Mitsui, A., and Cao, S., 1988, Isolation and culture of marine nitrogen-fixing unicellular cyanobacterium, *Synechococcus*, *Methods Enzymol.*, 167:105–113.
Mitsui, A., Kumazawa, S., Takahashi, A., Ikemoto, H., Cao, S, and Arai, T., 1986, Strategy by which nitrogen-fixing unicellular cyanobacteria grow photo-autotrophically, *Nature*, 323:720–722.
Nemoto, Y., 1995, Biological hydrogen photoproduction, in *Proceedings of the 1995 U.S. Department of Energy Hydrogen Program Review*, vol. II, National Renewable Energy Laboratory, Coral Gables, Florida, pp. 617–682.
Suda, S., Kumazawa, S., and Mitsui, A., 1992, Change in the H_2 photoproduction capability in a synchronously grown aerobic nitrogen-fixing cyanobacterium, *Synechococcus* sp. Miami BG 043511, *Arch. Microbiol.*, 158:1–4.

DEVELOPMENT OF SELECTION AND SCREENING PROCEDURES FOR RAPID IDENTIFICATION OF H_2-PRODUCING ALGAL MUTANTS WITH INCREASED O_2 TOLERANCE

Michael Seibert, Timothy Flynn, Dave Benson, Edwin Tracy, and Maria Ghirardi

National Renewable Energy Laboratory
Golden, Colorado 80401

Key Words

hydrogen production, green alga, *Chlamydomonas reinhardtii,* oxygen sensitivity, selection, hydrogenase

1. SUMMARY

Algal hydrogen photoproduction from water is catalyzed under anaerobic conditions by a reversible hydrogenase. However, the sensitivity of the enzyme to O_2 has precluded commercial applications. Two types of selective pressure have been developed to isolate mutants of the green alga *Chlamydomonas reinhardtii* that produce H_2 in the presence of O_2, and they depend on survival (or growth) of algal cells under either H_2-producing or H_2-uptake conditions in the presence of O_2. The application of H_2-production selective pressure yielded a variant of *C. reinhardtii* with 330% higher tolerance to O_2 (as measured by an increase in O_2 I_{50} for H_2 evolution), and, subsequently, a mutant with an additional 40% increase in O_2 I_{50}. Application of H_2-uptake selective pressure resulted in a mixture of survivors with a 50% increase in O_2 I_{50}. Our results demonstrate the efficacy of the two selective pressures in isolating desirable *C. reinhardtii* mutants; however, the selective pressures are not very specific. Thus, progress up until now has been limited due to the amount of time it takes to screen for O_2-tolerant, H_2-producing survivors. We now report a procedure based on the use of a H_2-sensitive chemochromic film that provides the potential to rapidly screen large numbers of colonies on agar plates for H_2 production capability.

BioHydrogen, edited by Zaborsky *et al.*
Plenum Press, New York, 1998

2. INTRODUCTION

Photobiological hydrogen production is catalyzed by the nitrogenase and hydrogenase enzyme systems that are present in photosynthetic bacteria, cyanobacteria, and green algae (Houchins, 1984). Nitrogenase-catalyzed reactions are quite energy-intensive, requiring adenosine triphosphate (ATP) input and a metabolic intermediate to drive either nitrogen fixation or H_2 photoproduction (Weaver et al., 1980). On the other hand, green algae photoevolve H_2 only through an inducible, reversible hydrogenase using water as the direct source of electrons (Schulz, 1996). From an overall thermodynamic perspective, algal production is potentially the most efficient photobiological process for generating H_2 (Ghirardi et al., 1996; Benemann, 1996). Two major challenges, however, currently preclude the use of algae in commercial H_2-producing systems. These are the sensitivity of hydrogenases to deactivation by O_2 produced by photosynthetic water-splitting and the low light saturation level of algal photosynthesis (Ghirardi et al., 1997). We have focused on the first challenge.

Our laboratory has developed two approaches to select for mutants of the green alga *Chlamydomonas reinhardtii* that produce H_2 in the presence of O_2. The two selective pressures depend on maintaining the reversible activities of the hydrogenase (i.e., the H_2-producing and H_2-uptake reactions) under conditions of O_2 stress. H_2-production selective pressure depends on the toxic effect of the drug metronidazole (MNZ) on algal photosynthesis in the absence of CO_2. When reduced, MNZ reacts with O_2 and releases superoxide radicals and hydrogen peroxide that kill the cells. However, hydrogenase can compete with MNZ for electrons available from light-reduced ferredoxin (Schmidt et al., 1977). If the hydrogenase is active, some of the electrons from ferredoxin will be used for H_2 production instead of MNZ reduction, and that may decrease the toxic effect of the drug (Ghirardi et al., 1997). The selective conditions that we developed require addition of O_2 to the system during MNZ treatment in order to inactivate wild-type (WT) hydrogenase. Survivors will have one of the following two phenotypes: (a) lower rates of photosynthetic electron transport to ferredoxin due to antenna or electron transport mutations (Schmidt et al., 1977), or (b) an O_2-tolerant hydrogenase that diverts electrons away from MNZ (Ghirardi et al., 1997). We have validated this selective pressure by investigating the effect of the following parameters on the number of survivors after treatment of algal cells with MNZ: drug concentration, light intensity, activation state of the hydrogenase, and imposed O_2 concentration (Ghirardi et al., 1997). Subsequently, we were able to isolate a H_2-producing variant of *C. reinhardtii*, D5, with increased tolerance to O_2 (measured as an O_2 I_{50} for H_2 evolution) by application of H_2-production selective pressure to a parental population of cw15 (cell-wall-less) algae in the presence of 2.8% added O_2.

Selection using H_2-uptake conditions was first applied by McBride et al. (1977), who subjected a population of WT cells to photoreductive conditions in the presence of controlled concentrations of O_2. Surviving algal cells grew by fixing CO_2 with electrons obtained from the oxidation of H_2 (using hydrogenase H_2-uptake activity) and ATP generated by cyclic electron transfer around photosystem I (Maione and Gibbs, 1986). Photosynthetic water oxidation (and consequently O_2 evolution) was blocked by the presence of the photosystem II inhibitor, DCMU. This selective pressure is much more specific for O_2-tolerant organisms than H_2-production selection. Unfortunately, the mutants isolated by McBride et al. (1977) had a higher O_2 I_{50} for both H_2 evolution and the back oxy-hydrogen reaction. The latter involves the dark uptake of H_2 and O_2 (Maione and Gibbs, 1986) and is certainly not desirable for an organism to be used in a photobioreactor for commercial H_2 production.

We have decided to apply both types of selective pressures to populations of *C. reinhardtii* in order to find the best technique (or combination of techniques) to obtain fully O_2-tolerant, H_2- producing mutants. This paper describes the application of a second round of H_2-production selective pressure to a mutagenized population of D5, optimization of the H_2-uptake selective pressure procedure, and the isolation of survivors from application of both types of selective pressures to cultures of cw15 and WT *C. reinhardtii* cells. We also describe preliminary rapid screening results for H_2 production capacity from individual algal isolates on agar plates using a procedure based on the chemochromic properties of a film of tungsten oxide.

3. EXPERIMENTAL

Wild-type *C. reinhardtii* (137c+) was a gift from Professor Susan Dutcher, University of Colorado, Boulder. The cell-wall-less strain, cw15, was obtained from the Chlamydomonas Genetics Center at Duke University. Algal cells were grown photoautotrophically either in Sager's minimal medium (WT) or in a modified Sueoka's high-salt medium (cw15, Ghirardi et al., 1997). Cultures were illuminated at 25 °C with cool white fluorescent light (8 W/m^2) and bubbled with a mixture of 1.7% CO_2 in air. Cells were harvested by centrifugation at 1000 g for 10 min and plated either in 1.5% (WT) or 0.8% (cw15) agar.

H_2 evolution was measured amperometrically with a Clark-type electrode (YSI 5331), poised at +0.40 V vs. an Ag/AgCl electrode. A second electrode, poised at -0.70 V vs. Ag/AgCl, was used to simultaneously determine the O_2 concentration in the assay chamber. Algal cells were anaerobically induced prior to the measurements, as described previously (Ghirardi et al., 1997). They were then injected anaerobically into the assay chamber containing assay buffer (50 mM MOPS, pH 6.8), which had been pre-adjusted to different initial concentrations of O_2. The cells were incubated in the dark for 2 min and then illuminated with saturating, heat-filtered incandescent light (Fiber-Lite High Intensity Illuminator, Model 170-D, Dolan-Jenner Industries) to induce H_2 evolution. Initial rates were calculated from the initial slopes of each curve, and gas concentrations were corrected for decreased solubility in aqueous solution at Golden, Colorado, located 1580 m above sea level.

Mutagenesis with nitrosoguanidine was done according to Harris (1989). An algal cell suspension (3×10^6 cells/mL) in citrate buffer (0.025M Na citrate, pH 5.0) was treated with nitrosoguanidine in the dark for different amounts of time. The cells were washed once with excess citrate buffer, twice with growth medium, and resuspended in about 50 mL of growth medium. This treated suspension was then incubated in the light for 2–3 days to allow chromosome segregation and expression of the mutations.

The H_2-production selective pressure was applied as described previously (Ghirardi et al., 1997), and survival rates were determined by counting the number of colonies detected on agar plates following the treatment. The H_2-uptake selective pressure was applied by treating a liquid culture of algal cells with DCMU (10–100 µM) to eliminate photosynthetic O_2 evolution, and incubating the cells in low light in anaerobic jars containing approximately 10% H_2, 1% CO_2, 8–10% O_2, and the balance N_2 for at least seven days. At the end of the incubation time, the cells were washed with growth medium and plated on minimal agar. Survivors were counted after three weeks of growth in the light. The screening procedure will be described in detail later in the paper and in the Figure 4 legend. It employs the chemochromic properties of a multi-layer, thin-film WO_3 sensor

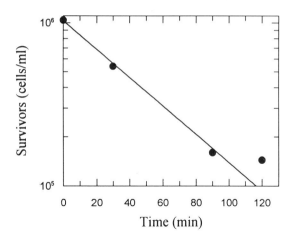

Figure 1. Semilogarithmic plot of the number of viable D5 *C. reinhardtii* cells as a function of incubation time with 1 μg/mL nitrosoguanidine. Treatments of 30–40 min killed 50% of the cells.

(Benson et al., 1996) that was adapted in a proprietary manner (U.S. patent application submitted) for use in direct contact with algal clones growing on agar plates.

4. RESULTS AND CONCLUSIONS

4.1. Mutagenesis

In order to generate *C. reinhardtii* mutants to be used with the two types of selective pressures described in section 2 (Introduction), we exposed the cultures to nitrosoguanidine, a chemical mutagen that induces random point mutations in algae (Harris, 1989; McBride et al., 1977). The algal cells were treated with the mutagen for different amounts of time, as described in section 3 (Experimental). Figure 1 shows a semilogarithmic plot of the number of survivors as a function of the incubation time. From this graph, we determined that a 30–40 min incubation treatment with the mutagen kills 50% of cell-wall-less cells. Subsequent mutagenesis experiments were done by exposing the cells to 1 μg/mL nitrosoguanidine for 30 min; however, cell-wall-less cells were more sensitive than WT cells (data not shown).

4.2. H_2-Production Selective Pressure

We have previously shown that a 20-min treatment of cw15 *C. reinhardtii* cells in the presence of 2.8% O_2 with 58 mM MNZ in the light killed more than 99.5% of the initial population (Ghirardi et al., 1997). The survivors from this initial H_2-production selection included the D5 isolate that had an O_2 I_{50} for H_2 evolution at least 330% higher than that of the original cw15 parental strain. We submitted the D5 isolate to mutagen treatment as described above, washed off residual mutagen, incubated cells in the light for 2–3 days, and exposed them to H_2-production selective pressure. The O_2 concentration during this second round of selective pressure was increased to 5%. The resultant survivors were plated and isolated colonies were transferred to a liquid medium. Hydrogen evolution rates were measured in the presence of different initial O_2 concentrations to determine an O_2 I_{50}. Figure 2 shows O_2 titration curves obtained with the initial cw15 parental strain (open circles), the D5 variant (closed circles), and the mutant IM6 (asterisks). As seen in

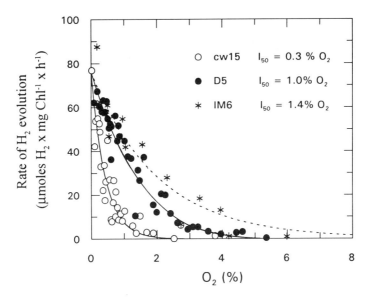

Figure 2. Oxygen titration of the rates of H_2 evolution by cw15 (■), the variant D5 (●), and mutant IM6 (✱). O_2 I_{50} for each organism is estimated by the concentration of added O_2 that lowers the rate of H_2 evolution to half the value measured in the absence of added O_2 (V_0, Y-axis intercept).

the inset, we were able to improve the O_2 tolerance of H_2 evolution almost 470% (from 0.3–1.4% O_2) by applying two rounds of H_2-production selective pressure. However, due to the lack of specificity of this selective pressure and the length of time required to do the I_{50} assays, improvements in identifying useful mutants must be developed.

4.3. H_2-Uptake Selective Pressure

We have optimized H_2-uptake selective pressure, based on the work described in McBride et al. (1977). Our protocol includes the following steps: (a) a 30-min treatment of an algal cell suspension with 1 µg/mL nitrosoguanidine in the dark (see Figure 1), followed by extensive washing of cells to remove residual mutagen and resuspension in growth medium; (b) 2–3 days of incubation of algal cells in the light to allow for chromosome segregation and expression of the mutations (this phase corresponds to 2–3 doubling times); (c) dark aerobic starvation of the cells for 2–3 days to deplete internal storage reserves (so that the cells grow only on H_2); and (d) exposure of the cells to the selective pressure in the presence of DCMU and O_2 for 5–7 days to eliminate WT and non-interesting phenotypes. Addition of O_2 to the selective pressure decreases the number of survivors by 3–4 orders of magnitude (data not shown), due to inactivation of the hydrogenase. If the cells are left in the selective pressure (+ O_2) for more than seven days, the number of cells increases again due to replication of the O_2-tolerant survivors.

Hydrogen-uptake selection was applied to a mutagenized population of WT *C. reinhardtii* cells. After a two-week incubation under the selective pressure in the presence of DCMU and 10% O_2, the suspended cells were assayed for H_2 evolution. The results, shown in Figure 3, indicate that the surviving population had a 50% increase in the O_2 I_{50} for H_2 evolution, confirming the efficacy of the selective pressure. However, as observed

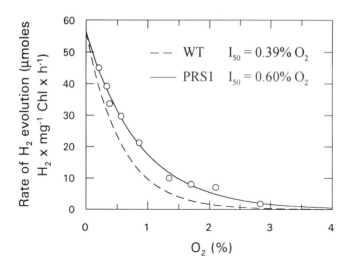

Figure 3. Oxygen titration of the rates of H_2 evolution by the surviving PRS1 population (○) from a H_2-uptake selective pressure experiment. Dashed curve represents the data for control WT cells.

previously, the assay for O_2 sensitivity is very time-consuming and has limited our ability to rapidly generate and screen for better mutants.

4.4. Screening

An individual 1-mm diameter algal colony on agar contains about 1 μg Chl. If one assumes that this colony can evolve H_2 at the maximum measurable rate under anaerobic conditions (see Figure 2) of about 80 μmol $H_2 \cdot$mg Chl$^{-1} \cdot$h^{-1} for about 2 min (before its hydrogenase becomes inactivated by photosynthetically produced O_2), then it should evolve 4 nmol of H_2. This number establishes a limit for the sensitivity of assays to be used for screening plated colonies for H_2 evolution capacity.

A very sensitive membrane detector for H_2 was developed at the National Renewable Energy Laboratory as part of a H_2 sensor system addressing H_2 vehicle safety issues (Benson et al., 1996). The detector (see Experimental section) changes color in the presence of H_2 evolved by anaerobically induced algal colonies on agar plates. Preliminary results are shown in Figure 4. Panel A shows an agar plate containing different algal colonies (that survived the selection conditions of Figure 3) covered with a piece of filter paper onto which a grid was placed. Panel B shows the chemochromic film placed on the filter paper and algae and then illuminated from the bottom of the agar plate with saturating light to induce H_2 production by the individual algal colonies. In panel C, the plate was removed, and the purple (seen as black) spots on the chemochromic film correspond to regions that changed color in the presence of H_2 evolved by the algal colonies. Finally, Panel D indicates the reversibility of the color reaction (and hence confirmation that H_2 was in fact detected) when the film was exposed to air for 5 min following the sensitization reaction. These results clearly demonstrate that the chemochromic film is sensitive enough to detect nanomol of H_2, and that this technique has the potential to rapidly screen a large number of colonies for H_2 production capacity. Preliminary experiments have also confirmed the relationship between the intensity of the spots and the amperometrically measured rates of H_2 evolution by the different colonies shown in Figure 4.

With the development of a rapid screen for H_2-producing mutants of *C. reinhardtii*, we are in a position to start combining the two selective pressure techniques discussed above in order to more rapidly isolate better mutants. Furthermore, it may also be useful to

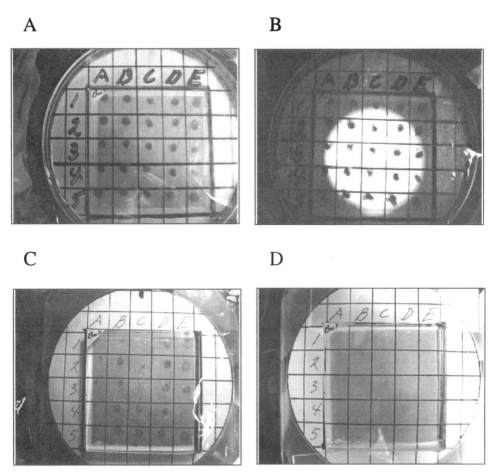

Figure 4. Chemochromic film screen for H_2-producing algal colonies. (A) Plate of individual algal colonies on agar. A grid was placed on filter paper covering the algal colonies to allow easy identification of individual colonies. (B) A chemochromic film was placed on top of the filter paper/grid and illuminated from the bottom of the plate to induce H_2 evolution by the algal colonies. (C) The chemochromic film and grid were removed from the plate containing the algal colonies for better observation of the purple spots on the film, which correspond to areas sensitized by H_2 evolved by the colonies. Note differences in intensity of the spots, which correspond to the amount of H_2 produced by the colony. (D) The sensitized film after exposure to air for 5 min. The spots disappeared, demonstrating that they were the result of exposure to H_2.

complement our classical mutagenesis work with a molecular biological approach to the hydrogenase O_2-sensitivity problem. Site-directed mutagenesis might be used to further increase the O_2 tolerance of the hydrogenase (McTavish et al., 1995) in mutants obtained by the classical mutagenesis approach.

ACKNOWLEDGMENTS

This work was sponsored by the Hydrogen Program, Office of Solar Thermal, Biomass Power, and Hydrogen Technologies, U.S. Department of Energy.

REFERENCES

Benemann, J.,1996, Hydrogen biotechnology: progress and prospects, *Nature Biotechnol.*, 14:1101–1103.

Benson, D.K., Tracy, C.E., and Bechinger, C., 1996, Design and development of a low-cost fiber-optic hydrogen detector, in *Proceedings of the 1996 U.S. Department of Energy Hydrogen Program Review*, vol. II, NREL/CP-21908430, pp. 605–624.

Ghirardi, M., Markov, S., and Seibert, M., 1996, Development of an efficient algal H_2-producing system, in *Proceedings of the 1996 U.S. Department of Energy Hydrogen Program Review*, vol. I, NREL/CP-21908430, pp. 285–302.

Ghirardi, M.L., Togasaki, R.K., and Seibert, M., 1997, Oxygen sensitivity of algal H_2-production, *Appl. Biochem. Biotechnol.*, 63: 141–151.

Harris, E., 1989, *The Chlamydomonas Sourcebook*, Academic Press, New York.

Houchins, J.P., 1984, The physiology and biochemistry of hydrogen metabolism in cyanobacteria, *Biochim. Biophys. Acta*, 768:227–255.

Maione, T.E., and Gibbs, M., 1986, Association of the chloroplastic respiratory and photosynthetic electron transport chains of *Chlamydomonas reinhardii* with photoreduction and the oxyhydrogen reaction, *Plant Physiol.*, 80:364–368.

McBride, A.C., Lien, S., Togasaki, R.K., and San Pietro, A., 1977, Mutational analysis of *Chlamydomonas reinhardi*: application to biological solar energy conversion, in *Biological Solar Energy Conversion*, Mitsui, A. et al. (eds.), Academic Press, New York, New York, pp. 77–86.

McTavish, H., Sayavedra-Soto, L.A., and Arp, D.A., 1995, Substitution of *Azotobacter vinelandii* hydrogenase small-subunit cysteines by serines can create insensitivity to inhibition by O_2 and preferentially damages H_2 oxidation over H_2 evolution, *J. Bacteriol.*, 177:3960–3964.

Schmidt, G.W., Matlin, K.S., and Chua, N.H., 1977, A rapid procedure for selective enrichment of photosynthetic electron transport mutants, in *Proceedings of the National Academy of Sciences (USA)*, 74:610–614.

Schulz, R., 1996, Hydrogenases and hydrogen production in eukaryotic organisms and cyanobacteria, *J. Mar. Biotechnol.*, 4:16–22.

Weaver, P.F., Lien, S., and Seibert, M., 1980, Photobiological production of hydrogen, *Solar Energy*, 24:3–45.

31

PHOTOSYNTHETIC HYDROGEN AND OXYGEN PRODUCTION BY GREEN ALGAE

An Overview

Elias Greenbaum and James W. Lee

Oak Ridge National Laboratory
P. O. Box 2008
Oak Ridge, Tennessee 37831

Key Words

hydrogen, oxygen, chlorophyll antenna size, Z-scheme, mutants, *Chlamydomonas*

1. SUMMARY

An overview of photosynthetic hydrogen and oxygen production by green algae in the context of its potential as a renewable chemical feedstock and energy carrier is presented. Beginning with its discovery by Gaffron and Rubin in 1942, then motivated by curiosity-driven laboratory research, studies were initiated in the early 1970s that focused on photosynthetic hydrogen production from an applied perspective. From a scientific and technical point of view, current research is focused on optimizing net thermodynamic conversion efficiencies represented by the Gibbs Free Energy of molecular hydrogen. The key research questions of maximizing hydrogen and oxygen production by light-activated water-splitting in green algae are: (1) removing the oxygen sensitivity of algal hydrogenases; (2) linearizing the light saturation curves of photosynthesis throughout the entire range of terrestrial solar irradiance—including the role of bicarbonate and carbon dioxide in optimization of photosynthetic electron transport; and (3) identifying the minimum number of light reactions that are required to split water to elemental hydrogen and oxygen. Each of these research topics is being actively addressed by the photobiological hydrogen research community.

2. INTRODUCTION

Photosynthetic hydrogen production by green algae was discovered in the pioneering experiments of Gaffron and Rubin (1942). This work was followed up by Gaffron and his colleagues in a series of seminal papers (Gaffron, 1960; Kaltwasser et al., 1969; Stuart and Gaffron, 1971, 1972; as well as many others). From the point of view of renewable fuels and chemical feedstock production, it is light-activated simultaneous photoproduction of hydrogen and oxygen that is of primary interest. The pioneering experiments in this field were performed by Spruit (1958), who developed a novel two-electrode polarographic technique for the simultaneous measurement of hydrogen and oxygen transients by the green alga *Chlorella*. The principal conclusion he reached was that hydrogen and oxygen metabolisms are closely related and they derived from water-splitting. Later work by Bishop and Gaffron (1963) indicated that light-dependent evolution of hydrogen appeared to require both photosystems.

Research on photosynthetic hydrogen production as a renewable energy source began in the 1970s (Gibbs et al., 1973; Lien and San Pietro, 1975; Mitsui et al., 1977). Using the two-electrode technique, Bishop et al. (1977) measured and interpreted hydrogen and oxygen production from a large group of green algae. However, due to the build-up of hydrogen and oxygen, with subsequent inhibition (*vide infra*) these reactions could be followed for only several minutes. Using a flow system that removed inhibitory oxygen, it was shown (Greenbaum, 1980) that sustained simultaneous photoproduction of hydrogen and oxygen could be observed for hours. In prior experiments using a glucose-glucose oxidase trap, Benemann et al. (1973) demonstrated hydrogen production from water by a chloroplast-ferredoxin-hydrogenase system. Measurement of the hydrogen analog of the Emerson and Arnold photosynthetic unit size (Greenbaum, 1977a, b) indicated that photogenerated reductant expressed as molecular hydrogen was derived from the mainstream of the photosynthetic electron transport chain. Direct measurement of the turnover time of photosynthetic hydrogen production (Greenbaum, 1979, 1982) demonstrated that this parameter was comparable to the turnover time of oxygen production. It was also shown (Greenbaum, 1988) that net conversion efficiencies of 5–10% could be achieved in the linear low-intensity region of the light saturation curve.

3. THE THREE PROBLEMS

Figure 1 is a schematic illustration of the minimum number of components required for light-activated hydrogen and oxygen production in green algae. The hydrogenase enzyme is synthesized de novo under anaerobic conditions. In normal photosynthesis, carbon dioxide is the preferred electron acceptor for photogenerated reductant from Photosystem I. However, direct kinetic competition between hydrogen evolution and the Calvin cycle can easily be observed (Graves et al., 1989; Cinco et al., 1993). The three scientific research problems associated with photosynthetic hydrogen and oxygen production are: (1) oxygen sensitivity of hydrogenase; (2) antenna size, bicarbonate, and the light saturation problem; and (3) the minimum number of light reactions required to split water to molecular hydrogen and oxygen.

3.1. Oxygen Sensitivity of Hydrogenase

In the application of intact unicellular green algae for hydrogen production, one is confronted with the problem of oxygen sensitivity of the hydrogenase enzyme. Hydrogenase is synthesized under anaerobic conditions and, at present, must be kept that way in order to preserve its functionality. As illustrated in Figure 1, oxygen and hydrogen by green algae are coproduced in the same volume. Therefore, a way must be found to pre-

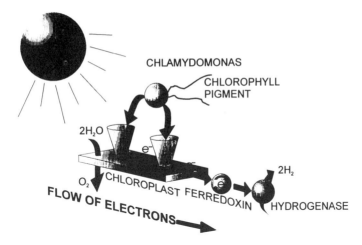

Figure 1. Schematic illustration of the process of photosynthetic hydrogen and oxygen evolution in green algae. The hydrogenase is synthesized under anaerobic conditions. In the absence of CO_2, the natural terminal electron acceptor, photogenerated reductant can be expressed as molecular hydrogen catalyzed the hydrogenase enzyme.

vent inhibition of hydrogenase activity by the photosynthetically produced oxygen. This challenging problem is the focus of research at the National Renewable Energy Laboratory (Ghirardi et al., 1997) and is reported elsewhere in these proceedings.

3.2. Antenna Size, Bicarbonate, and the Light Saturation Problem

Schematic illustrations of the relationship of the light saturation curves of photosynthesis and the light-harvesting antennas to the photosynthetic electron transport chains are presented in Figures 2 and 3. In full sunlight, ~1000 W/m^2, there exists a kinetic imbalance between the rate of photon excitation of the reaction centers and the ability of the thermally activated electron transport chains to process photogenerated electrons. Whereas the reaction centers can receive photoexcitations at the rate of ~2000 sec^{-1}, movement through the electron transport chain is of the order of 200 sec^{-1} or less (Gibbs et al., 1973). Therefore, normal photosynthesis saturates at much less that full sunlight, typically ~10%. This is illustrated in the lower parametric curve of Figure 2, labeled 200:1, implying 200 chlorophylls per reaction center. The actual number varies, of course, depending on alga and growth conditions (Melis, 1989; Melis et al., 1996). This is the situation that is illustrated schematically on the left-hand side of Figure 3, where 200 chlorophylls service a single reaction center.

Since there is little opportunity to increase the rate of thermally activated electrons through the photosynthetic electron transport chain, an alternate strategy is illustrated in the right-hand side of Figure 3. Kinetic balance between the rate of photon excitation and rate of photosynthetic electrons can, in principle, be balanced, even at full sunlight, by reducing the absolute antenna size per reaction center. Under optimum conditions, the idealized response is linear, illustrated by the parametric curve labeled 20:1. If such a response could be achieved in a real-world system, photosynthetic productivity on a per chlorophyll basis would increase and high solar irradiances would be converted to useful biomass energy. Linearization of the light saturation curve of photosynthesis was demonstrated by Herron and Mauzerall (1972). Melis et al. (1998) have demonstrated linearization of the

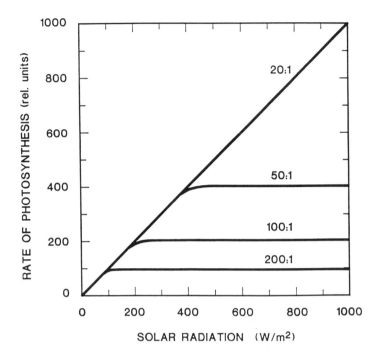

Figure 2. Schematic illustration of the relationship between photosynthetic antenna size and saturating light intensity. The parameters labeling each curve are the stoichiometric ratios of chlorophyll antenna pigment to photosynthetic reaction center. Please note that for Figures 2 and 3, the indicated ratios are representative of the range of possible values. For example, for Photosystem II the range is 500:1 to 40:1, while for Photosystem I the range can vary from 350:1 to 100:1.

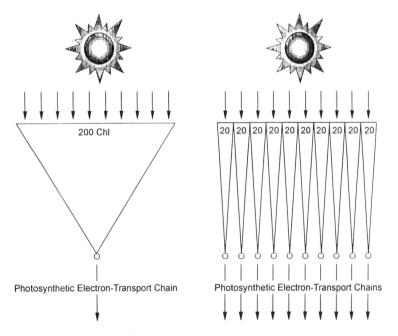

Figure 3. Schematic illustration of the relationship of chlorophyll antenna size, photosynthetic electron transport chains, and conversion efficiency.

light saturation curve for high-light grown cultures of *Dunaliella*. These results indicate that the concept is technically correct.

An additional complication of the light saturation problem involves the requirement of bicarbonate to optimize electron transport through Photosystem II. Since carbon dioxide/bicarbonate are the exclusive sink for photosynthetically generated reductant, they need to be removed so that the flow of electrons produces hydrogen rather than carbon dioxide fixation compounds. Complete removal of carbon dioxide, however, impairs electron transport in Photosystem II and further reduces the saturating light intensity for sustained simultaneous photoproduction of hydrogen and oxygen by about a factor of 10. Under these conditions, light saturation occurs at about 10 W/m^2. One strategy to overcome this limitation is to take advantage of the differential affinity of CO_2/bicarbonate between the Photosystem II binding site and the Calvin cycle. Such an approach has been explored by Cinco et al. (1993), in which light-activated hydrogen and oxygen evolution as a function of CO_2 concentration in helium were measured for the unicellular green alga *Chlamydomonas reinhardtii*. The concentrations were 58, 30, 0.8, and 0 ppm CO_2. The objective of these experiments was to study the differential affinity of CO_2/HCO_3^- for their respective Photosystem II and Calvin cycle binding sites vis-à-vis photoevolution of molecular oxygen and the competitive pathways of hydrogen photoevolution and CO_2 photoassimilation. The maximum rate of hydrogen evolution occurred at 0.8 ppm CO_2. The key result of this work was that the rate of photosynthetic hydrogen evolution can be increased, at least partially, by satisfying the Photosystem II CO_2/HCO_3^- binding site requirement without fully activating the Calvin-Benson CO_2 reduction pathway. These preliminary experiments suggest that mutants of *Chlamydomonas reinhardtii* that have a genetically engineered low CO_2 affinity for the Calvin cycle and relatively higher affinity for the PS II CO_2/HCO_3^- binding site may be good candidates to explore for relieving the CO_2 part of the light saturation constraint.

3.3. The Minimum Number of Light Reactions

It has recently been reported (Greenbaum et al., 1995; Lee et al., 1996) that mutants of the green alga *Chlamydomonas reinhardtii* that lack detectable levels of the Photosystem I (PS I) reaction center are capable of sustained simultaneous photoevolution of H_2 and O_2, CO_2 reduction, and photoautrophic growth. The data indicated that under some circumstances PS II alone is sufficient to generate reductant capable of driving hydrogen evolution and CO_2 fixation, not of course that the Z-scheme is universally wrong. Although the absence of PS I in mutants B4 and F8 for the data reported in the references was confirmed by physical, biochemical, and genetic techniques, subsequent analyses in our own laboratories, as well as those of colleagues to whom we have sent the mutants, indicate that there is variability in the PS I content of the cultures depending on growth conditions. While some strains retain undetectable levels of P700, others contain variable (up to 20%) amounts of wild-type P700. This property of mutants B4 and F8 has been communicated to the journals in which the results were initially published (Greenbaum et al., 1997a). The fact that mutants B4 and F8 can synthesize fluctuating levels of PS I raises the possibility that small amounts of this reaction center might exclusively generate reductant that can drive hydrogen evolution and carbon dioxide fixation.

In his analysis of this work, Boichenko (1996) postulated a "leaky" model of the Z-scheme, illustrated schematically in Figure 4. According to this model, under continuous high-intensity light, PS I is postulated to turn over with sufficient rapidity to accommodate reductant generated by multiple PS IIs, thereby preserving a key requirement of the Z-scheme: Only PS I is capable of generating low-potential reductant that can be used for

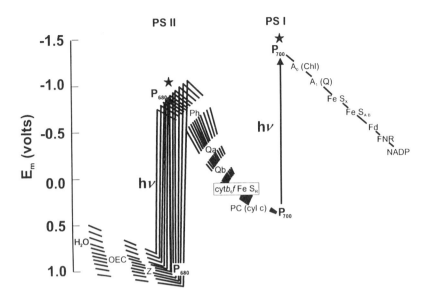

Figure 4. The "leaky" Z-scheme model of photosynthesis.

hydrogen evolution or carbon dioxide fixation. Based on comparative rates of photosynthesis with wild-type *Chlamydomonas reinhardtii* and known (Melis et al., 1996) properties of PS I antenna size, it was felt that the earlier results were not consistent with this model even if PS I were present, but undetectable. Preliminary data (Greenbaum et al., 1997b), testing the leaky Z-scheme model of Figure 4 on direct measurement of absolute single-turnover saturating flash yields of hydrogen and oxygen production in mutant Fud26, as well as photosynthesis measurements under light-limiting conditions (O_2 evolution with CO_2 as electron acceptor), suggest that the leaky Z-scheme model cannot quantitatively account for the measured amounts of photosynthesis. Therefore, in addition to the Z-scheme pathway, there is, in mutant Fud26, an alternative oxygenic photosynthetic pathway.

ACKNOWLEDGMENTS

The authors thank L.J. Mets, T.G. Owens, A. Melis, K. Redding, I. Vassieliev, and J. Golbeck for collaboration, advice, and discussion of this work. They also thank C.V. Tevault and S.L. Blankinship for technical support and L. Wagner for secretarial support. This research was supported by the U.S. Department of Energy. Oak Ridge National Laboratory is managed by Lockheed Martin Energy Research Corporation for the U.S. Department of Energy under contract DE-AC05-96OR22464.

REFERENCES

Bishop, N.I., and Gaffron, H., 1963, On the interrelation of the mechanisms for oxygen and hydrogen evolution in adapted algae, in *Photosynthetic Mechanisms in Green Plants*, Kok, B., and Jagendorf, A.T., (eds.), National Academy of Sciences and National Research Council, Washington, D.C., pp. 441–451.

Bishop, N.I., Fricke, M., and Jones, L.W., 1977, Photohydrogen production in green algae: water serves as the primary substrate for hydrogen and oxygen production, in *Biological Solar Energy Production*, Mitsui, A. et al. (eds.), Academic Press, New York, pp. 1–22.
Boichenko, V.A., 1996, Can photosystem II be a photogenerator of low potential reductant for CO_2 fixation and H_2 photoevolution? *Photosyn. Res.*, 47:291–292.
Cinco, R.M., MacInnis, J.M., and Greenbaum, E., 1993, The role of carbon dioxide in light-activated hydrogen production by *Chlamydomonas reinhardtii*, *Photosyn. Res.*, 38:27–33.
Gaffron, H., 1960, Energy storage: photosynthesis, *Plant Physiology*, IB:4–227.
Gaffron, H., and Rubin, J., 1942, Fermentative and photochemical production of hydrogen in algae, *J. Gen. Physiol.*, 26:219–240.
Ghirardi, M.L., Togasaki, R.K., and Seibert, M., 1997, Oxygen sensitivity of algal hydrogen production, *Appl. Biochem. Biotech.*, in press.
Gibbs, M., Hollaender, A., Kok, B., Krampitz, L.O., and San Pietro, A., 1973, in *Proceedings of the Workshop on Bio-Solar Conversion*, September 5–6, 1973, Bethesda, Maryland, Indiana University, National Science Foundation RANN Grant GI 40253.
Graves, D.A., Tevault, C.V., and Greenbaum, E., 1989, Control of photosynthetic reductant: the effect of light and temperature on sustained hydrogen photoevolution in *Chlamydomonas* sp. in an anoxic carbon dioxide containing atmosphere, *Photochem. Photobiol.*, 50:571–576.
Greenbaum, E., 1977a, The photosynthetic unit size of hydrogen evolution, *Science*, 196:878–879.
Greenbaum, E., 1977b, The molecular mechanisms of photosynthetic hydrogen and oxygen production, in *Biological Solar Energy Production*, Mitsui, A. et al. (eds.) Academic Press, New York, pp. 101–107.
Greenbaum, E., 1979, The turnover times and pool sizes of photosynthetic hydrogen production by green algae, *Solar Energy*, 23:315–320.
Greenbaum, E., 1980, Simultaneous photoproduction of hydrogen and oxygen by photosynthesis, *Biotechnol. Bioeng. Symp.*, 10:1–13.
Greenbaum, E., 1982, Photosynthetic hydrogen and oxygen production: kinetic studies, *Science*, 215:291–293.
Greenbaum, E., 1988, Energetic efficiency of hydrogen photoevolution by algal water splitting, *Biophys. J.*, 54:365–368.
Greenbaum, E., Lee, J.W., Tevault, C.V., Blankinship, S.L., and Mets, L.J., 1995, CO_2 fixation and photoevolution of H_2 and O_2 in a mutant of *Chlamydomonas* lacking photosystem I, *Nature*, 376:438–441.
Greenbaum, E., Lee, J.W., Blankinship, S.L., and Tevault, C.V., 1997b, Hydrogen and oxygen production in mutant Fud26 of *Chlamydomonas reinhardtii*, in *Proceedings of the 1997 U.S. Department of Energy Hydrogen Program Review*, May 21–23, 1997, Herndon, Virginia, pp. 1–10.
Greenbaum, E., Lee, J.W., Tevault, C.V., Blankinship, S.L., Mets, L.J., and Owens, T.G., 1997a, Photosystem I measurement in mutants B4 and F8 of *C. reinhardtii*, *Science*, 277:167–168, and *Nature*, 388:808.
Herron, H.A., and Mauzerall, D., 1972, The development of photosynthesis in a greening mutant of *Chlorella* and an analysis of the light saturation curve, *Plant Physiol.*, 50:141–148.
Kaltwasser, H., Stuart, T.S., and Gaffron, H., 1969, Light-dependent hydrogen evolution by *Scenedesmus*, *Planta* (Berlin), 89:309–322.
Lee, J.W., Tevault, C.V., Owens, T.G., and Greenbaum, E., 1996, Oxygenic photoautotrophic growth without Photosystem I, *Science*, 273:364–367.
Lien, S., and San Pietro, A., 1975, An inquiry into biophotolysis of water to produce hydrogen, Indiana University, National Science Foundation RANN report for grant GI 40253.
Melis, A., Neidhardt, J., Baroli, I., and Benemann, J.R., 1998, Maximizing photosynthetic productivity and light utilization in microalgae by minimizing the light-harvesting chlorophyll antenna size of the photosystems, in *Proceedings of BioHydrogen '97*, Plenum Publishers, in press.
Melis, A., 1989, Spectroscopic methods in photosynthesis: photosystem stoichiometry and chlorophyll antenna size, *Phil. Trans. R. Soc. Lond. B.*, 323:397–409.
Melis, A., Murakami, A., Nemson, J.A., Aizawa, K., Ohki, K., and Fujita, Y., 1996, Chromatic regulation in *Chlamydomonas reinhardtii* alters photosystem stoichiometry and improves the quantum efficiency of photosynthesis, *Photosyn. Res.*, 47:253–265.
Mitsui, A., Miyachi, S., San Pietro, A., and Tamura, S. (eds.), 1977, *Biological Solar Energy Production*, Academic Press, New York.
Spruit, C.J.P., 1958, Simultaneous photoproduction of hydrogen and oxygen by *Chlorella*. Mededel, *Landbouwhogeschool* (Wageningen/Nederland), 58:1–17.
Stuart, T.S., and Gaffron, H., 1971, The kinetics of hydrogen photoproduction by adapted *Scenedesmus*, *Planta* (Berlin), 100:228–243.
Stuart, T.S., and Gaffron, H., 1972, The mechanism of hydrogen photoproduction by several algae, *Planta* (Berlin), 106:91–100.

32

LIGHT-DEPENDENT HYDROGEN PRODUCTION OF THE GREEN ALGA *Scenedesmus obliquus*

Rüdiger Schulz,[1] Jörg Schnackenberg,[2] Kerstin Stangier,[1] Röbbe Wünschiers,[1] Thomas Zinn,[3] and Horst Senger[1]

[1]FB Biology/Botany
Philipps-University
Karl-von-Frisch-Str., D-35032 Marburg, Germany
[2]Max-Planck-Institut für Biochemie
 Abt. Strukturforschung, Am Klopferspitz 18A
 D-82152 Martinsried, Germany
[3]Macherey-Nagel GmbH & Co. KG
 Valencienner Str. 11, D-52355 Düren, Germany

Key Words

photohydrogen production, hydrogenase, light, photosynthesis, green alga, pilot bioreactor

1. SUMMARY

The unicellular green alga *Scenedesmus obliquus* is capable of producing molecular hydrogen in the light. The prerequisite is the activation of the responsible enzyme hydrogenase by anaerobic adaptation. Photohydrogen production is coupled to the photosynthetic electron transport chain of green algae. Results of biochemical and molecular biological investigations of a purified NiFe hydrogenase and an cloned "Fe-only" hydrogenase are presented, and thus the existence of two hydrogenases is shown for *Scenedesmus*. The cyanobacterium *Synechocystis* sp. PCC 6803 is introduced as a genetic tool for future manipulations of *S. obliquus* hydrogenase. Construction of a simple pilot bioreactor for photohydrogen production is explained and the general use of light-dependent hydrogen production by algae as a future energy source is discussed.

BioHydrogen, edited by Zaborsky *et al.*
Plenum Press, New York, 1998

2. INTRODUCTION

2.1. Occurrence and Characteristics of Hydrogenases

Since the beginning of the 20th century, it has been known that microorganisms are capable of metabolizing molecular hydrogen (Harden, 1901; Pakes and Jollyman, 1901). The responsible enzyme was named "hydrogenase" by Stephenson and Stickland (1931), and was subsequently found in many bacteria (Adams et al., 1981), in several algal species (Gaffron, 1939; Kessler, 1974; Boichenko and Hoffmann, 1994), in some trichomonades (Bui and Johnson, 1996), and in anaerobic fungi (Marvin-Sikkema et al., 1993). Some evidence for hydrogenase activity in mosses and higher plants is available in the literature (summarized in Schulz, 1996), but these are mainly single publications.

The enzyme catalyzes the reversible reductive formation of hydrogen from protons:

$$2H^+ + 2e^- \xleftrightarrow{hydrogenase} H_2$$

Some dozen prokaryotic hydrogenases are well-characterized presently (Voordouw, 1992; Hahn and Kück, 1994; Boichenko and Hoffmann, 1994), but only two hydrogenases of green algae have been isolated (Happe and Naber, 1993; Schnackenberg et al., 1993). Although the enzyme catalyzes the chemically simple reduction of protons to hydrogen and oxidation of hydrogen, hydrogenases are a very heterogenic class of enzymes. They greatly differ in their molecular masses, subunit compositions, specificity for electron donors and acceptors, cofactors, specific activities, oxygen sensitivities and metal contents at the active site. According to the latter, hydrogenases can be classified into four distinctive groups:

- "Fe-only" hydrogenases: The active site of these hydrogenases is built up by an unknown type of FeS-cluster (Adams, 1990).
- NiFe hydrogenases: These contain nickel (Graf and Thauer, 1981). Recently, the active site of one was resolved by x-ray diffraction spectroscopy at a resolution of 2.85 Å (Volbeda et al., 1995). It contains a sulfur-coordinated hetero-dimer of a Ni and Fe ion (Fontechilla-Camps, 1996).
- NiFeSe hydrogenases: The active site of these hydrogenases resembles the NiFe type. One Ni-coordinating cysteine is replaced by a selenocysteine (Yamazaki, 1982).
- Metal-free hydrogenases: This novel class of hydrogenases was found in methanogenic archaea like *Methanobacterium thermoautotrophicum* (Zirngibl et al., 1990). The enzyme turned out to be a H_2-forming methylenetetrahydromethanopterin dehydrogenase (Thauer et al., 1996).

With respect to the biological production of hydrogen gas, three main responsible metabolic pathways are of interest:

- Fermentation: In fermentative bacteria, metabolism leads to the formation of H_2 and CO_2 from organic compounds. The energy is derived from the organic compound itself.
- Anoxygenic photosynthesis: In phototrophic bacteria, organic or reduced sulfur substrates are converted to H_2 and CO_2 by the use of light energy.
- Oxygenic photosynthesis: In green algae and cyanobacteria, water is split by the use of solar radiation to yield O_2 and, by hydrogenase activity, H_2. The coupling

of photohydrogen production and water-splitting in the photosynthetic electron transport chain of the oxygenic photosynthesis in algae and cyanobacteria and the avoidance of CO_2 production, make this system particularly interesting for the production of molecular hydrogen by sunlight.

2.2. Photohydrogen Evolution by Hydrogenases in Green Algae

In contrast to the biochemically and genetically well-characterized hydrogenases from prokaryotic organisms, there are much less data available for the eukaryotic enzymes (for review: Hahn and Kück, 1994; Boichenko and Hoffmann, 1994; Schulz, 1996).

About 50 years ago, Gaffron (1939, 1940; Gaffron and Rubin, 1942) revealed that the unicellular green alga *Scenedesmus obliquus* is capable of either light-dependent production of molecular hydrogen (photoproduction of hydrogen) or its uptake (photoreduction of CO_2), under anaerobic conditions dependent on the partial pressure of H_2 and CO_2. Subsequently, hydrogenase activity was found in a great variety of algae, mainly Chlorophyceae, Euglenophyceae, Phaeophyceae, and Rhodophyceae (Kessler, 1974; Boichenko and Hoffmann, 1994).

Both photosystems of the green algal photosynthetic apparatus were shown to be involved in photohydrogen evolution by serving as electron donors for proton reduction in the hydrogenase reaction (Randt and Senger, 1985) (Figure 1). In addition, it was shown that the electrons and protons are derived from the photosynthetic water-splitting process (Senger, 1980). However, when photosystem II or the water-splitting complex is blocked, the reducing power derived from the fermentation of organic substrates can be used (Senger and Bishop, 1979).

In the present paper, we would like to present part of our efforts to investigate the hydrogen producing machinery of the green alga *Scenedesmus obliquus* and to apply this system for the production of hydrogen as an energy source on a technical scale.

3. EXPERIMENTAL

Culturing of *Scenedesmus obliquus*, its anaerobic adaptation, purification of the hydrogenase enzyme, and measuring photohydrogen production and enzyme activity were performed as described elsewhere (Senger and Bishop, 1979; Schnackenberg et al., 1993; Urbig et al., 1993; Zinn et al., 1994). Molecular biological methods were performed ac-

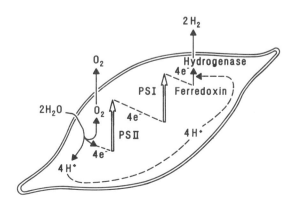

Figure 1. Mechanism of photohydrogen production in *Scenedesmus*. The coupling of hydrogenase to the photosynthetic apparatus for photohydrogen production in chloroplasts of the green alga is shown schematically.

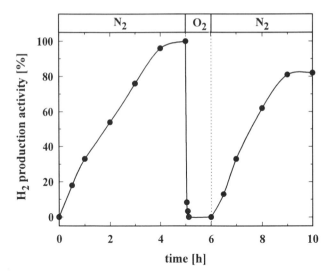

Figure 2. Activation and reactivation of hydrogenase activity in whole cells of *Scenedesmus obliquus*. By flushing the algal cell suspension with nitrogen in darkness, anaerobic adaptation was performed. This led to an increase of hydrogenase activity, shown here over a period of 5 h. Very rapid inactivation of hydrogenase enzyme activity within minutes occurred by incubation with oxygen. Reactivation of hydrogenase activity from the 6–10th hour of the experiment was possible for *Scenedesmus* in whole cells only.

cording to lab manual protocols, such as Sambrook et al. (1989) or detailed in Appel and Schulz (1996).

4. RESULTS, DISCUSSION, AND CONCLUSIONS

4.1. Anaerobic Adaptation, Activation, and Inactivation

After adaptation of algal cell cultures to an anaerobic atmosphere, full hydrogenase activity develops within several hours in *Scenedesmus* (Senger and Bishop, 1979). All attempts to activate the hydrogenase in a cell-free extract fail (Urbig et al., 1993). When whole cells or cell-free extracts of adapted cells are exposed to even traces of oxygen, the activity is lost within minutes. After such treatment, the enzyme can be reactivated in whole cells only (Figure 2).

4.2. Purification and Characterization of a NiFe Hydrogenase from *Scenedesmus*

After anaerobic adaptation, an active hydrogenase was purified from *Scenedesmus obliquus* cells (Schnackenberg et al., 1993). All manipulations of the enzyme have to be performed under strict anaerobiosis. The enzyme consists of at least two subunits of 36 and 55 kDa.

Evidence of nickel in the active site was first shown by incorporation of radioactive ^{63}Ni in the enzyme protein and by the effects of different nickel concentrations in the growth medium on the enzyme activity (Zinn et al., 1994). Recently, preliminary results were obtained by the application of EPR (electron paramagnetic resonance), TRXF (total reflection x-ray fluorescence), and AAS (atomic absorption spectroscopy), confirming the conclusion that the purified hydrogenase is of the NiFe type.

Western blot experiments with an antibody directed against the large NiFe hydrogenase subunit of *Bradyrhizobium japonicum* (kindly provided by Dr. Arp and Dr. Sayavedra-Soto, Corvallis, Oregon, United States) gave evidence that at least the 55-kDa subunit of the NiFe hydrogenase is constitutively expressed in *S. obliquus*. Southern blot analysis with genomic, chloroplastic, and mitochondrial DNA of *S. obliquus* hybridized with cloned hydrogenase DNA fragments from *Scenedesmus* as homologous probes (see below) clearly shows that the hydrogenase is encoded in the nucleus. But, by inhibition of the cytoplasmic or plastidic transcription and translation apparatus during anaerobic adaptation by cycloheximide, chloramphenicol, and other inhibitors, the involvement of the chloroplastic transcription and translation apparatus in hydrogenase activation was shown. Thus, we assume the existence of a chloroplastic activation factor. This hypothetical activation factor has not been identified yet.

More is known about the inactivation of hydrogenase activity in crude protein extracts of *Scenedesmus*. Preliminary results show that reduced thioredoxin (kindly provided by Dr. Follmann, Kassel, Germany) has an inhibitory effect on hydrogenase activity of *Scenedesmus*. Thioredoxin is a typical regulatory component of photosynthetic reactions. Mostly, it activates metabolic enzymes (Follman and Häberlein, 1996). Thus, this protein is discussed as a natural regulator of the hydrogenase.

4.3. Strategies to Clone the NiFe Hydrogenase of *Scenedesmus*

We first tested heterologous DNA probes and antisera against NiFe hydrogenases from several prokaryotic sources for their applicability in screening procedures with DNA libraries from the green alga *Scenedesmus*. Although there were some reactions in Southern and Western blot experiments as described before, none of the probes was specific enough to react in the heterologous screening procedures. At the moment, three different cloning strategies are in progress:

1. The amino acid sequence at the N-terminus of the small 36-kDa NiFe hydrogenase subunit of *S. obliquus* was determined in collaboration with Dr. Linder (Giessen, Germany) and an oligonucleotide derived from this sequence was used to screen a cDNA library of the green alga. Work in progress includes sequencing of the positive-reacting clones.
2. An antiserum against an oligopeptide, synthesized according to the N-terminal amino acid sequence of the 36-kDa subunit, was generated. This antiserum reacts very well with the 36-kDa protein from *Scenedesmus* and is in use for an immunoscreening of expression libraries from the green alga.
3. DNA primers were created by use of amino acid sequences of conserved prokaryotic Ni-binding sites of NiFe hydrogenases (Przybyla et al., 1992). These primers were used for PCR (polymerase chain reaction) experiments with genomic DNA of *S. obliquus*.

From these screening experiments, we isolated several PCR products and cDNAs, which were sequenced and the DNA sequences were compared with DNA and protein databanks. Until now none of the clones shows either significant homologies to known sequences of NiFe hydrogenases or to any other known protein sequence. Work in progress focuses on further screening experiments and on investigations to identify the already-isolated DNA clones. Experiments are being performed at the moment to learn more about the transcriptional and translational regulation of the isolated DNA clones and about their enzyme activity by heterologous expression in *E. coli* and *Synechocystis*.

4.4. Biochemical and Physiological Evidence for an "Fe-Only" Hydrogenase

Recently, ferredoxin, the terminal electron acceptor of the photosynthetic electron transport chain, was purified from *Scenedesmus* to homogeneity and identified by amino acid sequencing and UV/VIS and EPR spectroscopy. The functional role of reduced ferredoxin for hydrogen evolution was shown by measuring hydrogenase activity in cell-free extracts of *S. obliquus*. The function of reduced ferredoxin as an electron donor for the hydrogenase is a typical feature of "Fe-only" hydrogenases.

Parallel to the isolation of the NiFe hydrogenase from *Scenedesmus* in our lab (Schnackenberg et al., 1993), a hydrogenase from another green alga, *Chlamydomonas reinhardtii*, was purified by another group (Happe and Naber, 1993) and identified as an "Fe-only" hydrogenase (Happe et al., 1994). The antibody directed against the latter enzyme gave a cross reaction with a 46-kDa protein (the typical size of "Fe-only" hydrogenases) from *S. obliquus* (this unpublished experiment was performed by Zinn and Schulz in collaboration with Dr. Naber and Dr. Happe, Bochum, Germany).

4.5. Cloning of an "Fe-Only" Hydrogenase from *Scenedesmus*

To investigate the existence of an "Fe-only" hydrogenase in *S. obliquus*, two oligonucleotide primers were synthesized according to amino acid sequences of the 46-kDa α subunit from the *Desulfovibrio vulgaris* (strain Hildenborough and Monticello) "Fe-only" hydrogenase (Voordouw and Brenner, 1985; Voordouw et al., 1989). Conserved regions of the FeS-binding, cysteine-rich parts of the sequence were used. The sequence of the resulting PCR product from *Scenedesmus* shows significant similarities and identities to "Fe-only" hydrogenases in DNA and protein databanks. Using the PCR product as a hybridization probe, cDNAs were isolated, which show all sequence characteristics of "Fe-only" hydrogenases.

Thus, as a conclusion, the existence of two hydrogenases, one of the NiFe type and the other of the "Fe-only" type, is postulated for the green alga *Scenedesmus obliquus*.

4.6. Pilot Bioreactor for Photohydrogen Production

The main long-term goal of our work is the production of biohydrogen on a technical scale. With this goal in mind, we built a small pilot bioreactor (Figure 3) composed of three functional units: a flat plastic container for the cell culture (volume 250 mL), a PEM (polymer electrolyte membrane) fuel cell, and an electric motor.

After anaerobic adaptation of the algal cells in the flat plastic container and addition of Na-dithionite for recapturing the oxygen evolved by the photosynthetic apparatus, the cell container is exposed to light. Without any improvements, a hydrogen production rate of at least 1 mL/h by 30 g algae (wet weight) was observed. Gas chromatographic analysis of the produced gas gave no evidence of any gaseous impurities besides hydrogen. After injection of the evolved hydrogen into a fuel cell (kindly provided by Dr. Lehmann, Stralsund, Germany), the electric power produced by the fuel cell was sufficient to drive the small electric motor.

4.7. Future Work

The main future goal of our research is the application of hydrogenases in photohydrogen production on a technical scale (Schulz and Borzner, 1996). Due to the coupling of hydrogenases to photosynthesis, algae are capable of using light energy for photohydrogen

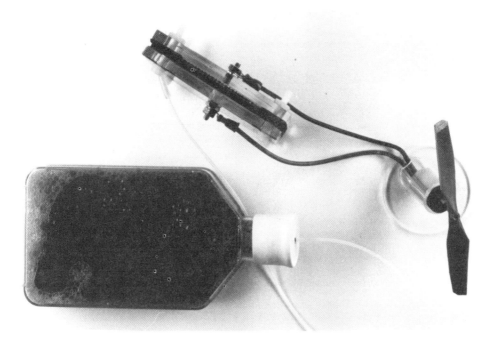

Figure 3. Pilot bioreactor system for photohydrogen production by the green alga *Scenedesmus obliquus*. On the left side of the figure, a 250-mL plastic flask is shown, which is illuminated by sunlight or a slide projector. The flask carries anaerobic adapted green alga cells in a medium containing Na-dithionite for recapturing the oxygen evolved by the photosynthetic apparatus. In the upper part of the figure, a fuel cell and an electric motor coupled with a small propeller is shown. The algal cell container is connected to the fuel cell by a plastic tube, while the fuel cell and the electric motor are in contact by electric cables.

production directly and in addition avoiding production of CO_2. To take advantage of this efficient system, further details of the process have to be collected. The main problem that remains to be solved is the extreme sensitivity of the algal hydrogenase to oxygen. With the help of the cyanobacterium *Synechocystis* as a genetic tool (see paper of Appel et al. in this issue), we should be able to manipulate the enzyme toward oxygen insensitivity genetically.

ACKNOWLEDGMENTS

This work was supported financially by BMFT (Bundesministerium für Forschung und Technologie der Bundesrepublik Deutschland, "Biologische Wasserstoffgewinnung"), DFG (Deutsche Forschungsgemeinschaft, Graduierten-Kolleg "Enzymchemie" and DFG-Project "Hydrogenasen oxygener Mikroorganismen"), COST Action 8.18 (European Co-Operation in the Field of Scientific and Technical Research, "Hydrogenases and their Biotechnical Applications"), Fonds der Deutschen Chemischen Industrie, Deutsche Bundesstiftung Umwelt, and Studienstiftung des Deutschen Volkes.

REFERENCES

Adams, M.W.W., 1990, The metabolism of hydrogen by extremely thermophilic, sulfur-dependent bacteria, *FEMS Microbiol. Rev.*, 75:219–238.
Adams, M.W.W., Mortenson, L.E., and Chen, J.S., 1981, Hydrogenase, *Biochim. Biophys. Acta*, 594:105–176.

Appel, J., and Schulz, R., 1996, Sequence analysis of an operon of a NAD(P)-reducing nickel hydrogenase from the cyanobacterium *Synechocystis* sp. PCC 6803 gives additional evidence for direct coupling of the enzyme to NAD(P)H-dehydrogenase (complex I), *Biochim. Biophys. Acta*, 1298:141–147.

Boichenko, V.A., and Hoffmann, P., 1994, Photosynthetic hydrogen production in procaryotes and eucaryotes: occurrence, mechanism, and function, *Photosynthetica*, 30:527–552.

Bui, E.T.N., and Johnson, P.J., 1996, Identification and characterization of [Fe]-hydrogenases in the hydrogenosome of *Trichomonas vaginalis*, *Mol. Biochem. Parasitology*, 76:305–310.

Follmann, H., and Häberlein, I., 1996, Thioredoxins: universal, yet specific thiol-disulfide redox cofactors, *BioFactors*, 5:147–156.

Fontecilla-Camps, J.C., 1996, The active site of Ni-Fe hydrogenases: model chemistry and crystallographic results, *JBIC*, 1:91–98.

Gaffron, H., 1939, Über auffallende Unterschiede in der Physiologie nahe verwandter Algenstämme, nebst Bemerkungen über die Lichtatmung, *Biol. Zentralblatt*, 59:302–312.

Gaffron, H., 1940, Carbon dioxide reduction with molecular hydrogen in green algae, *Am. J. Bot.*, 27:273–283.

Gaffron, H., and Rubin, J., 1942, Fermentative and photochemical production of hydrogen in algae, *J. Gen. Physiol.*, 26:219–240.

Graf, E.G., and Thauer, R.K., 1981, Hydrogenase from *Methanobacterium thermoautotrophicum*, a nickel-containing enzyme, *FEBS Lett.*, 136:165–169.

Hahn, D., and Kück, U., 1994, Biochemical and molecular genetic basis of hydrogenases, *Proc. Biochem.*, 29: 633–644.

Happe, T., and Naber, J.D., 1993, Isolation, characterization, and N-terminal amino acid sequence of hydrogenase from the green alga *Chlamydomonas reinhardtii*, *Eur. J. Biochem.*, 214:475–481.

Happe, T., Mosler, B., and Naber, D., 1994, Induction, localization, and metal content of hydrogenase in the green alga *Chlamydomonas reinhardtii*, *FEBS Lett.*, 222:769–774.

Harden, A., 1901, The chemical action of *Bacillus coli communis* and similar organisms on carbohydrates and allied compounds, *J. Chem. Soc.*, 79:610–628.

Kessler, E., 1974, Hydrogenase, photoproduction and anaerobic growth, in *Algal Physiology and Biochemistry*, Stewart, W.D.P. (ed.), Blackwell Scientific, Oxford, pp. 456–473.

Marvin-Sikkema, F.D., Kraak, M.N., Vennhuis, M., Gottschal, J.C., and Prins, R.A., 1993, The hydrogenosomal enzyme hydrogenase from the anaerobic fungus *Neocallimastix* sp. L2 is recognized by antibodies, directed against the C-terminal microbody protein targeting signal SKL, *Eur. J. Cell Biol.*, 61:86–91.

Pakes, W.C.C., and Jollyman, W.H., 1901, The bacterial decomposition of formic acid into carbon dioxide and hydrogen, *J. Chem. Soc.*, 79:386–391.

Przybyla, A.E., Robbins, J., Menon, N., and Peck Jr., H.D., 1992, Structure-function relationships among the nickel-containing hydrogenase, *FEMS Microbiol. Rev.*, 88:109–136.

Randt, C., and Senger, H., 1985, Participation of the two photosystems in light dependent hydrogen evolution in *Scenedesmus*, *Photochem. Photobiol.*, 42:553–557.

Sambrook, J., Fritsch, E.F., and Maniatis, T., 1989, *Molecular Cloning: A Laboratory Manual*, Cold Spring Harbor Laboratory, Cold Spring Harbor, New York.

Schnackenberg, J., Schulz, R., and Senger, H., 1993, Characterization and purification of a hydrogenase from the eukaryotic green alga *Scenedesmus obliquus*, *FEBS Lett.*, 327:21–24.

Schulz, R., 1996, Hydrogenases and hydrogen production in eukaryotic organisms and cyanobacteria., *J. Mar. Biotechnol.*, 4:16–22.

Schulz, R., and Borzner, S., 1996, Wasserstoff produzierende Hydrogenasen - Schlüsselenzyme zur Lösung des Energieproblems? *Chemie in Labor und Biotechnik*, 47:496–500.

Senger, H., 1980, Is water the source of photohydrogen? in *Biological Applications of Solar Energy*, Gnanam, A. et al. (eds.), Maximillan Company of India, Ltd., Madras, Bombay, Calcutta, Delphi, pp. 196–198.

Senger, H., and Bishop, N.I., 1979, Observations on the photohydrogen producing activity during the synchronous cell cycle of *Scenedesmus obliquus*, *Planta*, 145:53–62.

Stephenson, M., and Stickland, L.H., 1931, Hydrogenase: a bacterial enzyme activating molecular hydrogen, *Biochem. J.*, 25:205–214.

Thauer, R.K., Klein, A.R., and Hartmann, G.C., 1996, Reactions with molecular hydrogen in microorganisms: evidence for a purely organic hydrogenation catalyst, *Chem. Rev.*, 96:3031–3042.

Urbig, T., Schulz, R., and Senger, H., 1993, Inactivation and reactivation of the hydrogenases of the green algae *Scenedesmus obliquus* and *Chlamydomonas reinhardtii*, *Z. Naturforsch.*, 48c:41–45.

Volbeda, A., Charon, M.H., Piras, C., Hatchikian, E.C., Frey, M., and Fontecilla-Camps, J.C., 1995, Crystal structure of the nickel-iron hydrogenase from *Desulfovibrio gigas*, *Nature*, 373:580–587.

Voordouw, G., 1992, Evolution of hydrogenase genes, *Adv. Inorg. Chem.*, 38:397–422.

Voordouw, G., and Brenner, S., 1985, Nucleotide sequence of the gene encoding the hydrogenase from *Desulfovibrio vulgaris* (Hildenborough), *Eur. J. Biochem.*, 148:515–520.

Voordouw, G., Strang, J.D., and Wilson, F.R., 1989, Organization of the genes encoding [Fe] hydrogenase in *Desulfovibrio vulgaris* subsp. oxamicus Monticello, *J. Bacteriol.*, 171:3881–3889.

Yamazaki, S., 1982, A selenium-containing hydrogenase from *Methanococcus vannielii*, *J. Biol. Chem.*, 257:7926–7929.

Zinn, T., Schnackenberg, J., Haak, D., Römer, S., Schulz, R., and Senger, H., 1994, Evidence for nickel in the soluble hydrogenase from the unicellular green alga *Scenedesmus obliquus*, *Z. Naturforsch.*, 49c:33–38.

Zirngibl, C., Hedderich, R., and Thauer, R.K., 1990, N^5-N^{10}-methylenetetra-hydromethanopterin dehydrogenase from *Methanobacterium thermoautotrophicum* has hydrogenase activity, *FEBS Lett.*, 261:112–116.

33

ASSOCIATION OF ELECTRON CARRIERS WITH THE HYDROGENASE FROM *Scenedesmus obliquus*

Jörg Schnackenberg,[1] Wolfgang Reuter,[2] and Horst Senger[3]

[1]National Institute of Bioscience and Human Technology
Higashi 1-1
Tsukuba-shi, Ibaraki-ken, Japan
[2]Max-Planck-Institut für Biochemie
Am Klopferspitz 18a
D-82152 Planegg-Martinsried, Germany
[3]Philipps Universität Marburg
Fachbereich Biologie
Karl-von Frisch-Straße
D-35032 Marburg/Lahn, Germany

Key Words

green algae, photosynthetic hydrogen production, hydrogenase, multiprotein complexes, ferredoxin, cytochrome c_6

1. SUMMARY

The hydrogenase from *Scenedesmus obliquus* has been purified as a multiprotein complex. The occurrence of the hydrogenase as a multiprotein complex has been confirmed by its purification via alternative isolation methods. The subunit composition reveals two proteins in addition to the previously described large and small hydrogenase subunits. These additional polypeptides could be identified as ferredoxin and cytochrome c_6 by spectroscopic and molecular analyses. Comparative kinetic studies with methylviologen and purified ferredoxin revealed a significantly higher affinity of hydrogenase toward ferredoxin. Whereas the close association between hydrogenase and ferredoxin has previously been postulated, the significant amount of cytochrome c_6 as part of the active hydrogenase complex requires new considerations about the electron pathway in photohydrogen metabolism.

2. INTRODUCTION

Hydrogenases are classified by the international enzyme commission as E.C.1.12.-.-. and catalyze the turnover of molecular hydrogen. The enzymes of procaryotes were initially described by Stephenson and Stickland (1931). Analogous enzymes were detected in green algae (Gaffron, 1939; Kessler, 1974; Adams et al., 1981; Biochenko and Hoffmann, 1994). Algal hydrogenases are not active under aerobic conditions; expression of the activity has to be induced by an anaerobic adaptation period (Bishop, 1966; Kessler, 1974; Ben-Amotz, 1979). Recently, the hydrogenases from *Chlamydomonas reinhardtii* (Happe and Naber, 1993) and *Scenedesmus obliquus* (Schnackenberg et al., 1993) have been purified and characterized. Although these organisms belong to the same algal group, the hydrogenases from *C. reinhardtii* and *S. obliquus* exhibit significant differences in their polypeptide composition and metal content. Whereas the hydrogenase from *C. reinhardtii* is a monomeric enzyme with M_r 48000 and belongs to the group of "Fe-only"-hydrogenases (Happe and Naber, 1993; Happe et al. 1994), the *S. obliquus* hydrogenase comprises two protein subunits with molecular masses of M_r 55000 and 36000, respectively (Schnackenberg et al., 1993). Additionally, the presence of nickel within the protein complex could be shown (Zinn et al., 1994). The hydrogenase from *C. reinhardtii* was found to be exclusively located in the chloroplast (Maione and Gibbs, 1986). In principle, the close interaction of hydrogenase with photosynthetic centers (Efimtsev et al., 1976) suggests that the algal hydrogenase is loosely bound to the thylakoid membrane (Weaver et al., 1980) and forms binary complexes of hydrogenase and ferredoxin at anaerobic conditions, similar to the ferredoxin : $NADP^+$ reductase complexes (Apley and Wagner, 1988; Knaff, 1991). Formation of molecular hydrogen (H_2) requires the transfer of two electrons via ferredoxin. It could be shown that a single turnover of the photosynthetic reaction centers, that is, the splitting of one molecule of H_2O mediated by the oxygen-evolving complex (OEC), provides the hydrogenase with the two electrons necessary for the formation of H_2 (Greenbaum, 1988). Consequently, one molecule of hydrogenase must either simultaneously or sequentially react with two molecules of ferredoxin. Besides cyanobacteria, green algae represent the predominant group of photosynthetic, hydrogen-producing organisms (Boichenko and Hoffmann, 1994). From cyanobacteria, some strategies for exclusion of high intracellular oxygen concentrations, which would inhibit hydrogen and nitrogen turnover, are known (Tandeau de Marsac and Houmard, 1993; Brass et al., 1994). Similar adaptation mechanisms for green algae are yet uncertain. A solution to the problem of the high oxygen sensitivity of their hydrogenases is required for realization of biological hydrogen production with green algae. Screening for organisms or mutants with oxygen-tolerant hydrogenase-systems may overcome this problem (Gogotov, 1986). Nevertheless, accompanying biochemical and genetic studies, as well as investigations on the structure-function relation in hydrogen-producing systems, can provide the necessary information, enabling the improvement of the stability of hydrogenase either by genetic manipulation or by culture conditions. In addition, understanding of the mechanism by which the hydrogenase is inhibited by oxygen would provide valuable information that could lead to the development of new strategies for efficient photohydrogen production.

Previous research showed that the partial inhibition of the water-splitting complex of PS II by a high pH environment leads to an extended duration of H_2 evolution (Schnackenberg et al., 1996), even though the amount of accumulated hydrogen is rather low compared with the initial rates of photohydrogen evolution in a photosynthetically active state of the cells. There is no doubt that in *S. obliquus*, both photosystems are required for efficient photohydrogen production (Randt and Senger, 1985). The contribution of photosynthesis, in particular of the O_2-evolving PS II, enhances the problem of deactivation of algal hydrogenases by molecular oxygen. Hence, understanding of the mechanism of inactivation of hydrogenase by oxygen is of prime importance. Based on our present knowledge, this phenomenon is not only due to the oxidation of metal ions in the active site of the enzyme. Additional structural changes which, in vitro, are irreversible must be suggested. To obtain more insight into this problem, we are developing new methods to purify stable hydrogenase complexes from *S. obliquus* in sufficient quantities. In the present study, we introduce a new method for the purification of hydrogenase and discuss the role of electron mediators.

3. EXPERIMENTAL

3.1. Growth of *Scenedesmus obliquus* and Anaerobic Adaptation

Cells of *Scenedesmus obliquus* D3 (Gaffron, 1939) were cultivated heterotrophically at 30 °C in the dark. The growth medium (KNO_3 8.0×10^{-3} M, NaCl 8.0×10^{-3} M, $Na_2HPO_4 \times 2 H_2O$ 1.0×10^{-3} M, $NaH_2PO_4 \times 2 H_2O$ 3.0×10^{-3} M, $CaCl_2 \times 2 H_2O$ 1.0×10^{-4} M, $MgSO_4 \times 7 H_2O$ 1.0×10^{-3} M, $Fe_2(SO_4)_3 \times 6 H_2O$ 7.5×10^{-3} M, Na-Citrat $\times 2 H_2O$ 5.5×10^{-4} M) according to Kessler et al. (1957) has been previously modified by Bishop and Senger (1971) and was supplemented with the trace elements H_3BO_3 4.5×10^{-5} M, $MnCl_2 \times 6 H_2O$ 8.0×10^{-6} M, $ZnSO_4 \times 6 H_2O$ 7.0×10^{-7} M, $CuSO_4 \times 6 H_2O$ 3.0×10^{-7} M, and MoO_3 1.0×10^{-7} M (Kratz and Myers, 1955). For heterotrophical growth, the cells were additionally supplied with 0.5% (wt/v) glucose and 0.25% (wt/v) yeast extract (Bishop and Wong, 1971).

The cells were harvested by centrifugation (1400 x g, 10 min), washed twice with 50 mM bis-Tris buffer, pH 7.5, and adjusted to a packed cell volume of 500 µl mL^{-1}. The algal suspension was anaerobically adapted by sequential evacuation and flushing with nitrogen followed by a dark adaptation period of 4 h at 32 °C (Strecker and Senger, 1981).

3.2. Biochemical Procedures

Adapted cells were disrupted in a vibrogen cell mill (Schnackenberg et al., 1993) and purified from cell debris by centrifugation at 48000 x g. To remove non-soluble particles, this protein extract was centrifuged at 300000 x g for 1 h at 4 °C. The supernatant, usually containing near 95% of hydrogenase activity, was used for further experiments.

Corresponding to their electrical charge, the proteins were separated on a Fractogel TSK DEAE-650 (Merck, Darmstadt, Germany) column (29 × 150 mm) equilibrated with bis-Tris buffer (50 mM, pH 7.5). Bound proteins were eluted stepwise (50 mM) with a NaCl-gradient from 50–400 mM (in bis-Tris, 50 mM, pH 7.5).

Hydrogenase-active fractions were separated on different non-denaturing agarose and polyacrylamide gels. Gel electrophoreses were performed in a Tris/boric acid buffer system, comprising 100 mM Tris, 73 mM boric acid, and 7% (wt/v) sucrose at pH 8.5 (Reuter and Wehrmeyer, 1988).

The subunit compositions of the hydrogenase complexes were determined on SDS-PAGE as previously described by Reuter and Wehrmeyer (1988).

Non-denaturing and SDS polyacrylamide gels were stained with Serva Blue R-250 (Reuter and Wehrmeyer, 1988).

Absorbance of the fractions was recorded with a Hitachi U 3200 spectrophotometer.

The activity of hydrogenase in agarose and polyacrylamide gels was detected according to the method of Ackrell et al., (1969). In solution, hydrogen evolution was measured amperemetrically with a Clark-type electrode (Schnackenberg et al., 1993). The enzyme assay consisted of 10 mM Na-dithionite and 5 mM methylviologen in Tris/HCl buffer (50 mM, pH 6.3).

4. RESULTS AND DISCUSSION

After anaerobic adaptation, disruption, and ultracentrifugation, the entire hydrogenase activity was exclusively found in the water-soluble centrifugation supernatant (Schnackenberg et al., 1993). This protein fraction was used for further purification under strict anaerobic conditions. Anion exchange chromatography revealed two hydrogenase-active fractions eluted at NaCl concentrations of 200 and 300 mM NaCl. Although the absorbance spectra of the two fractions show significant differences (Figures 1a and 1b), both reveal a high absorbance level around 325 nm, indicating the presence of hydrogenase (Schneider and Schlegel, 1976; Hatchikian et al., 1978). Subsequent experiments with non-denaturing gel electrophoresis proved this method to be suitable to achieve further purification of the hydrogenase. The separation of the active fractions, either on agarose or polyacrylamide tube gels, revealed no significant differences in the molecular composition of the colored hydrogenase fractions (Figure 2). Preparative non-denaturing electrophoresis on agarose was chosen as the second purification step. Spectral analyses of the further purified hydrogenase fractions (DEAE-200 and DEAE-300) gave initial evidence of the additional presence of cytochrome in the DEAE-200 fraction and ferredoxin in the DEAE-300 fraction. Furthermore, an increase of the absorbance at 325 nm relative to that of the proteins at 280 nm can be observed (Figures 1c and 1d). The hydrogenase-active protein bands obtained by separation on agarose tube gels were eluted and subsequently separated on a non-denaturing polyacrylamide slab gel (Figure 3). In the electrophoresis, a red-colored (lane a, b; band 2) and a brown-colored protein band (lane c, d; band 1) are clearly visible. Protein staining (Figure 3) showed only one additional protein band (band 3). The experiment also demonstrated that the presence of oxygen in aerobically isolated hydrogenase fractions has no influence on the stability of the protein complexes (lane b, d). After the second gel electrophoretic purification step, the two virtual hydrogenase complexes derived from the DEAE-200 and DEAE-300 fraction were eluted and their absorbance spectra recorded. The purified DEAE-200 fraction showed the characteristic absorbance spectrum of a reduced c-type cytochrome, whereas the purified DEAE-300 fraction shows, besides the hydrogenase-specific absorbance at 325 nm, two additional maxima that could be attributed to ferredoxin (Figure 1e and 1f). After electrophoresis, remarkable amounts of aggregated protein could be detected on the gel surface and were eluted for further investigation. This aggregation behavior is probably due to reduced solubility by strong protein-protein interactions and/or strong surface activity of these proteins at high concentrations.

All results suggest that the hydrogenase from *S. obliquus* is part of one multiprotein complex. The occurrence of the two different hydrogenase complexes is caused by the conditions during anion exchange chromatography. Consequently, it could be omitted by direct electrophoretical separation of the soluble protein fraction on stabilizing polyacrylamide gels. Under these conditions, only one colored hydrogenase-active protein band appeared (Figures 4a and 4b). The dissociation of the multiprotein complex into two sub-complexes is probably due to the dissociating effect of the high concentration of Cl^- ions

Association of Electron Carriers with the Hydrogenase from *Scenedesmus obliquus*

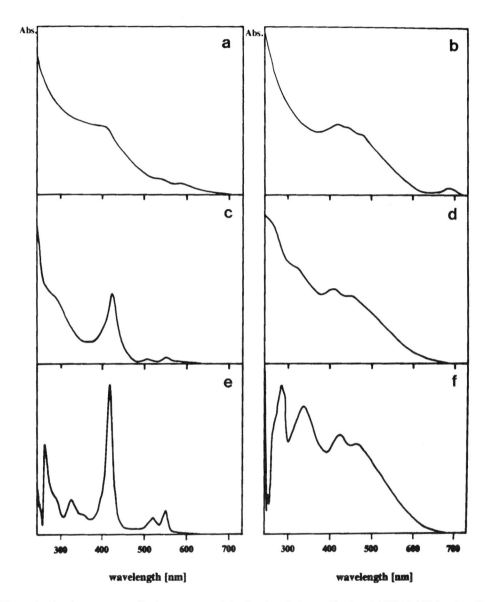

Figure 1. Absorbance spectra of hydrogenase containing fractions during purification: (a) DEAE-200 fraction; (b) DEAE-300 fraction; (c) DEAE-200 fraction purified on agarose tube gel (1.2%); (d) DEAE-300 fraction further purified on agarose gel (1.2%); (e) final purification of the DEAE-200 fraction, eluted from agarose tube gels, on a non-denaturing polyacrylamide gel (9%); (f) final purification of the DEAE-300 fraction eluted from agarose tube gels, on a non-denaturing polyacrylamide gel (9%).

Figure 2. Separation of hydrogenase-active DEAE-fractions on agarose/polyacrylamide gels: (a) hydrogenase-active DEAE-200 fraction separated on a polyacrylamide tube gel (3% polyacrylamide); (b) hydrogenase-active DEAE-300 fraction separated on a polyacrylamide tube gel (3% polyacrylamide); (c) hydrogenase-active DEAE-200 fraction separated on an agarose tube gel (1.2% agarose); (d) hydrogenase-active DEAE-300 fraction separated on an agarose tube gel (1.2% agarose).

during anion exchange chromatography. This hypothesis is supported by the results of the polypeptide analysis of the two hydrogenase subcomplexes (Figure 5). Both purified DEAE-fractions show the two hydrogenase subunits with M_r 55000 and 36000, as reported earlier (Schnackenberg et al., 1993). The DEAE-300 fraction (Figures 5A, b) contains an additional polypeptide with a molecular mass of approximately M_r 23000 identified as ferredoxin, whereas the DEAE-200 fraction (Figures 5A, c) shows the low molecular weight polypeptide identified as cytochrome c. The precipitated proteins mainly consist of hydrogenase polypeptides (Figures 5B, a, b). Precipitation of the hydrogenase proteins on the surface of the polyacrylamide gels is obviously caused by the dissociation of the protein complexes. A similar effect has been observed during anion exchange chromatography on Mono-Q material (Pharmacia, Freiburg, Germany) consisting of a polystyrol matrix (Schnackenberg et al., 1993). Only the association with the electron carriers cytochrome and/or ferredoxin prevents the precipitation of the hydrogenase.

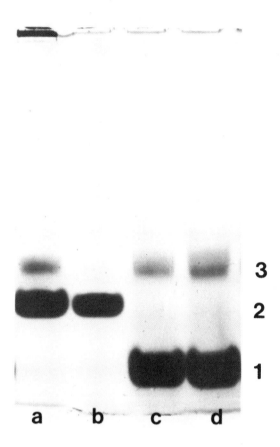

Figure 3. Final purification of hydrogenase-active fractions on non-denaturing polyacrylamide gels. After purification of the hydrogenase containing DEAE-200 and DEAE-300 fractions on agarose tube gels, final purification of the hydrogenase complexes was achieved on a non-denaturing polyacrylamide gel (9% polyacrylamide). (a) DEAE-200 fraction, aerobically isolated; (b) DEAE-200 fraction, anaerobically isolated; (c) DEAE-300 fraction, aerobically isolated; (d) DEAE-300 fraction, anaerobically isolated.

For characterization of the c-type cytochrome, the polypeptide-enriched subcomplex was eluted from the non-denaturing polyacrylamide gel at non-stabilizing conditions. The dissociated proteins were electrophoretically separated on a non-denaturing polyacrylamide gel. The resulting purified cytochrome was again eluted and absorbance spectra were recorded (Figure 6). The cytochrome showed absorbance maxima at 553 nm (α), 522 nm (β), 416 nm (γ), and 319 nm (δ), typical for cytochrome c_6. Any attempts to reduce the purified cytochrome failed, whereas the addition of ferricyanide (40 pM) led to the typical changes in the absorbance maxima reported for cytochrome c_6 in its oxidized state (Bendall, 1968). At slightly alkaline pH values, the reduced cytochrome c_6 appears very stable against oxidation; under acidic pH, both the reduced and oxidized form are present and the oxidized form increases with time.

This study clearly demonstrates the close association of the hydrogenase from *S. obliquus* with the electron carriers ferredoxin and cytochrome c_6, forming a multiprotein complex. The function of ferredoxin as an electron donor for the formation of molecular

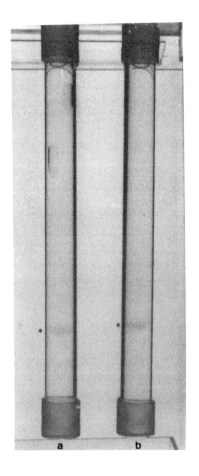

Figure 4. Direct separation of a hydrogenase containing water-soluble protein extract from *S. obliquus*. After induction of hydrogenase activity, an extract of water-soluble protein was prepared under (a) anaerobic or (b) aerobic conditions and directly separated on polyacrylamide tube gels (3% polyacrylamide). Colored protein bands are indicated by asterisks; only the anaerobically isolated protein band expressed hydrogenase activity.

hydrogen via hydrogenase has been reported earlier (Happe and Naber, 1993) and is supported by the presented results. The determined K_M-value of 28.5 μM for ferredoxin as the electron donor for the hydrogenase-mediated evolution of molecular hydrogen is approximately 80 times lower than the K_M-value determined for the artificial electron donor methylviologen (Schnackenberg, unpublished data). This fact points out the high affinity of hydrogenase for reduced ferredoxin, leading to the conclusion that ferredoxin is the natural electron donor for the formation of molecular hydrogen.

The occurrence of cytochrome c_6 in association with hydrogenase raises a number of questions. In principle, the participation of c-type cytochromes in hydrogen evolution and in the oxidation of molecular hydrogen is well-characterized. In *Desulfovibrio vulgaris* strain MIYAZAKI, cytochrome c_3 acts as an electron donor for light-induced hydrogen evolution (Hiraishi et al., 1996), whereas in *D. vulgaris*, cytochrome c_3 is involved in the hydrogenase mediated oxidation of molecular hydrogen (R'zaigui et al., 1980). In *S. obliquus*, neither cytochrome c_3 nor c_6 acts as an electron donor for light-induced hydrogen evolution (Schnackenberg, unpublished data) but the potential role of cytochrome c_6 in hydrogen oxidation has not yet been investigated.

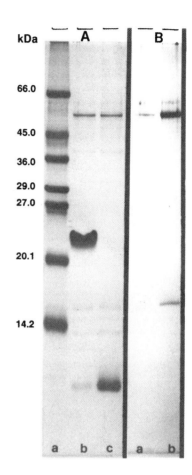

Figure 5. SDS gel electrophoresis of the purified hydrogenase complexes. (A) DEAE-200 and DEAE-300 fraction after separation on agarose tube gels and non-denaturing polyacrylamide slab-gels: (a) marker proteins (MW SDS-70L, SIGMA, München, Germany); (b) DEAE-300 fraction; (c) DEAE-200 fraction. (B) Analysis of the proteins precipitated on the surface of non-denaturing polyacrylamide slab gels [Figure 3]; (a) DEAE-200 fraction; (b) DEAE-300 fraction.

Based on the present knowledge, synthesis of either cytochrome c_6 or its analogous molecule plastocyanin only depends on regulation by the relative availability of iron and copper during growth (Merchant and Bogorad, 1987). Both proteins primarily function as mobile electron carriers between cytochrome b_6/f and PS I complexes. Hence, in the photosynthetic state of S. obliquus, they are assumed to be exclusively located within the thylakoid lumen. Cytochrome c_6 is encoded by the nuclear petJ gene and the polypeptide is synthesized in the cytoplasm as a preapocytochrome, which contains a transit peptide for translocation into the chloroplast and/or the thylakoid lumen (Merchant and Bogorad, 1987; Bovy et al., 1992; Howe and Merchant, 1994). Nevertheless, the proof for its exclusive position in the thylakoid lumen has not yet been adduced. The present knowledge about the location and function of cytochrome c_6 does not exclude its presence on both sides of the thylakoid membrane. Ferredoxin is an electron carrier located in the stroma of the chloroplast interacting with hydrogenase and/or $NADP^+$:reductase. Both cytochrome c_6 and ferredoxin can be isolated as part of a common hydrogenase-active multiprotein complex. This result suggests the location of both proteins at one side of the thylakoid membrane. Therefore, depending on the anaerobic metabolism of S. obliquus, cytochrome c_6 may act on both sides of the thylakoid membrane and probably represent a bifunctional

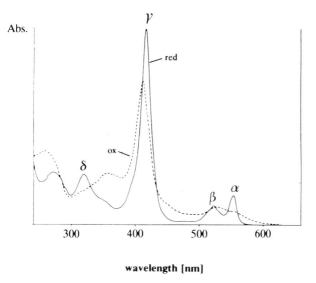

Figure 6. Absorbance spectra of cytochrome c_6 from *S. obliquus*: After dissociation of the cytochrome containing hydrogenase complex (DEAE-200 fraction), the cytochrome was separated on a non-denaturing polyacrylamide gel (9% polyacrylamide). The eluted cytochrome shows the characteristic absorbance spectra of a reduced *c*-type cytochrome. Upon the addition of Na-dithionite (10 pM), no changes in the absorbance maxima could be detected, whereas the addition of ferricyanide (40 pM) induced oxidation of the cytochrome, indicated by the characteristic changes of the absorbance maxima.

electron carrier molecule. However, the meaning of cytochome c_6 within the anaerobic adaptation and metabolism can be clarified by extensive studies of the synthesis and the activation mechanism of the hydrogenase complex.

ACKNOWLEDGMENTS

This study was supported by the Bundesministerium für Forschung und Technologie (Biologische Wasserstoffgewinnung) and by grants from the Deutsche Forschungsgemeinschaft (RE 1131-2/1, 2/2). The authors would like to thank Ms. C. Nickel-Reuter for her help with the figures and manuscript.

REFERENCES

Ackrell, B.A.C., Asato, R.N., and Mowers, H.F., 1969, Multiple forms of bacterial hydrogenases, *J. Bacteriol.*, 92(4):828–838.

Adams, M.W.W., Mortenson, L.E., and Chen, J.S., 1981, Hydrogenase, *Biochim. Biophys. Acta*, 594:105–176.

Apley, E.C., and Wagner, R., 1988, Chemical modification of the active site of ferredoxin-$NADP^+$ reductase and conformation of the binary ferredoxin/ferredoxin-NAD^+ reductase complex in solution, *Biochim. Biophys. Acta*, 936:269–279.

Ben-Amotz, A., 1979, Hydrogen metabolism, in *Encyclopedia of Plant Physiology*, vol. 6, Gibbs, M., and Latzko, E., (eds.), Springer Verlag, Berlin, pp. 497–506.

Bendall, D.S., 1968, Oxidation reduction potentials of cytochromes in chloroplasts from higher plants, *Biochem. J.*, 109:46–47.

Bishop, N.I., 1966, Partial reactions of photosynthesis and photoreduction, *Ann. Rev. Plant Physiol.*, 17:185–208.

Bishop, N.I., and Senger, H., 1971, Preparation and photosynthetic properties of synchronous cultures of *Scenedesmus*, in *Methods in Enzymology*, vol. 23a, San Pietro, A., (ed.), Academic Press, New York, pp. 53–66.

Bishop, N.I., and Wong, J., 1971, Alternate fates of the photochemical reducing power generated in photosynthesis: hydrogen production and nitrogen fixation, *Curr. Topics Bioenergetics*, 8:3–31.

Boichenko, V.A., and Hoffmann, P., 1994, Photosynthetic hydrogen production in procaryotes and eucaryotes: occurrence, mechanisms, and functions, *Photosynthetica*, 30(4):527–552.

Bovy, A., de Vrieze, G., Borrias, M., and Weisbeek, P., 1992, Isolation and sequence analysis of a gene encoding a basic cytochrome c-553 from the cyanobacterium *Anabaena* SP. PCC 7937, *Plant Mol. Biol.*, 19:491–492.

Brass, S., Westermann, M., Ernst, A., Reuter, W., Wehrmeyer, W. and Böger, P., 1994, Utilization of light for nitrogen fixation by a new *Synechocystis* strain is extended by its low photosynthetic efficiency, *Appl. Environ. Microbiol.*, 60:2575–2583.

Efimtsev, E.I., Boichenko, V.A., Zatolokin, N.E., and Litvin, F.F., 1976, Investigation of primary processes of photoinduced hydrogen evolution in *Chlorella* under flash illumination, *Dokl. Akad. Nauk.* (in Russian), SSSR 227:731–734.

Gaffron, H., 1939, Reduction of carbon dioxide with molecular hydrogen in green algae, *Nature*, 143:204–205.

Gogotov, I.N., 1986, Hydrogenases of phototrophic microorganisms, *Biochemie*, 68:181–187.

Greenbaum, E., 1988, Hydrogen production by algal water splitting, in *Algae and Human Affairs*, Lembi, C.A., and Waaland, J.R. (eds.), Cambridge University Press, Cambridge, pp. 283–304.

Happe, T., and Naber, J.D., 1993, Isolation, characterization, and N-terminal amino acid sequence of hydrogenase from the green alga *Chlamydomonas reinhardtii*, *Eur. J. Biochem.*, 214:475–481.

Happe, T., Mosler, B., and Naber, D.J., 1994, Induction, localization, and metal content of hydrogenase in the green alga *Chlamydomonas reinhardtii*, *Eur. J. Biochem.*, 222:769–774.

Hatchikian, E.C., Bruschi, M., and LeGall, J., 1978, Characterization of the periplasmic hydrogenase from *Desulfovibrio gigas*, *Biochem. Biophys. Res. Comm.*, 82(2):451–461.

Hiraishi, T., Kamachi, T., and Okura, I., 1996, Kinetic studies of electron transfer on photoinduced hydrogen evolution with hydrogenase, *J. Photochem. Photobiol. A. Chem.*, 101(1):45–47.

Howe, G., and Merchant, S., 1994, Role of heme in the biosynthesis of cytochrome c_6, *J. Biol. Chem.*, 269(8):5824–5832.

Kessler, E., 1974, Hydrogenase, photoproduction and anaerobic growth, in *Algal Physiology and Biochemistry*, Stewart, W.D.P. (ed.), Blackwell Scientific, Oxford, pp. 456–473.

Kessler, E., Arthur, W., and Brugger, J.E., 1957, The influence of manganese and phosphate on delayed light emission, fluorescence, photoreduction, and photosynthesis in algae, *Arch. Biochem.*, 71:326–335.

Knaff, D.B., 1991, Ferredoxin-dependent enzymes, *Biochim. Biophys. Acta*, 1056:93–125.

Kratz, W.A., and Myers, F., 1955, Nutrition and growth of several blue-green algae, *Am. J. Bot.*, 42:282–285.

Maione, T.E., and Gibbs, M., 1986, Hydrogenase-mediated activities in isolated chloroplasts of *Chlamydomonas reinhardtii*, *Plant Phyiol.*, 80:360–363.

Merchant, S., and Bogorad, L., 1987, The Cu(II)-repressible plastidic cytochrome c, *J. Biol. Cem.*, 262:9062–9067.

Randt, C., and Senger, H., 1985, Participation of the two photosystems in light-dependent hydrogen evolution in *Scenedesmus*, *Photochem. Photobiol. J.*, 42(5):553–557.

Reuter, W., and Wehrmeyer, W., 1988, Core substructure in *Mastigocladus laminosus* phycobilisomes. I. microheterogeneity in two of three allophycocyanin core complexes, *Arch. Microbiol.*, 150:534–540.

R'zaigui, M., Hatchikian, E.C., and Benlian, D., 1980, Redox reactions of the hydrogenase-cytochrome c_3 system from *Desulfovibrio gigas* with the synthetic analogue of ferredoxin active sites $[Fe_4S_4(S-CH_2-CH_2 OH)_4]^{2-}$, *Biochem. Biophys. Res. Comm.*, 92(4):1258–1265.

Schnackenberg, J., Schulz, R., and Senger, H., 1993, Characterization and purification of a hydrogenase from the eukaryotic green alga *Scenedesmus obliquus*, *FEBS Lett.*, 327:21–24.

Schnackenberg, J., Ikemoto, H., and Miyachi, S., 1996, Photosynthesis and hydrogen evolution under stress conditions in a CO_2-tolerant marine green alga, *Chlorococcum littorale*, *J. Photochem. Photobiol. D/B. Biol.*, 34:59–62.

Schneider, K., and Schlegel, H.G., 1976, Purification and properties of soluble hydrogenase from *Alcaligenes eutrophus* H16, *Biochim. Biophys. Acta*, 452:66–80.

Stephenson, M., and Stickland, L.H., 1931, Hydrogenase: a bacterial enzyme activating molecular hydrogen, *Biochem. J.*, 25:205–214.

Strecker, U., and Senger, H., 1981, Photohydrogen evolution, photoreduction of CO_2, and hydrogenase activity during the cell cycle of *Scenedesmus obliquus*, in *Photosynthesis VI, Photosynthesis and Productivity, Photosynthesis and Environment*, Akoyunoglou, G. (ed.), Balaban International Science Service, Philadelphia, pp. 681–688.

Tandeau de Marsac, N., and Houmard, J., 1993, Adaptation of cyanobacteria to environmental stimuli: new steps towards molecular mechanisms, *FEMS Microbiol. Rev.*, 104:119–190.

Weaver, P.F., Lien, S., and Seibert, M., 1980, Photobiological production of hydrogen, *Solar Energy*, 24:3–45.

Zinn, T., Schnackenberg, J., Haak, D., Römer, S., Schulz, R., and Senger, H., 1994, Evidence for nickel in the soluble hydrogenase from the unicellular green alga *Scenedesmus obliquus*, *Z. Naturforsch.*, 49c:33–38.

34

ALGAL CO_2 FIXATION AND H_2 PHOTOPRODUCTION

Akiko Ike, Ken-ichi Yoshihara, Hiroyasu Nagase, Kazumasa Hirata, and Kazuhisa Miyamoto

Environmental Bioengineering Laboratory
Graduate School of Pharmaceutical Sciences
Osaka University
1-6 Yamadaoka, Suita
Osaka 565, Japan

Key Words

microalgae, CO_2 fixation, lactic acid fermentation, photosynthetic bacteria, H_2 production

1. SUMMARY

A combined system for recovering air pollutants and converting them to H_2 was studied. CO_2 and nitric oxide (NO) were removed simultaneously from a model flue gas using *Dunaliella tertiolecta* and *Chlamydomonas reinhardtii* cultures in a tubular photobioreactor. The algal biomass was converted to a substrate suitable for bacterial H_2 production by the starch-hydrolyzing lactic acid bacterium, *Lactobacillus amylovorus*. The photosynthetic bacterium *Rhodobium marinum* produced H_2 from lactic acid fermentation products, and a conversion yield of 6 mol H_2 per mole of starch-glucose in the algal biomass was observed in a system starting from biomass of *D. tertiolecta*. Such a combined system would give additional value and thereby improve practical efficiency of solar energy conversion to H_2.

2. INTRODUCTION

In recent decades, studies of H_2 metabolism have identified the roles of the photosynthetic apparatus and enzymes in H_2 evolution and uptake. In an effort to develop efficient and stable biological systems for H_2 production, we have combined CO_2 fixation by

BioHydrogen, edited by Zaborsky *et al.*
Plenum Press, New York, 1998

microalgae and H_2 production by photosynthetic bacteria using lactic acid fermentation as a mediator.

Various systems of biological H_2 production have been proposed and investigated (Miyamoto, 1994). Reported data on the conversion efficiency of light to hydrogen energy is compared below: Very high efficiencies (6–24% based on absorbed photosynthetically active radiation) were observed with green algae in a specially designed experimental apparatus (Greenbaum, 1988). A high conversion efficiency (2.5%) was obtained with cyanobacterial cultures in a simple tubular reactor operated for 15 days (Miyamoto, 1995). However, the conversion efficiencies observed under outdoor conditions were much lower than those under artificial light conditions in the laboratory (Miyamoto, 1995). These facts point to the need to continue basic and engineering research on photosynthetic microorganisms and photobioreactors.

On the other hand, many chemical, physical, and biological systems for CO_2 fixation have been investigated to address greenhouse warming. In particular, biological systems such as planting and microalgal cultivation are considered to be environmentally friendly, energy-saving, and cost-effective. Compared to planting, microalgae are generally superior with respect to the CO_2 fixing rate. In addition, salty water and arid fields unsuitable for plant cultivation can be used for microalgal mass cultivation; it would not, therefore, compete with agriculture. Microalgal CO_2 fixation systems, however, would produce enormous amounts of waste biomass when applied to solving the global warming problem. Such algal biomass would easily return to CO_2, and it is therefore important to develop the means to effectively recycle such reclaimed CO_2.

We have proposed the use of microalgae for the recycling of air pollutants such as CO_2 and NO_x (Yoshihara et al., 1996; Nagase et al., 1997). In the present study, this system for the simultaneous recovery of CO_2 and NOx could be combined with a bacterial H_2 production system, in which algal biomass is pre-treated by lactic acid fermentation and thereafter converted to H_2 with the aid of photosynthetic bacteria.

3. EXPERIMENTAL

3.1. Microorganism and Cultivation Conditions

The green alga *Dunaliella tertiolecta* (ATCC30929) was cultivated in a modified f/2 seawater medium (Guillard and Ryther, 1962) in a vertical, tubular photobioreactor (culture volume, 4 L), as described by Yoshihara et al. (1996). *D. tertiolecta* was precultivated in the reactor at 25 °C under illumination from white fluorescent lamps (38 W/m^2) until the optical density of the cell culture at 680 nm (OD_{680}) reached 2.5. The cell concentration was monitored by measurement of the OD_{680}. The OD_{680} was correlated with the ash-free dry weight (AFDW) of the cell culture, which was determined as described by Nagase et al. (1997). For preculture, 15% CO_2 in air was supplied at the bottom of the reactor through a glass-ball filter (diameter, 10 mm; grain size, 40–50 µm) at a flow rate of 150 mL/min. After preculture, *D. tertiolecta* was cultivated continuously, keeping the OD_{680} at 2.5 by supplying a new medium and withdrawing the culture liquid at the same speed (turbidstat method). In continuous cultivation, 15% CO_2 in N_2 was supplied for two days and then a model flue gas (100 ppm NO, 15% CO_2 in N_2) was supplied at a flow rate of 150 mL/min.

For the lactic acid fermentation experiments, the freshwater green alga *Chlamydomonas reinhardtii* (IAM C-238), grown photoautotrophically in a modified Bristol medium (Miura et al., 1982), and a marine green alga, *D. Trtiolecta*, also grown photoautotrophically in f/2 medium, were used. The algae were cultivated at 30 °C for 10 days in a cylindrical glass bottle (working volume, 2 L; diameter, 10 cm) with 1% CO_2 gas supplied continuously under illumination (10 W/m^2, fluorescent lamp).

The lactic acid bacterium *Lactobacillus amylovorus* (ATCC 33620) was grown anaerobically in Lacto Bacii MRS broth (Difco, United States) at 37 °C. The photosynthetic bacterium *Rhodobium marinum* (ATCC 35675) was grown anaerobically in a basal medium (Ike et al., 1997) with 3.92 g sodium lactate, 1.32 g ammonium sulfate, and 1.0 g yeast extract per 1 L, at 30 °C under illumination (170 W/m^2, tungsten lamps).

3.2. Lactic Acid Fermentation of Algal Biomass

C. reinhardtii and *D. tertiolecta* were harvested by centrifugation (13000 × g, 10 min) and biomass was prepared with a concentration 100 times as dense as that of the original algal culture. This concentrated and freeze-thawed (-30 °C) biomass was used for fermentation. Four milliliters of an actively growing culture of *L. amylovorus* (OD_{600} = ca. 9) was centrifuged (17000 × g, 10 min) and a cell pellet was added to 40 mL of algal biomass suspension. Inoculated samples were incubated at 37 °C for 2–4 days in plastic bottles (working volume, 50 mL). To prevent a cessation of fermentation occurring as a result of the low pH caused by the production of lactic acid, 10 N NaOH was added to the plastic bottles every 6 h throughout the fermentation as a neutralizer. The lactic acid fermentation by *L. amylovorus* from starch can be expressed by the following equation:

$$(C_6H_{12}O_6)_n \rightarrow 2nC_3H_6O_3$$

The lactic acid yield from the initial starch was calculated using this equation.

3.3. H_2 Production from Fermentates of Algal Biomass

The fermentates of *C. reinhardtii* and *D. tertiolecta* biomass were diluted with sterilized water to yield a lactic acid concentration of 1.8 g/L (20 mM). The medium components contained in the basal medium used for the photosynthetic bacteria (detailed above), plus 10 mM sodium glutamate, 0.15% $NaHCO_3$, and 1% NaCl, were added to the diluted fermentate. The diluted fermentate (60 mL) was degassed and placed in a glass tube (working volume, 70 mL; diameter: 2.5 cm), provided with a gas outlet and sampling nozzle. The actively growing cultures of photosynthetic bacteria were harvested by centrifugation (6300 × g, 10 min) and resuspended in the diluted fermented to make the calculated OD_{660} of the bacteria ca. 1. The inoculated samples were incubated at 30 °C under illumination (330 W/m^2, tungsten lamps) for 5 days. The H_2 gas produced was trapped in a 100-mL glass cylinder above a water bath. H_2 production of photosynthetic bacteria from lactic acid can be expressed as:

$$C_3H_6O_3 + 3H_2O \rightarrow 6H_2 + 3CO_2$$

The H_2 yield from the initial lactic acid was calculated using this equation.

Assay methods The NO and O_2 concentrations in the outlet gas were measured as described by Nagase et al. (1997). Assays of volatile fatty acids (the sum of acetic, propionic, butyric, and valeric acids) and ammonia were done by methods described by Ike et al. (1996). Assays of starch, total sugar, total lipid, protein, lactic acid, and evolved H_2 gas were done by methods described by Ike et al. (1997).

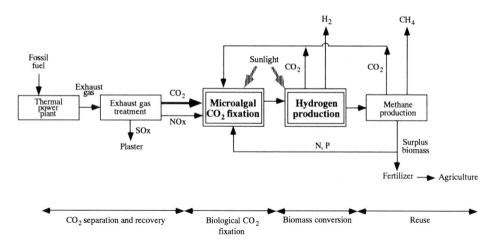

Figure 1. Total system for energy production based on microalgal CO_2 fixation.

4. RESULTS AND DISCUSSIONS

4.1. Total System for CO_2 Recycling

The recycling of materials and thus the minimizing of wastes should be basic goals in the coming 21st century. In terms of quantity, the most waste at present is CO_2, and therefore the recycling of CO_2 seems to be the most pressing issue. We propose the use of photosynthetic microalgae to fix CO_2, and thereby produce energy and chemicals as one approach to CO_2 recycling (Figure 1). The system for CO_2 recycling consists of three main steps. The first stage is designed for the recovery of CO_2 originating from a large point source such as an electric power plant, using various established technologies such as chemical absorption. In the second step, recovered CO_2 is photosynthetically fixed by microalgal cultures, which would produce a large amount of biomass. However, algal biomass itself is regarded as a low-grade fuel because of its high moisture content. Although we could make a powdered fuel of dried algae, it takes much energy in harvesting and drying to produce such a fuel. In addition, direct burning of biomass would produce nitrogen oxides (NO_x). These disadvantages would limit the use of microalgae as solid fuel. Algae, however, can be used to produce modern gaseous and liquid fuels in which lipids, carbohydrates, or proteins would be converted by chemical or biological processes to liquid fuels, methane, or hydrogen. Therefore, for the third stage, we propose a combined system where algal biomass is converted by biological or physicochemical treatment to prepare substrates for H_2 production promoted by photosynthetic or anaerobic bacteria. If we can produce additional useful materials such as foods and fertilizers, it would increase the efficiency of the total system.

4.2. Simultaneous Recovery of CO_2 and NO

In addition to a large amount of CO_2, the combustion of fossil fuels generates sulfur and nitrogen oxides (SO_x and NO_x). Emissions of these oxides and their derivatives cause acid rain, which disrupts the balance of the natural ecosystem. Existing technologies (e.g., wet limestone processes for desulfurization) have difficulty in removing nitric oxide (NO),

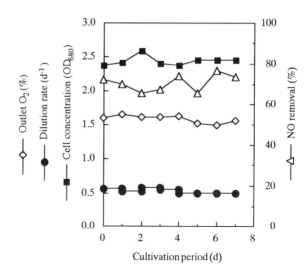

Figure 2. Continuous cultivation of *D. tertiolecta* to remove CO_2 and NO. The medium was continuously supplied to the culture to maintain the cell concentration at OD_{680} of 2.5. The model flue gas (15% CO_2 and 100 ppm NO in N_2) was supplied at a flow rate of 150 mL/min. One OD_{680} unit corresponds to a cell concentration of 0.28 g AFDW/L.

the dominant species of NO_x in flue gas, because NO is virtually insoluble in water. Therefore, a biological process for the simultaneous recovery of CO_2 and NO was investigated.

Previously, we found that some strains of microalgae could grow at inlet NO concentrations up to 500 ppm in an atmosphere of 15% CO_2 and remove 60–70% of NO, irrespective of the inlet NO concentration in a 2-m vertical tubular reactor (Yoshihara et al., 1996; Nagase et al., 1997). Figure 2 shows data for the continuous cultivation of *D. tertiolecta*, aerating 15% CO_2 and 100 ppm NO in N_2. For 7 days, the dilution rate was kept at 0.50–0.55 d^{-1} by monitoring the cell concentration. Thus, the CO_2 fixing rate was kept constant for seven days. As one OD_{680} unit corresponds to 0.28 g AFDW/L, the cell concentration of OD_{680} at 2.5 corresponds to 0.7 g/L. From the reactor volume of 4 L and dilution rate of 0.5 d^{-1}, the production rate of algal biomass, in which the carbon content of dried cells was taken as 50% (w/w) on the ash-free base, the amount of CO_2 fixed by the cells was calculated at 0.7 g C/day, that is, 2.6 g CO_2/day.

No evident inhibition of cell growth by NO in the model flue gas was observed. Almost 70% of the NO was eliminated constantly for seven days. On the other hand, 4.4% of supplied CO_2 was fixed by *D. tertiolecta* in this reactor system, as 58.3 g CO_2/day was supplied under conditions of 15% CO_2 and at 150 mL/min.

4.3. Lactic Acid Fermentation of Algal Biomass

We found that algal biomass is a good substrate for lactic acid fermentation (Ike et al., 1997), and therefore lactic acid fermentation of algal biomass was subsequently studied. *D. tertiorecta* biomass was tested for its value as a substrate for lactic acid fermentation by the starch-hydrolyzing lactic acid bacterium, *Lactobacillus amylovorus*, (Nakamura, 1981). *C. reinhardtii* biomass was also tested because this alga accumulated up to 50% starch in the cells. When we used the authentic starch as a substrate for fermentation by *L. amylovorus*, addition of nutrients was essential. In the case of algal biomass, however, fermentation proceeded much faster without any additional nutrients. Both *D. tertiorecta* and *C. reinhardtii* biomass were effectively converted to lactic acid with high yields of 98 and 93%, respectively.

The lactic acid fermentates were analyzed to determine the contents of total sugar (starch, in large part), glycerol, total lipid, protein, and lactic acid. Figure 3 compares the composition of the original biomass and the lactic acid fermentates. Starch (80% of total

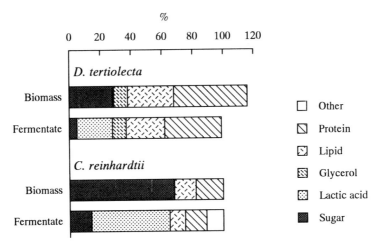

Figure 3. Lactic acid fermentation by *L. amylovorus*. The condensed (100 times as dense as the original algal culture) and freeze-thawed algal biomass were inoculated with *L. amylovorus* and incubated for 2–4 days at 37 °C.

sugar in *D. tertiorecta* and 85% in *C. reinhardtii* biomass) was reduced by fermentation and lactic acid content was increased correspondingly; no significant decrease was observed for any other component. Although *D. tertiorecta* is known to accumulate much glycerol as an osmoregulatory solute (Borowitzka and Brown, 1974), glycerol and lipid content was almost unchanged after fermentation.

4.4. H_2 Photoproduction from Lactic Acid Fermentate

Without any further treatment, fermentate of algal biomass could be used as a good substrate for H_2 production by a photosynthetic bacterium, *R. marinum*, which was se-

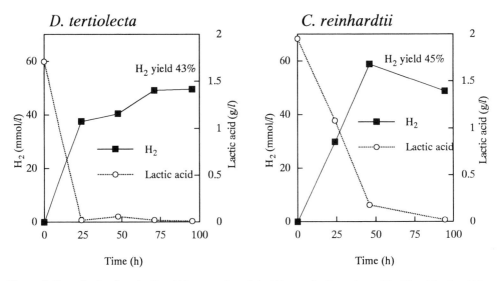

Figure 4. H_2 production from lactic acid fermentation of algal biomass by *R. marinum*. The diluted lactic acid fermentate of algal biomass containing 20 mM lactic acid was inoculated with *Rhodobium marinum* and incubated under illumination of 330 W/m² and with 1% NaCl at 30 °C.

lected as the best H_2 producer from the lactic acid fermentate of algal biomass. Figure 4 shows the time-course of consumption of lactic acid and production of H_2. *R. marinum* consumed lactic acid completely and evolved 50 and 60 mmol of H_2 per 1 L of fermentate of *D. tertiorecta* and *C. reinhardtii* biomass, with almost the same H_2 yield. After optimization for NaCl concentration, the yield could be improved up to 53% for *D. tertiorecta* biomass and 71% for *C. reinhardtii* biomass.

A multi-functional system for recovering pollutants and converting them to useful materials should be strategically designed to improve total conversion efficiency. We proposed here to use lactic acid fermentation as a mediator between algal CO_2 fixation and H_2 photoproduction. Algal biomass of *D. tertiolecta* and *C. reinhardtii* was successfully converted to H_2 with a good yield by the successive use of a lactic acid bacterium, *L. amylovorus*, and a photosynthetic bacterium, *R. marinum*. The molar yield of H_2 from starch accumulated in the *D. tertiorecta* biomass was 6.2 mol H_2/mol starch-glucose under salty conditions. The combined use of *D. tertiolecta* and *R. marinum* would make possible the use of seawater both in the steps of CO_2 (and NO_x) fixation and H_2 production, promising additional benefits for the system.

REFERENCES

Borowitzka, L.J., and Brown, A.D., 1974, The salt relation of marine and halophilic strains of unicellular green alga, *Dunaliella*; the role of glycerol as a compatible solute, *Archives of Microbiology*, 96:37–52.

Greenbaum, E., 1988, Energetic efficiency of hydrogen photoevolution by algal water splitting, *Biophysical Journal*, 54:365–368.

Guillard, L., and Ryther, J.H., 1962, Studies of marine planktonic diatoms, *Can. J. Microbiol.*, 8:229–239.

Ike, A., Saimura, C., Hirata, K., and Miyamoto, K., 1996, Environmentally friendly production of H_2 incorporating microalgal CO_2 fixation, *J. Mar. Biotechnol.*, 4:47–51.

Ike, A., Toda, N., Hirata, K., and Miyamoto, K., 1997, Hydrogen photoproduction from CO_2-fixing microalgal biomass: application of lactic acid fermentation by *Lactobacillus amylovorus*, *J. Ferment. Bioeng.*, 84:428–433.

Miura, Y., Yagi, K., Shoga, M., and Miyamoto, K., 1982, Hydrogen production by a green alga *Chlamydomonas reinhardtii* in an alternating light/dark cycle, *Biotechnol. Bioeng.*, 24:1555–1563.

Miyamoto, K., 1994, Hydrogen production by photosynthetic bacteria and microalgae, in *Recombinant Microbes for Industrial and Agricultural Applications*, Murooka, Y., and Imanaka, T. (eds.), Marcel Dekker, Inc., New York, Basel, and Hong Kong, pp. 771–786.

Miyamoto, K., 1995, Biological H_2 production, in *Bioremediation: The TOKYO '94 Workshop*, OECD Documents, pp. 625–630.

Nagase, H., Yoshihara, K., Eguchi, K., Yokota, Y., Matsui, R., Hirata, K., and Miyamoto, K., 1997, Characteristics of biological NO_x removal from flue gas in a *Dunaliella tertiolecta* culture system, *J. Ferment. Bioeng.*, 83:465–469.

Nakamura, L.K., 1981, *Lactobacillus amylovorus*, a new starch-hydrolyzing species from cattle waste-corn fermentations, *Int. J. Syst. Bacteriol.*, 31:56–63.

Yoshihara, K., Nagase, H., Eguchi, K., Hirata, K., and Miyamoto, K., 1996, Biological elimination of nitric oxide and carbon dioxide from flue gas by marine microalga NOA-113 cultivated in a long tubular photobioreactor, *J. Ferment. Bioeng.*, 82:351–354.

35

HYDROGEN PRODUCTION BY FACULTATIVE ANAEROBE *Enterobacter aerogenes*

Shigeharu Tanisho

Faculty of Engineering
Yokohama National University
79-2, Tokiwadai
Hodogaya-ku, Yokohama 240-8501, Japan

Key Words

hydrogen, *Enterobacter aerogenes*, dissolved oxygen concentration, fermentation, NADH, FAD

1. SUMMARY

Enterobacter aerogenes is a facultative anaerobic bacterium. It grows under both aerobic and anaerobic conditions. In our experiments, it assimilated glucose at a rate as high as 17 mmol glucose/(g-dry cell h) and consequently evolved hydrogen at near-equal speed. The yields of hydrogen, though, were small, at 1.0 mol from 1 mol glucose and 2.5 mol from 1 mol sucrose. The optimum temperature for bacterium growth was about 40 °C and the optimum pH for hydrogen evolution was roughly 5.7, though 7.0 was the optimum pH for growth. This bacterium immediately restored the respiratory function to its aerobic condition when cultivation was changed from anaerobic to aerobic conditions. Based on the mechanism of hydrogen evolution through a NADH (Nicotinamide Adenine Dinucleotide, reduced form) pathway, two methods were proposed to improve the yield of hydrogen from glucose. One was to inhibit the electron transport chain at the sites of the NADH dehydrogenase complex or cytochrome b-c_1 complex and the other was to cultivate in the range of pH 5 to 4 under anaerobic conditions.

2. INTRODUCTION

There are two ways of evolving hydrogen using microorganisms: photobiological evolution and fermentative hydrogen evolution. The former only functions in the daylight

BioHydrogen, edited by Zaborsky *et al.*
Plenum Press, New York, 1998

Table 1. Representative rates of hydrogen evolution of microorganisms

Category	Doubling time (h)	H_2 evolution rate	
		mmol H_2/L·h	mmol H_2/g·h
A. Photochemical evolution			
1. Oxygenic photosynthetic organisms	7~25		
Oscillatoria sp. Miami BG7		0.4	0.4
Anabaena cylindrica	25	1.2	1.3
2. Anoxygenic photosynthetic organisms	2.2~9		
Rhodopseudomona capsulata		5.3	5.3
Rhodosprillum rubrum		3.0	2.5
Rhodobacter spaeroides 8703		—	10.4
B. Fermentative evolution	0.16~2		
1. Strict anaerobe			
Clostridium butyricum		—	7.3
Clostridium beijerinckii AM21B		17	25
2. Facultative anaerobe			
Citrobacter intermedius		11	9.5
Enterobacter aerogenes E.82005	0.25	36	17

because the photosynthetic microorganisms utilize solar energy as their energy source. The latter functions not only in the daylight but also in darkness because the fermentative microorganisms utilize organic substrates as their energy source. Biomass as an organic substrate is synthesized by plants using sunlight; therefore, fermentative production of hydrogen is also a production technique utilizing solar energy. The latter microbes generally grow faster then the former, as shown in Table 1, and also evolve hydrogen faster.

Enterobacter aerogenes strain E.82005 is a fermentative bacterium that was isolated from leaves of a flower grown in the university garden (Tanisho et al., 1983). It is one of the very common bacteria living in the intestines of humans and animals and is a facultative anaerobe; therefore, it grows under both aerobic and anaerobic conditions. Fundamental properties of this strain are shown in Table 2. One of its advantages is its short doubling time and consequent fast rate of hydrogen evolution (Tanisho et al., 1987). The yield of hydrogen from glucose, however, was smaller at 1 mol than other bacteria, such as *Clostridium butyricum,* which yielded 2.3 mol from 1 mol glucose (Wood, 1961).

Table 2. Properties of *Enterobacter aerogenes* st. E 82005

Metabolism	Facultative anaerobe
Suitable pH	
For growth	7.0
For H_2 evolution	5.5–6.0
Temperature suitable for growth	37–40°C
Yield of H_2	
From glucose	1.0 mol H_2/mol
From sucrose	2.5 mol H_2/mol
Evolution rate of H_2	
Batch cultivation	17 mmol/g dry cell/h
	21 mmol/L culture/h
Continuous cultivation	36 mmol/L culture/h

Calculations of the metabolites' mass balance with data from continuous cultivation lasting 42 days clearly showed that this bacterium evolves hydrogen through the NADH pathway (Tanisho et al., 1995). The mechanism of this evolution was also proposed from the electrochemical point of view (Tanisho et al., 1989). The NADH (Nicotinamide Adenine Dinucleotide, reduced form) is oxidized at an active site of the membrane-bound hydrogenase opened to the cytoplasm side (inside) of the cell, and protons accept electrons transported to the periplasm side (outside) through the hydrogenase to become molecular hydrogen.

$$\text{Inside: } NADH + H^+ \rightarrow NAD^+ + 2H^+ + 2e \tag{1}$$

$$\text{Outside: } 2H^+ + 2e \rightarrow H_2 \tag{2}$$

$$\text{Total: } NADH + H^+ \rightarrow NAD^+ + H_2 \tag{3}$$

The purpose of this reaction for the cell seems to be to reoxidize residual NADH and also reduce the surrounding proton concentration. Therefore, there is a potential to increase the yield of hydrogen by increasing the amount of residual NADH.

In the metabolic reaction of glucose under aerobic conditions, it is well-known that 10 mol NADH and 2 mol $FADH_2$ (Flavin Adenine Dinucleotide, reduced form) produced through the glycolysis and TCA cycle (tricarboxylic acid cycle) are re-oxidized by transporting electrons to oxygen through the electron transport chain.

$$\text{Glycolysis: } C_6H_{12}O_6 + 2NAD^+ \rightarrow 2CH_3COCOOH + 2NADH + 2H^+ \tag{4}$$

$$\text{TCA cycle: } 2CH_3COCOOH + 8NADH^+ + 2FAD + 6H_2O \rightarrow \\ 6CO_2 + 8NADH + 8H^+ + 2FADH_2 \tag{5}$$

$$\text{Electron transport chain: } 10NADH + 10H^+ + 2FADH_2 + 6O_2 \rightarrow 10NAD^+ \\ + 2FAD + 12H_2O \tag{6}$$

$$\text{Total: } C_6H_{12}O_6 + 6O_2 \rightarrow 6H_2O + 6CO_2 \tag{7}$$

Equation 5 and Figure 1 give a very interesting hint that inhibiting the electron transport chain at the sites of the NADH dehydrogenase complex or cytochrome b-c_1 complex leaves much unoxidized NADH and, consequently, the inhibition may promote the yield of hydrogen as follows (Tanisho et al., 1992):

$$C_6H_{12}O_6 + 6H_2O \rightarrow 12H_2 + 6CO_2 \tag{8}$$

This paper presents experimental results on hydrogen evolution under aerobic conditions and a plan for improving the yield of hydrogen.

3. EXPERIMENTAL

The microorganism used in this experiment was *Enterobacter aerogenes* strain E.82005. The medium of pre-cultivation consisted of 15 g glucose, 5 g peptone, 14 g

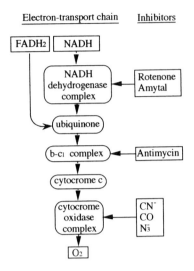

Figure 1. Electron transport chain and sites of action of inhibitor.

K_2HPO_4, 6 g KH_2PO_4, 2 g $(NH_4)_2SO_4$, 1 g sodium citrate, and 0.2 g $MgSO_4 \cdot 7H_2O$ in 1 L sterilized water. After 20 h cultivation at 38 °C, 50 mL of the cultivated liquid was inoculated into the experimental culture liquid, which contained the same amount of nutrients as the pre-cultivation, with the exception of peptone.

The volume of fermenter was 2.5 L and the initial pH of 1 L culture liquid was adjusted with 50 mL of acetic acid buffer; thus, the total volume of the culture liquid was 1.1 L. The pH of the culture liquid was controlled at 5.1 or 4.0 with 0.2 M acetic acid buffer by an automatic controller during the experiment.

Nitrogen or air was introduced continuously through a porous sparger into the culture liquid at a speed of 1.8 mL/s and the dissolved oxygen concentration was measured with a DO meter (DO 1A, TOA Electronics, Ltd.) The concentration of hydrogen in the discharged gas was analyzed by gas chromatography using molecular sieves (Molecular Sieve 5A, Gasukuro Kogyo, Inc.) as packing. Cultivation temperature was kept at 38 °C.

4. RESULTS

Figure 2 shows concentrations of dissolved oxygen during cultivations at pH 5.1, where nitrogen gas or air was introduced through a sparger at a rate of 1.8 mL/s. The saturated condition of dissolved oxygen (7.12 mg/L) was immediately broken by an inoculation of cultivated liquid even though air was still introduced in the form of fine bubbles. The dissolved O_2 concentration decreased to about 5.8 mg/L within 10 min; however, the concentration was maintained thereafter in the peptone-free culture. In a culture where nitrogen gas was introduced, the dissolved O_2 concentration of about 0.4 mg/L was also decreased to about 0 mg/L in an error range immediately after an inoculation of cultivated liquid.

Because the dissolved O_2 concentration of the pre-cultivated liquid also indicated about 0 mg/L in an error range when it was measured before the inoculation, bacteria inoculated had surely been in an anaerobic condition. These bacteria grown under anaerobic

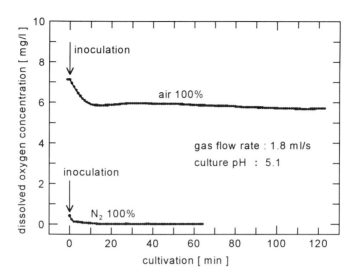

Figure 2. Changes of dissolved oxygen concentration by cultivation.

conditions, in general, were considered not to have the TCA cycle and the electron transport chain, and construct them only after they perceived oxygen. The bacteria's quick response to the dissolved oxygen suggests the existence of the TCA cycle and the electron transport chain even under anaerobic conditions.

Figure 3 shows hydrogen concentrations contained in discharged gas under pH 5.1 and 4.0. According to Tanisho et al. (Tanisho et al., 1987), pH 5.1 was a critical condition for bacterial growth, under which the productivity of cell mass was only 25% that under

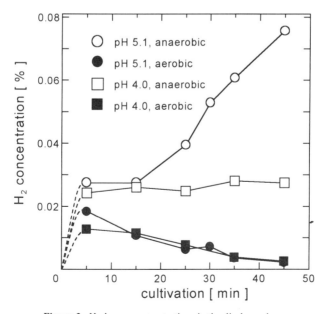

Figure 3. Hydrogen concentrations in the discharged gas.

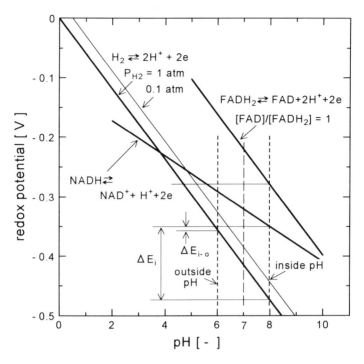

Figure 4. Redox potential of H_2, NAD, and FAD in relation to pH.

pH 7.0. Consequently, the rate of hydrogen evolution was also decreased to 70% of the maximum rate. pH 4.0 was more critical than pH 5.1, where survival was considerably difficult for this bacterium. At time 0, the concentration was naturally 0, but 5 min after the inoculation, approximately 0.02% of hydrogen was detected in the discharged gas at both aerobic and anaerobic conditions. This means that the activity of hydrogen evolution under aerobic cultivation was as high as under anaerobic cultivation even after 5 min of exposure to air. Under anaerobic cultivation, the H_2 concentration at pH 5.1 increased significantly and even at pH 4.0 it increased gradually. In contrast, under aerobic cultivation, the H_2 concentration gradually decreased both pH 5.1 and 4.0 and reached to a trace level after 45 min.

The activity of hydrogen evolution, as seen above, increases under anaerobic cultivation and decreases under aerobic cultivation; thus, hydrogenase surely loses its activity by exposure to oxygen. But considering the high activities under the constant concentration of dissolved oxygen after 10 min (see Figure 2), and particularly the detectable activities after 45 min, the loss of the activity may be caused by the change of the NADH oxidizing path from the H_2 evolution path to the respiratory path and may not be necessarily caused by the inhibition of hydrogenase by oxygen.

5. CONCLUSION

As seen in Equation 5, 2 mol pyruvate made by the glycolysis are decomposed to 8 mol NADH and 2 mol $FADH_2$, in addition to 6 mol CO_2 through the TCA cycle. To increase the hydrogen yield, therefore, the TCA cycle has to be functioning even under anaerobic cultivation. Because it was shown in this experiment that the TCA cycle was suspended under anaerobic condition, the reason the cycle did not function under the condition was considered from the electrochemical point of view, as follows: NADH can be oxidized by hydrogen evolution in the range of pH 6 to 5 (see Figure 4); however, since the standard redox potential of FAD (E_0' = -220 mV) is 100 mV higher than that of NAD (E_0' = -320 mV), $FADH_2$ may be accumulated in the cell and, consequently, the TCA cycle may stop its function. Thus, if the cycle is devised to function under anaerobic condition, that is, if a means of oxidizing $FADH_2$ under anaerobic conditions can be found, residual NADH may increase in the cell and, consequently, the H_2 yield may also increase.

In our laboratory, we are examining whether the inhibition of NADH dehydrogenase complex under aerobic conditions (see Figure 1) contributes to the increment of H_2 concentration in the discharged gas. In addition, since the redox potential of FAD in the cell is nearly equal to or lower than the redox potential of H_2 at pH below 5, as seen in Figure 4, we are examining whether the cells evolve H_2 from $FADH_2$ directly by cultivating the cells in the range of pH 5 to 4 under anaerobic condition and measuring the H_2 yield.

REFERENCES

Tanisho, S., Wakao, N., and Kosako, Y., 1983, Biological hydrogen production by *Enterobacter aerogenes*, *J. Chem. Eng. Japan*, 16:529–530.

Tanisho, S., Suzuki, Y., and Wakao, N., 1987, Fermentative hydrogen evolution by *Enterobacter aerogenes* strain E.82005, *Int. J. Hydrogen Energy*, 12:623–627.

Tanisho, S., Kamiya, N., and Wakao, N., 1989, Hydrogen evolution of *Enterobacter aerogenes* depending on culture pH, *Biochem. Biophys. Acta*, 973:1–6.

Tanisho, S., Ishiwata, Y., and Takahashi, K., 1992, Hydrogen production by fermentation and a trial for improvement on the yield of hydrogen, *Hydrogen Energy Progress IX*, Proceedings of the Ninth World Hydrogen Energy Conference, Paris, pp. 583–590.

Tanisho, S. and Ishiwata, Y., 1995, Continuous hydrogen production from molasses by fermentation using urethane foam as a support of flocks, *Int. J. Hydrogen Energy*, 20:541–545.

36

ARTIFICIAL BACTERIAL ALGAL SYMBIOSIS (PROJECT ArBAS)

Sahara Experiments

Ingo Rechenberg

Bionics and Evolutiontechnique
Technical University Berlin
Ackerstraße 71-76
D-13355 Berlin, Germany

1. SUMMARY

The blue alga *Nostoc muscorum* is the model for a two-stage compound reactor that splits water. In the first stage, green algae (*Chlamydomonas oblonga*) produce oxygen and excrete carbohydrates by imitating the vegetative cells of *Nostoc*. The excreted matter of the algae enters the second stage (the analogon of the heterocysts of *Nostoc*), where purple bacteria (*Rhodobacter capsulatus*) decompose the carbohydrates into hydrogen and carbon dioxide. Carbon dioxide is fed back into the algal reactor to get reloaded with hydrogen. This process has been tested since 1988 in a small dune region at the edge of the Sahara in southern Morocco. The reactor is cooled by passing the heat through a cooling tube imbedded deep into the dune. Field experiments have demonstrated that the rate of hydrogen production is significantly increased if the reflected solar radiation from the dunes is used; that fluorescent laser dyes will further amplify hydrogen formation; and that contamination of the algal reactor by microorganisms, which consume the excreted carbohydrates, is the main problem during outdoor experiments.

2. INTRODUCTION

Experiments began in 1979 to produce biohydrogen with purple bacteria. *Rhodobacter capsulatus* (strain DSM 1710 from the German collection of microorganisms, Brunswick) was the pilot organism. Although progress was made in the experiments, one major question arose: Why do purple bacteria need so much light energy to split lactate into carbon dioxide and hydrogen? This was not necessary from a thermodynamic point of view, but may have been due to the fact that nitrogen fixation (ammonia synthesis) needs high activation energy, and hydrogen release is a mistake made by the nitrogenase enzyme when nitrogen is not available. This was the motivation to launch a world-wide screening of purple bacteria.

Another unsatisfactory situation was that purple bacteria do not split water. Purple bacteria need carbohydrates as electron donors, while it was our goal to realize the biophotolysis of water. The thrust of the work was therefore changed to combine bacterial and algal photosynthesis. A simple consideration shows that the total efficiency of the combined process is

$$\frac{1}{\eta_{total}} = \frac{1}{\eta_{bact}} + \frac{1}{\eta_{alga}}$$

and not

$$\eta_{total} = \eta_{bact} \cdot \eta_{alga}$$

One other challenge remained, that is, the use of purple bacteria outside the laboratory. Evolution has resulted in most purple bacteria living under low light intensities in muddy water. Many experiments with purple bacteria are conducted under the (weak) light of incandescent lamps. As such, the extrapolation of the measured indoor hydrogen production rates to outdoor conditions seemed questionable. Much work needed to be done to obtain the same results under the strong solar radiation of the Sahara as in the laboratory.

3. EXPERIMENTAL RESULTS

3.1. Bacteria Screening

Purple bacteria are found nearly everywhere on earth. We have succeeded in isolating purple bacteria even in Antarctica (on King George Island) and 1800 km from the north pole (in Ny Alesund, Spitsbergen). However, these polar strains evolve less than 50 mL hydrogen per 100 mL bacteria suspension, and are not listed in Table 1. The best bacteria are from Morocco and evolve more than 500 mL hydrogen per 100-mL suspension.

Our standardized 100-mL bacterial culture contained 0.36 mg lactic acid. No effort was made to produce an anaerobic culture solution. The standardized production test was as follows: Cultivation under the light of incandescent lamps at 10,000 lux and 37 °C. No isolation technique was applied for the first hydrogen production test. All water samples were first grown in a 250-mL flask containing the standardized nutrition. Then, 5 mL of this enriched bacteria culture was inserted into the standard test tube. The total amount of the evolved gas, which normally contained 92% hydrogen, is given in Table 1.

3.2. Directed Evolution of Purple Bacteria

The focus of this research has not been bioengineering; rather, the "Evolution Strategy" has occupied our main effort (Rechenberg, 1989, 1994). The Evolution Strategy imitates rules of biological evolution in order to optimize technical systems. Proponents have, for example, developed an airfoil, truss bridge, supersonic nozzle, galvanic bath, and even a coffee composition by means of this Evolution Strategy. It seemed reasonable to apply

Table 1. Purple bacteria: wild strains vs. DSM laboratory strains (shaded cells) (numerical values=volume of hydrogen produced by a 100-mL test tube)

Moulay Idriss	Morocco	530	Sidi el Yamani	Morocco	433
Ain Defali	Morocco	522	Tiffrit	Algeria	433
Ouarzazate	Morocco	518	Tit Tamanrasset	Algeria	431
Ouarzazate III	Morocco	509	Chefchaouen	Morocco	429
Tetouan	Morocco	509	Ait Boukha	Morocco	428
Ain Defali II	Morocco	507	Talioune	Morocco	428
S. el A. Ayacha	Morocco	495	Ksar Ljir	Morocco	427
Rabat	Morocco	494	Meski Ziz	Morocco	427
Oued Moulouya	Morocco	493	R. capsulatus	DSM 1710	427
El Hajeb	Morocco	487	Oued Mellah	Morocco	425
Mdiq	Morocco	487	Laayoun	West Sahara	423
Zagora I	Morocco	485	Meski I	Morocco	422
Tozeur	Tunisia	473	Timahdite II	Morocco	420
Takoradi	Ghana	469	Agadez	Niger	419
Erfoud	Morocco	468	Olivia	Spain	419
Ouezzane	Morocco	464	Ait Ridi	Morocco	418
Restinga Smir	Morocco	463	Ouarzazate II	Morocco	415
Armeskroud	Morocco	460	Benguerir	Morocco	414
Lago di Lucrino	Italia	458	Chopr	Russia	412
Meski	Morocco	455	Mdiq IV	Morocco	411
Lago di Bolsena	Italia	450	Oasis Jorf	Morocco	411
Mdiq II	Morocco	450	Oued Tensift	Morocco	410
Zagora Draa	Morocco	450	Meknes	Morocco	408
Draa el Asef	Morocco	448	Oued Ouenha	Morocco	406
Tazlida	Morocco	448	Boumaine	Morocco	403
Marrakech	Morocco	445	Carretera	Colombia	402
Marina Smir	Morocco	445	Lake Kaweah	USA	402
Ouarzazate	Morocco	445	Morhane	Morocco	398
Transikht	Morocco	444	Dar Ben Sadouk	Morocco	393
Ait Ourir	Morocco	442	Oued Touil	Algeria	393
Lago d'Averno	Italia	441	Ain Elati	Morocco	393
Ouezzane II	Morocco	439	Hawaii	USA	392
Oed Korifla	Morocco	438	Ouezzane	Morocco	390
Mdiq I	Morocco	437	Berriane	Algeria	389
Mdiq II	Morocco	435	Ait Bou Krir	Morocco	388
Guelta Hirhafok	Algeria	434	Souk el Arba	Morocco	388
Taliouine	Morocco	434	Freuae	Algeria	386
Oued Zem	Morocco	433	Hoggar	Algeria	386

Table 1. (*Continued*)

Agouim Atlas	Morocco	384	Tamarrakechte	Morocco	341
Tiouine	Morocco	384	Pont du Loukkos	Morokko	338
Oued Aoufouss	Morocco	382	Baie Mahault	Guadeloupe	334
Pantano Grande	Italia	381	Hassi Elabad	Algeria	333
Youb	Algeria	380	Mdiq V	Morocco	331
Casablanca	Morocco	378	St. Anne	Guadeloupe	330
Boudenib	Morocco	376	Rich Ziz	Morocco	330
Ifri Ziz	Morocco	374	Fes el Bari	Morocco	325
Aflou	Algeria	372	Zadar	Yugoslavia	324
Lago Biviere	Italia	372	S. Ali Ou Brahim	Morocco	322
Bouznika	Morocco	369	Volubilis	Morocco	322
Albuquerque	USA	368	Tofliht	Morocco	321
Palau	Italia	367	Zagora II	Morocco	321
Ghardaia II	Algeria	361	Volvisee	Greece	315
Opuzen	Yoguslavia	361	Interstate Baker	USA	311
Niamey	Niger	360	Marrakech I	Morocco	310
El Goléa	Algeria	359	Lake Russel	Australia	309
Benguerir I	Morocco	358	Aoufouss	Morocco	306
Jorf	Morocco	358	Jorf Rheris	Morocco	305
Benguerir II	Morocco	354	Laghourat	Algeria	302
Rich	Morocco	354	L.Alimini Grande	Italia	302
Ravnie Chaude	Guadeloupe	353	Khemisset	Morocco	297
Saida	Algeria	351	Ladrilleros III	Colombia	296
Sidi Bel Abbes	Algeria	350	Panke	Berlin	295
Horray	Great Britain	349	Midelt	Morocco	294
R. rubrum	DSM 467	349	Rabat Pont Fida	Morocco	293
Ifrane	Morocco	348	Tarkwa	Ghana	290
Lago di Pergusa	Italia	348	R. sphaeroides	DSM 2340	289
Edinburgh I	Great Britain	347	Bundaberg	Australia	287
Marrakech II	Morocco	347	Libertad I	Peru	287
Afilal Guelta	Algeria	346	Dayet Sri	Morocco	284
Lago di Vico	Italia	346	Tegeler See	Berlin	282
Accra	Ghana	345	Pinios	Greece	280
L. di Ciancardi	Italia	345	Malinalco II	Mexico	278
Imlaoulaouene	Algeria	344	Takhmet	Algeria	278
Tazrout	Morocco	343	Atlas	Morocco	275
Timahdite I	Morocco	342	Evora	Portugal	275
Ain Defali III	Morocco	341	Edinburgh II	Great Britain	274

Table 1. (*Continued*)

Pamplemousses	Mauritius	274	Tegeler Fließ	Berlin	170
Spiggle	Great Britain	270	Arbaoua	Morocco	169
Béchar	Algeria	269	Detroit I	USA	168
M. B. Ksiri	Morocco	265	Heiligensee	Berlin	165
See Lübars	Berlin	257	In Sallah II	Algeria	158
Rinnoji-Koen	Japan	255	Khenifra	Morocco	157
Taddert	Morocco	251	L. di Montecosaro	Italia	155
Ladrilleros	Colombia	250	Pointe-à-Pitre	Guadeloupe	152
Malinalco I	Mexico	248	L. Alimini Piccolo	Italy	151
Mottuji-Ko	Japan	248	Teufelssee	Berlin	144
Chancana	Peru	247	Bacina	Yugoslavia	138
Isabella Lake	USA	247	Rio Nanay	Peru	137
Io-San	Japan	244	Tinerhir	Morocco	136
Libertad II	Peru	241	Lago Varano	Italy	133
Britzer Dorfteich	Berlin	237	Jorf Elmelha	Morocco	132
Fes	Morocco	234	Schäfersee	Berlin	130
Matata	New Zealand	233	Pamplemousses I	Mauritius	124
Trou aux Biches	Mauritius	231	Bogazkale	Turkey	121
Lake Genezareth	Israel	228	Dudley Creek	Australia	119
Derkaoua	Morocco	218	Fes el Bari	Morocco	118
Mirror Lake	USA	214	Hubertussee	Berlin	110
Tomb	Great Britain	212	Krumme Lanke	Berlin	110
Zagora	Morocco	211	Teltowkanal	Berlin	110
Buckingham P.	Great Britain	208	Marcel Bay	St. Maarten	109
Laderillos I	Colombia	208	Prinoth	Italy	104
Lerwick	Great Britain	208	Chuzonji-Ko	Japan	97
Nordhafen	Berlin	197	Rendezvous Bay	Anguilla	86
Spree	Berlin	193	Loch Ness	Great Britain	83
Golan	Israel	191	St. Louis Reservoir	USA	83
Rio Finco	Colombia	190	Tarikt	Morocco	83
Iquitos	Peru	189	El Jadida	Morocco	82
Kairo Nil	Egypt	189	Death Valley I	USA	67
Sellerjoch	Italy	186	Ile Fourche	Martinique	58
Hikone	Japan	181	L. di Bolsano	Italy	58
Stößensee	Berlin	180	Montfort	Israel	58
Aberdin	Great Britain	177	Orkney	Great Britain	56
El Ksiba	Morocco	177			
Detroit II	USA	174			

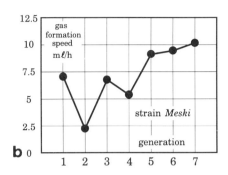
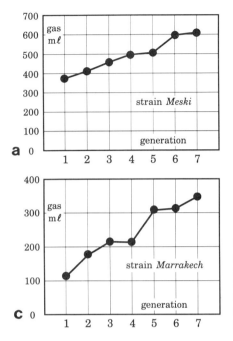

Figure 1. (a) Directed evolution with (3, 9)-Evolution Strategy. Goal: maximum total gas volume. Test tube: 100 mL. H-donor: Lactat (40 mmol/L). (b) Parallel rise of the speed of gas formation. Experimental conditions in (a). (c) Directed evolution with the (3, 9)-Evolution Strategy. Goal: maximum total gas volume. Test tube: 100 mL. H-donor: Saccharose (40 mmol/L).

the Evolution Strategy to improve the amount of hydrogen released by purple bacteria. Gerald Koch (Koch and Schwessinger, 1996) applied a (3, 9) Evolution Strategy, a nomenclature meaning three parents producing nine offspring, from which the three best become parents of the next generation.

The experiments began with the Moroccan strain *Meski*, which was the best available at the time. Nine isolated colonies from an agar shake of *Meski* were placed in nine test tubes. The three best colonies producing the highest quantities of hydrogen were placed together in a new agar shake culture, from which nine randomly isolated colonies produced the offspring for the second generation. Again, the the best colonies were selected. The evolution experiment was halted after seven generations. For generation 7, the 100-mL test tube produced 607 mL of gas (Figure 1a), very near to the theoretical maximum. Figure 1b shows for the same experiment the increase of the speed of hydrogen evolution (measured in milliliters hydrogen per milliliter culture liquid an hour). Koch conducted a similar experiment to adapt purple bacteria to saccharose (Figure 1c). He stopped after seven generations.

A critical observation: To modify the nitrogenase for the new task of hydrogen release, the speed of hydrogen evolution is the proper measure. In this case, the evolution experiment may need a hundred or more generations to show an effect. In addition, in keeping with the Evolution Strategy, the mutation size must be controlled. We did not have the financial support to undertake such a prolonged experiment. However, the simple pilot experiments may demonstrate that the algorithm of the Evolution Strategy may fulfill the future dream of directed evolution.

3.3. Erg Chebbi: Experiments at the Planning Stage

Erg Chebbi is a small sand desert at the edge of the Sahara in South Morocco. We arrived in 1982 to collect purple bacteria. The fundamental question was whether purple

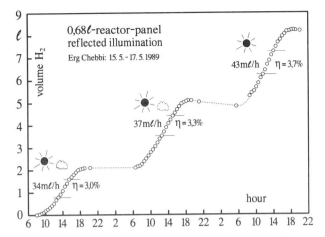

Figure 2. Reflected light from the dunes gives the best solar efficiency.

bacteria would produce hydrogen under these extreme conditions. Erg Chebbi thus became the field laboratory to test hydrogen-producing photobioreactors.

The first task was developing an apparatus for the Sahara that would enable the bacteria to withstand the heat. The use of an electric-driven thermostat was considered unrealistic. It was necessary to find a means of dispersing the heat in the depth of the dunes. In the autumn of 1987, this idea was tested over a period of 10 days using a dummy reactor filled with red colored water. When the sun was at its zenith, the temperature on the surface of the dunes measured 60 °C but only 30 °C at a depth of 1 m. The maximum temperature of the dummy bacteria suspension was 54 °C. Due to the low heat conductivity of the Sahara sand, the idea of ground cooling was not pursued. (In the summer of 1994, ground cooling was used to great success).

In 1988, an expedition car was driven to the same experimental site. It was equipped with a small 140-mL reactor, which was cooled by a solar-driven Peltier element. Because it broke, it was changed to simple evaporation cooling. This nomadic cooling principle was very effective, and within 16 days, 0.6 L of biohydrogen was produced.

3.4. Advanced Experiments

Field-testing continued at the site in the spring and autumn of 1989. The new bioreactor design had an industrial evaporation cooler. The low relative humidity of the desert air made this cooling system very effective. Two reactor tanks, with a total volume of 1.4 L, were designed like solar panels which could be aligned to the sun. The results were disappointing. Hydrogen production dropped hourly, presumably because the purple bacteria were damaged by the strong radiation. As a final test, the reactors were directed to the dunes. The change was dramatic, as the reflected light of the dunes proved to be much more conducive to hydrogen production (Figure 2).

3.5. Intermezzo: The Project ArBAS

Purple bacteria require an organic substrate for hydrogen production. Lactic acid was used in the Sahara experiments. The primary goal, however, has been to split water

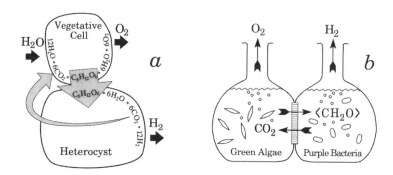

Figure 3. Cell linkage of *Nostoc muscorum* (a) and composed bioreactor (b).

into hydrogen and oxygen. All green plants do this, but the hydrogen becomes non-existent. Plants do not generate molecular hydrogen with one exception—in the process of nitrogen fixation.

The blue alga *Nostoc muscorum* was selected as the biological model to design a bioreactor that splits water (Figure 3). The vegetative cell of *Nostoc* corresponds with the flask containing green algae. Here, the hydrogen of the water will be contained in carbohydrates. The carbohydrates are transported through a special membrane into the heterocyst, which corresponds with the flask containing purple bacteria. The carbohydrates are broken into CO_2 and H_2 by a complex process including the nitrogenase. The process in the blue alga *Nostoc*, as well as in the reactor, is driven by light energy. This process works only if the green algae excrete carbohydrates. In the literature (Fogg, 1966), algal strains are identified that excrete up to 50% of the total photosynthetic production into their surroundings. In addition, CO_2 works as a carrier molecule which is loaded with hydrogen and then unloaded. This idea was formulated in 1981 (Rechenberg, 1981, 1987); others may have also done so. This is mentioned because biological evolution has not found a more intelligent method to get hold of hydrogen. Preservation and transportation of hydrogen in carbohydrates might be a practical idea.

The Artificial Bacterial Algal Symbiosis (ArBAS) project was established to test the concept. In 1989, students supervised by Koch first produced hydrogen from a bacterial algal linkage in a batch operation. The alga *Chlamydomonas oblonga* (known to emit carbohydrates) grows under strong light and produces 3% CO_2 in the inlet gas. After 200 hours, the culture was inoculated with purple bacteria (Figure 4). In this experiment, algae and bacteria worked in one tank at different times. Later, we were able to do the same with the filtrated carbohydrates excreted by *C. oblonga*.

It must be noted that there is no hope that an ordinary algal culture, inoculated with purple bacteria, will produce hydrogen. Much time was wasted between 1981 and 1989 because of two factors: (1) Every effort had to be made to obtain an exceptionally high carbon/nitrogen ratio in the growing algal culture and (2) precautions had to be made to prevent parasitic bacteria from consuming the excreted carbohydrates.

3.6. High C/N Ratio

Algae need a nitrogen source to grow. But nitrogen inhibits the hydrogen evolution of purple bacteria. This means that as long as the algae are growing, no hydrogen will occur, even if the algae excrete carbohydrates. Figure 5 (Koch-Schwessinger, 1996) shows

Artificial Bacterial Algal Symbiosis (Project ArBAS)

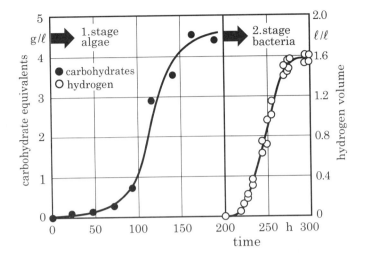

Figure 4. Biophotolysis of water by a bacterial algal linkage.

two clearly separated culture phases of *C. oblonga*. The phase of "growing cells" is represented by an inclining line. The concentration of carbohydrates in the algal liquid grows proportionally to the density of the culture. The phase of "resting cells" is represented by the vertical line. Constant extinction of the light means constant density of the culture. Along the vertical line, the resting algal cells permanently produce carbohydrates. The best result was 8.65 g carbohydrate equivalents per liter algal liquid. The change from the inclined line to the vertical line occurs when the nitrogen source is exhausted. This point comes about earlier with a lower nitrogen concentration in the nutrition. A biotechnologi-

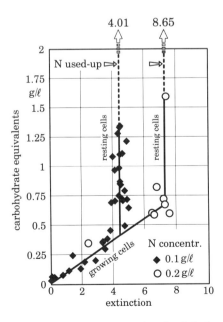

Figure 5. Growing and resting cells. Accumulation of carbohydrates (*C. oblonga*).

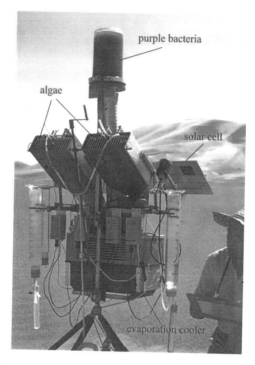

Figure 6. Bioreactor design, 1989–1992. Algae and purple bacteria in the Sahara.

cal process, in which algal cells excrete carbohydrates for as long as possible without growing, was a goal of the ArBAS project.

3.7. Algae and Bacteria in Erg Chebbi

Success with the bacterial algal composition encouraged us to repeat the experiment. The flat reactor vessels contained the algae while the purple bacteria were grown in a cylindrical glass column (Figure 6). This columnar reactor exploited the radiation from the sun and dunes. The experiment used algae, purple bacteria, and nutrient solution in a solar-driven refrigerator, small gas cylinders with CO_2, a pH meter, photometer, solarimeter, distilled water, and so forth.

The experiment ran well. The algae became dark green with a small yellow tinge at the end. There seemed to be no difference with the laboratory culture. However, the first biological water split remained elusive. After inoculation with purple bacteria, no small hydrogen bubble could be detected. We concluded that as a result of the day and night cycle, parasitic bacteria had enough time to consume the produced carbohydrates so that the carbon/nitrogen ratio could not exceed the required threshold.

3.8. Laser Dye Bioreactors

The frustration over the unsuccessful algae experiment pushed the test with the purple bacteria into the background. At this time, a breakthrough ocurred. In the column-shaped reactor tank, the lactate-fed *Meski*-purple bacteria produced liters of hydrogen. *Meski* is the name of our best purple bacteria strain, named after the small Moroccan oasis

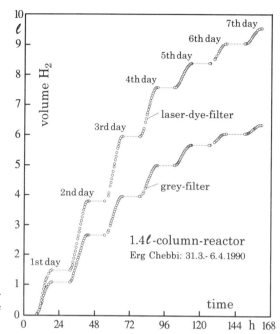

Figure 7. Step-up of hydrogen evolution by a laser dye filter. Color filter and gray filter have the same energy absorption.

of Meski in which it was found. The first test with the unprotected reactor in solar radiation was a failure. It was the second experiment that became so successful. For this experiment, the reactor on the sunlit side was covered with orange-colored foil.

The new bioreactor design was made of two glass flasks with transparent cooling jackets, which were not used for cooling. To study the filter effect, the outer ring volume was filled with various colored liquids. In Figure 7, a laser dye filter is compared to a grey filter. Both have the same energy absorption. The optimum laser dye has an absorption range from 420–520 nanometers, which corresponds with the absorption range of the carotenoids. These red and yellow pigments protect cells from damaging sun rays. In addition, hydrogen production may be enhanced because the laser dye transforms the absorbed wavelengths into longer ones, which are more effective in photosynthesis.

4. RESULTS AND CONCLUSIONS

4.1. Solved and Unsolved Problems

The world-wide screening of purple bacteria and a directed evolution experiment in the laboratory resulted in the isolation of the strain *Meski* for maximum total hydrogen evolution. For the Sahara experiments, it was important that *Meski* work at 40 °C. The strain *Meski* has has been preserved as *Rhodobacter sphaeroides* DSM 9483 in the German Collection of Microorganisms, D-38124 Braunschweig.

The experiment with the Evolution Strategy gives rise to the idea of remodeling the nitrogenase into an enzyme that evolves hydrogen with a higher efficiency.

The experiments in the Sahara have shown a serious problem, that is, purple bacteria are not able to make use of the high concentrated energy of the sun. A substitute of carote-

Figure 8. Heliomites; photobioreactors (300 L) in an art exhibition.

noid pigments must protect the bacteria from the damaging sun rays. But even then, the efficiency will not exceed 2% in the bright sun. The radiation is too strong at the surface and too low in the interior of the reactor. We have not devised a bioreactor design that dilutes the sun's rays uniformly into the volume.

The cultivation of algae (here, *C. oblonga*) under the Sahara sun was less problematic than expected. However, the algal culture could not be sufficiently sterilized in the experimental surroundings of the desert. Parasitic bacteria had enough time to consume the excreted carbohydrates of the algae. As a consequence, the ArBAS project could not be completed in the desert.

4.2. A Futuristic Vision

Between 1988 and 1990, 124.31 L of hydrogen was produced in the Sahara. This is the energy-equivalent of 0.04 L of gasoline. How can the bioreactor be enlarged? For an art exhibition in 1986, we designed a purple bacteria bioreactor in the form of a cone. The exhibit was named "Heliomites in the Sahara." The three heliomites with 300 L volume operated under floodlights for several weeks (Figure 8). The discovery that the reflected radiation from the dunes is so effective gave rise to the idea of building heliomites for use in the Sahara. The design in Figure 9 has a volume of 1500 L. With a production rate of 190 mL H_2 per liter bacteria suspension per hour (which was the average production rate of the 1990 Sahara reactor), one heliomite could produce the equivalent of 1 kW. The 100 heliomite energy farm shown in Figure 9 could produce 100 kW.

There is no doubt that hydrogen production by an algal bacterial compound is uneconomical today. But intensive research should increase the efficiency. Algae and bacte-

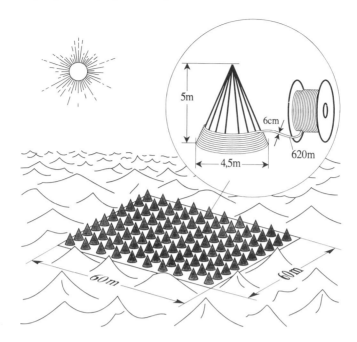

Figure 9. Heliomites in the Sahara. Futuristic view of 100-kW biological hydrogen farm.

ria utilize different wavelengths of the light spectrum. They can be arranged one upon the other without stealing each other's light. Furthermore, there is potential to improve photobiological efficiency, which is low (2%) because the light intensity is too high at the surface and too low in the depth of the reactor. Solar radiation has to be scattered uniformly into the reactor volume. Wavelength transformer should also increase the efficiency.

It is the objective of bionics (or biomimetics) to use artificial systems in place of the evolved systems found in nature. It eventually may be possible to imitate the Z-scheme of photosynthesis by an artificial system. For biophotolysis of water, the hydrogen would be bound to a carbohydrate analogue in the first stage. The hydrogen complex, unable to rereact with oxygen, would then move to a second stage, where the hydrogen is released. The blank carrier molecule would move back to the first stage and be reloaded with hydrogen.

But what would happen if carbon dioxide were replaced by silicon dioxide? A glance into a chemical textbook proves that in some cases a homologous silicon chemistry is possible. This may be nonsense; what would happen if a trace gas like carbon dioxide acted as a carrier molecule for hydrogen? At present, a solution à la bionics is not in sight. The substitution of the process by biological components should be explored.

REFERENCES

Benemann, J.R., Miyamoto, K., and Hallenbeck, P.C., 1980, Bioengineering aspects of biophotolysis, *Enzyme Microb. Technol.*, 2:103–111.
Fogg, G.E., 1966, Extracellular products of algae, *Oceanogr. Mar. Biol. Ann. Rev.*, 4:195–212.
Koch-Schwessinger, G., 1996, Wasserstoffproduktion durch Purpurbakterien, Werkstatt Bionik und Evolutionstechnik Band 3, Frommann Holzboog, Stuttgart.

Miyake, J., and Kawamura, S., 1987, Efficiency of light energy conversion to hydrogen by the photosynthetic bacterium *Rhodobacter sphaeroides*, *Int. J. Hydrogen Energy*, 12(3):147–149.

Rechenberg, I., 1981/1987, Wasserstofferzeugung mit Purpurbakterien. Wissenschafts magazin TU-Berlin (1), 36–43 / Schweizerische Laboratoriums-Zeitschrift, 44(11):417–425.

Rechenberg, I., 1989, Evolution strategy—nature's way of optimization, in Optimization: Methods and Applications, Possibilities and Limitations, Bergmann, H.W. (ed.), DLR lecture notes in engineering, Springer, Berlin, 47:106–128.

Rechenberg, I., 1994, Evolutionsstrategie '94, Werkstatt Bionik und Evolutionstechnik Band 1, Frommann Holzboog, Stuttgart.

Rechenberg, I, 1994, Photobiologische Wasserstoffproduktion in der Sahara, Werkstatt Bionik und Evolutionstechnik Band 2, Frommann Holzboog, Stuttgart.

Watanabe, Y., de la Noüe, J., and Hall, D.O., 1995, Photosynthetic performance of a helical tubular photobioreactor incorporating the cyanobacterium *Spirulina platensis*, *Biotechnol. Bioeng.*, 47:261–269.

Zaborsky, O.R., Mitsui, A., and Black, C.C. (eds.), 1982, *Handbook of Biosolar Resources*, parts 1 and 2, CRC Press, Boca Raton.

37

THE EFFECT OF *Halobacterium halobium* ON PHOTOELECTROCHEMICAL HYDROGEN PRODUCTION

Vedat Sediroglu,[1] Meral Yucel,[2] Ufuk Gunduz,[2] Lemi Turker,[3] and Inci Eroglu[1]

[1]Department of Chemical Engineering
[2]Department of Biology
[3]Department of Chemistry
Middle East Technical University
06531, Ankara, Turkey

Key Words

H. halobium, cellulose acetate membranes, electrochemical hydrogen production, photoelectrochemical reactor, bacteriorhodopsin

1. SUMMARY

A bio-photoelectrochemical (bio-PEC) reactor containing the photosynthetic bacterium *Halobacterium halobium* has been constructed. *H. halobium* contains a retinal protein, bacteriorhodopsin, in its purple membrane that gives protons upon illumination. In this system, the biocatalytic effect of bacteriorhodopsin on photoelectrochemical hydrogen production was determined.

The bio-PEC reactor has a 1-L volume, contains *H. halobium* cells in 20% NaCl solution and houses a Pt electrode as the cathode, an Ag electrode as the anode, a combined pH electrode, and a temperature probe. The minimum voltage required for hydrogen production in this system was -0.88 V. The photoelectrochemical hydrogen production experiments were performed by using free *H. halobium* packed cells, or material immobilized on cellulose acetate membranes either in the form of packed cells or purple membrane fragments. The results were compared with experiments performed without *H. halobium*. It was concluded that the presence of light and bacteriorhodopsin enhanced hydrogen production.

2. INTRODUCTION

The energy sources used today are mainly fossil fuels and nuclear power. However, fossil fuels are limited and the deleterious environmental impact associated with fossil fuels and nuclear energy has forced humanity to search for alternative energy sources. One of these alternative sources is based on hydrogen, which is a potentially renewable, clean energy source. Fossil-derived hydrogen has attained an important position as a fuel and basic raw material for chemical industries.

The most economical methods of producing hydrogen are through chemical processing, particularly steam-reforming of methane and other hydrocarbons, coal gasification, and electrochemical processes such as water electrolysis.

In addition, rapid advancements in biotechnology in recent years have drawn attention to technologies for hydrogen production using microorganisms and solar energy, called photobiological hydrogen production. There is much research on photobiological hydrogen production by photosynthetic bacteria, green algae, cyanobacteria, and cell-free systems (Markov et al., 1995; Benemann, 1997). However, technical problems currently limit the yield and duration of the systems and therefore it is necessary to develop alternative practical photobiological hydrogen-producing systems for the future.

The halophilic archaebacterium of *Halobacterium halobium* is a photosynthetic bacterium that lives in 4-M salt solutions (Oesterhelt and Stoeckenius, 1971). The cell membrane of *H. halobium* has been subdivided into five sections: red, brown, white, yellow, and purple. Although they have different functions in the organism, 80% of these fragments correspond to purple membrane fragments. The purple membrane (PM) contains a single protein called bacteriorhodopsin (BR). When BR absorbs light, it ejects protons from the cell, thus generating an electrochemical gradient across the cell membrane. Accordingly, there are attempts to make use of this natural proton pump for the transduction of light energy into a more immediately useful form (Singh and Caplan, 1980). Even if the cell dies, the retinal protein BR is very stable and functions continuously (Trissl, 1990). Although *H. halobium* pumps protons upon illumination, it lacks a hydrogenease enzyme to convert protons into molecular hydrogen. Therefore, a proton reduction system must be coupled with *H. halobium*. The photoproduction of hydrogen has been achieved by *H. halobium* coupled to a salt-tolerant *Escherichia coli* (Khan and Bhatt, 1989; Patel and Madamwar, 1994; Kaya et al., 1996) or by *H. halobium* coupled to electrochemical systems (Khan and Bhatt, 1990, 1992).

The electrochemical system using *H. halobium* as a bio-catalyst was found to be very promising, as was the H_2-evolution bio-system using *H. halobium* coupled to *E. coli*. The advantage of the electrochemical system is that it does not require any nutrient supply and the electrolyte (NaCl) used in the system is very cheap and abundant. However the mechanism and consumption rate of the expensive silver are not known.

The METU BR Research Group has been working on long-term illumination of BR, studying the photoresponse of purple membrane fragments (Yucel et al., 1995), BR reconstituted on liposomes (Eroglu et al., 1991), BR immobilized on polyacrylamide gel membranes (Eroglu et al., 1994), and BR immobilized on cellulose acetate membranes (CAM) (Sediroglu et al., 1996). BR is quite stable and keeps its photoactivity for two years on CAM.

The objective of this study is to develop a new immobilization technique for *H. halobium* packed cells on CAM, to investigate its effect on photoelectrochemical hydrogen gas production, and to determine the efficiency of such a system. For this purpose, a bio-photoelectrochemical reactor (bio-PEC) was designed and constructed. The experiments were performed by using *H. halobium* packed cells, immobilized packed cells, and immobilized purple membrane fragments. The results were compared with the experiments performed with *H. halobium* free systems.

3. EXPERIMENTAL

3.1. Materials

The common reagents used were all of reagent quality and purchased from Merck (Germany). The *H. halobium* S-9 strain was a gift of Prof. Dieter Oesterhelt of the Max Plank Institute (Germany).

3.2. Growth of *H. halobium*

H. halobium cells (strain S-9) were grown as described by Oesterhelt and Stoeckenius (1974) at 39 °C in a sterilized medium containing 250 g NaCl, 20.0 g $MgSO_4 \cdot 2H_2O$, 3.64 g trisodium citrate 5.5 H_2O, 2.0 g KCl, and 10 g bacteriological peptone L-37 in 1000 mL of distilled water at pH 7.0 in a shaking incubator. Bacterial growth was followed spectrophotometrically by absorbance measurement at 660 nm, and the biosynthesis of BR was followed by an increase in absorbance value at 560 nm. The cells were grown for seven days.

3.3. Preparation of *H. halobium* Packed Cells and Purple Membrane Fragments

After seven days, the cells were collected by centrifugation at 6600 g for 20 min in a Sorvall centrifuge (GSA rotor). Then, *H. halobium* packed cells were obtained and suspended in 4 M NaCl solution. The purple membrane fragments were prepared according to Yucel et al. (1995). Cell pellets were suspended in 20 mL of basalt solution overnight at 20 °C, then 0.15 mL of DNAase (2000 units/mL) was added and the solution dialyzed at 4 °C against 2 L of 0.1 M NaCl for three days. The lysate was centrifuged at 2500 g (in SS-34) at 4 °C for 20 min. The sediment containing small pellets was discarded. The colored supernatant was centrifuged at 43000 g (in SS-34) for 60 min at 4 °C. The purple pellets were collected and then suspended in 0.1 M NaCl. The centrifugation step was repeated once more. The pellets, recollected and suspended in 5ml of distilled H_2O, were stored at -20 °C until used.

3.4. Measurement of Photoactivity of *H. halobium* Packed Cells and PM Fragments

The photoactivity of *H. halobium* packed cells and PM fragments was measured as pH vs. time by using a combined pH electrode connected to a Nel pH transmitter at 25 °C, as described by Yucel et al. (1995). Continuous illumination was achieved by a projector lamp (1000 W/m^2).

3.5. Design of Bio-PEC

The bio-PEC reactor was made of glass having an inner volume of 1 L (Figure 1). The upper part of the reactor had four openings for the electrode assembly, combined pH electrode, temperature probe, and H_2 gas outlet. The electrode assembly system contained

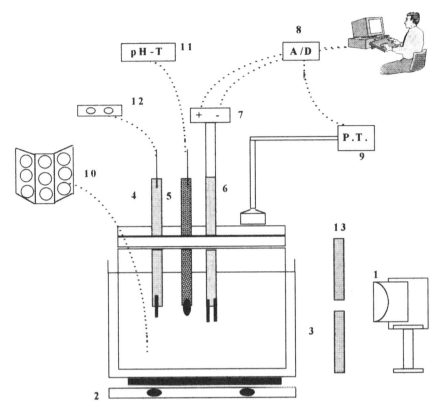

Figure 1. Experimental set-up: (1) light source, (2) magnetic stirrer, (3) water bath, (4) temperature probe, (5) pH electrode, (6) anode and cathode, (7) power supply, (8) analog-to-digital converter, (9) pressure transducer, (10) immobilization assembly, (11) pH transmitter, (12) potentiometer, (13) black plate.

a platinum electrode as the cathode (surface area, 0.8 cm^2) and a silver electrode as the anode (surface area, 4.5 cm^2), located 1 cm apart. The electrodes were connected to a constant voltage power supplier. Voltage and current were recorded with a voltmeter that was connected to an analog-to-digital converter. The combined pH electrode was connected to a pH transducer and the temperature probe was connected to a potentiometer. All H$_2$ gas produced was collected in a 1.4-L closed vessel and increases in pressure were recorded by a pressure transducer (Cole Parmer, J-7352 Series) connected to an analog-to-digital converter. The reactor was placed in a constant temperature water bath made of glass and stirred with a magnetic stirrer at 100 rpm.

3.6. Immobilization of *H. halobium* Packed Cells and Purple Membrane Fragments on Cellulose Acetate Membranes

Cellulose acetate membranes (CAM) having a pore size and thickness of 0.2 μm and 0.15 mm, respectively, were placed into an immobilization assembly that had nine 2-cm diameter holes. The phospholipid solution was prepared by dissolving egg yolk phosphatidyl cholin (lecithin) in chloroform at a concentration of 10 mg/mL. At the selected lecithin-to-BR ratio, the required amount of phospholipid solution was applied on one side of the CAM and

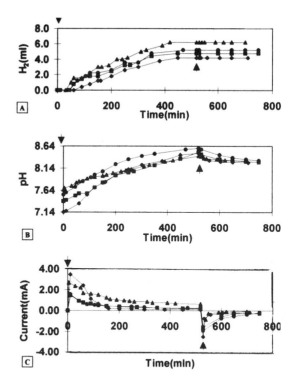

Figure 2. Effect of NaCl concentration on photoelectrochemical hydrogen production (A) total H_2 produced as volume versus time, (B) change of pH versus time, (C) change of current versus time (↓ = Power ON-Light ON; ↑ = Power OFF-Light OFF; [♦] 0.1 M, [■] 1 M, [●] 2 M, [▲] 3.5 M).

allowed to dry. Afterwards, packed cells or PM fragments with a known amount of BR were spread onto the lecithin-impregnated membrane surface. All the membranes were kept in 4 M NaCl solution for at least 24 h at 4 °C in the dark before the experiments.

3.7. Photoelectrochemical Hydrogen Production

The control experiments without *H. halobium* were carried out for the optimization of the operating conditions. As a result of control experiments, the stirring rate (100 rpm), NaCl concentration (20%), temperature (25 °C) and voltage applied (-0.88 V) were set. Fresh salt solution was used and both Pt and Ag electrodes were regenerated before each experiment.

Illumination was at a light intensity of 1000 W/m². All experiments were carried out under anaerobic conditions. During the experiments, voltage, current, pressure change, and pH were recorded with respect to time and all the measurements were calibrated. Experiments were repeated with *H. halobium* either as free packed cells or immobilized on CAM under the same conditions. The purity of hydrogen gas was determined by gas chromatography (Hewlett Packard 5890, Series II).

4. RESULTS AND CONCLUSIONS

4.1. Results of Control Experiments

The effect of NaCl concentration on photoelectrochemical hydrogen production is shown in Figure 2. The NaCl concentration of the solution was varied from 0.1 M to 3.5

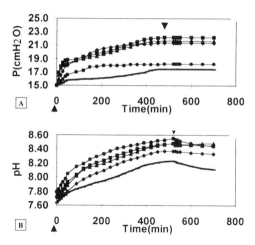

Figure 3. Effect of *H.halobium* packed cells on the rate of hydrogen production (A) in terms of pressure, (B) pH change with respect to time. (↑ = Power ON-Light ON; ↓ = Power OFF-Light OFF; [●] control, [▫] without stirring, [♦] BR = 0.098 μmol, [▲] BR = 0.76 μmol, [■] BR = 1.6 μmol.)

M. As seen in Figure 2a, the rate of hydrogen production and the amount of H_2 produced increased as the NaCl concentration of the solution was increased. Maximum H_2 production was achieved at 3.5 M (20%) NaCl solution. The reaction taking place on the electrodes was expected to be as noted below:

At the cathode: $2H_2O + 2e^- \rightarrow H_2 + 2OH^-$, $E° = -0.83$ V

At the anode: $Ag + Cl^- \rightarrow AgCl + e^-$, $E° = -0.22$ V

and/or: $2Ag + 2OH^- \rightarrow Ag_2O + H_2O + 2e^-$, $E° = -0.34$ V

The surface analysis of silver electrodes by scanning electron microscopy showed the formation of Ag_2O on silver electrodes after all photoelectrochemical hydrogen production experiments (Sediroglu, 1997). A lower hydrogen production rate observed at low NaCl concentrations was attributed to the accumulation of Ag_2O on the silver electrode surface and low conductivity. The solubility of silver chloride in water is very limited, but its solubility increases with increasing NaCl concentration. In addition, alkali chlorides convert silver oxides into silver chloride (Mellor, 1963). Therefore, at high NaCl concentrations, chloride ions covering the silver surface may dissolve silver oxide and silver chloride formed on the anode, and with the photoreactions suggested by Mellor (1963), might help the regeneration of the Ag electrode. Hydrogen production ceasing after 400 min might be attributed to Ag_2O accumulation on the anode; however, it should be emphasized that H_2 could not be produced by replacing the Ag electrode with a new one if the Pt electrode was not regenerated.

Figure 2b illustrates the change of pH versus time in the same set of experiments. As H_2 was produced, pH of the solution increased. The current decreased during H_2 production and reversed its direction when the power and the light were turned off (Figure 2c). This might indicate galvanic reactions taking place in the system.

4.2. Effect of Free *H. halobium* Packed Cells on H_2 Production

Figure 3 illustrates the effect of BR on photoelectrochemical hydrogen production when free *H. halobium* packed cells were used in the bio-PEC. As the amount of BR was

Figure 4. Effect of immobilized *H. halobium* on CAM on photoelectrochemical hydrogen production. Immobilization area: 28.2 cm² (↑ = Power ON-Light ON; ↓ = Power OFF-Light OFF; [-] control; [▭] CAM [BR = 0.033 µmol]; [■] CAM [BR = 0.067 µmol]; [●] CAM [BR = 0.098 µmol]; [◆] CAM-PM [BR = 0.067 µmol]; [▲] H.H$_{free}$ [BR = 0.098 µmol].)

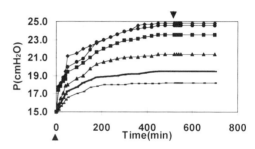

increased (Figure 3a), the hydrogen production rate also increased. Although there was a slight increase in total hydrogen production with increasing BR content, the difference with BR compared to the results of the control experiments lacking BR was quite significant. These results demonstrate the effect of BR on photoelectrochemical hydrogen production.

Stirring also had an important role in photoelectrochemical hydrogen production. The lowest rate of hydrogen production was attained in experiments performed without stirring and BR. Moreover, the rate of hydrogen production was much greater during the initial time periods (ca. 20 min) compared to long-term periods.

The pH change with respect to time during the hydrogen production in these experiments is shown in Figure 3b. The rate of increase in pH in control experiments was much higher than that in experiments containing *H. halobium* packed cells. This result indicates the buffering effect of BR on the solution. According to the Nerst equation, the required voltage for the hydrogen production increases with increased OH⁻ concentration. Therefore, as pH increases, H_2 production should decrease. Upon illumination, BR released protons and thus enhanced hydrogen production.

4.3. Effect of Immobilization of *H. halobium* Packed Cells and PM Fragments on CAM

The results of experiments carried out with *H. halobium* packed cells immobilized on CAM having different initial amounts of BR (0.033 to 0.098 µmol) is given in Figure 4. This figure also shows the results with free *H. halobium* packed cells (BR=0.098 µmol), together with control experiments. As seen in Figure 4, increasing the amount of BR increases the H_2 production rate. This enhancement is much more pronounced than that with the free system. This might be due to the synergetic behavior of BR in the immobilized system. Another explanation could be that immobilization might increase the utilization of light energy effectively since all BR is oriented when immobilized. Photoelectrochemical hydrogen production was also carried out with PM fragments containing 0.067 µmol BR immobilized on CAM. In this case, however, the rate of hydrogen production was higher than that seen with the same amount of BR in *H. halobium* packed cells.

In order to find the effect of immobilization area of CAM on photoelectrochemical hydrogen production, additional experiments were performed using three different BR amounts (0.033, 0.067, and 0.098 µmol) loaded on two different areas of CAM (18.84 and 28.2 cm²) shown in Figure 5. At high BR amounts, decreasing the surface area of CAM decreased the total hydrogen production. However, at the low BR amount, decreasing the surface area increased the total hydrogen production. This might indicate that there is an optimum utilization efficiency of BR on immobilized CAM. In other words, the CAM sur-

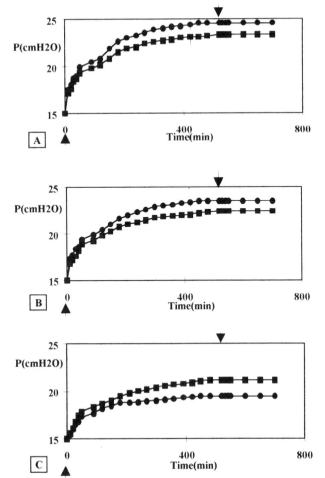

Figure 5. Effect of immobilization area of CAM on rate of hydrogen production; in terms of pressure. (A) BR = 0.098 μmol; (B) BR = 0.067 μmol; (C) BR = 0.033 μmol (↑ = Power ON-Light ON; ↓ = Power OFF-Light OFF; [■] 18.84 cm^2, [●] 28.2 cm^2).

face has a certain capacity for the immobilization of BR. If more BR is immobilized on the CAM surface, some of it might leave the surface or come over to each other, thereby interfering with their activity.

Figure 6A illustrates the H$_2$ production ratio (HPR), which is defined as the ratio of the difference between the total number of moles of H$_2$ produced in the presence and absence of BR within a certain period of time, divided by the number of moles of BR present in the system. It has been concluded that the optimum conditions were obtained with immobilized packed cells containing 0.067 μmol BR on 28.2 cm^2 CAM.

Figures 6B illustrates the initial rates obtained from experiments with immobilized and free systems. As seen in Figure 6B, initial rates are almost two times higher in the case of the former compared to the latter. Moreover, the initial rate of PM fragments is higher than any of the others. Figure 6C shows the power requirement based on hydrogen production. The minimum power requirement is achieved with PM fragments immobilized on CAM. As seen from this figure, power requirements are lowered by BR compared to the control experiments.

In conclusion, the immobilization of *H. halobium* enhances hydrogen production in a salt solution. Studies on the different immobilized systems for the improvement of sta-

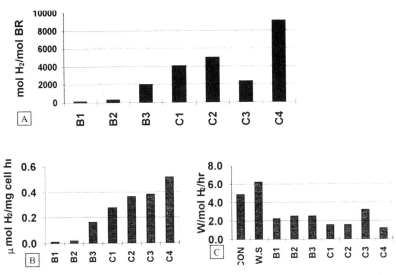

Figure 6. A) Comparison of hydrogen production ratio; (B) comparison of initial rates; (C) comparison of power requirements. B1-B3 denote free *H.halobium* packed cell, C1-C3 denote immobilized *H.halobium* packed cell on CAM, C4 denotes immobilized PM on CAM; B1 (BR = 1.6 μmol), B2 (BR = 0.76 μmol), B3 (BR = 0.098 μmol), C1 (BR = 0.098 μmol), C2 (BR = 0.067 μmol), C3 (BR = 0.033 μmol), C4 (BR = 0.067 μmol), CON = control experiments, W.S = control experiments without stirring).

bility and hydrogen production efficiency are continuing. Continuous H_2 production needs further research. It must be carried out in order to understand the mechanism of the system, silver consumption, and its applicability to continuous H_2 production.

ACKNOWLEDGMENTS

This research is supported by the METU Research Fund, project number AFP 96-07-02-02, and by the Turkish Scientific and Research Council, project number TBAG 15 35.

REFERENCES

Benemann, J.R., 1997, Feasibility analysis of photobiological hydrogen production, *Int. J. Hydrogen Energy*, 22:979–987.

Eroglu, I., Zabut, B.M., and Yucel, M., 1991, Modeling and kinetics of light induced proton pumping of bacteriorhodopsin reconstituted liposomes, *J. Membrane Sci.*, 61:325–336.

Eroglu, I., Aydemir, A., Turker, L. and Yucel, M., 1994, Photoresponse of bacteriorhodopsin immobilized in polyacrylamide gel membranes, *J. Membrane Sci.*, 86:171–179.

Kaya, B., Gunduz, U., Eroglu, I., Turker, L., and Yucel, M., 1996, Hydrogen gas production using coupled enzyme systems from *E. coli* and *H. halobium*, *Proceedings of 11th World Hydrogen Energy Conference*, Stuttgart, Germany, June 23–28, pp. 2595–2600.

Khan, M.M.T., and Bhatt, J.P., 1989, Light dependent hydrogen production by *H. halobium* MMT_{22} coupled to *E. coli*, *Int. J. Hydrogen Energy*, 14:643–645.

Khan, M.M.T., and Bhatt, J.P., 1990, Photoelectrochemical studies on *H. halobium* or continuous production of hydrogen, *Int. J. Hydrogen Energy*, 15:477–80.

Khan, M.M.T., and Bhatt, J.P., 1992, Large-scale photobiological solar hydrogen generation using *H. halobium* MMT$_{22}$ and silicon cell, *Int. J. Hydrogen Energy*, 17:93–95.

Markov, S.A., Bazin, M.J., and Hall, D.O., 1995, The potential of using cyanobacteria in photobioreactors for hydrogen production, in *Advances in Biochemical Engineering Biotechnology*, vol. 52, Springer-Verlag, Berlin Heidelberg, pp. 59–86.

Mellor, J.W., *Supplement to Mellor's Comprehensive Treatise on Inorganic and Theoretical Chemistry*, vol. 2, Longmans, London, 1963, p. 390.

Oesterhelt, D., and Stoeckenius, J.W., 1971, Rhodopsin-like protein from the purple membrane of *H. halobium*, *Nature New Biol.*, 233:149–154.

Oesterhelt, D., and Stoeckenius, J.W., 1974, Isolation of the cell membrane of *H. halobium* and its fraction into red and purple membrane, *Methods in Enzymology*, 31:667–671.

Patel, S., and Madamwar, D., 1994, Photohydrogen production from a coupled system of *H. halobium* and *P. valderianum*, *Int. J. Hydrogen Energy*, 19:733–738.

Sediroglu, V., Aydemir, A., Gunduz, U., Yucel, M., Turker, L., and Eroglu, I., 1996, Modeling of long-term photoresponse of bacteriorhodopsin immobilized on cellulose acetate membranes, *J. Membrane Sci.*, 113:65–71.

Sediroglu, V., 1997, *Photoelectrochemical hydrogen production by using Halobacterium halobium*, Ph.D. thesis, Middle East Technical University, Ankara, Turkey.

Singh, K., and Caplan, R.S., 1980, The purple membrane and solar energy conversion, *TIBS*, March, pp. 62–64.

Trissl, H.W., 1990, Photoelectric measurements of purple membranes, *Photochem. Photobiol.*, 51:793–801.

Yucel, M., Zabut, B.M., Eroglu, I., and Turker, L., 1995, Kinetic analysis of light induced proton dissociation and association of bacteriorhodopsin in purple membrane fragments under continuous illumination, *J. Membrane Sci.*, 104:65–72.

PHOTOSYNTHETIC BACTERIAL HYDROGEN PRODUCTION WITH FERMENTATION PRODUCTS OF CYANOBACTERIUM *Spirulina platensis*

Katsuhiro Aoyama,[1] Ieaki Uemura,[1] Jun Miyake,[2] and Yasuo Asada[3]

[1]Frontier Technology Research Institute
Tokyo Gas Co., Ltd.
1-7-7 Suehiro-cho
Tsurumi-ku, Yokohama, 230-0045, Japan
[2]National Institute for Advanced Interdisciplinary Research
1-1-4 Higashi
Tsukuba, Ibaraki, 305-0046, Japan
[3]National Institute of Bioscience and Human-Technology
1-1 Higashi
Tsukuba, Ibaraki, 305-0046, Japan

Key Words

Spirulina, cyanobacterium, photosynthetic bacterium, hydrogen, organic acid

1. SUMMARY

The cyanobacterium *Spirulina platensis* accumulates glycogen photoautotrophically in a nitrogen-deficient medium. Under anaerobic conditions in the dark, the glycogen degrades into organic compounds. As molecular hydrogen also evolves in this process, hydrogenase participation is suggested in this metabolism. We investigated the several conditions necessary for the evolution of hydrogen and production of organic compounds. The effects of cell concentration, initial pH, and concentration of the buffer were determined.

These fermentation products were then converted into molecular hydrogen by using the photosynthetic bacterium *Rhodobacter sphaeroides* RV with light energy. The composition of the evolved gas was mainly hydrogen and carbon dioxide. This photosynthetic bacterial production of hydrogen was caused by a nitrogenase-dependent mechanism. Combining this system with photosynthesis of cyanobacteria resulted in the production of hydrogen by splitting water.

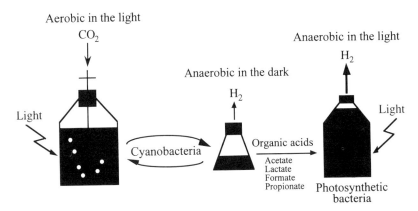

Figure 1. Hydrogen production by combining cyanobacteria with photosynthetic bacteria.

2. INTRODUCTION

Hydrogen gas has been receiving much attention as a clean and renewable energy carrier that does not produce carbon dioxide in combustion. Using biotechnology as a means of environmentally acceptable hydrogen production has been a goal of researchers throughout the world.

Ever since Gaffron and Rubin (1942) discovered light-dependent hydrogen production by a green algae, *Scenedesmus* sp., hydrogen production by green algae and cyanobacteria have been extensively studied (Miyamoto et al., 1993). Miura and Miyamoto (1992) demonstrated an alternating light-dark cycle as a model of the day-night cycle on combining green algae, *Chlamydomonas* sp. Hydrogenase is one of the key enzymes in this metabolism. Cyanobacterial hydrogenase can also be active in the evolution of molecular hydrogen, particularly in the dark during fermentative metabolism (Asada and Kawamura, 1984, 1986). Here, we describe hydrogen production and related fermentatives by *S. platensis* and photosynthetic bacterial hydrogen production by fermentation products of *S. platensis*.

3. MATERIALS AND METHODS

3.1. Strains and Cultivation

The non-nitrogen-fixing and filamentous cyanobacterium, *Spirulina platensis* NIES-46, was cultivated at 30 °C in SOT medium (Ogawa and Terui, 1970) under continuous illumination by fluorescent lamps (5 klux). ("Nitrogen-free" indicates the absence of sodium nitrate in the SOT medium.)

A photosynthetic bacterium, *Rhodobacter sphaeroides* RV, was cultivated at 30 °C in a basal medium (Miyake et al., 1984) containing 10 mM ammonium sulfate, 36 mM sodium succinate, and 0.1% yeast extract (pH 6.8) under continuous illumination by tungsten lamps.

3.2. Cyanobacterial Fermentation

S. platensis was harvested in the nitrogen-free medium at the late logarithmic growth phase, collected by filtration using filter paper, and resuspended in a sodium phosphate buffer

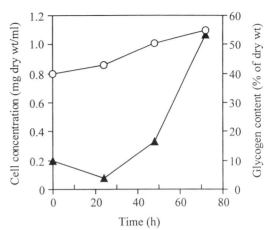

Figure 2. Accumulation of glycogen in *S. platensis* under nitrogen-starved conditions. *S. platensis* cells were harvested in the late logarithmic growth phase and transferred into the nitrogen-free medium. ○ = cell concentration; ▲ = glycogen content.

(20 mM, pH 7.0, unless otherwise stated). Twenty milliliters of the cell suspension were sealed in an Erlenmeyer flask (60 mL total volume) with a rubber stopper. The gas phase was replaced with nitrogen gas. The flasks were shaken overnight in the dark at 30 °C.

The fermentation broth was transferred to the immobilized photosynthetic bacterial hydrogen production system (Figure 1) instead of the gL medium (Miyake et al., 1984).

3.3. Hydrogen Production by Photosynthetic Bacterium

R. sphaeroides RV was immobilized in a basal medium containing 4% (w/v) Bacto-Agar (DIFC) in a Roux bottle. The gL medium for hydrogen evolution consisted of a basal medium containing 10 mM L-glutamic acid monosodium salt, 50 mM D,L-lactic acid sodium salt, and 0.15% sodium hydrogen carbonate. The Roux bottle was illuminated with halogen lamps. The light intensity was measured by a Radiometer Model 4090 (Springfield Jarco Instruments).

3.4. Analysis Methods

Glycogen content in *S. platensis* cells was determined according to Ernst et al. (1984).
Organic acids excreted by *S. platensis* were analyzed by high-performance liquid chromatography (Shimadzu LC-8A system) using a Shim-Pack SCR-101H (Shimadzu) column.

The evolved gas was analyzed by gas chromatography (GC-8A, Shimadzu Corp.). A molecular sieve 5A column (GL Sciences, Inc.) and Porapak Q column (GL Sciences, Inc.) were used to determine hydrogen and carbon dioxide, respectively.

4. RESULTS AND CONCLUSION

4.1. Production of Hydrogen Gas and Organic Compounds by *Spirulina*

S. platensis accumulated glycogen when photoautotrophically incubated in a nitrogen-free medium (Figure 2). The cells incubated in the nitrogen-free medium were demonstrated to have higher hydrogen production activity than those grown in an ordinary culture medium (Figure 3). In Figure 3, the rate and sustainability of hydrogen production

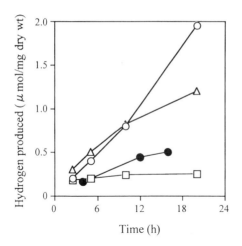

Figure 3. Effects of cell concentration and nitrogen starvation on hydrogen production by *S. platensis*. Nitrogen-starved cells of *S. platensis* prepared by incubation for 72 h as described in Figure 2 were suspended in 20 mM sodium phosphate, pH 7.0, at different cell concentrations. The following symbols indicate nitrogen-starved cells: ○ = 1.62; △ = 4.06; □ = 8.70 (mg dry wt/mL); ● = non-nitrogen-starved cells at a concentration of 1.62 mg dry wt/mL.

are inversely associated with cell concentration. One explanation could be product inhibition or pH decrease; fermentation products or pH might reach an inhibitory level earlier in a dense suspension than in the thin suspension. The main organic compounds produced by the nitrogen-starved cells were ethanol and acetic, lactic, and formic acids (Figure 4).

It is shown in Figure 5 that productivities were lower in acidic pH, that a higher buffer concentration enhanced hydrogen production, and the organic acids might suggest the occurrence of inhibition of fermentation by acidic products.

The following reactions were deduced. The glycogen degraded into ethanol via pyruvate. In this pathway, ferredoxin is reduced and the cyanobacterial hydrogenase mediates the hydrogen production. At the same time, several organic acids are produced by way of acetyl-CoA.

4.2. Hydrogen Production by Photosynthetic Bacteria Using the Fermentation Products of *Spirulina*

The fermentation medium of *S. platensis* was used as the substrate for hydrogen production by *R. sphaeroides*. During the first 10 h, both the gL medium and the fermentation

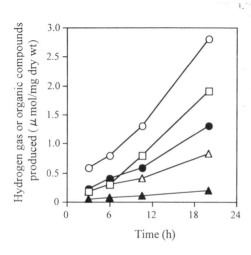

Figure 4. Production of hydrogen and organic compounds by nitrogen-starved cells of *S. platensis*. □ = hydrogen gas; ○ = acetic acid; △ = formic acid; ▲ = lactic acid; ● = ethanol.

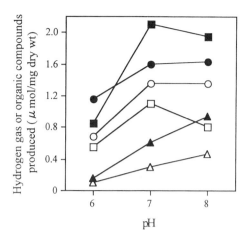

Figure 5. Effects of pH and buffer concentration on production of hydrogen and organic acids. Nitrogen-starved cell suspension of *S. platensis* (4.06 mg) in 20 or 100 mM sodium phosphate buffer, at pH 6.0, 7.0, 8.0: ☐, ■ = hydrogen gas; ○, ● = acetic acid; △, ▲ = formic acid (open and filled symbols indicate 20 and 100 mM sodium phosphate buffer, respectively).

medium of *S. platensis* showed the same hydrogen evolution rate as the substrate (data not shown). However, the rate of hydrogen evolution slowed rapidly due to the consumption of lactic, acetic, and propionic acids. The composition of the gas was mainly hydrogen and carbon dioxide (78 and 10%, respectively). The total amount of hydrogen evolved was nearly equal to the sum of the stoichiometrical estimation based on each organic acid degraded.

ACKNOWLEDGMENTS

This work was performed under the management of RITE as a part of the Research and Development Project on Environmentally Friendly Technology for the Production of Hydrogen, supported by the New Energy and Industrial Technology Development Organization (NEDO).

REFERENCES

Asada, Y., and Kawamura, S., 1984, Hydrogen evolution by *Microcystis aeruginosa* in darkness, *Agric. Biol. Chem.*, 48:2595–2596.
Asada, Y., and Kawamura, S., 1986, Screening for cyanobacteria that evolve molecular hydrogen under dark and anaerobic conditions, *J. Ferment. Technol.*, 64:553–556.
Ernst, A., Kirschenlohr, H., Diez, J., and Boeger, P., 1984, Glycogen content and nitrogenase activity in *Anabaena variabilis*, *Arch. Microbial.*, 140:120–125.
Miura, Y., Saitoh, C., Matsuoka, S., and Miyamoto, K., 1992, Stably sustained hydrogen production with high molar yield through a combination of a marine green alga and a photosynthetic bacterium, *Biosci. Biotech. Biochem.*, 56:751–754.
Miyake, J., Mao, X.Y., and Kawamura, S., 1984, Photoproduction of hydrogen from glucose by a co-culture of a photosynthetic bacterium and *Clostridium butyricum*, *J. Ferment. Technol.*, 62:531–535.
Miyamoto, K., 1993, Hydrogen production by photosynthetic bacteria and microalgae, in *Recombinant Microbes for Industrial and Agricultural Applications*, Murooka, Y., and Imanaka, T. (eds.), Marcel Dekker, New York, pp. 771–786.
Ogawa, T., and Terui, G., 1970, Studies on the growth of *Spirulina platensis*, I. on the pure culture of *Spirulina platensis*. *J. Ferment. Technol.*, 48:361–367.

39

HYDROGEN PHOTOPRODUCTION FROM STARCH IN CO_2-FIXING MICROALGAL BIOMASS BY A HALOTOLERANT BACTERIAL COMMUNITY

Akiko Ike, Naohumi Toda, Tomoko Murakawa, Kazumasa Hirata, and Kazuhisa Miyamoto

Environmental Bioengineering Laboratory
Faculty of Pharmaceutical Sciences
Osaka University
1-6 Yamadaoka
Suita, Osaka 565-0871, Japan

Key Words

hydrogen photoproduction, algal biomass, photosynthetic bacteria, heat-HCl treatment, lactic acid fermentation, bacterial community

1. SUMMARY

To produce H_2 from CO_2-fixing algal biomass, two-step systems including heat-HCl treatment or lactic acid fermentation followed by H_2 production, have been attempted. An alternate system involving direct conversion of algal starch to H_2 is proposed. Such a single-step conversion has been achieved using a halotolerant bacterial community selected from night soil treatment sludge. H_2 yield from starch-glucose was lowest in the process via heat-HCl treatment, while the process with lactic acid fermentation followed by H_2 photoproduction of *Rhodobacter sphaeroides* RV gave the highest yield of 4.6 mol/mol starch-glucose from *Chlamydomonas reinhardtii* biomass. Although the single-step system gave a lower H_2 yield at present, it produced H_2 directly from starch in *C. reinhardtii* biomass.

2. INTRODUCTION

Among the many global environmental problems facing society today, greenhouse warming is widely recognized as being one of the most serious. CO_2 fixation by microal-

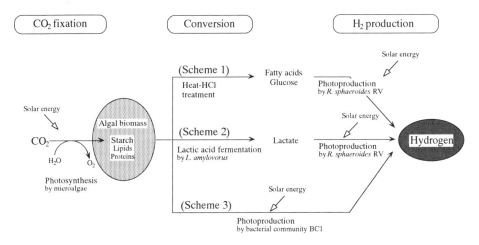

Figure 1. Various pathways of H_2 production from algal biomass.

gae is considered to be an environmentally friendly, energy-saving means of addressing this problem. However, microalgal systems produce large amounts of waste biomass that would easily be converted to CO_2. Therefore, it is necessary to establish processes by which algal biomass can be recycled to resource production.

H_2 is one of the most promising energy media because it is easily converted to electricity and is cleanly combustible. We have proposed three systems to convert algal biomass to H_2 using photosynthetic bacteria (Figure 1). Since photosynthetic bacteria cannot utilize raw algal biomass as an efficient substrate for H_2 production, pretreatment is required to transform it into suitable substrates. As described in Scheme 1, we attempted to decompose algal biomass by heat-HCl treatment to produce fatty acids, which then served as substrates for H_2 production by photosynthetic bacteria (Ike et al., 1996). We also attempted to convert algal starch, a main storage compound in *Chlamydomonas reinhardtii*, to lactic acid by means of fermentation. H_2 was then produced by a photosynthetic bacterium from the fermentate (Scheme 2) (Ike et al., 1997).

However, in terms of cost-savings, environmental effects, and simplicity, a single-step biological conversion system would be more desirable than the two-step systems described above. An example of such a single-step conversion would be the production of H_2 from starch by photosynthetic bacteria. This topic has received little attention to date due to the limited ability of photosynthetic bacteria to use starch as an electron donor for H_2.

In the present work, we propose a new system, described as Scheme 3 (Figure 1), in which algal biomass is converted to H_2 in a single step by a bacterial community. As a first step in constructing this system, we obtained a microbial community, consisting of photosynthetic bacteria and other microorganisms, which could convert starch to H_2 in a single-step culture system.

3. EXPERIMENTAL

3.1. Algal Biomass

The freshwater green algae *C. reinhardtii* (IAM C-238) and *Chlorella pyrenoidosa* (IAM C-212) were grown photoautotrophically in modified Bristol medium (Miura et al.,

1982). A marine green alga, *Dunaliella tertiolecta* (ATCC 30929), was also grown photoautotrophically in modified f/2 medium (Guillard et al., 1962). Algae were cultivated at 30 °C for 10 days in a cylindrical glass bottle (working volume 2 L; diameter 10 cm) with 1% CO_2 gas supplied continuously under illumination (10 W/m^2, fluorescent lamp). Algal cells were harvested and concentrated a hundredfold by means of centrifugation (13000 × g, 10 min).

3.2. Heat-HCl Treatment of Algal Biomass

To optimize the conditions of heat-HCl pretreatment, concentrated algal biomass was autoclaved (120 °C, 118 kPa) in the presence of 0–0.5 N HCl for 0–80 min. Extracts were then treated with the cation exchange resin Dowex 50W-8X (Dow Chemical Company, United States) to remove ammonia and diluted three times with sterile water, after which pH was adjusted to 7.0 with 10 N NaOH.

3.3. Lactic Acid Fermentation of Algal Biomass

Lactobacillus amylovorus, a lactic acid bacterium capable of hydrolyzing starch (Nakamura, 1981), was precultivated in Lacto Bacilli MRS broth (Difco Corporation, Ltd., United States), and 2.5 mL of an actively growing cell suspension culture (OD_{600} = ca. 9) was harvested by centrifugation (17000 × g, 10 min). The cells were washed once with sterile water and added to 25 mL of concentrated intact algal biomass. Inoculated samples were incubated at 37 °C for 2–4 days in 50-mL screw-cap plastic bottles. To prevent lactic acid production from stopping the fermentation process, either 500 mg of $CaCO_3$ was added before the fermentation, or 10 N NaOH was added every 6 h during fermentation to adjust pH to 6.0. Thus, lactic acid yield was calculated based on the equation

$$(C_6H_{12}O_6)_n \rightarrow 2nC_3H_6O_3$$

The fermentate obtained was diluted with sterile water to give a calculated lactic acid concentration of 2.7 g/L (30 mM).

3.4. Selection of Bacterial Community

For screening of a halotolerant bacterial community capable of directly converting starch to H_2, activated sludge was used from the M night soil treatment plant, located in the southern area of Osaka Prefecture, and which employs a seawater dilution system to control the temperature in the reactor. Medium 1, containing 1.64 g sodium acetate, 1.92 g sodium propionate, 1.84 g glycerol, 3.60 g glucose, 1.32 g ammonium sulfate, 0.2 g yeast extract, and 24 g synthetic seawater mixture Aqua Marine S (Yashima Chemical Co., Japan) per 1 L of basal medium (Ike et al., 1997) was used for enrichment cultures. Medium 2 with 5.40 g/L succinic acid (disodium salt), 30 g/L NaCl and 1.5% agar replacing glycerol, glucose, and Aqua Marine S in Medium 1 was used to select for a bacterial community, designated BC1. Succinic acid was employed because it is efficiently used by photosynthetic bacteria.

3.5. H_2 Production

Mineral components of the basal medium (Ike et al., 1997), plus 10 mM sodium glutamate and 0.15% $NaHCO_3$, were added to the degassed heat-HCl treatment extract or lac-

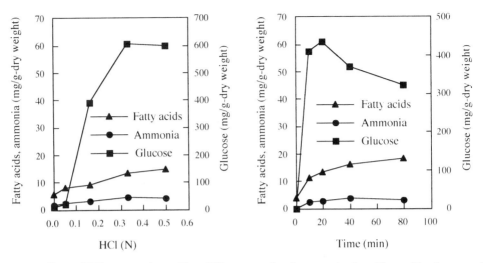

Figure 2. Effects of HCl concentration and heat-HCl treatment duration on production of fatty acids, glucose, and ammonia from *C. reinhardtii* biomass. *C. reinhardtii* cells were harvested by centrifugation (13000 × g, 10 min), concentrated a hundredfold, and autoclaved (120 °C, 118 kPa) with 0–0.5 N HCl for 0–80 min.

tic acid fermentate. The extract (25 mL) and fermentate (60 mL) were placed in glass tubes (volume 30 and 70 mL, diameter 1.0 and 2.5 cm, respectively) provided with a gas outlet and sampling nozzle. Aliquots (5 and 12 mL) of an actively growing *Rhodobacter sphaeroides* RV culture (OD_{660} = ca. 5 corresponding to ca. 5 mg dry wt/mL) were centrifuged (6300 × g, 10 min), and the cells were resuspended into the extract and the fermentate, respectively. The inoculated samples were incubated at 30 °C under 330 W/m² tungsten illumination for five days. The H_2 gas produced was trapped in a 100-mL glass cylinder above a water bath.

Hydrogen evolution by the bacterial community BC1 was tested using glucose, starch, cellobiose, and cellulose. Each H_2 production medium contained one substrate at 25 mM (polysaccharide and disaccharides concentrations are expressed as the equivalent of their constituent monomers), the mineral components of the basal medium, 10 mM sodium glutamate, 0.15% $NaHCO_3$, and 3% NaCl. After degassing, 65 mL of each medium was placed in a glass tube (volume 70 mL, diameter 2.5 cm) provided with a gas outlet and sampling nozzle. An actively growing BC1 culture (OD_{660} = ca. 2.5) was harvested by centrifugation (6300 × g, 10 min) and resuspended in each H_2 production medium to give a final OD_{660} of 0.6. The inoculated samples were incubated in the same manner as the extracts and fermentates described above.

Hydrogen production from algal biomass was done in a similar manner as described above. *C. reinhardtii* cells were harvested by means of centrifugation (13000 × g, 10 min) and diluted to give a final starch concentration of 25 mM (expressed as glucose). An actively growing BC1 culture was added to the algal cell suspension to test H_2 production from starch.

3.6. Analytical Methods

Volatile fatty acids (the sum of acetic, propionic, butyric, and valeric acid), glucose, and ammonia were assayed according to Ike et al. (1996). Lactic acid and H_2 were measured as described previously (Ike et al., 1997).

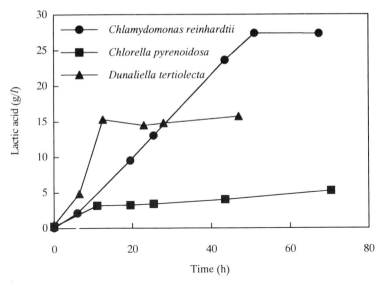

Figure 3. Lactic acid fermentation of algal biomass by *L. amylovorus*. *C. reinhardtii*, *C. pyrenoidosa*, and *D. tertiolecta* cells were harvested by centrifugation (13000 × g, 10 min), concentrated a hundredfold, inoculated with *L. amylovorus*, and incubated at 37 °C for 2–4 days with 20 g/L $CaCO_3$.

4. RESULTS

4.1. Heat-HCl Treatment of Algal Biomass and H_2 Production from the Extract

With heat and/or acid treatment, algal cells could be converted to organic compounds, including fatty acids which are favorable substrates for H_2 production by photosynthetic bacteria. However, under optimal treatment conditions, only 20 mg of fatty acids per gram dry weight of algal biomass was produced, although 400–600 mg/g dry weight of glucose was produced (Figure 2).

The use of cation exchange resin and threefold dilution lowered ammonia concentration to ca. 30 mg/L, while fatty acid concentration was ca. 600 mg/L. From this diluted extract, *R. sphaeroides* RV evolved 3.4 mmol of H_2 per liter within 65 h.

4.2. Lactic Acid Fermentation of Algal Biomass

The freshwater or marine green algal biomass samples (*C. reinhardtii* and *C. pyrenoidosa*, or *D. tertiolecta*) were fermented by *L. amylovorus*. The results (Figure 3) show that lactic acid was produced from all three algal biomass types, and that high lactic acid yields (70 and 80%) were obtained from the biomass of *C. reinhardtii* and *D. tertiolecta*, respectively. The yield reached almost 100% when freeze-thawed algal cells were used as substrate.

The fermentate of *C. reinhardtii* was diluted and the lactic acid concentration was adjusted to 2.7 g/L (30 mM). *R. sphaeroides* RV evolved 75 mmol of H_2 from 1 L of the diluted fermentate within 100 h.

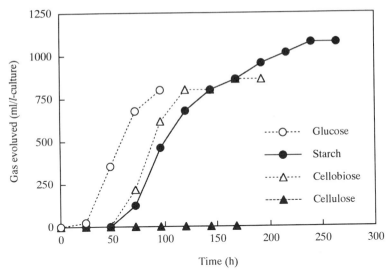

Figure 4. H_2 production by the bacterial community BC1. Each substrate was supplied at 25 mM (polysaccharide and disaccharides are expressed in terms of their constituent monomers), and the gas evolution was measured. The H_2 concentration in the evolved gas was ca. 70% when analyzed at the end of each experimental run.

4.3. Selection of Bacterial Community for Single-Step H_2 Production

The seed sludge was diluted two- to a hundredfold with Medium 1 and then incubated under tungsten illumination (170 W/m^2) at 20–25 °C with or without slow agitation. After one month, cultures appearing pinkish in color were spread on agar plates of Medium 2 and incubated for one week. Red, orange, or pink colonies were transferred to the hydrogen production medium and their growth and H_2 productivity were evaluated. One colony exhibited good growth and the resulting bacterial community produced H_2 from starch as well as from various fatty acids and sugars. We named this community BC1 and used it in subsequent experiments. It was found that BC1 contains at least two bacterial strains that are responsible for H_2 production from starch.

Figure 4 shows that BC1 could produce H_2 from starch, glucose, and cellobiose, but not from cellulose. BC1 evolved 28.3 mmol H_2 from 1 L of culture containing 25 mM starch (expressed as glucose). As shown in Figure 5, BC1 produced H_2 directly from starch in algal cells, and the amount of H_2 produced was much higher from algal biomass than from authentic starch, suggesting that some algal cell constituents would stimulate bacterial growth and/or H_2 production.

5. DISCUSSION

In the heat-HCl treatment system (Scheme 1), only 20 mmol of H_2 per mole of algal starch-glucose was produced. Although low molecular weight fatty acids seem to be good substrates for H_2 production by photosynthetic bacteria, production of such fatty acids from algal biomass by heat treatment was not sufficient. Glucose, which is produced from starch by heat-HCl treatment, was not a good substrate for H_2 production by *R.*

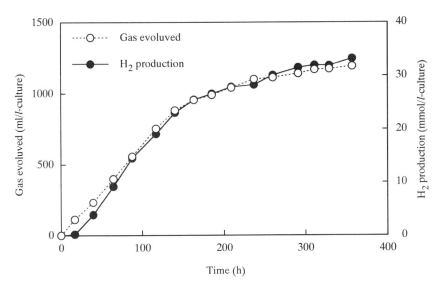

Figure 5. H_2 production from *C. reinhardtii* biomass by the bacterial community BC1. Algal biomass was supplied to bacterial cell suspension of BC1 and the gas (containing mainly H_2 and CO_2) evolution and H_2 production were measured.

sphaeroides RV. Other problems include the energy-consuming nature of the treatment and the need for subsequent ammonia removal.

We succeeded in the fermentative conversion of algal biomass of *C. reinhardtii* and *D. tertiolecta* to lactic acid. Starch, the main photosynthetic storage component in *C. reinhardtii*, was completely consumed by *L. amylovorus*, and good H_2 yields were subsequently achieved from the resulting lactic acid. In this system (Scheme 2), 4.6 mol of H_2 per mole of algal starch-glucose was produced by successive use of *L. amylovorus* and *R. sphaeroides* RV. However, two bacterial cultivation steps were required. In terms of cost-savings, environmental effects, and simplicity, a single-step biological cultivation system (Scheme 3) is more desirable.

To establish such a system, we selected the bacterial community BC1, which could convert both starch and cellobiose directly to H_2. The sludge from the M night soil treatment plant, from which BC1 was selected, is ordinarily exposed to various waste components and high salinity. Therefore, BC1 would have wide biological diversity and halotolerance.

Although H_2 yield in Scheme 3 was relatively lower (1.3 mol of H_2 per mole of starch-glucose) than that of Scheme 2 (H_2 production via lactic acid fermentation), it could be improved by optimization of culture conditions such as pH and temperature, or by reconstruction of the members of BC1. We could demonstrate production of H_2 from algal biomass by means of BC1. This system (Scheme 3) will have the advantages of energy efficiency (compared to Scheme 1) and simplicity (compared to Scheme 2).

REFERENCES

Guillard, L., and Ryther, J.H., 1962, Studies of marine planktonic diatoms, *Can. J. Microbiol.*, 8:229–239.
Ike, A., Saimura, C., Hirata, K., and Miyamoto, K., 1996, Environmentally friendly production of H_2 incorporating microalgal CO_2 fixation, *J. Mar. Biotechnol.*, 4:47–51.

Ike, A., Toda, N., Hirata, K., and Miyamoto, K., 1997, Hydrogen photoproduction from CO_2-fixing microalgal biomass: application of lactic acid fermentation by *Lactobacillus amylovorus*, 84:428–433.

Miura, Y., Yagi, K., Shoga, M., and Miyamoto, K., 1982, Hydrogen production by a green alga *Chlamydomonas reinhardtii* in an alternating light/dark cycle, *Biotechnol. Bioeng.*, 24:1555–1563.

Nakamura, L.K., 1981, *Lactobacillus amylovorus*, a new starch-hydrolyzing species from cattle waste-corn fermentations, *Int. J. Syst. Bacteriol.*, 31:56–63.

40

HYDROGEN PRODUCTION BY PHOTOSYNTHETIC MICROORGANISMS

Yoshiaki Ikuta,[1] Tohru Akano,[2,3] Norio Shioji,[1,4] and Isamu Maeda[1,4]

[1]Mitsubishi Heavy Industries, Ltd.
[2]Kansai Electric Power Co.
[3]RITE Amagasaki 2nd and Nankoh Laboratory
Hyogo, Japan
[4]Osaka University
Osaka, Japan

1. SUMMARY

Hydrogen is a clean energy alternative to fossil fuels. We have developed a stable system for the conversion of solar energy into hydrogen using photosynthetic microorganisms. Our system consists of the following three stages: (1) photosynthetic starch accumulation in green algae (400 L x 2); (2) dark anaerobic fermentation of the algal starch biomass to produce hydrogen and organic compounds; and (3) further conversion of the organic compounds to produce hydrogen using photosynthetic bacteria.

We constructed a test plant for this process at Kansai Electric Power Company's Nankoh power plant in Osaka, and conducted a series of tests using CO_2 obtained from a chemical absorption pilot plant. The photobiological hydrogen production process used a combination of a marine alga, *Chlamydomonas* sp., and the marine photosynthetic bacterium *Rhodovulum sulfidophilum* sp. WIS. The dark anaerobic fermentation of algal starch biomass was also investigated.

2. INTRODUCTION

The sustainable development of human activities in harmony with the global environment can be achieved by increasing the recycling of the resources necessary for those activities. Hydrogen is a clean, renewable form of energy. It can be converted to electric energy with high efficiency. Hydrogen production by the biophotolysis of water based on microalgal photosynthesis in the carbon cycling system is an ideal solar energy conversion system.

Based on studies that have been conducted in small-scale tests, we constructed our bench-scale test apparatus at the Nankoh power plant of Kansai Electric Power Company in Osaka.

BioHydrogen, edited by Zaborsky *et al.*
Plenum Press, New York, 1998

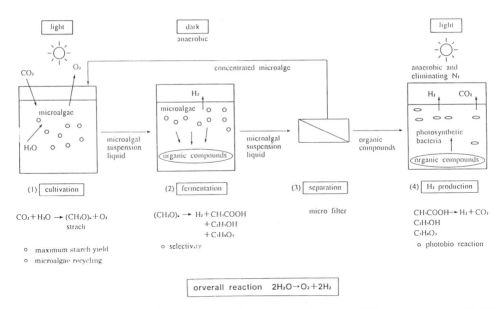

Figure 1. Process of photobiological hydrogen production by a combination of *Chlamydomonas* sp. MGA 161 and *Rhodovulum sulfidophilum sp.* WIS.

Our photobiological hydrogen production system is shown in Figure 1. This system consists of the following four processes:

1. Cultivation of microalgae: In this process, starch is made from microalgal photosynthesis. The microalga marine *Chlamydomonas sp. MGA 161* was used.
2. Fermentation of microalgae: Organic products, such as acetic acid, ethanol, and glycerol, are produced from the dark anaerobic reaction of the starch in the microalgae produced in the prior process.
3. Separation of microalgae: After the fermentation is completed, the microalgal suspension liquid is treated by a microfilter. Concentrated microalgae are recycled to the algae cultivation pond and a permeate containing organic compounds is transferred to the H_2 production process.
4. Hydrogen production by photosynthetic bacteria: Organic products, such as acetic acid, ethanol, and glycerol, produced in the prior process are converted to hydrogen using photosynthetic bacteria in light. The photosynthetic bacterium marine *Rhodovulum sulfidophilum sp.* WIS was used.

In this hydrogen production system, it is necessary to recycle the microalgae to the algae cultivation pond after fermentation.

3. CULTIVATION OF MICROALGAE

3.1. Test Apparatus

Figure 2 shows the raceway-type reactor used in the cultivation of *Chlamydomonas* MGA 161. The specifications of this reactor are as follows: material, acrylic resin; size, 2800 × 800 × 250 mm; area, 2 m^2; capacity, 400 L; and agitator, paddle-type.

Figure 2. Raceway-type cultivator.

3.2. Test Results

Figure 3 shows algal productivity of *Chlamydomonas* MGA 161 in a raceway-type cultivator, using CO_2 obtained from a chemical absorption pilot plant treating actual power plant flue gas. The maximum productivity of the starch is 15 $g/(m^2/day)$.

The relationship between the productivity and content of starch is shown in Figure 4. The productivity of starch is maximized in the range of starch content from 20~30 wt % (dry). Therefore, microalgae (the same microalgae are used repeatedly between the cultivation and fermentation processes) should be subjected to the fermentative part of this process when the starch content reaches this range.

4. FERMENTATION OF MICROALGAE

4.1. Test Apparatus

A photograph of the fermentator used in the test is shown in Figure 5. The specifications of this fermentator are as follows: material, polyvinyl chloride; size, 450 mm 9 n diameter × 1250 mm; and capacity, 155 L.

4.2. Test Results

The relationship between productivity of organic products (y) and starch decomposition (x) is shown in Figure 6. Theoretically, the relationship between y and x should be y =

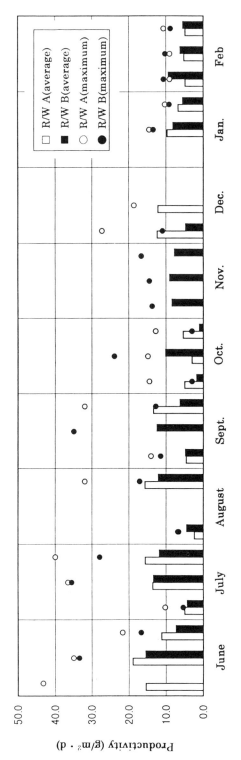

Figure 3. Algal productivity of *Chlamydomonas* sp. MGA 161 in raceway-type cultivator (June 1996 through February 1997).

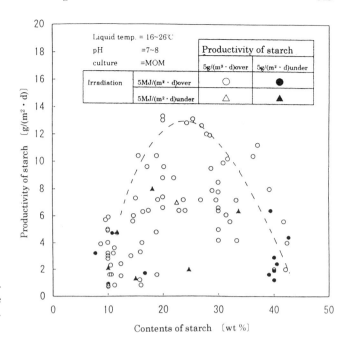

Figure 4. Relationship between productivity and content of starch in the cultivation of *Chlamydomonas* sp. MGA161.

2x, but in this test, y = 1.6x was obtained. The reason is that other products, such as lactic acid, are produced in addition to acetic acid, ethanol, and glycerol.

The relationship between the productivity of organic products and concentration of microalgae is shown in Figure 7. This relationship is depicted as a straight line up to the microalgae concentration of 5000 mg/L, suggesting that in order to increase the organic compound concentration during fermentation, the microalgal suspension liquid should be concentrated.

5. HYDROGEN PRODUCTION BY PHOTOSYNTHETIC BACTERIA

Figure 8 illustrates the parallel plate-type photobioreactors, which were used in a combined test with a microalgae cultivator and microalgae fermentator. The specifications of this reactor are as follows: material, acrylic resin; plate size, 2000 × 50 × 500 mm; number of plates, 6; spacing of plates, 120 mm; optical area, 1.8 m^2; and culture volume, 150 L.

6. MICROALGAE RECYCLING TEST

In this hydrogen production system, it is necessary to recycle the microalgae to the algae cultivation pond after fermentation is completed. The photosynthetic starch accumulation and fermentative production of organic compounds through starch degradation were carried out during an alternating light-dark cycle. Stable starch accumulation was repeatedly observed in the light period of this cycle. The fermentative production of acetic acid, ethanol, and glycerol was stable during the dark period of this cycle.

Figure 5. Microalgae fermentator.

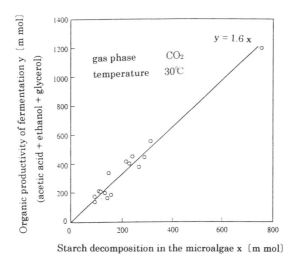

Figure 6. Relationship between organic productivity and starch decomposition in the fermentator.

Hydrogen Production by Photosynthetic Microorganisms

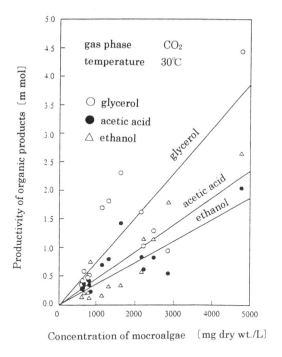

Figure 7. Relationship between productivity of organic products and concentration of microalgae.

Figure 8. Parallel plate photobioreactor.

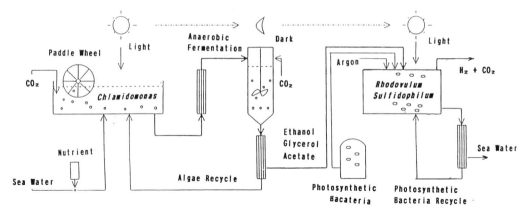

Figure 9. Bench-scale test plant flow.

7. BENCH-SCALE TEST PLANT

7.1. Bench-Scale Test Flow

The bench scale test plant flow is shown in Figure 9. The bench-scale plant consists of two raceway-type reactors, one fermentator, and a series of the parallel plate-type photobioreactors. The process is as follows:

- Microalgae produced during the day are directed to anaerobic fermentation through a microfilter. CO_2 is fed to a raceway-type reactor under pH control.
- The starch accumulated in microalgae is anaerobically fermented and acetic acid, ethanol, and glycerol are produced.
- After fermentation, microalgae are recycled to the raceway-type cultivator through a microfilter.
- Acetic acid, ethanol, and glycerol are sent to the parallel plate-type photobioreactors, where hydrogen is evolved using photosynthetic bacteria.

7.2. Test Results

The results of the continuous operation of the bench-scale test conducted from October 5 through October 28, 1997, are shown in Figure 10. All apparatus, the raceway-type reactor, fermentator, and the series of parallel plate-type photobioreactors are operated concurrently.

The performances of the test are summarized below:

- 23 days of continuous operation were performed for the raceway-type reactor and three times for the recycle operation of the parallel plate-type photobioreactors without decreasing hydrogen evolution;
- after 10 days of operation, hydrogen evolution decreased drastically because of the degradation of photosynthetic bacteria from contamination by other bacteria; and
- the accumulated amount of hydrogen evolved during these 23 days was 78 N L (normal liters).

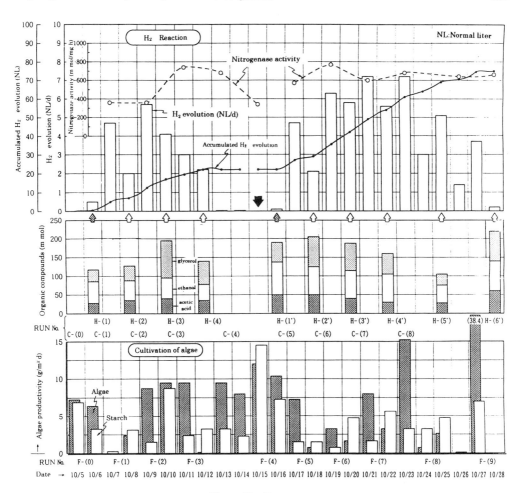

Figure 10. Test results.

8. CONCLUSIONS

1. The productivity of starch is maximized in the range of 20~30 wt %. Therefore, microalgae (the same microalgae are used repeatedly between the cultivation and fermentation process) should be subjected to the fermentative part of this process when the starch content reaches this range.

2. A microalgae recycle test was conducted. Photosynthetic starch accumulation and fermentative production of organic compounds through starch degradation occurred during an alternating light-dark cycle without any damage to the microalgae.

3. Twenty-three days of continuous operation were performed for the raceway-type reactor and three times for the recycle operation of the parallel plate-type photobioreactors without decreasing hydrogen evolution.

The accumulated amount of hydrogen evolved during these 23 days is 78 N L.

REFERENCES

Akano, T., Miura, Y., Fukatsu, K., Miyasaka, H., Ikuta, Y., Matsumoto, H., Shioji, N., Miziguchi, T., Yagi, K., and Maeda, I., 1996, Hydrogen production by photosynthetic microorganism, *Applied Biochemistry and Biotechnology*, 56/57:677–688.

Miura, Y., Saitoh, C., Matsuoka, K., and Miyamoto, K., 1992, Stability sustained hydrogen production with high molar yield through a combination of a marine green alga and a photosynthetic bacterium, *Biosci. Biotechnol., Biochem.*, 56:751–754.

41

DEVELOPMENT OF EFFICIENT LARGE-SCALE PHOTOBIOREACTORS

A Key Factor for Practical Production of Biohydrogen

James C. Ogbonna,[1,2] Toshihiko Soejima,[1] and Hideo Tanaka[1,2]

[1]Institute of Applied Biochemistry
University of Tsukuba
1-1-1 Tennodai
Tsukuba, Ibaraki 305-8572, Japan
[2]CREST
Japan Science and Technology Corporation
Japan

Key Words

photobioreactor design, light-distribution coefficient, light-supply coefficient, stirred-tank photobioreactor, scale-up, internal illumination.

1. SUMMARY

This work was aimed at developing an efficient stirred-tank photobioreactor for the large-scale cultivation of photosynthetic cells. A quantitative method of evaluating light conditions inside photobioreactors was first investigated and a light-supply coefficient, which is a product of a light-distribution coefficient and light energy supplied per unit volume, was proposed as an engineering parameter for design and scale-up of photobioreactors. Using this parameter, a method of designing and scaling-up internally illuminated photobioreactors was proposed. A photobioreactor was considered as consisting of units, and an optimum unit size was defined as a reactor volume that is optimally illuminated by a centrally located single light source. A large-scale photobioreactor with the optimum light-supply coefficient can thus be constructed by determining the optimum unit size for the target process and then increasing the number of units in three dimensions. Based on this concept, an optimum unit was constructed and then scaled-up to 20.0 L while maintaining a constant light-supply coefficient. With a 20-W fluorescent lamp, the unit size (diameter) that gave the optimum light-supply coefficient for the cultivation of *Chlorella* was 0.075 m. Carbon dioxide fixation by *Chlorella pyrenoidosa* and α-tocopherol production by *Euglena gracilis* cells in the 20.0-L photobioreactor were the same as those of the single unit.

2. INTRODUCTION

Various methods of increasing the efficiency of biohydrogen production are being widely investigated and many promising results have accumulated. However, translation of such results into production of biohydrogen as a substitute to fossil fuel requires development of efficient large-scale bioreactors.

Currently, open ponds are used for almost all commercial production of algae, but it is difficult to obtain high productivities with the open ponds because both the temperature and light intensities vary throughout the day and around the year. Furthermore, aside from their low productivity, production and collection of gaseous products using open cultivation ponds are technically very difficult. As an alternative to open cultivation ponds, various closed photobioreactors have been proposed for microalgae cultivation. These include vertical or horizontal tubular photobioreactors (Pirt et al., 1983), helical (serpentine) photobioreactors (Hoshino et al., 1991), and inclined or horizontal thin panel photobioreactors (Tredici et al., 1991). Although some of these photobioreactors are efficient at the laboratory scale, most of them may not work at all, or at best are very inefficient when scaled-up. The need to keep the S/V ratio high sets limits as to the extent the diameters of the tubes or the depths of the panels can be increased. Furthermore, mixing in such narrow tube or thin-panel photobioreactors is often very low with consequent problems, such as cell adhesion to, and growth on, the reactor walls and accumulation of photosynthetically produced oxygen and generated heat inside the photobioreactor (Tredici and Materassi, 1992). These problems are more pronounced in dense cultures of filamentous strains that show high viscosity of non-Newtonian behavior (Torzillo et al., 1993). Also, many small-scale, internally illuminated photobioreactors have been reported (Yakunin et al., 1986; Mori, 1985; Javanmardian and Palsson, 1991), but neither the quantitative evaluation of the light conditions in the reactors nor the design criteria and scale-up method have been discussed.

There is thus a need for large-scale closed photobioreactors in order to fully tap the potential of microalgae, especially for the production of biohydrogen. In order to construct a photobioreactor for efficient large-scale cultivation of photosynthetic cells, the following points must be considered:

  a. Light is the most important factor in photoautotrophic cultures and efficient light supply to photobioreactors is the most important challenge in photobioreactor design and construction. Many small-scale photobioreactors with good light supply have been constructed but most are extremely difficult to scale-up. Ease of scale-up for practical application should be a basic goal in designing a photobioreactor.
  b. Utilization of solar energy is very desirable because it is abundant and free. At present, solar energy utilization in outdoor open ponds seems to be the only means by which some cheap algae-derived products can be produced commercially. It is thus necessary to design a photobioreactor that can easily be illuminated by both solar and artificial light sources.
  c. Although it is necessary to have good mixing inside the photobioreactor, many photosynthetic cells have no cell walls while some are mobile or filamentous, which make them very fragile and sensitive to shear stress. It is therefore desir-

able to provide the desired degree of mixing but keep the hydrodynamic stress as low as possible.

d. The risk of contamination by heterotrophic microorganisms is low when there is no organic carbon source in the medium. However, within facilities where many other phototrophic cells are cultivated, contamination by other phototrophs can be a serious problem. Depending on the product, it may be necessary to operate under sterile conditions. Thus, cultivation under sterile conditions should be possible.

We have been working on the design and construction of a photobioreactor for the large-scale cultivation of photosynthetic cells. We are proposing a new concept for photobioreactor design where scale-up is a primary design criterion. Our aim is to design a photobioreactor that: (1) can be easily and efficiently scaled-up, (2) can be illuminated by both solar and artificial light sources, (3) has good mixing properties but low hydrodynamic stress, and (4) can be operated under sterile conditions.

3. MATERIALS AND METHODS

3.1. Light Supply Index

Since light is the most important factor affecting photoautotrophic cultivation, rational design and scale-up of photobioreactors must be based on the light characteristics inside the photobioreactors. The first step in this study is therefore to determine an index for the quantitative evaluation of light conditions inside photobioreactors; consequently, the reliability of various light parameters was investigated.

3.1.1. Incident Light Intensities. In the case of externally illuminated photobioreactors, the incident light intensities are the light intensities on the inner surface of the illumination side. For internally illuminated photobioreactors, the light intensities on the surface of the cold light pipe or the glass tube housing the fluorescent lamp were taken as the incident light intensities.

3.1.2. Average Light Intensities. The average light intensities in the cuboidal photobioreactors were calculated from Equation 1 (Camacho-Rubio et al., 1985). Descriptions of the cuboidal and other photobioreactors used for the basic experiments are given in Ogbonna et al. (1995c).

$$I_{ave} = I_o[1 - \exp(-EXL)]/EXL \tag{1}$$

where I_O = incident light intensity ($\mu mol/m^2 \cdot s$), E = light extinction coefficient (m^2/kg), X = cell concentration (kg/m^3), and L = depth of the photobioreactor (m). The cell concentrations during the transition from the exponential to the linear growth phases were used for the calculations.

3.1.3. Light Energy Per Unit Volume. The unit of light intensity used in this study is $\mu mol/m^2 \cdot s$, where 1 μmol = 0.2176 J. The light energy supplied per unit volume is thus given by Equation 2.

$$E_t/V = 0.2176 I_o S_A/1000V \tag{2}$$

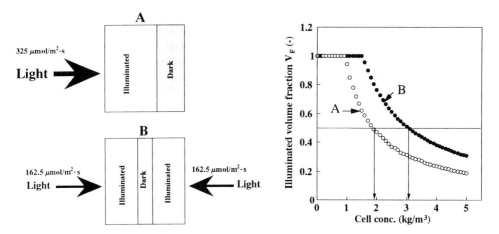

Figure 1. Effect of cell concentration on illuminated volume fraction of a cuboidal photobioreactor either illuminated from one surface at light intensity of 325 μmol/m²·s (A) or from two surfaces at light intensities of 162.5 μmol/m²·s (B). The arrows indicate the Kiv for the reactors.

where E_t = total light energy supplied to the reactor (kJ/s), S_A = illumination surface area (m²), and V = culture broth volume (m³).

3.1.4. Light Distribution Inside the Photobioreactor. When a photobioreactor containing a high cell concentration is illuminated from the surface, light is absorbed rapidly by the cells so that the light intensity decreases sharply from the illumination surface. As a result, the photobioreactor can be divided, on the basis of light intensity, into illuminated and non-illuminated volume fractions. The illuminated volume fraction (V_F) is defined as the ratio of the photobioreactor volume that receives sufficient light for cell growth (i.e., the fraction of the reactor volume where light intensity [I] is higher than the critical light intensity [I_C]) to the total volume of the photobioreactor. At the initial stage, when the cell concentration in the photobioreactor is very low, V_F may be unity (the entire photobioreactor volume is illuminated); however, as shown in Figure 1, this parameter decreases as the cell concentration in the photobioreactor increases (as the cultivation progresses). We proposed the use of a light-distribution coefficient (Kiv) as an index of the light distribution within the photobioreactor. This is defined as the cell concentration at which the illuminated volume fraction is reduced to 0.5.

The method for calculating Kiv depends on the shape of the photobioreactor as well as on the method of illumination. For the internally illuminated cylindrical photobioreactor, the central part of the photobioreactor is illuminated while the region close to the photobioreactor surface is dark (Figure 2). By assuming that the light source is either cylindrical or housed in a cylindrical glass tube, the liquid volume in the photobioreactor can be expressed by Equation 3:

$$V = \pi h\{(D/2)^2 - (d/2)^2\} \quad (3)$$

where d = diameter of the light source (glass tube housing the light source) (m). If the photobioreactor is visualized as consisting of two concentric compartments (cylinders) of equal volume (Figure 2), the entire photobioreactor volume can be expressed in terms of

Figure 2. Schematic diagram of light distribution in internally illuminated cylindrical photobioreactor. I = light intensity (μmol/m²·s); I_o = incident light intensity (μmol/m²·s); I_C = critical light intensity below which no photosynthetic growth can occur (μmol/m²·s); L_L = distance from the illumination surface to a point where I = I_C (m); d = diameter of the light source (m); D = diameter of the photobioreactor (m); D_I = distance from the surface of the light source to the surface of the inner cylinder (m).

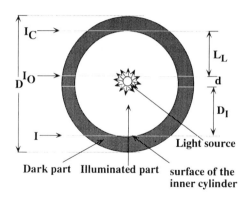

Dark part   Illuminated part   surface of the inner cylinder

the volume of the inner cylinder, as shown in Equation 4, and the distance (D_I) from the surface of the light source to the surface of the inner cylinder can be calculated from Equation 5.

$$\pi h \left\{ \left(\frac{d}{2}\right)^2 - \left(\frac{d}{2}\right)^2 \right\} = 2\pi h \left\{ \left(D_I + \frac{d}{2}\right)^2 - \left(\frac{d}{2}\right)^2 \right\} \quad (4)$$

$$D_I = \frac{-d + \sqrt{d^2 + 2\left\{\left(\frac{D}{2}\right)^2 - \left(\frac{d}{2}\right)^2\right\}}}{2} \quad (5)$$

Assuming that light absorption and dispersion can be neglected (no cells inside the photobioreactor), the total light energy on the surface of the light source is the same as the total light energy on the surface of the concentric cylinder (diameter = $2D_I + d$). This is represented by Equation 6 and the relationship between the light intensity on the surface of the inner cylinder (I) and the incident light intensity (I_o) is given by Equation 7.

$$I_o \pi dh = I \pi h (2D_I + d) \quad (6)$$

$$I = I_o d/(2D_I + d) \quad (7)$$

$$I = I_o \exp(-EXD_I) \quad (8)$$

In the presence of suspended cells in the photobioreactor, light attenuation inside the photobioreactor due to light absorption by the cells can be assumed to obey the Lambert-Beer law. Thus, the light intensity (I) at a distance (D_I) from the illumination surface is given by Equation 8.

Combining Equations 7 and 8 yields Equation 9. When 50% of the entire volume (only the inner compartment) receives enough light for photosynthetic growth, the light intensity on the surface of the inner cylinder, I = I_C, D_I = L_L, and X = Kiv. Thus, for an internally illuminated cylindrical photobioreactor, Kiv can be calculated from Equation 10. Equations for calculation of Kiv for externally illuminated cylindrical and cuboidal photobioreactors have been presented by Ogbonna et al. (1995c).

$$I = \frac{I_0 d}{2D_I + d} \exp(-EXD_I) \qquad (9)$$

$$Kiv = \frac{\ln\left[\frac{I_0 d}{\{I_C(2L_L + d)\}}\right]}{EL_L} \qquad (10)$$

3.2. Micoorganisms

Chlorella pyrenoidosa C-212, *Euglena gracilis* Z strain, and *Spirulina platensis* from the algal collection of the Institute of Applied Microbiology, University of Tokyo, Japan, were used in this study.

3.3. Correlation of Growth Rates and Various Light Parameters

Effects of various light parameters on linear growth rates of *Chlorella pyrenoidosa* C-212 and *Spirulina platensis* were investigated using both externally illuminated and internally illuminated photobioreactors. These include various sizes of cuboidal photobioreactors that were illuminated uniformly from one surface by either a 500-W halogen lamp or seven daylight-type fluorescent lamps, a cylindrical photobioreactor that was uniformly illuminated from the entire surface by eighteen 10-W daylight-type fluorescent lamps arranged vertically around the reactor, an internally illuminated cylindrical photobioreactor with a cold light pipe (Shibata Hario Co. Ltd., Tokyo), and cylindrical photobioreactors that were internally illuminated by a 4-W daylight-type fluorescent lamp inserted in a centrally fixed glass tube. Detailed descriptions of these photobioreactors are given in Ogbonna et al. (1995b). The medium composition and cultivation conditions were as described by Ogbonna et al. (1995a).

3.4. Determination of Optimum Unit Size

Prototype photobioreactor units with various diameters (Figure 3) were used to determine the optimum unit size for the new photobioreactor. A glass tube with an external

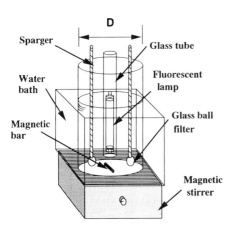

Figure 3. Schematic diagram of a single-unit prototype photobioreactor used to determine the optimum unit size. The diameters of the reactors (D) used for the experiments were 0.055, 0.065, 0.075, 0.095, and 0.124 m.

Development of Efficient Large-Scale Photobioreactors

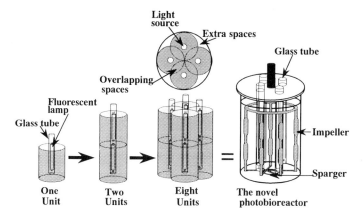

Figure 4. Schematic diagram of the design concept and the novel photobioreactor.

diameter of 2.4 cm was fixed in the center of the unit, and a daylight-type fluorescent lamp was inserted into the glass tube for illumination. Each reactor was inoculated with a pre-culture cell suspension and immersed in a water bath for temperature control. Mixing was achieved by sparging 5% CO_2 in air through two glass ball filters at a rate of 0.6 vvm and by means of magnetic stirrers. The incident light intensity was the same in all the reactors and, consequently, the light-supply coefficient varied with the diameter of the reactor vessel. The optimum unit diameter was determined from the plot of the light-supply coefficient (with the corresponding vessel diameter) against the linear growth rates.

3.5. Construction of a 20-L Prototype Reactor

A prototype photobioreactor, consisting of eight units, was constructed with Pyrex transparent glass (Figure 4). The diameter of each unit was 0.075 m while the total volume of the reactor was 20.0 L. The photobioreactor was 140 cm high and 15.3 cm in internal diameter. Each unit was equipped with a centrally fixed glass tube (diameter = 4 cm), into which two 20-W lamps or a 40-W fluorescent lamp were inserted. Since the lamps were not mechanically fixed and could easily be replaced, only the reactor was heat-sterilized (where necessary) and, after cooling, the lamps were inserted, thus making it possible to cultivate under sterile conditions. A modified impeller was installed for mixing. This impeller had very low shear stress but good mixing capacity. Aeration was done through a ring sparger with 12 holes. The diameter of each hole was 0.5 mm. The glass housing units served as baffle plates in breaking the gas bubbles, thus increasing the kLa. This 20-L reactor differed from the 3-L reactor (Ogbonna et al., 1996) in that in the 3-L reactor, 4-W fluorescent lamps were used for illumination while a paddle-type impeller was used for mixing.

3.6. Cultivation in the Novel Photobioreactor

Chlorella pyrenoidosa C-212 and *Euglena gracilis* Z strain were cultivated in the novel photobioreactor. The media and cultivation conditions for Chlorella were as described previously (Ogbonna et al., 1996). A modified Hutner medium was used for cultivation of *Euglena gracilis*.

3.7. Analytical Methods

The light intensities were measured by analogue photometer (LI-185B, Licor, Nebraska, United States). The cell concentrations and linear growth rates were determined as described previously by Ogbonna et al. (1995a, b, c). The biomass yield coefficient ($Y_{X/E}$) and the tocopherol yield per supplied light energy ($Y_{t/E}$) were calculated from Equations 11 and 12, respectively. Here, X_{final} = final cell concentration (kg/m^3) and K_p = α-tocopherol productivity (μg/m^3·d). The photosynthetic efficiency (PE) was calculated from Equation 13, where E_c = calorific value of the cells (24 kJ/g-cell, (Hirata et al., 1996)), E_t = total energy supplied to the reactor (kJ/s), and V = culture broth volume. Extraction and measurement of α-tocopherol concentration was determined according to the method described by Shigeoka et al. (1986).

$$Y_{X/E} = X_{final}/(E_t/V) \qquad (11)$$

$$Y_{t/E} = K_p/(E_t/V) \qquad (12)$$

$$PE = \frac{dX}{dt} \cdot 100 E_c \cdot \frac{V}{E_t} \qquad (13)$$

4. RESULTS AND DISCUSSION

4.1. Incident and Average Light Intensities

The reliability of incident light intensity and average light intensity as indices of light conditions inside photobioreactors was investigated in cuboidal photobioreactors of various sizes (Ogbonna et al., 1995c), using the linear growth rate as the growth index (Ogbonna et al., 1995b). Although for a photobioreactor of the same size there was a good relationship between the linear growth rates and the incident or average light intensities, there was no correlation between them when data from photobioreactors of different sizes were considered. This implies that the apparently good correlation reported between the photosynthetic growth rates and the incident-light intensities is probably because low cell concentrations were used in such studies (Ogawa et al., 1971). This means that neither the incident nor average light intensities is a good index of light-supply efficiency of photobioreactors and thus cannot be used for meaningful evaluation of light conditions inside photobioreactors.

4.2. Light Energy Supplied Per Unit Volume

The relationships between the light energy supplied per unit photobioreactor volume and the linear growth rates are shown in Figure 5A. Although the relationship approaches a linear one, there is much scatter of the data near the curve. At a given light energy supplied per unit volume (e.g., 1.2 kJ/m^3·s as indicated by an arrow in Figure 5A), the linear growth rates decreased with an increase in depth of the photobioreactors, indicating that the light distribution inside the photobioreactor must be considered for the rational design and scale-up of photobioreactors. In comparison to a fairly homogenous light distribution

Development of Efficient Large-Scale Photobioreactors

Figure 5. Correlations between A: light energy supplied per unit volume (Et/V), B: light distribution coefficient (Kiv), or C: light supply coefficient ($E_t/V \cdot Kiv$) and the linear growth rates (K) of *Chlorella pyrenoidosa*. In A and B, various sizes of cuboidal photobioreactors were used. The depths of the photobioreactors were (○), 0.02 m; (●), 0.04 m; (▲), 0.06 m; (□), 0.08 m and (■), 0.16 m. In C, the types of photobioreactors were (○), five sizes of externally illuminated cuboidal photobioreactors; (●), an externally illuminated cylindrical photobioreactor; (■), a cylindrical photobioreactor illuminated internally by a cold light pipe, and (□), two cylindrical photobioreactors illuminated internally by daylight-type fluorescent lamps.

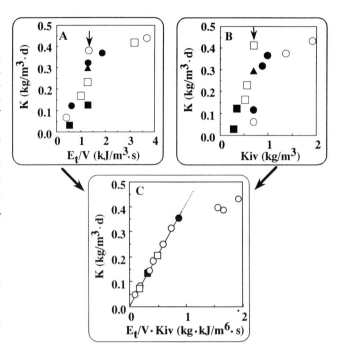

inside a very shallow photobioreactor, there would be spatial heterogeneity in the light intensities inside the deep photobioreactors.

4.3. Light-Distribution Coefficient

The effects of light-distribution coefficients on the linear growth rates *of Chlorella pyrenoidosa* were therefore investigated as shown in Figure 5B. The results showed that the linear growth rates increased with an increase in the light-distribution coefficient. However, as in the case of light energy supplied per unit volume, the data were scattered, indicating that the light-distribution coefficient alone is not a sufficient index of light-supply efficiency of photobioreactors. Nevertheless, it is a good parameter of light-distribution inside the reactor. As shown in Equation 10, it incorporates the properties of the reactor (the size), the incident-light intensities (I_o), and the characteristics of the cells (the critical-light intensity (I_C) and the light-extinction coefficient [E]).

The significance of Kiv can be seen from a comparison of the Kiv values of two 0.02-m deep cuboidal photobioreactors (Figure 1). Photobioreactor A is illuminated from one surface at an incident light intensity of 325 µmol/m²·s, while photobioreactor B is illuminated from two surfaces at incident-light intensities of 162.5 µmol/m²·s. By assuming a critical light intensity (I_C) of 7.65 µmol/m²·s, and a light-extinction coefficient of 200 m²/kg (Ogbonna et al., 1995b), the effect of cell concentration on the illuminated volume fraction was calculated as shown in Figure 1. Although the total light energy supplied per unit volume in the two photobioreactors is the same, photobioreactor B is more uniformly illuminated and, consequently, Kiv (indicated by the arrows in Figure 1) for photobioreactor B (3.1 kg/m³) is higher than that for photobioreactor A (1.9 kg/m³).

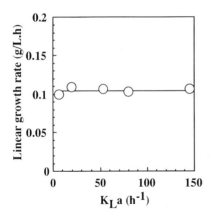

Figure 6. Effect of volumetric oxygen transfer coefficient (K_La) on the photo-autotrophic linear growth rate of *Chlorella pyrenoidosa*.

4.4. Light-Supply Coefficient

At a constant light-distribution coefficient (Kiv), a linear relationship was observed between the linear growth rate and the light energy supplied per unit volume (Et/V). Similarly, when the Et/V was held constant, there was a good correlation between Kiv and the linear growth rate. A light-supply coefficient, defined as the product of the light energy supplied per unit volume and the light distribution coefficient (Et/V·Kiv), was therefore proposed as an index of the light-supply efficiency of photobioreactors. There was a linear relationship between the light-supply coefficient and the linear growth rates of both *Chlorella pyrenoidosa* and *Spirulina platensis* in cuboidal photobioreactors of various sizes (Ogbonna et al., 1995c). As shown in Figure 5C, when various other types of both internally illuminated and externally illuminated cylindrical photobioreactors were used, good correlation was found between the linear growth rates of *Chlorella* and the light-supply coefficient.

4.5. Scale-Up Parameter

Most microbial fermentation processes are significantly affected by the degree of mixing within the bioreactor and, consequently, various parameters that directly or indirectly describe mixing behavior in the bioreactor have been used as bioreactor design criteria. Presently, the volumetric mass-transfer coefficient is the most widely used index for bioreactor scale-up (Tanaka et al., 1991; Ju and Chase, 1992; Sumino et al., 1993). The effect of the volumetric mass-transfer coefficient (k_La) on the photo-autotrophic cultivation of *Chlorella pyrenoidosa* was investigated using the 3-L photobioreactor reported previously (Ogbonna et al., 1996a), and it was found that variation of k_La between 6 and 145 h^{-1} had no significant effect on the linear growth rate (Figure 6).

In photobioreactors, mixing help: (1) keep the cells in suspension, (2) distribute both the nutrients and the generated heat within the photobioreactor, (3) improve CO_2 transfer into the photobioreactor, (4) degas the photosynthetically produced O_2, (5) improve mass transfer between the cells and the liquid broth, and (6) facilitate the movement of cells in and out of the illuminated part of the photobioreactor. The degree of mixing required in photobioreactors depends on the importance of the above factors. With paddle-type impeller used in this study, very low agitation speed (k_La) is sufficient to achieve objectives 1, 2, and 5 while under our experimental conditions (aeration with 5% CO_2 in air and a light-supply coefficient of 0.374 kJ·kg/m^6·s), the growth rate of *Chlorella* would not be limited by the rate of CO_2 transfer (Märkl, 1977). Furthermore, although the dissolved oxygen (DO) concentration inside the photobioreactor was not measured, it is expected that in the stirred-tank reactor, the DO would not increase to levels that inhibit the growth of *Chlorella*. Although these experimental results are not enough to make a generalized conclusion, for most photosynthetic cells, CO_2 limitation or O_2 inhibition becomes important only when the light supply is high enough to support high growth rates.

On the other hand, as discussed above, irrespective of the cell type, photobioreactor type, and size, there was a linear relationship between the linear growth rates and the light-supply coefficient (Et/V·Kiv). This coefficient is, therefore, a good index for design and scale-up of photobioreactors. It was therefore adapted as a scale-up parameter in this study.

4.6. Scale-Up Method

It was assumed that data obtained with a small-scale photobioreactor can be reproduced for a large-scale photobioreactor if both photobioreactors have the same light-supply coefficient. As a method of constructing a large photobioreactor having the same light supply coefficient as a smaller one, a new concept in photobioreactor design and scale-up was proposed. The photobioreactor is conceptualized as consisting of units. One unit consists of a reactor volume (space) that is illuminated from the center by a single lamp. An optimum unit size for a process (a unit volume that is efficiently illuminated by the single lamp, and thus has an optimum light-supply coefficient) is experimentally determined, and a larger reactor with the same light-supply coefficient can be constructed by simply increasing the number of units in three dimensions (Figure 4). Depending on the size of the reactor (number of units), different unit arrangements can be used to achieve this, but it is easiest if 4^n units (n = integer) are used as the base, and the desired reactor size is then obtained by increasing the height (increasing the number of units in a vertical direction). For example, to construct a reactor comprising 20 units, four units (n = 1) can be used as the base and the height increased five times, or 16 units (n = 2) can be used as the base and the height increased 1.25 times.

4.7. Optimum Light-Supply Coefficient

The effects of the light-supply coefficient, corresponding to various unit sizes (diameters), on the linear growth rate and biomass yield coefficient during the cultivation of *Chlorella* are shown in Figure 7. The linear growth rate increased with the light-supply coefficient (decreasing unit diameter). The highest linear growth rate is obtained at a high light-supply coefficient corresponding to very narrow unit diameters.

In the case of large unit sizes, which mean very low light-supply coefficients, the illuminated volume fraction of the unit is low (a large portion of the unit is in the dark). This would lead to increased maintenance energy and thus a decrease in the productivity and yield coefficient (Ogbonna et al., 1995a; Ogbonna and Tanaka, 1996). The yield coefficient increased with decreasing unit diameter because at moderate light-supply coefficients, most of the supplied light energy is absorbed and efficiently used for cell growth and product formation. On the other hand, at very high light-supply coefficients, more light is supplied than the cells can efficiently absorb and utilize. Consequently, the yield coefficients at high light-supply coefficients were low due to photoinhibition and/or en-

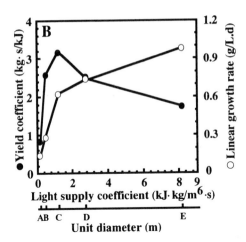

Figure 7. The effects of light-supply coefficient on the linear growth rates and biomass yield per supplied light energy. The light intensity was kept constant and the light-supply coefficient was varied by changing the diameter of the unit. The diameters of the units for points A, B, C, D, and E were 0.055, 0.065, 0.075, 0.095, and 0.124 m, respectively.

ergy loss in the form of heat. There is thus an optimum light-supply coefficient for maximum yield from the supplied light energy.

With a given light intensity, the optimum unit size depends on both the type of cell and the process economy. If light represents a significant percentage of the total production cost, then greater importance should be attached to efficiency of light utilization and the unit size giving the highest yield coefficient should be selected. However, if the cost of light is relatively cheap (e.g., if solar energy is used), then the design criterion should be to obtain the highest productivity. In this case, a high light-supply coefficient is desirable, provided that the light intensity is not too high as to cause photoinhibition. In most cases, a compromise is made between the productivity and yield coefficient.

4.8. CO_2 Fixation by *Chlorella pyrenoidosa*

The effects of scaling-up the reactor on carbon dioxide fixation by *Chlorella pyrenoidosa* is shown in Table 1. By using the method proposed in this study, one unit reactor (2.5 L) was successfully scaled up to eight units (20 L) while maintaining the light-supply coefficient constant at 1.6 $kJ \cdot kg/m^6 \cdot s$. Consequently, the linear growth rate, final cell concentration, and thus the volumetric CO_2 fixation rates were the same in all the reactors. Because of variations in the climatic and growth conditions and growth characteristics of the algae used, as well as methods of determining and reporting productivities, it is extremely difficult to compare the productivities of various algae cultures. However, the final cell concentrations and the linear growth rates (productivity) obtained in this

Table 1. Effect of photobioreactor scale on cell growth and carbon dioxide fixation by *Chorella pyrenoidosa*

Reactor volume (L)	Linear growth rate (g/L·day)	Final cell conc. (g/L)	Volumetric CO_2 fixation rate (g-CO_2/L·day)	CO_2 fixation rate (g-CO_2/L·day))	Photosynthetic efficiency (%)[a]
2.5	0.62	6.9	1.11	2.78	6.94
5.0	0.61	7.0	1.10	5.5	6.83
20.0	0.62	6.8	1.11	22.2	6.93

[a]Calculated based on the total light energy supplied. A cell calorific value of 24 kJ/g-cell was used.

Table 2. Scale-up of α-tocopherol production by *Euglena gracilis* strain Z

Reactor volume (L)	Linear growth rate (g/L·day)	Final cell conc. (g/L)	Volumetric α-tocopherol productivity (μg/L·day)	α-tocopherol productivity (μg/day)	Yield per supplied energy (μg/kJ)
2.5	0.076	1.46	179	358	0.84
20.0	0.075	1.50	187	3180	0.87

study are high when compared to the values reported for other photoautotrophic algae cultures (Terry and Raymond, 1985; Richmond et al., 1993; Miyamoto et al., 1988; Tredici et al., 1991; Tredici and Materassi, 1992). The daily carbon dioxide fixation rate increased linearly with the reactor volume, an indication of successful scale-up. The photosynthetic efficiency based on the total light intensity was about 6.9% in each reactor. In comparison with other systems, this photosynthetic efficiency is very high (Watanabe et al., 1995; Hirata et al., 1996) and shows that efficient light utilization is achieved by using the method proposed in this study for photobioreactor design.

4.9. α-Tocopherol Production by *Euglena gracilis* Z

A comparison of α-tocopherol production in 2.5-L and 20-L photobioreactors is shown in Table 2. As in the case of *Chlorella*, both the linear growth rates and the volumetric α-tocopherol productivity were almost the same in both reactors. *Chlorella* was used for determination of the optimum unit size and as a result, the light-supply coefficient in the photobioreactors was higher than the optimum value for *Euglena gracilis* cells. The linear growth rates were thus relatively low, but the final α-tocopherol contents of the cells (1600 μg/g) were higher than the values obtained at the optimum light intensity for cell growth.

5. CONCLUSION

The above results have shown that the proposed light-supply coefficient is a good parameter for photosynthetic process scale-up. In other words, by keeping this coefficient constant, results obtained in a small reactor can be reproduced in a large-scale reactor. We are optimistic that by using the above method (determining the optimum unit size for the process and increasing the number of units in three dimensions), efficient industrial-scale stirred tank photobioreactors can be constructed. Since the lamps are not mechanically fixed to the reactor, the conventional method of bioreactor sterilization can be employed and the reactor is illuminated by inserting the lamps into the housing glass tubes after sterilization.

For most products such as hydrogen, it is desirable to use solar light for illumination. We are currently investigating the use of a light collection device consisting of fresnel lenses to collect the light and distribute it inside the reactor through optical fibers. Since the position of the sun changes continuously, the device is equipped with a light-tracking sensor so that the lenses rotate with the position of the sun. One of the problems with this system is that it is currently expensive. However, as the demand increases (especially in the housing industries), the prices are expected to drop sharply. The efficiency of light collection and transmission is still low (38%) but more efficient systems are currently being developed.

6. NOMENCLATURE

D = diameter of the photobioreactor (m) (see Figure 2)
D_I = distance from the surface of the light source to the surface of the inner cylinder (m) (see Figure 2)
d = diameter of the light source (or glass tube housing the light source) (m)
E = light extinction coefficient (m^2/kg)
E_c = calorific value of the cells (24 kJ/g-cell)
E_t = total light energy supplied to the reactor (kJ/s)
I = light intensity ($\mu mol/m^2 \cdot s$)
I_C = critical light intensity below which no photosynthetic growth can occur ($\mu mol/m^2 \cdot s$)
I_O = incident light intensity ($\mu mol/m^2 \cdot s$)
Kiv = light distribution coefficient (kg/m^3) (defined as the cell concentration at which the illuminated volume fraction is reduced to 0.5)
K_P = α-tocopherol productivity ($\mu g/m^3 \cdot d$)
L = depth of the photobioreactor (m) (see Figure 2)
L_L = distance from the illumination surface to a point where $I = I_C$ (m) (see Figure 2)
PE = photosynthetic efficiency
S_A = illumination surface area (m^2)
V = culture broth volume (m^3)
V_F = illuminated volume fraction (-) (defined as the ratio of the photobioreactor volume that receives sufficient light for cell growth to the total volume of the photobioreactor)
X = cell concentration (kg/m^3)
X_{final} = final cell concentration (kg/m^3)
$Y_{t/E}$ = tocopherol yield per supplied light energy ($\mu g/kJ$)
$Y_{X/E}$ = biomass yield coefficient ($kg \cdot s/kJ$)

REFERENCES

Camacho-Rubio, F., Padial-Vico, A., and Martinez-Sancho, M.E., 1985, The effect of the mean intensity of light on the cultivation of *Chlorella pyrenoidosa*, *Inter. Chem. Eng.*, 25:283–288.

Hirata, S., Hayashitani, M., Taya, M., and Tone, S., 1996, Carbon dioxide fixation in batch culture of *Chlorella* sp. using a photobioreactor with a sunlight collection device, *J. Ferment. Bioeng.*, 81:470–472.

Hoshino, K., Hamochi, M., Mitsuhashi, S., and Tanishita, K., 1991, Measurements of oxygen production rate in flowing *Spirulina* suspension, *Appl. Microbiol. Biotechnol.*, 35:89–93.

Javanmardian, M., and Palsson, B.O., 1991, High-density photoautotrophic algal cultures: design, construction, and operation of a novel photobioreactor system, *Biotechnol. Bioeng.*, 38:1182–1189.

Ju, L.K., and Chase, G.G., 1992, Improved scale-up strategies of bioreactors, *Bioproc. Eng.*, 8:49–53.

Märkl, H., 1977, CO_2 transport and photosynthetic productivity of a continuous culture of algae, *Biotechnol. Bioeng.*, 19:1851–1862.

Mori, K., 1985, Photoautotrophic bioreactor using solar rays condensed by fresnel lenses, *Biotechnol. Bioeng. Symp.*, 15:331–345.

Miyamoto, K., Wable, O., and Benemann, J.R., 1988, Vertical tubular reactor for microalgae cultivation, *Biotechnol. Lett.*, 10:703–708.

Ogawa, T., Kozawa, H., and Terui, G., 1971, Studies on the growth of *Spirulina platensis* (II) growth kinetics of an autotrophic culture, *J. Ferment. Technol.*, 50:143–149.

Ogbonna, J.C., Yada, H., and Tanaka, H., (1995a), Effect of cell movement by random mixing between the surface and bottom of photobioreactors on algal productivity, *J. Ferment. Bioeng.*, 79:152–157.

Ogbonna, J.C., Yada, H., and Tanaka, H., 1995b, Kinetic study of light-limited batch cultivation of photosynthetic cells, *J. Ferment. Bioeng.*, 80:259–264.

Ogbonna, J.C., Yada, H., and Tanaka, H., 1995c, Light supply coefficient—a new engineering parameter for photobioreactor design, *J. Ferment. Bioeng.*, 80, 369–376.

Ogbonna, J.C., Yada, H., Masui, H, and Tanaka, H., 1996, A novel internally illuminated stirred tank photobioreactor for large scale cultivation of photosynthetic cells, *J. Ferment. Bioeng.*, 82:61–67.

Ogbonna, J.C., and Tanaka, H., 1996, Night biomass loss and changes in biochemical composition of cells during light/dark cyclic culture of *Chlorella pyrenoidosa, J. Ferment. Bioeng.*, 82:558–564.

Pirt, J.S., Lee, Y.K., Walach, M.R., Pirt, M.W., Balyuzi, H.H.M., and Bazin, M.J., 1983, A tubular bioreactor for photosynthetic production of biomass from carbon dioxide: design and performance, *J. Chem. Tech. Biotechnol.*, 33B:35–58.

Richmond, A., Boussiba, S., Vonshak, V., and Kopel, R., 1993, A new tubular reactor for mass production of microalgae outdoors, *J. Appl. Phycol.*, 5:327–332.

Shigeoka, S., Onishi, T., Nakano, Y., and Kitaoka, S., 1986, The contents and subcellular distribution of tocopherols in *Euglena gracilis, Agric. Biol. Chem.*, 50:1063–1065.

Sumino, Y., Sonoi, K., and Doi, M., 1993, Scale-up of purine nucleoside fermentation from a shaking flask to a stirred-tank fermentor, *Appl. Microbiol. Biotechnol.*, 38:581–585.

Tanaka, H., Ishikawa, H., Nobayashi, H., and Takagi, Y., 1991, A new scale-up method based on the effect of ventilation on aerated fermentation systems.

Terry, K.L., and Raymond, L.P.,1985, System design for the autotrophic production of microalgae, *Enzyme Microb. Technol.*, 7:474–487.

Torzillo, G., Carlozzi, P., Pushparaj, B, Montaini, E., and Materassi, R., 1993, A two-plane photobioreactor for outdoor culture of *Spirulina, Biotechnol. Bioeng.*, 42:891–898.

Tredici, M.R., Carlozzi, P. Zittelli, G.C., and Materassi, R., 1991, A vertical alveolar panel (VAP) for outdoor mass cultivation of microalgae and cyanobacteria, *Biores. Technol.*, 38:153–159.

Tredici, M.R., and Materassi, R., 1992, From open ponds to vertical alveolar panels: the Italian experience in the development of reactors for the mass cultivation of phototrophic microorganisms, *J. Appl. Phycol.*, 4:221–231.

Watanabe, Y., Noüe, J., and Hall, D.O., 1995, Photosynthetic performance of a helical tubular photobioreactor incorporating the cyanobacterium *Spirulina platensis, Biotechnol. Bioeng.*, 47:261–269.

Yakunin, A.F., Tsygankov, A.A., Gogotov, I.N., and L'vov, N. P., 1986, Dependence of the nitrogenase and nitrate reductase activities on the molybdenum concentration in the medium and assimilation of nitrate in cells of *Rhodopseudomonas capsulata, Mikrobiologiya*, 55:564–569.

42

LIGHT PENETRATION AND WAVELENGTH EFFECT ON PHOTOSYNTHETIC BACTERIA CULTURE FOR HYDROGEN PRODUCTION

Eiju Nakada,[1] Satoshi Nishikata,[1] Yasuo Asada,[2] and Jun Miyake[3]

[1]Fuji Electric Corporate R&D, Ltd.
2-2-1 Nagasaka
Yokosuka, 240-0194, Japan
[2]National Institute of Bioscience & Human-Technology
Higashi
Tsukuba, 305-0046, Japan
[3]National Institute of Advanced Interdisciplinary Research
1-1-4 Higashi
Tsukuba, 305-0046, Japan

Key Words

photobioreactor, light penetration, hydrogen production, *Rhodobacter sphaeroides*

1. SUMMARY

The penetration of light into a photobioreactor and its relation to hydrogen production were analyzed using the photosynthetic bacterium, *Rhodobacter sphaeroides*. Two photobioreactor types were used to examine hydrogen evolution at various light penetration depths: the A-type was comprised of four compartments of bacterial suspension and the B-type consisted of two cell-immobilized gels. A large portion of the incident light energy was absorbed in the upper part of the reactor. In the first compartment (0–5 mm) of the A-type reactor, 69% of the incident light energy was absorbed; 21% was absorbed in the second (5–10 mm). About 75% of the incident light energy was entrapped in the front gel of the B-type reactor. The cells in the deep parts of both reactors showed higher light energy conversion efficiencies to hydrogen, although the light spectrum reaching the cells was altered and the intensity was low. Alteration of the light spectrum and light intensity upon passage of light through the reactor greatly affected hydrogen production. Excess light energy in the shallow region reduced the total efficiency of the reactor. Light reaching the deep part of the reactor (600–780 nm) could be used effectively for hydrogen production.

2. INTRODUCTION

Application of solar energy technology is desirable to prevent environmental problems (e.g., the greenhouse effect, acid rain) caused by fossil energy sources. Hydrogen is a well-known, high weight density energy carrier, and its combustion does not pollute the environment. Photosynthetic bacteria can evolve hydrogen gas under solar illumination and could be applied for the commercial production of storage fuel from sunlight (Miyamoto, 1994).

Among various photosynthetic microorganisms, photosynthetic bacteria have high hydrogen production rates (Miyake et al., 1987). The highest conversion efficiency of light to hydrogen yet recorded in this class of organism was ca. 7% using light from a solar simulator (Miyake et al., 1987, 1989). Impressive as this number appears, improvements in the efficiency are still needed.

Application of molecular genetic approaches and redesign of the photobioreactor to improve efficiency have been studied (Jahn et al., 1994; Tsygankov et al., 1994). We have been examining the light conditions as a key factor to enhance the conversion efficiency of hydrogen production, particularly the effects of light penetration on photosynthetic bacteria culture.

Detailed analysis is required to understand the distribution of light in a photobioreactor and the relationship between this distribution and hydrogen production. We used two kinds of photobioreactors to analyze the relationship between hydrogen production and light penetration in a suspension culture and cell-immobilized gel. The A-type photobioreactor incorporates four compartments of bacterial suspension and the B-type contains two layers of immobilized gel.

Furthermore, we examined the relationship between light wavelength and hydrogen production. The C-type photobioreactor (200-mL flat glass bottle) was used in the experiments and two kinds of light sources were used to examine the wavelength dependence of hydrogen production. The profiles of light penetration into the photobioreactor and the light energy conversion efficiency were discussed.

3. EXPERIMENTAL

3.1. Organisms and Culture Conditions

Rhodobacter sphaeroides RV was used in these experiments (Miyake et al., 1984). The growth medium (aSy) was comprised of a basal medium (inorganic salts and vitamins), 0.1% yeast extract, 10 mM ammonium sulfate, and 75 mM sodium succinate (Miyake et al., 1984; Mao et al., 1986). Cells were grown for three days with tungsten light (30 W/m^2) at 30 °C.

3.2. Reactors

The A-type photobioreactor made up of four equal compartments is shown in Figure 1A. Each compartment had a 5-mm light path and a 37.5-cm^2 irradiation area. The

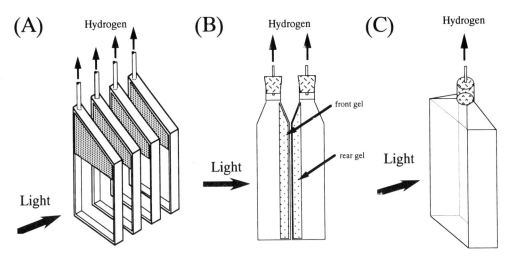

Figure 1. Schematic of the photobioreactors. (A) A-type reactor with four compartments. Compartment windows made of transparent polyacryl resin; upper part of each window covered with aluminum tape to obtain an irradiation area of 37.5 cm^2; light path of each compartment 0.5 cm. (B) B-type reactor with two cell-immobilized gels; two flat 200-mL flat glass bottles containing cell-immobilized gels combined to simulate the front and rear section of a thick gel; each gel irradiation area of 76 cm^2 and thickness of 0.53 cm; side of the reactor covered with aluminum tape to shield gel from light. (C) C-type reactor (200-mL flat glass bottle); irradiation area of 76 cm^2 and thickness of 2.5 cm.

compartments were aligned along the light path with a 3-mm gap between compartments for the circulation of water to control the temperature. Each compartment was made of polyacryl resin (5 mm thick) using transparent resin for the windows and black resin for the frame.

The B-type photobioreactor consisting of two cell-immobilized gel layers is shown in Figure 1B. Each gel had an irradiation area of 76 cm^2 and thickness of 0.53 cm. Both cell-immobilized gels were placed in 200-mL flat glass bottles. The sides of the bottles were covered with aluminum tape to shield the light.

The C-type photobioreactor (200-mL flat glass bottle) is shown in Figure 1C. It had an irradiation area of 76 cm^2 and thickness of 2.5 cm.

3.3. Hydrogen Production

The hydrogen production medium (gL) was prepared with the basal medium, 75 mM sodium D,L-lactate, 10 mM sodium glutamate, and 18 mM sodium bicarbonate (Miyake et al., 1987). Cell suspensions (100 mL) were mixed with freshly prepared gL medium (300 mL). The cells were cultivated in a 400-mL flat glass bottle with tungsten light (100 W/m^2) at 30 °C for 24–48 h until the hydrogen production rate became constant. The cell suspension (13 mg dry wt mL^{-1}) was then transferred to the compartments of the A-type and C-type reactors.

When the B-type reactor was used, the cell suspension (70 mL, ca. 1.5 mg dry wt mL^{-1}) was mixed with an equal volume of sterilized basal medium that contained 4% of Bacto-Agar (DIFCO) in a horizontally placed 200-mL flat glass bottle. After solidifying for 1 h at room temperature, freshly prepared gL medium was also carefully poured into the B-type reactor, carefully. Two flat glass bottles were then aligned as seen in Figure 1B.

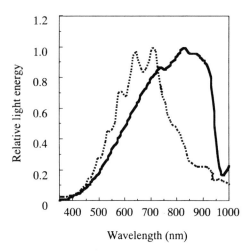

Figure 2. Spectra of light sources. (—) infrared depleted tungsten light; (-----) infrared depleted halogen light (with wavelength-selective transparent mirror).

The reactors were immersed in a water bath to keep the temperature at 30 °C. The cell density of the photosynthetic bacteria was measured using a spectrophotometer (Hitachi, Model 110, Japan). Hydrogen gas was collected in glass syringes to measure the evolution rates as reported by Miyake et al. (1982).

3.4. Light Sources

A tungsten lamp of 150 W (color temperature of 2900 K) was used as the light source when A- and B-type reactors were used. In addition to the tungsten lamp, a halogen lamp of 110 W with a wavelength-selective transparent mirror was used when a C-type reactor was used.

A water vessel was placed in front of the reactor to filter out infrared. The light intensity at the surface of the reactor (only light passed through the water filter) was adjusted to 720 W/m^2 (A- and B-type) or 500 W/m^2 (C-type). It was set close to the intensity of full sunlight, although the experimental conditions could not simulate the sunlight exactly due to technical difficulties. The spectra at the surface of the reactor are illustrated in Figure 2.

A radiometer (YSI-Kettering, Model 65A, United States) was used to measure total light energy. The wavelength dependence of light penetrating into the reactor was measured by a spectrophotometer (Hitachi, Model 330, Japan) with an integrating sphere.

4. RESULTS AND CONCLUSIONS

4.1. Light Penetration into the Photobioreactor

4.1.1. Suspension, A-Type. The light penetration profiles in the A-type reactor are illustrated in Figure 3A. Light around the absorption maxima (at 800 and 850 nm) decreased rapidly during passage through the cell culture. The spectrum of light was significantly altered in the deep part of the reactor where the remaining light was at

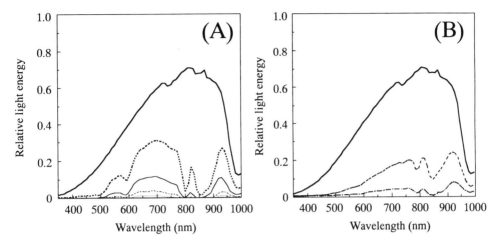

Figure 3. Spectrum alteration of light passing though a photobioreactor. (A) Spectra at various depths in the A-type reactor (suspension). Depths were (—) 0 cm (at surface of first compartment); (-----) 0.5 cm (at surface of second compartment); (—) 1 cm (at surface of third compartment); (-----) 1.5 cm (at surface fourth compartment); (—) 2 cm (after passing through the fourth compartment). (B) Spectra in B-type reactor (immobilized); positions were (—) surface of front gel; (-----) surface of rear gel; (- • - • -) after passing real gel.

500–800 nm. Absorption by soret band (about 400 nm) did not affect the profile much because light from the tungsten lamp did not emit much energy at this wavelength.

The total energy distribution along the light path was calculated based on the transmittance data of Figure 3A. The amount of incident light energy absorbed in the first compartment was 69%. The second compartment absorbed 21%, the third 7.0%, and the fourth 2.0%. The light energy absorbed in each compartment decreased exponentially. The incident light was absorbed predominantly by the cells in the first compartment. Cells located deep in the reactor were only poorly illuminated.

4.1.2. Immobilized, B-Type. The light energy and the spectrum in the B-type reactor also altered with the depth of light penetration. The spectrum of incident light at the surface of the gel is shown in Figure 3B. All spectra were measured as soon as the cells were immobilized in the gel. Light energy decreased with passage through the gels, especially light energy around 800 and 850 nm as before. The amount of incident light energy absorbed in the front gel was 75%, while that in the rear gel was 20%. About 95% of the incident light energy was absorbed in 10 mm of gel.

4.2. Hydrogen Production in the Various Parts of the Reactors

4.2.1. Suspension, A-Type. The hydrogen evolution rates from each compartment of the A-type reactor (Figure 4A) were recorded for up to 6 h. The rates of hydrogen evolution became constant after a transient period of up to 2 h, and the cells grew only slowly in the compartments regardless of the illumination conditions. Hydrogen evolution from each compartment represents productivity from various depths in this reactor. Although the major part of the energy was absorbed in the first compartment (0–5 mm), the hydrogen evolution rate was nearly equivalent to that of the second compartment (5–10 mm). Hydrogen was also evolved from the third compartment (10–15 mm) although only a lim-

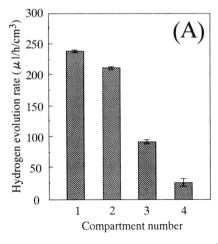 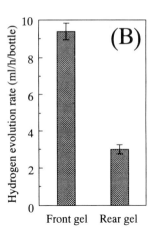

Figure 4. Hydrogen evolution rates at 720 W/m^2. (A) A-type reactor, (B) B-type reactor.

ited amount of energy reached it. Hydrogen was evolved in the fourth compartment, and the rate was unstable.

4.2.2. Immobilized, B-Type. Figure 4B shows the hydrogen evolution rates from the front and rear gels of the B-type reactor. The rates of hydrogen evolution became constant after a transient period of up to 4 h, and they were recorded for up to 18 h. The hydrogen evolution rate from the front gel was about 3 times higher than the rear gel. Hydrogen was evolved from the rear gel although only a limited amount of energy reached it.

4.3. Light Energy Conversion Efficiency to Hydrogen in the Various Parts of the Reactors

The light energy conversion efficiency to hydrogen was calculated according to the following equation:

Efficiency (%) = (combustion enthalpy of hydrogen) / (absorbed light energy) × 100

The absorbed light energy in each compartment or gel was calculated based on the transmittance in the wavelength range of 350–1000 nm in Figure 3. The other wavelength regions were not included in the calculation because the light sources did not emit much energy at the wavelength below 350 nm and light over 1000 nm did not contribute to hydrogen evolution.

The efficiencies calculated from the above equation ignore the energy of the substrate used by bacteria to evolve hydrogen because only D,L-lactate was used as a substrate and the concentration of substrate was kept approximately constant in experiments.

4.3.1. Suspension = A-Type. The light energy conversion efficiencies are plotted versus the depth (compartment number) in Figure 5A. The efficiency did not correlate linearly with the absorbed energies. The efficiency in the first compartment (0–5 mm) was the lowest, even though most of the light energy was absorbed in this compartment. The high

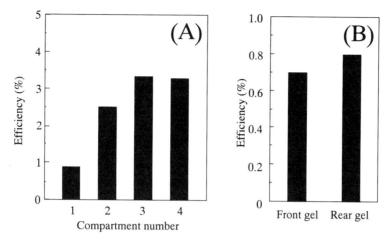

Figure 5. Light energy conversion efficiency to hydrogen at 720 W/m^2. (A) A-type reactor, (B) B-type reactor.

efficiencies were observed in the third and fourth compartments (10–20 mm), although only a limited amount of energy reached them.

4.3.2. Immobilized = B-Type. The light energy conversion efficiency to hydrogen in the B-type reactor is shown in Figure 5B. The efficiency of the front gel was lower, although most of the light energy was absorbed in it. A higher efficiency was observed in the rear gel although only limited energy reached it.

5. LIGHT PENETRATION EFFECT ON LIGHT ENERGY CONVERSION EFFICIENCY

A common observation was that higher efficiencies were recorded in the deep parts of both types of reactors. Cells located deep in the reactor were poorly illuminated. The effect of light intensity on light energy conversion efficiency has been investigated by Miyake et al. (1987), and they confirmed that the efficiency became high at low light intensity.

We also considered that alteration of the light spectrum upon passage of light through the reactor contributed to higher efficiencies in the deep parts of reactors. The light spectra reaching cells in these deep parts of the reactor were substantially different from that of the light source. Only light around 500–800 nm (mainly 600–780 nm) reached deep into the reactors. Nevertheless, the cells utilized the light for hydrogen production even though they did not contain much energy in the absorption maxima of the bacteriochlorophylls. This is inconsistent with the results of Nogi et al. (1985) that light around 600–780 nm was not effective for hydrogen evolution.

To clarify the relationship between light wavelength and hydrogen production, the hydrogen evolution rates from the C-type reactor were measured under illumination with two kinds of light sources. The light energy conversion efficiencies to hydrogen were 1.1% under illumination with infrared-depleted tungsten light that mainly radiated the energy around 750–900 nm, and 1.3% under illumination with infrared-depleted halogen light that mainly radiated the energy around 600–750 nm. This data suggest that light around 600–750 nm that can reach deep into the rectors is effective for hydrogen evolution rather than light around 800–900 nm. It is necessary to analyze the wavelength dependence of the light energy conversion efficiency to hydrogen in more detail. Further work is in progress to investigate the relationship with light wavelength and hydrogen evolution by using monochromatic light or band passed light.

In this paper, we have reported that alteration of light intensity and light spectrum upon passage of the light through the reactor affected light energy conversion efficiencies to hydrogen. To enhance the total conversion efficiency of light to hydrogen in a photobioreactor, light energy should be equally distributed in every part of the reactor and light should be supplied at low intensity. The use of optical fibers is a way to redistribute the light evenly in the reactor. However, such a system has the disadvantage of prohibitively increasing cost.

ACKNOWLEDGMENTS

This work was performed under the management of RITE as a part of the Research and Development Project on Environmentally Friendly Technology for the Production of Hydrogen supported by New Energy and Industrial Technology Development Organization (NEDO).

REFERENCES

Garcia, A., Vernon, L.P., Ke, B., and Mollenhauer, H., 1968, Some structural and photochemical properties of *Rhodopseudomonas palustris* subchromatophore particles obtained by treatment with Triton X-100, *Biochemistry*, 7:319–325.

Hirayama, O., Uya, K., Hiramatsu, Y., Yamada, H., and Moriwaki, K., 1986, Photoproduction of hydrogen by immobilized cells of a photosynthetic bacterium, *Rhodospirillum rubrum* G-9 BM, *Agric. Biol. Chem.*, 50:891–897.

Jahn, A., Keuntje, B., Dorffler, M., Klipp, W., and Oelze, J., 1994, Optimizing photoheterotrophic H_2 production by *Rhodobacter capsulatus* upon interposon mutagenesis in the *hup*L gene, *Appl. Microbiol. Biotechnol.*, 40:687–690.

Mao, X.Y., Miyake, J., and Kawamura, S., 1986, Screening photosynthetic bacteria for hydrogen production from organic acid, *J. Ferment. Technol.*, 64:245–249.

Miyake, J., Tomizuka, N., and Kamibayashi, K., 1982, Prolonged photo-hydrogen production by *Rhodospirillum ruburum*, *J. Ferment. Technol.*, 60:199–203.

Miyake, J., Mao, X.Y., and Kawamura, S., 1984, Photoproduction of hydrogen from glucose by a co-culture of a photosynthetic bacterium and *Clostridium butyricum*, *J. Ferment. Technol.*, 62:531–535.

Miyake, J., and Kawamura, S., 1987, Efficiency of light energy conversion to hydrogen by the photosynthetic bacterium *Rhodobacter sphaeroides*, *Int. J. Hydrogen Energy*, 39:147–149.

Miyake, J., Asada, Y., and Kawamura, S., 1989, Nitrogenase, in *Biomass Handbook*, Hall, C.W., and Kitani, O. (eds.), Gordon and Breech Scientific Publishers, New York, pp. 362–370.

Miyamoto, K., 1994, Hydrogen production by photosynthetic bacteria and microalge, in *Recombinant Microbes for Industrial and Agricultural Applications*, Murooka, Y., and Imanaka, T. (eds.), Marcel Dekker, New York, pp. 771–786.

Nogi, Y., Akiba, T., and Horikoshi, K., 1985, Wavelength dependence of photoproduction of hydrogen by *Rhodopseudomonas rutila*, *Agric. Biol. Chem.*, 49:35–38.

Shneour, E.A., 1962, Carotenoid pigment conversion in *Rhodopseudomonas sphaeroides*, *Biochem. Biophys. Acta*, 62:534–540.

Tsygankov, A.T., Hirata, Y., Miyake, M., Asada, Y., and Miyake, J., 1994, Photobioreactor with photosynthetic bacteria immobilized on porous glass for hydrogen photoproduction, *J. Ferment. Bioeng.*, 77:575–578.

43

CYLINDRICAL-TYPE INDUCED AND DIFFUSED PHOTOBIOREACTOR

A Novel Photoreactor for Large-Scale H_2 Production

Reda M. A. El-Shishtawy, Shozo Kawasaki, and Masayoshi Morimoto

Kajima Laboratory
Project for Biological Production of Hydrogen
Research Institute of Innovative Technology for the Earth
2-19-1 Tobitakyu
Chofu, Tokyo 182, Japan

Key Words
induced and diffused photobioreactor, scale-up, hydrogen

1. SUMMARY

A novel photobioreactor based on a polyacrylate light-receiving face and modified polyester diffusion sheet is presented. With it, the photosynthetic bacterium *Rhodobacter sphaeroides* RV produced H_2 from lactate and glutamate as carbon and nitrogen sources, respectively. The maximum efficiency of energy conversion to H_2 was 9.23% using a 1-cm culture width with illumination of 300 W/m² from a halogen lamp. The reactor showed good proportionality between the rate of gas evolution and incident light energy. A 20-cm deep cylindrical-type photobioreactor that had previously failed to produce H_2 was made efficient for H_2 production after being modified with induced plates to create a cylindrical induced and diffused photobioreactor (20-cm IDPBR). The productivity of this 20-cm IDPBR was 5.03 mM H_2/mM lactate, while that of the unmodified photobioreactor was 1.62 mM H_2/mM lactate and that of a 5-cm deep version used as the standard was 3.07 mM H_2/mM lactate under light of 300 W/m². Stirring effects decreased the productivity of the unmodified photobioreactor to 0.20 mM H_2/mM lactate. Our novel system could be used to study other light-dependent bioreactions.

2. INTRODUCTION

Photobiological H_2 production (Weaver et al., 1980) by photosynthetic bacteria is a promising method of solar energy conversion. Many photosynthetic bacteria produce H_2

under suitable conditions (Sasikala et al., 1993). Since light is the main driving force for H_2 production, its efficient utilization by photosynthetic bacteria is a prerequisite for an efficient photobioreactor (PBR). Various PBRs have been proposed for H_2 production. A typical PBR is illuminated either from the outside through a transparent lid (Sasikala et al., 1992), or from the inside by a bulb at the center (Stevens et al., 1983). The main design criterion in these PBRs is to maximize the illumination surface-to-volume ratio. Consequently, the culture thickness of these PBRs is very thin and thus impractical for scale-up. Mignot et al. (1989) reported a successful approach of using optical fibers to guide light to the reactor. However, these optical fibers are very expensive, and many technical problems must be solved before their practical application.

Recently, we introduced (Morimoto et al., 1996) a design concept for an induced and diffused photobioreactor (IDPBR). The design harvests the solar energy and distributes the light homogeneously inside the PBR. We have reported (El-Shishtawy et al., 1997) a prototype IDPBR and successfully obtained efficient H_2 production.

In this paper, we describe the effect of light intensity on H_2 production using our prototype IDPBR. We also present comparisons between a large-scale, 20-cm deep, cylindrical-type induced and diffused photobioreactor, an unmodified version and a 5-cm standard model.

3. EXPERIMENTAL

3.1. Microorganism and Culture Conditions

Rhodobacter sphaeroides RV from the National Institute of Bioscience and Human Technology in Tsukuba, Japan, was pre-cultured in basal medium (inorganic salts and vitamins) containing 0.1% (w/v) yeast extract, 10 mM ammonium sulfate, and 60 mM sodium succinate (Mao et al., 1986; Miyake et al., 1984). The strain was grown anaerobically in a completely full Roux flask of 6 cm width and 272 cm^2 area under illumination (43 W/m^2 from a tungsten lamp) for three days at 30 °C. The H_2 production medium (HPM) was then made from a 1/3 (v/v) ratio of the preculture and a freshly prepared basal medium containing 72 mM sodium D,L-lactate and 10 mM sodium glutamate (Nakada et al., 1995), respectively. HPM (initial bacterial concentration was about 85 mg-dw/L) was transferred to the IDPBRs for H_2 photoproduction by continuous illumination using a halogen lamp (300 W/m^2) for the time indicated at 30 °C without stirring.

Also, HPM was transferred to the cylindrical plane reactors (490 cm^2) of 5- and 20-cm depths and the 20-cm deep IDPBR for H_2 photoproduction by continuous illumination (300 W/m^2) for the time indicated at 30 °C without and with stirring (600 rpm).

Figure 1 shows the concept of the IDPBR, which is based on a polyacrylate light receiving face and modified polyester reflection sheet, and the actual 20-cm deep IDPBR. The effect of culture width and light intensity on the production of H_2 (Figures 2 and 3) were investigated using rectangular IDPBRs. The rectangular IDPBRs with different culture widths (0.5–2.5 cm) were placed in water bath at 30 °C to have the edge face in front of the light source. The incident light intensity at the surface (0.002 m^2, the light receiving area) was 300 W/m^2 (Figure 2) or as indicated in Figure 3. The front face of the culture was covered with black tape to avoid direct illumination and the scattered light from the base of water bath was ignored.

The 20-cm deep cylindrical plane reactor was changed to a 20-cm deep IDPBR after modification with three layers of cylindrical induced plates of about 16 cm in height and 2

Cylindrical-Type Induced and Diffused Photobioreactor 355

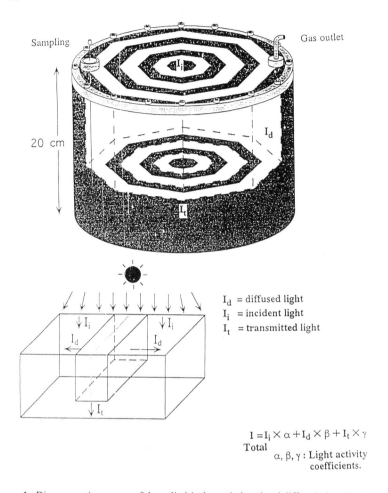

$$I_{Total} = I_i \times \alpha + I_d \times \beta + I_t \times \gamma$$

α, β, γ : Light activity coefficients.

Figure 1. Diagrammatic concept of the cylindrical type induced and diffused photobioreactor.

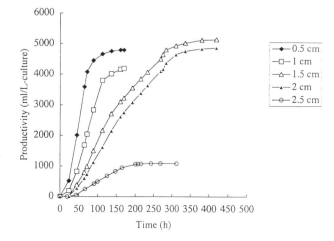

Figure 2. Effect of IDPBR culture widths on the productivity of H_2 per culture volume. Conditions: halogen lamp (300 W/m^2), 0.5 cm (150 mL), 1 cm (300 mL), 1.5 cm (450 mL), 2 cm (600 mL), 2.5 cm (750 mL), 20 cm^2, no stirring, 30 °C.

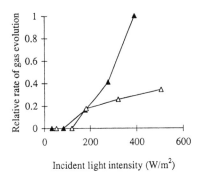

Figure 3. Rate of H_2 evolution at different light intensities using different widths of IDPBR. Conditions: halogen lamp, 1 cm (300 mL), 2 cm (600 mL), 20 cm², with stirring (600 rpm), 30 °C. 1 cm width (▲), 2 cm width (△).

cm in thickness to make cylindrical layers of culture of about 1.6 cm thickness and clearance of about 4 cm at the bottom of the reactor where a magnetic bar was placed for stirring.

Cell growth was determined either photometrically by measuring the optical density of the culture at 660 nm or gravimetrically by weighing the amount of dry cell after overnight desiccation at 120 °C. A linear relationship between OD_{660} and cell dry weight values was observed and the cell dry weight was calculated from the graph (1 OD_{660} = 0.34 mg [dry weight]/mL).

3.2. Analytical Methods

The light intensity at the surface of the reactor was measured with a photometric sensor (LI-COR Inc., Model LI-189). The volume of gas evolved was measured by displacement of acidic water at room temperature, and the H_2 content was analyzed by Sensidyne detector tubes (Gastec Corp.) and/or gas chromatography. Organic acid excretion and lactate consumption in the medium were measured by HPLC using a Shim-Pack SCR-102H column (8 x 300 mm, Shimadzu) and a conductivity detector (CCD-6A, Shimadzu).

The efficiency of light energy conversion (Miyake and Kawamura, 1987) was calculated according to the following equation:

Efficiency (%) = (gross combustion heat of the evolved H_2)/(input light energy) × 100

Gross combustion heat of H_2 = 3.05 Kcal/L (Miyake and Kawamura [1987] adopted net combustion heat).

The conversion efficiency of the organic acid substrate (lactate) to H_2 was calculated on the basis of the following (Vincenzini et al., 1981):

$$CH_3CH(OH)COOH + 3H_2O = 6H_2 + 3CO_2$$

4. RESULTS AND DISCUSSION

We have reported (Morimoto et al., 1996) that the diffused light intensity that might be directly received by the culture decreases in accordance with the following equation:

Diffused light intensity = (inlet light intensity × inlet area)/(outlet area) × α

where α is the energy loss parameter (<1).

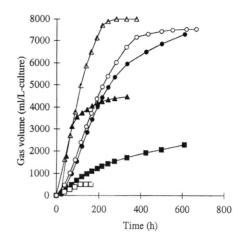

Figure 4. Stirring effect on the course of H_2 production of various PBRs under continuous irradiation (300 W/m²). Conditions: halogen lamp (300 W/m²), 5 cm (2.45 L), 20 cm (9.8 L), 20 cm IDPBR (5.33 L), 490 cm², stirring (600 rpm), 30 °C. (▲) 5 cm, (■) 20 cm, (●) 20 cm IDPBR, (△) 5 cm (with stirring), (○) 20 cm IDPBR (with stirring).

As shown in Figure 2, the productivity per culture volume using IDPBRs of different culture widths (El-Shishtawy et al., 1997) under 300 W/m² illumination is more or less the same for all widths except that of the 2.5-cm model, which has the lowest production. These results indicate that diffused light in an IDPBR is quite effective for H_2 production up to 2 cm wide, over which it is no longer effective but negative due to the increased effect of mutual shading that prevents penetration of light. The maximum productivity observed was 7812 mL/m²·h at 300 W/m², which is equivalent to 9.23% light energy conversion to H_2.

Having the optimum culture widths of 1 and 2 cm, it was desirable to study the effect of light intensity on H_2 production using 1- and 2-cm culture width IDPBRs. As expected, H_2 production could be obtained only under high incident energy (Figure 3). However, the rate of gas evolution was proportional to incident light energy. This result reflects the absence of photoinhibition and the increase of light saturation energy that makes the IDPBR a suitable device for solar energy conversion.

It has become clear that other modifications for enhancing the energy inside the IDPBR and for large-scale cultivation are necessary. Therefore, a cylindrical-type IDPBR is proposed to make use of all forms of light, that is, incident, transmitted, and diffused light, as depicted in Figure 1.

Figure 1 shows the principle of a modified IDPBR that would make use of all light conditions, and thus scaled-up in cylindrical or rectangular form. We have (RITE, Japan) conducted several experiments for H_2 production using simple cylindrical-type PBRs (plane PBR) of different depths and concluded that the 5-cm depth should be considered the standard for H_2 production. Consequently, we started with the cylindrical form and eventually designed the cylindrical-type induced and diffused photobioreactor.

As shown in Figure 4, the productivity of H_2 using a 20-cm IDPBR (5.97 mL/L-culture) was much higher than that of the 5-cm (3.64 mL/L-culture) and 20-cm PBRs (1.92 mL/L-culture). Upon stirring, the productivity of the 5-cm version became more or less similar to that of the 20-cm IDPBR, while that of the 20-cm PBR decreased to a level of 0.24 mL/L-culture. The results indicate that stirring has no effect on the 20-cm IDPBR, which means that light energy inside the 20-cm IDPBR is quite homogenous, owing not only to the better distribution of light inside the reactor but also to the self-mixing of the culture by virtue of the H_2 bubbles that rise strongly from the bottom of the reactor.

While stirring enhanced the productivity of the 5-cm PBR, it lessened that of the 20-cm PBR. In the case of the 5-cm PBR without stirring, the incident energy (300 W/m²)

Table 1. Substrate and energy conversion to H_2 using cylindrical PBRs

Cylindrical PBRs	Light conversion (maximum %)[a]		Substrate conversion (mM H_2/mM lactate)[b]	
	Without stirring	With stirring	Without stirring	With stirring
5 cm	3.71	2.61	3.07	5.32
20 cm	2.07	1.07	1.62	0.20
20 cm IDPBR	2.78	2.85	5.03	4.80

[a]Values were obtained with an initial concentration of 53 mM D,L-lactate in the HPM using the final accumulated H_2 volume. No organic acid substrate was detected at the end of H_2 production in the culture broth of the 5-cm PBR and 20-cm IDPBR; however, 14.61 and 39.37 mg/L of organic acids were detected in the culture broth of the 20-cm PBR in both conditions of without and with stirring, respectively.

[b]Maximum light energy conversion efficiencies were obtained from their corresponding maximum H_2 evolution rates.

was too strong on the cells and caused photoinhibition that led to low productivity. However, stirring distributed light nicely inside the PBR and this led to high productivity.

In the case of the 20-cm PBR, stirring lowered the productivity because the light energy was not sufficient before stirring, and so stirring decreased the proportion of the reactor volume under light. As such, this condition would lead to no growth, and endogenous respiration and cell death could take place in the dark zone. As shown in Table 1, the substrate and energy conversion efficiencies are in good agreement with the above discussion. As result of sufficient light energy in the 5-cm PBR and the 20-cm IDPBR, the light conversion efficiency in both PBRs is similar. Furthermore, in these PBRs, there was no detection of organic acids in the culture broth at the end of H_2 production. However, in case of the 20-cm PBR, without and with stirring, amounts of 14.61 and 39.37 mg/L of organic acids were detected in the culture broth, respectively. This result indicates that a cylindrical 20-cm IDPBR is a promising PBR for wastewater H_2 production process.

ACKNOWLEDGMENTS

R.M.A. El-Shishtawy, on leave from the National Research Centre, Cairo, Egypt, was supported by a Japanese the New Energy and Industrial Technology Development Organization Fellowship Award to work under the management of the Research Institute of Innovative Technology for the Earth. The authors thank Miss Sakamoto for her analytical work.

REFERENCES

El-Shishtawy, R.M.A., Kawasaki, S., and Morimoto, M., 1997, *Biotechnol. Tech.*, 11:403–407.
Mao, X.Y., Miyake, J., and Kawamura, S., 1986, *J. Ferment. Technol.*, 64:245–249.
Mignot, L., Junter, G.A., and Labbe, M., 1989, *Biotechnol.Tech.*, 299–304.
Miyake, J., and Kawamura, S., 1987, *Int. J. Hydrogen Energy*, 12:147–149.
Miyake, J., Mao, X.Y., and Kawamura, S., 1984, *J. Ferment. Technol.*, 62:531–535.
Morimoto, M., Kawasaki, S., El-Shishtawy, R.M.A., Ueno, Y., and Kunishi, N., 1996, in *Proceedings of the 11th World Hydrogen Energy Conference*, Stuttgart, Germany, pp. 2761–2766.
Nakada, E., Asada, Y., Arai, T., and Miyake, J., 1995, *J. Ferment. Bioeng.*, 80:53–57.
Sasikala, K., Ramana, Ch.V., and Raghuveer Rao, P., 1993, in *Advances in Applied Microbiology*, vol. 38, Academic Press, Inc., pp. 211–295.
Sasikala, K., Ramana, Ch.V., and Raghuveer Rao, P., 1992, *Int. J. Hydrogen Energy*, 17:23–27.
Stevens, P., van der Sypt, H., De Vos, P. and De Ley, J., 1983, *Biotechnol. Lett.*, 5:369–374.
Vincenzini, M., Balloni, W., Mannelli, D., and Florenzano, G., 1981, *Experientia*, 37:710–711.
Weaver, P.F., Lien, S., and Seibert, M., 1980, *Sol. Energy*, 24:3–45.

44

ANALYSIS OF COMPENSATION POINT OF LIGHT USING PLANE-TYPE PHOTOSYNTHETIC BIOREACTOR

Yoji Kitajima,[1] Reda M. A. El-Shishtawy,[1] Yoshiyuki Ueno,[2] Seiji Otsuka,[1] Jun Miyake,[3] and Masayoshi Morimoto[1]

[1]Kajima Laboratory
Project for Biological Production of Hydrogen
Research Institute of Innovative Technology for the Earth
2-19-1 Tobitakyu
Chofu, Tokyo 182, Japan
[2]Marine Biotechnology Institute
Heita Kamaishi
Iwate 026, Japan
[3]National Institute of Bioscience and Human-Technology
Higashi
Tsukuba, Ibaraki 305, Japan

Key Words

bioreactor, hydrogen, uptake, compensation, *Rhodobacter sphaeroides*

1. SUMMARY

Five cylindrical, plane-type, photosynthetic bioreactors of different depths (1, 3, 5, 10, and 20 cm) were used to examine the effect of reactor depth with agitation on hydrogen production using *Rhodobacter sphaeroides* RV and artificial waste water. This batch test was performed with simulated solar energy using halogen lamps. The irradiation pattern was a sine curve of light$_{12h}$/dark$_{12h}$ cycle with a peak value of 1 kW/m^2. Investigation of the effects of depth showed that the rate of hydrogen production decreased and organic acid concentration increased as reactor depth increased. This was probably due to the increasing effect of the dark zone. Under dark conditions and with strong agitation of the gas/liquid phase in the 1-cm deep reactor, the uptake rate of hydrogen was 100 mL H$_2$/g dry cell weight/day. Hydrogen production and uptake were occurring and balancing in the photosynthetic bioreactor. The light compensation point using a plane-type photosynthetic bioreactor with agitation was about 102 MJ/m^3 culture/day. Thus, an important factor to be considered for scaling-up photosynthetic bioreactors is the penetration of light into the cell culture and the absence of a dark zone in the reactor.

2. INTRODUCTION

Hydrogen production by microorganisms can be divided into two main categories: anaerobic non-photosynthetic bacteria producing both hydrogen and organic acids, and photosynthetic bacteria consuming organic acids and producing hydrogen. Anaerobic microflora convert cellulose to hydrogen with high efficiency (Ueno et al., 1996). However, when attempting to scale-up the experimental apparatus and perform experiments, the phenomenon of rapid reduction in the rate of hydrogen production has puzzled researchers (Kitajima et al., 1996).

In this report, we analyze this reduction in hydrogen production by mass cultures and establish a new methodology for analyzing hydrogen production. We introduce concepts that include the presence of hydrogen uptake reactions and the presence of an equilibrium point between hydrogen production and uptake reactions depending on illumination intensity, which we call the light compensation point.

3. EXPERIMENTAL

3.1. Microorganism and Culture

Rhodobacter sphaeroides RV (Miyake et al., 1984) was provided by the National Institute of Bioscience and Human Technology (NIBH). The strain was pre-cultured in an ASY medium consisting of a basal medium (inorganic salts and vitamins), 0.1 w/v % yeast extract, 10 mM $(NH_4)_2SO_4$, and 35 mM sodium succinate. The GL medium used as artificial waste water for hydrogen production consisted of a basal medium containing 50 mM sodium D,L-lactate and 10 mM sodium glutamate (Nakada et al., 1995). Cells in the ASY medium were grown anaerobically in a 6-cm deep Roux flask under illumination (40 W/m^2 from tungsten lamps) for 3 days at 30 °C.

3.2. Hydrogen Production in Light

Five cylindrical, plane-type, photosynthetic bioreactors with different depths (1, 3, 5, 10, and 20 cm) were constructed to examine the effect of reactor depth with agitation on the course of hydrogen production. Each reactor had a 25-cm diameter and a 490-cm^2 light-receiving area (Figure 1a). A cell suspension, pre-cultured in ASY medium, was mixed with the diluted GL medium at a ratio of 1:3 to create a suspension with an initial cell concentration of about 100 mg dw/L and transferred to the reactor. The reactor was immersed in a water bath at 30 °C. This batch test was performed under simulated solar energy using a halogen lamp unit (Solar Simulator, Marubishi Bioengineering Co., Ltd.). The irradiation pattern was a sine curve of light$_{12h}$/dark$_{12h}$ cycle with a peak value of 1 kW/m^2 at the reactor surface.

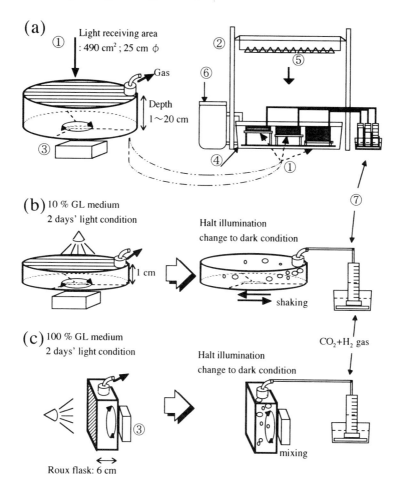

Figure 1. Schematic diagram of experimental arrangement: (a) H_2 produced in reactors of various depths under light conditions with agitation; (b) and (c) H_2 consumption under dark conditions with agitation; lactate absent and present. (1) Plane-type photosynthetic bioreactor, (2) solar simulator, (3) magnetic stirrer, (4) water bath, (5) halogen lamp unit, (6) cooler, (7) gas cylinder.

3.3. Hydrogen Uptake in the Dark

For measurement under dark conditions in the absence of lactate, a 1-cm deep cylindrical plane reactor and a diluted GL medium with lactate and glutamate reduced to one-tenth were used. A cell suspension, pre-cultured in ASY medium, was mixed with the diluted GL medium at a ratio of 1:3 to create a suspension with an initial cellular concentration of about 100 mg dw/L, and 500 mL of the suspension was poured into a 1-cm deep bioreactor. The bioreactor was illuminated at 300 W/m² (halogen lamp), and bacteria that produce hydrogen by photosynthesis were allowed to grow. After two days, when hydrogen production had ceased due to substrate depletion, the illumination was halted and the gas in the reactor was replaced with a mixture of H_2 and CO_2 (H_2:CO_2 = 91:9). Broth was continued with strong shaking of gas and medium under dark conditions in the absence of lactate (Figure 1b).

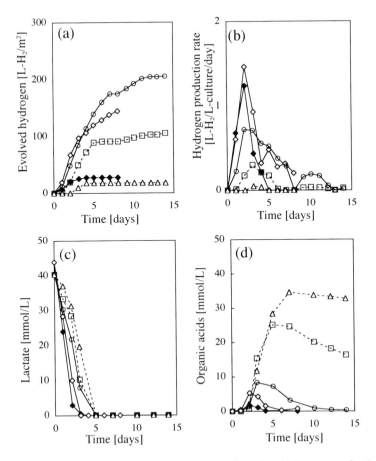

Figure 2. Time-course of hydrogen production and concentration of organic acids in reactors of various depths. Irradiation pattern was a sine curve of light$_{12h}$/dark$_{12h}$ cycle with a peak value of 1 kW/m^2. (a) Total hydrogen production by light-receiving area; (b) hydrogen production rate by culture medium; (c) lactate consumption; (d) production of lower fatty acids (acetate + propionate + butyrate). Depth (and volume) of reactor: ♦ = 1 cm (0.5 L); ◇ = 3 cm (1.5 L); ○ = 5 cm (2.5 L); ☐ = 10 cm (5 L); △ = 20 cm (10 L).

For measurement under dark conditions in the presence of lactate, a cell suspension pre-cultured in ASY medium was mixed with standard GL medium at a ratio of 1:3 to create a suspension with an initial cellular concentration of about 100 mg dw/L. A total of 1000 mL of this suspension was poured into a Roux flask 6 cm deep, the flask was kept at 30 °C, agitated, and illuminated at 300 W/m^2. After two days, while hydrogen production was continuing and some organic acid substrate remained, illumination was halted. The gas in the flask was replaced with a mixture of H_2 and CO_2 (H_2:CO_2 = 84:16) and measurements of gas quantity and composition were performed under dark conditions in the presence of lactate (Figure 1c).

3.4. Analytical Methods

The light intensity was measured with a radiation sensor (LI-COR, Inc., Model LI-189). The gas volume was measured by displacement of acidic water, and the hydrogen and carbon dioxide content was analyzed by Sensidyne detector tubes (Gastec Corp.) and a gas chromatograph (Shimadzu, GC-9A). The amounts of D,L-lactate and lower (C2~C5) fatty acids in the medium were measured by HPLC. Cellular dry weight (g-dw) was obtained by the filtration method (pore size: 0.45 μm)

4. RESULTS AND CONCLUSIONS

4.1. Hydrogen Production in Light

Hydrogen production per area (L H_2/m^2) increased as the depth of the reactor increased from 1 to 3 to 5 cm. When the depth was 10 or 20 cm, hydrogen production decreased (Figure 2a). The rate of hydrogen production per volume of culture (L H_2/L culture/day) was higher for shallower reactors with a depth of 1 or 3 cm (Figure 2b). It was expected that hydrogen would be produced at high rates in the surface layers of deeper reactors with depths of 10 or 20 cm because their surface layers received similar (and sufficient) amounts of light energy, as did shallower reactors with depths of 1 or 3 cm. However, cumulative hydrogen production (L H_2/m^2) in the deeper reactors was smaller, and greatest for reactors with a depth of 5 cm. In reactors of all depths, lactate was entirely consumed within days after the start of the tests (Figure 2c). On the other hand, in deeper reactors, the production and accumulation of other organic acids such as acetate, propionate, and butyrate became apparent (Figure 2d). Based on these findings, it was proposed that in deeper reactors, side reactions occur that consume hydrogen produced and cause organic acids to accumulate, in addition to the main reaction of lactate consumption and hydrogen production.

4.2. Hydrogen Uptake in the Dark

When the culture was illuminated at 300 W/m^2 for 24 h, the rate of hydrogen production was as high as 1000 mL (40 mmol) H_2/L culture/day under normal light conditions with lactate (Figure 3a, left). We presumed the following regarding the concentration of organic acids: within the first five days of culturing, during which the hydrogen production rate is high, lactate disappears as do major organic acids such as acetate, propionate, and butyrate, which are produced later. After the lactate concentration has decreased, organic acids produced in the reactor later are re-absorbed as substrate, so hydrogen production proceeds efficiently (Figure 3a, right). In another experiment, a GL medium diluted one-tenth was introduced as a substrate to a 1-cm deep cylindrical plane-type bioreactor and exposed to light. After two days, illumination was halted because the organic acid substrate had disappeared and the culture bacteria were starving. High levels of hydrogen uptake occurred temporarily just after illumination was halted, and hydrogen was afterward consumed steadily at the rate of 100 mL (4 mmol) H_2/g dw/day (Figure 3b, left). Moreover, organic acids (which had earlier disappeared) were produced and their concentrations gradually increased (Figure 3b, right). These results demonstrate that the hydrogen uptake and organic acid production assumed to have occurred in earlier tests using reactors of different depths can occur. After illuminating a GL medium in the presence of an organic acid substrate for two days, illumination was halted while some lactate remained. On the left in Figure 3c is the time course of the hydrogen production rate and cell concentration, and on the right in Figure 3c is the time course of lactate, acetate, propionate, and butyrate concentrations. Under culture conditions described in Figure 3a,

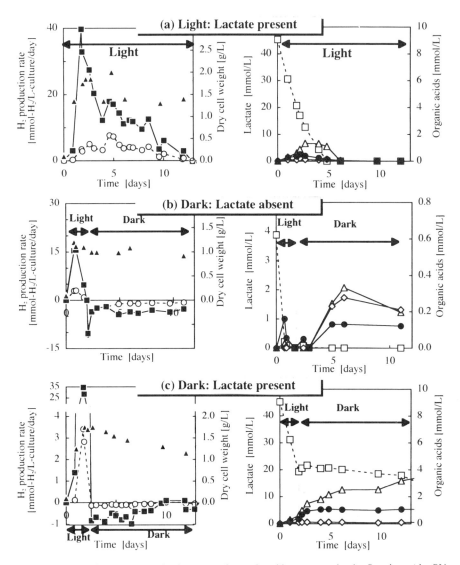

Figure 3. Time-course of hydrogen production rate and organic acids concentration by *R. sphaeroides* RV under various conditions. Light condition was adjusted to 300 W/m^2. Dark condition was prepared after light was continuously irradiated for two days. (a) Control: continuous light condition at 300 W/m^2 in presence of lactate using Roux flask 6-cm deep (culture volume: 1000 mL). (b) Dark condition in absence of lactate using plane-type reactor of 1 cm depth (culture volume: 500 mL). (c) Dark condition with presence of lactate using Roux flask 6 cm deep (culture volume: 1000 mL). ■ = H$_2$ gas; ○ = CO$_2$ gas; ▲ = dry cell weight; □ = lactate; △ = acetate; ◇ = propionate; ● = butyrate.

the illumination of a culture flask was halted when organic acid substrate components such as lactate were present. The rate of hydrogen production was high during the period of illumination, but there was continuous uptake of hydrogen after illumination was halted. After the initial two days of culturing, during which the rate of hydrogen production was high, half the initial quantity of lactate remained and major organic acids such as acetate, propionate, and butyrate were also present. Under these circumstances, illumina-

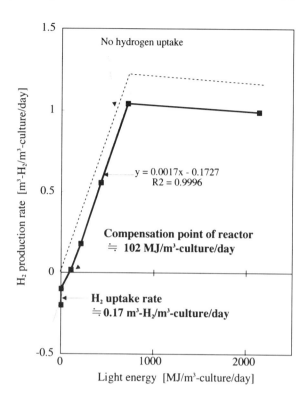

Figure 4. Relationship between light energy and hydrogen production rate from lactate using plane-type photosynthetic bioreactor cultured *R. sphaeroides* RV with agitation. The irradiation pattern was a sine curve of light$_{12h}$/dark$_{12h}$ cycle with peak value of 1 kW/m^2. The light energy indicated in this figure was recalculated without heat ray radiation (>approximately 1000 nm) because the light energy was measured from behind the 5-cm water phase. The hydrogen production rate was averaged throughout the period until the lactate decreased over 95%. Practical data is represented by the solid line. Broken line assumes no hydrogen uptake.

tion was halted so the consumption of lactate stopped, and quantities of lactate, acetate, propionate, and butyrate increased slightly. This indicates that hydrogen was consumed in the dark even when the substrate remained. Moreover, since the cell concentration in this situation was about 1.5 g dw/L, it appears that in the dark the bacteria could not grow using H$_2$, CO$_2$, or organic acids as the substrate.

4.3. Analysis of Light Compensation

The rate of hydrogen production in five bioreactors while illuminated, and the rate of hydrogen uptake in the dark, are expressed schematically with respect to energy from simulated sunlight per day per unit volume of culture in Figure 4. For each case, the average rates of hydrogen production during the time until more than 90% of the lactate had been consumed were used as the rate of hydrogen production under illumination. Rates of hydrogen production decreased as amounts of light energy per unit of reactor volume decreased, but the hydrogen production changed to uptake even before the light energy diminished to zero. The light compensation point using the plane-type, photosynthetic bioreactor cultured strain *R. sphaeroides* RV with agitation was about 102 MJ/m^3 culture/day. The average cell concentration obtained was about 1–2 g dw/L. The rate of hydrogen production at zero light obtained from the curve in Figure 4 was -0.17 m^3 H$_2$/m^3 culture/day, very close to the rate of hydrogen uptake obtained from experimental data collected under dark conditions. Figure 5a shows light irradiation energy per unit area as calculated using the procedures mentioned above (Figure 4) for assumed reactor depths of 1, 3, 5, 10, and 20 cm, and the corresponding expected rate of hydrogen production per unit illuminated area (L H$_2$/m^2/day). However, in even deeper reactors, the volume of culture per unit area becomes larger and the uptake of

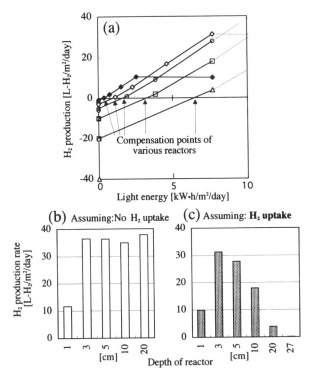

Figure 5. Estimated effect of hydrogen uptake on hydrogen production rate from lactate by *R. sphaeroides* RV in reactors with agitation and various depths (plotted data calculated from the experimental data shown in Figure 4). Based on the estimation, no apparent production of hydrogen would be observed if a reactor is 27 cm deep under the lighting conditions, as the rates of hydrogen production and uptake are balanced in the reactor (c). (a) Calculated hydrogen production rate on the basis of light energy per light-receiving area and the compensation points. (b) Calculated hydrogen production rates in reactors with various depths assuming no hydrogen uptake (light energy = 27.7 MJ/m^2/day). (c) Observed hydrogen production rates in reactors with various depths (light energy = 27.7 MJ/m^2/day). Depths of reactor : ◆ = 1 cm; ◇ = 3 cm; ○ = 5 cm; □ = 10 cm; △ = 20 cm.

hydrogen per volume of culture increases gradually to levels which are not negligible; thus there is a possibility that hydrogen production will stop if illumination energy decreases even slightly, such as from cloudy or rainy weather. Furthermore, it is assumed from Figure 4 that the number of light compensation points for a given reactor volume is one, and Figure 5a shows that the number of light compensation points per area differs depending on the average optical path length (depth) of the reactor. Figures 5b and c show calculated hydrogen production rates, assuming plane-type reactors of different depths are agitated identically while being illuminated at 7.7 kWh/m^2/day (27.7 MJ/m^2/day). This is regarded as the highest outdoor illumination energy under clear skies, based on Figure 5a. The chart on the right assumes the presence of the hydrogen uptake phenomenon described above, while the chart on the left shows results derived from the broken line in Figure 4, which assumes no hydrogen uptake. If light energy and hydrogen production by photosynthesis have a linear relationship through the origin, as the reactor depth increases the rate of hydrogen production per unit illuminated area must approach a constant value (Figure 5b). However, as indicated by the chart on the right Figure 5c, although the rate of hydrogen production per area increases with reactor depth until a certain depth (under conditions of the present experiment, about 3 cm), the rate of hydrogen production per area decreases as reactor depth becomes greater than that depth. In our experiment, illumination was supplied by a solar simulator in order to generate results which could be compared to the results of outdoor experiments.

In conclusion, it is suggested that the cause is a hydrogen uptake reaction in the *R. sphaeroides* RV strain. These concepts of hydrogen uptake reaction and light compensation point have the potential to explain without any inconsistencies the reduction of hydrogen production in deeper reactors. It is anticipated that the values of light compensation points or rates of hydrogen production and uptake will differ depending on the strain of bacteria, substrate composition, and, in actual sewage, the coloration of the sewage.

ACKNOWLEDGMENTS

This work was performed under the management of the Research Institute of Innovative Technology for the Earth (RITE) as part of the Development Project of Environmentally Friendly Technology for Hydrogen Production supported by the New Energy and Industrial Technology Development Organization (NEDO). The authors thank Miss Ikuko Sakamoto and Mr. Shozo Kawasaki for their analytical work.

REFERENCES

Kitajima, Y., Morimoto, M., Otsuka, S., Sakamoto, I., Kunishi, N., and Ueno, Y., 1996, Environmentally friendly technology for hydrogen production, *Kajima Annual Report*, 44:295–299.

Miyake, J., Mao, X.Y., and Kawamura, S., 1984, Photoproduction of hydrogen from glucose by a co-culture of photosynthetic bacterium and *Clostridium butyricum*, *J. Ferment. Technol.*, 62:531–535.

Nakada, E., Asada, Y., Arai, T., and Miyake, J., 1995, Light penetration into cell suspensions of photosynthetic bacteria and relation to hydrogen production, *J. Ferment. Bioeng.*, 80:53–57.

Ueno, Y., Otsuka, S., and Morimoto, M., 1996, Hydrogen production from industrial waste water by anaerobic microflora chemostat culture, *J. Ferment. Bioeng.*, 82:194–197.

45

HYDROGEN PRODUCTION BY A FLOATING-TYPE PHOTOBIOREACTOR

Toshi Otsuki, Shigeru Uchiyama, Kiichi Fujiki, and Sakae Fukunaga

Research Institute
Ishikawajima-Harima Heavy Industries Co., Ltd.
1, Shin-Nakahara-cho
Isogo-ku, Yokohama 235, Japan

Key Words

photobioreactor, photosynthetic bacteria, *Rhodopseudomonas palustris* R-1, H_2 production

1. SUMMARY

We have succeeded in continuously producing H_2 over three months using a floating-type photobioreactor, artificial light (halogen lamp), artificial raw waste water (identical to the effluent of an anaerobic pre-treatment process), and *Rhodopseudomonas palustris* R-1. The goal was to determine the specifications of artificial raw waste water, whose constituents compel bacteria to produce H_2. We eliminated nutrients such as P, K, N, Fe, Ca, vitamin, and yeast-extract, which caused bacteria to grow instead of producing H_2. The raw waste water contained 800 mg/L each of acetate, propionate, and butyrate and 400 mg/L of ethanol, which are the suitable electron donors for H_2 photoproduction.

A thermostat controlled liquid temperature at 35 °C, and 24 halogen lamps illuminated the reactor surface for 24 h at 434 W/m^2. The raw waste water was fed at 0.8 L/day to the reactor. The bacteria separated from the pre-culture was added to the reactor several times during the experiment.

We accumulated 85 L of gas during 1583 h. The H_2 content in the gas was nearly 85% and CH_4 was less than 1%. N_2 was less than 6% and O_2 was not detected.

The gas production rate was 100–300 mL/L•day during the early stage and 300–500 mL/L•day in the later stage. The efficiency of light conversion to H_2 was 0.308%.

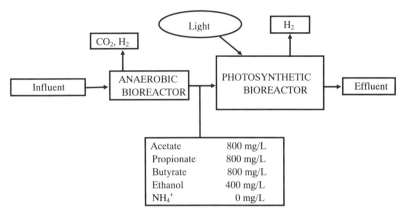

Figure 1. H_2 production from sugary waste water.

2. INTRODUCTION

Methods for large-scale H_2 production by photosynthetic bacteria were studied together with the pre-treatment of organic materials used as substrates for H_2 production. Glucose contained in a sugary waste water was used as organic material. The waste water was treated with two bioreactors: The first anaerobic bioreactor produced organic acids from glucose by anaerobes and the second photobioreactor produced H_2 from organic acids by photosynthetic bacteria. As the waste water was treated in the anaerobic stage, glucose changed to such organic acids as acetate, propionate, and butyrate, which are good substrates for photosynthetic bacteria in the following stage (Figure 1). The photobioreactor that we constructed had the following characteristics:

- floating-type reactor constructed for use on sea or lake;
- liquid in the reactor easily mixed by waves;
- no energy required for cooling the reactor; and
- little land space required.

We found *Rhodopseudomonas palustris* R-1 to be a suitable photosynthetic bacteria for H_2 production from waste water.

There are few experimental reports on the issues (Kim, 1982), but many concerning the physiology of photosynthetic bacteria (e.g., Sasikala, 1993).

3. EXPERIMENTAL

The conditions of the experiment are listed in Table 1.

The reactor was made of acrylic plate and had a volume of 6 L, depth of 50 mm, and a triangle roof, moving on a seesaw-like bed at 5 rpm, similar to the reactor we envision for at-sea application. The temperature and concentration of liquid in the reactor were kept constant using a small magnet pump. This pump circulated the liquid between the ORP and pH electrodes in the monitoring box and reactor. A thermostat controlled liquid temperature at 35 °C, and 24 halogen lamps illuminated the reactor for 24 h at 434 W/m² on the surface of the roof.

Table 1. Conditions of the experiment, indoors, 6-L reactor

Bacteria type	*Rhodopseudomonas palustris* strain R-1
Reactor type	Triangle roof on moving bed
Reactor volume	6 L
Water depth	50 mm (excluding roof height)
Moving rate	5 rpm (like a seesaw)
Light source	24 halogen lamps
Irradiation area	0.107 m^2
Irradiation energy	434 W/m^2
Mixed-liquor circulation rate	7 L/min
Raw water	Artificial anaerobic effluent
Feed rate	0.8 L/day (continuous, 0.14/day)
Temperature	35 °C
Separation of bacteria from pre-culture	Yes

We fed the raw waste water at 0.8 L/day to the reactor using a roller pump. We daily checked the pH, SS (suspended solids), OD (660 nm), NH_4^+, and organic acids of the effluent, as well as amounts of gas and concentrations of five constituents (H_2, O_2, N_2, CH_4, CO_2) in the produced gas. The produced gas was collected in the cylinder and the volume measured.

Rhodopseudomonas palustris R-1 was cultured for two days using the medium for growth and then added to the reactor several times after separation from the medium by centrifugation (about 10000 rpm, 15 min) during the experiment (Figure 2).

The raw waste water contained 800 mg/L of each sodium acetate, sodium propionate, and sodium butyrate, 400 mg/L of ethanol, and other small amounts of nutrients (Table 2).

As we describe later, whole constituents were used at the beginning of the experiment, but we eliminated the nutrients (below the $NaHCO_3$ in Table 2) in the early stage.

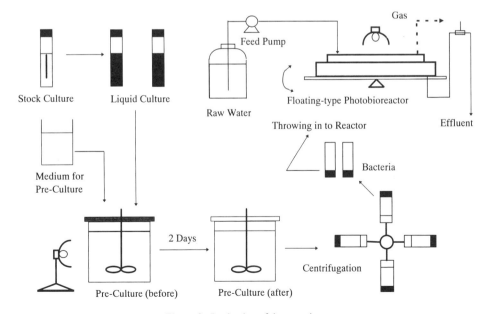

Figure 2. Production of the experiment.

Table 2. Composition of artificial anaerobic effluent

Sodium acetate	0.8 g
Sodium propionate	0.8 g
Sodium butyrate	0.8 g
Ethanol	0.4 g
NaHCO$_3$	2.0 g
NH$_4$Cl	0.04 g
K$_2$HPO$_4$	0.15 g
FeSO$_4$·7H$_2$O	0.03 g
MgCl$_2$·6H$_2$O	0.82 g
CaCl$_2$·2H$_2$O	0.1 g
Biotin	0.01 mg
Thiamine-HCl	0.2 mg
Yeast extract	0.1 g
Distilled water	1 L
pH	5~6
Sodium glutamate (temporary)	0.2 g

One large floating-type bioreactor was constructed and tested on the sea surface near the Research Institute during the summer. The reactor had a volume of 800 L (2.0 m wide ∞ 4.0 m long ∞ 0.1 m deep), with light-penetrating acrylic plate, buoys for floatation, and a foothold for maintenance. We checked the motion (pitch, roll) of the reactor and measured the temperature of sea water, atmosphere, and liquid in the reactor.

4. RESULTS AND DISCUSSION

We obtained 85 L of gas, which contained about 85% H$_2$, during 1583 h (66 days) (Figure 3). The gas production rate was 100–300 mL/L·day during the early stage and 300–500 mL/L·day in the latter stage. At the beginning of this reactor operation, we generated only small amounts of gas; moreover, CH$_4$ was dominant. However, as the testing progressed, efficiency of light conversion to H$_2$ reached 0.308% for the long-term (66 days)

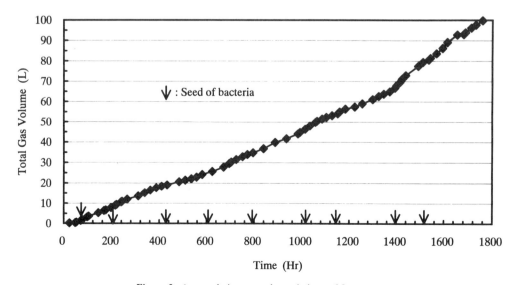

Figure 3. Accumulative gas volume, indoors, 6-L reactor.

Table 3. Efficiency of light energy conversion to H_2. Continuous light irradiation and continuous raw water feed, indoors, 66 days, 6-L reactor

Irradiation energy of lamp	434.2 W/m^2 (measured) = 591386 kcal/m^2
Irradiation area	0.107 m^2
Volume of gas production	84.87 L/66 days (35 °C, 1 atm) = 75.23 L/66 days (0 °C, 1 atm)
H_2 content	85%
Volume of H_2 production	63.94 L H_2/66 days
H_2 production per irradiation area	597.7 L H_2/m^2 (66 days)
Combustion energy of H_2	68.3 kcal/mol
Combustion energy of produced H_2	1822 kcal/m^2 (66 days)
Efficiency	0.308% (66 days)
	0.551% (in the latter stage)

and 0.551% in the latter stage (8 days) (Table 3). The efficiency was expressed as the ratio of the combustion energy of H_2 in the produced gas to the light energy illuminated.

The H_2 content of the gas was nearly 85%, and CH_4 decreased to 1%. N_2 was less than 6% and O_2 was not detected. CH_4 is a problem that we cannot avoid in our non-disinfecting system, but we have produced more H_2 and very little CH_4.

During the experiment, the pH and ORP of the liquid were monitored, and showed their proper values, pH 6.6~7.8 and ORP –400~600 mV, respectively. The OD (660 nm) and the SS of the effluent were 0.1~3.0 and 100~400 mg/L, respectively. The mean removal efficiency of organic acids was 46.2% (for acetate), 85.6% (for propionate), and 58.7% (for butyrate). It showed that the propionate decomposed easily.

The reason we have succeeded in producing much H_2 is the changing of constituents of the raw waste water. We eliminated the nutrients when making the raw waste water, so it contained only four constituents (acetate, propionate, butyrate, and ethanol). The nutrients, P, K, N, Fe, Ca, vitamin, and yeast-extract caused bacteria to grow instead of producing H_2. The new raw waste water should compel bacteria to produce H_2 because there is no nitrogen source and excess reduced energy in the cell accumulated.

Furthermore, we confirmed the cooling effect of surrounded water for the large reactor (800 L) on the sea surface in summer. The temperatures of liquid in the reactor and sea water were 28 and 25°C, respectively, while the atmospheric temperature was 36 °C. The bacteria becomes extinct when the liquid temperature continues to be over 40 °C, and it often occurs if no cooling system is attached on hot days. Inasmuch as a conventional cooling method, like a cooling tower, wastes much energy, we should use an economical one, and recommend using natural water (sea water or lake water). The sea water has another advantage because the liquid in the reactor is mixed well by wave motion.

Some problems still remain unsolved, for example, what condition will stimulate more nitrogenase activity? We have much work to do to achieve stable H_2 production, and we have not yet produced H_2 under natural conditions. We also need engineering data, such as suitable hydraulic retention time and light energy intensity.

ACKNOWLEDGMENTS

This work was performed under the management of the Research Institute of Innovative Technology for the Earth as a part of the Development of Environmentally Friendly Technology for the Production of Hydrogen Project, supported by the New Energy and Industrial Technology Development Organization.

REFERENCES

Kim, J.S, Ito, K, and Takahashi, H., 1982, Production of molecular hydrogen on outdoor batch cultures of *Rhodopseudomanas sphaeroides*, *Agric. Biol. Chem.*, 46:937–941.

Sasikala, K., Ramana, C.V., Rao, P.R., and Kovacs, K.L., 1993, Anoxygenic phototrophic bacteria: physiology and advances in hydrogen production technology, *Adv. Appl. Microb.*, 38:211–298.

46

PHOTOHYDROGEN PRODUCTION USING PHOTOSYNTHETIC BACTERIUM *Rhodobacter sphaeroides* RV

Simulation of the Light Cycle of Natural Sunlight Using an Artificial Source

Tatsuki Wakayama,[1] Akio Toriyama,[2] Tadaaki Kawasugi,[2] Takaaki Arai,[3] Yasuo Asada,[4] and Jun Miyake[5]

[1]Research Institute of Innovative Technology for the Earth (RITE)
1-1 Higashi, Tsukuba, Ibaraki, 305-0046, Japan
[2]Kubota Corp.
5-6 Koyodai, Ryugasaki, Ibaraki, 301-0852, Japan
[3]Department of Industrial Chemistry
College of Industrial Technology, Nihon University
Narashino, Chiba, 275-0006, Japan
[4]National Institute of Bioscience & Human-Technology (NIBH)
[5]National Institute for Advanced Interdisciplinary Research (NAIR)
Agency of Industrial Science and Technology
Ministry of International Trade and Industry
1-1 Higashi, Tsukuba, Ibaraki, 305-0046, Japan

Key Words

Rhodobacter sphaeroides, hydrogen production, photosynthetic bacteria, simulation of sunlight light cycle

1. SUMMARY

Production of hydrogen by photosynthetic bacteria under sunlight was studied. In August 1996, we conducted experiments on hydrogen production by natural sunlight. The hydrogen levels produced by the photosynthetic bacteria depended on the irradiation intensity of the sunlight; the maximum production rate was 4.0 L/m^2/h and the conversion efficiency was relatively high, exceeding 2%. A laboratory hydrogen production experiment was also conducted using artificial light and the results were compared with those of the outdoor experiment.

2. INTRODUCTION

Hydrogen production using photosynthetic bacteria is a promising method of obtaining energy from sunlight (Weaver et al., 1979), and is being investigated at many research laboratories and the Research Institute of Innovative Technology for the Earth (RITE) using waste water as the substrate and sunlight as the energy source (Ikemoto et al., 1984; Kim et al., 1982, 1987; Tassinnto et al., 1990).

One challenge in developing this energy conversion method is that the profile of the sunlight varies from season to season, including the irradiation angle and intensity, and that the frequency spectrum changes throughout a single day. Moreover, there is a daily light-dark cycle. Therefore, to obtain the technical data needed for designing related equipment, such as outdoor reactors, it is vital to collect data on the strength of sunlight and on the change of angle. It is also essential to determine the relationship between incident solar energy and hydrogen photoproduction.

Despite this need, hydrogen production studies with photosynthetic bacteria and using natural sunlight are very rare in comparison to laboratory experiments using artificial sunlight. In this work, therefore, we measured hydrogen production by photosynthetic bacteria under various natural sunlight conditions, and compared these results with laboratory measurements of hydrogen production using artificial sunlight.

3. EXPERIMENTAL

For these studies, we used a *Rhodobacter sphaeroides* RV strain, a photosynthetic bacterium that exhibits high, stable production of hydrogen (Miyake et al., 1984).

The natural sunlight experiments were conducted on the rooftop of the National Institute of Bioscience and Human-Technology in Tsukuba, Ibaraki Prefecture, Japan. Three sets of three-day experiments were completed in August 1996 (Figure 1). We measured the illuminance (Lux), energy intensity ($W/m^2/s$), photons ($\mu E/mol$), outdoor temperature (°C), and temperature in the reactor (°C). We studied two different light-receiving systems to assess which system had the highest sunlight utilization efficiency: (1) a reactor covered with aluminum foil on the bottom and sides and (2) an uncovered reactor.

We placed the reactors in an aquarium facing south at an angle of 30°. The aquarium was filled with water and acted as an incubator. For the reactors, we used Roux flasks (Eagle, Ltd.), whose light path length was 4.5 cm, volume 700 mL, and irradiation area 150 cm^2. The depth of water in the incubator was 4 cm.

The hydrogen was collected in an upside-down graduated cylinder filled with water. We determined the hydrogen yield by reading the volume on the graduated cylinder and then calculating the production per unit area (L/m^2) and production rate ($L/m^2/h$).

The efficiency of the conversion of light energy to hydrogen was calculated as:

$$\text{Light conversion efficiency (\%)} = 100 \times \frac{\text{hydrogen combustion energy}}{\text{absorbed light energy}}$$

For the laboratory experiments, we used an M-26 halogen bulb (Kondo-Sylvania, Ltd.) as the light source, which is the standard bulb used for hydrogen photoproduction experiments in the RITE Hydrogen Project. When the M-26 halogen bulb is used with a

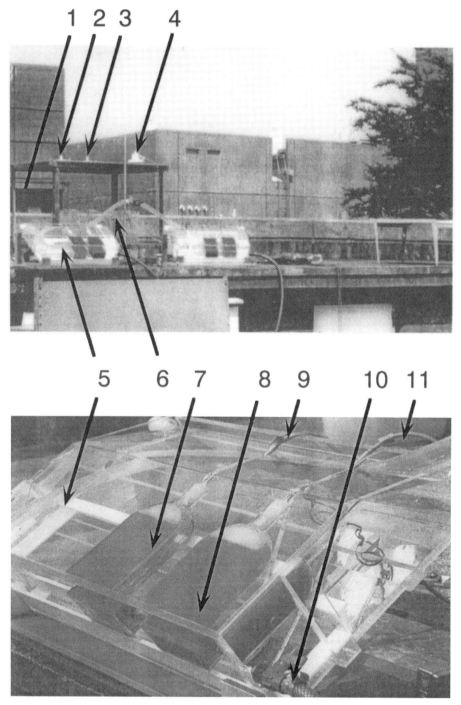

Figure 1. Hydrogen production in outdoor batch cultures for simulation of the light cycle of sunlight using the photosynthetic bacterium *Rhodobacter sphaeroides* RV: (1) rooftop of the NIBH building, (2) illuminance sensor (Lux), (3) photon sensor ($\mu E/m^2/s$), (4) energy sensor (W/m^2), (5) incubator (faced south, tilted 30°), (6) H_2 collector, (7) reactor 2 (covered), (8) reactor 1 (uncovered), (9 and 11) cooling system (30 °C), and (10) reactor temperature sensor (°C).

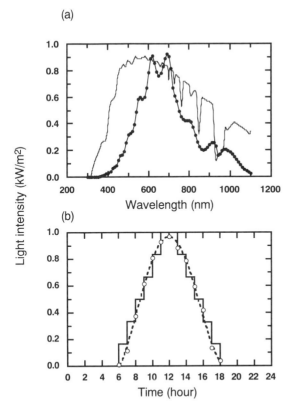

Figure 2. Characterization of light sources: (a) spectrum of sunlight (ERDA AM1.5) and M-26, (b) daily change of light intensity. (a) l = M-26, — = ERDA AM 1.5, (b)·O·· = sunlight, — = M-26.

wavelength selective dichroic mirror, the infrared spectrum can be eliminated, thus generating a spectrum similar to sunlight (Figure 2a) (ERDA and NASA, 1977).

According to the daily variation in sunlight intensity observed on clear sunny days, we determined that the maximum irradiation was 1 kW/m^2, integral daily irradiation energy was 7 kWh/m^2 and the irradiation period was 12 h.

To reproduce the observed daily variations in sunlight intensity, over a period of 6 h we increased the irradiation intensity of artificial sunlight in six discrete steps to a maximum of 1.0 kW/m^2 and then over a subsequent period of 6 h we decreased it in six discrete steps (Figure 2b).

The *Rba. sphaerides* RV strain was subcultured in aSy broth, which is a medium recommended by Miyake et al. (Mao, et al.,1986). Bacteria were precultured by the prescribed protocol and then anaerobically cultured in a gL medium with 50 mM sodium lactate as the carbon source and 10 mM sodium glutamate as the nitrogen source (30 °C, 15000 Lux). These bacteria were stored in darkness for 12 h prior to use in hydrogen photoproduction experiments.

4. RESULTS

The outdoor experiments were conducted on August 15, 1996. The sky was partly cloudy and a high hydrogen production rate was recorded. The light intensity and total irradiation energy during the experimental period fluctuated by up to 60%, depending on the degree of reflection and absorption by clouds (Figure 3a). Although the total irradiation energy for each day differed, the daily total irradiation energy ranged from 6 to 7 kWh/m^2.

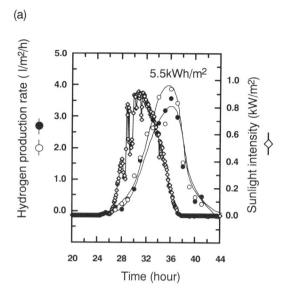

Figure 3. Outdoor batch culture and hydrogen production: (a) time course of hydrogen production, (b) total volume of hydrogen and light conversion efficiency. ○ = uncovered reactor, ● = covered reactor, ◇ = sunlight intensity.

(b) Total volume of H_2 and light conversion efficiency.

	Volume (l/m^2)	Efficiency (%)
Uncovered	40.0	2.2
Covered	25.4	1.4

The hydrogen production rate and total volume produced were high for the reactors facing south. Hydrogen production was observed 4 h after sunrise in the uncovered reactor. The rate increased with increasing irradiation intensity, and the maximum production rate of 4.0 L/m²/h was reached 4 h after the irradiation peak. The total daily production volume was about 40 L/m², and the photoconversion efficiency was 2.2% (Figure 2b).

The hydrogen production rate varied with light intensity, with a time lag of 4 h. The total hydrogen production was about 60% lower in the reactor whose bottom and sides were covered. This indicates that the distribution of light in the reactor influences bacterial activity. Because all faces of the uncovered reactor were open, the reactor received moderate energy during the day, which enhanced the total hydrogen production. When the irradiation exceeded a threshold level, the hydrogen production rate by the photosynthetic bacteria reached a saturation level.

For the laboratory experiments using artificial sunlight, the variation in sunlight irradiation intensity was simulated using the insolation data obtained from the outdoor experiments, with the intensity being varied in six discrete steps over a period of 6 h (Figure 4).

Hydrogen production was observed 2 h after the radiation onset and reached a maximum of 4.5 L/m²/h 4 h after the radiation peak, just as in the outdoor experiments (Figure 4a). Throughout the three-day experiment, daily changes in hydrogen production showed similar patterns as those in the outdoor experiments. Total hydrogen production was about 25 L/m²/day, similar to the yield from the outdoor experiment. Photoconversion efficiency was 1.1%, indicating the artificial light source sufficiently reproduced the natural light source, despite differences in the light spectrum. Because the daily maximum irradiation level of 1.0 kW/m² was shown to be too strong for photosynthetic bacteria, the low con-

(a)

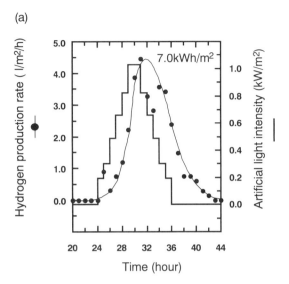

(b)
Total volume of H$_2$ and light conversion efficiency.

	Volume (l/m²)	Efficiency (%)
6 steps	25.3	1.1

Figure 4. Simulation of sunlight using artificial light for hydrogen production by *Rba. sphaeroides* RV: (a) time course of hydrogen production, (b) total volume of hydrogen and light conversion efficiency.

version rate, in comparison to the outdoor experiment, was probably caused by the light intensity reaching a saturation level (data not shown).

The outdoor results indicate that photosynthetic bacteria achieve relatively high production rates of hydrogen with natural sunlight. We achieved acceptable hydrogen production rates with a system that had a fixed reactor facing south and that required no power supply or solar-tracking system.

Our measurements with artificial sunlight suggest the possibility of simulating sunlight with artificial light. However, continued effort is required to develop a simpler device for producing artificial sunlight with the proper light spectrum. We are also currently improving the system to better simulate oblique irradiation, which imitates the morning and evening sunlight angles.

ACKNOWLEDGMENTS

This work was done under the management of RITE as a part of the Research and Development Project on Environmentally Friendly Technology for the Production of Hydrogen, supported by NEDO.

REFERENCES

ERDA and NASA, 1977, Terrestrial photovoltaic measurement procedures, NASA TM 73702.
Ikemoto, H., and Mitsui, A., 1984, Continuous hydrogen photoproduction from sulfide by an immobilized marine photosynthetic bacterium,*Chromatium* sp. MIAMI PBS 1071, *Advances in Photosynthesis Research*, pp. 789–792.
Kim, J.S., Ito, K., and Takahashi, H., 1982, Production of molecular hydrogen in outdoor batch culture of *Rhodopseudomonas sphaeroides*, *Agric. Biol. Chem.*, 46(4):937–941.

Kim, J.S., Ito, K., Izaki, K., and Takahashi, H., 1987, Production of molecular hydrogen by a semi-continuos outdoor culture of *Rhodopseudomonas sphaeroides*, *Agric. Biol. Chem.*, 51(4):1173–1174.

Mao, X., Miyake, J., and Kawamura, S., 1986, Screening photosynthetic bacteria for hydrogen production from organic acid, *Journal of Fermentation Technology*, 64(3):245–249.

Miyake, J., Mao, X., and Kawamura, S., 1984, Photoproduction of hydrogen from glucose by a co-culture of a photosynthetic bacteria and *Clostridium butyricum*, *Journal of Fermentation Technology*, 62(6):531–535.

Tassinnto, G., Capolino, E., Sili, C., and Vincenzini, M., 1990, Outdoor hydrogen photoevolution by immobilized cells of *Rhodopseudomonau palstris*, in *Nitrogen Fixation*, Kluwer Academic Publishers, pp. 591–592.

Weaver, P.F., Lien, S., and Seibert, M., 1979, Photobiological production of hydrogen, *Solar Energy*, 24:3–45.

47

BIOREACTORS FOR HYDROGEN PRODUCTION

Sergei A. Markov

Biological Faculty
M.V. Lomonosov Moscow State University
Moscow, Russia
National Renewable Energy Laboratory
1617 Cole Boulevard
Golden, Colorado 80401

Key Words

bioreactors, molecular hydrogen, cyanobacteria, purple bacteria

1. SUMMARY

This mini-review surveys data on bioreactors for H_2 production, with particular emphasis on the author's own data for the last eight years. Two groups of bioreactors can be distinguished based on the nature of H_2-evolving reactions: photobioreactors with H_2 photoproduction (cyanobacteria and purple bacteria) and "dark" bioreactors based on dark anaerobic H_2 production (phototrophic purple bacteria and chemotrophic anaerobic bacteria). Three types of cyanobacterial photobioreactors are described: glasstube photobioreactors, membrane photobioreactors, and spiral tubular photobioreactors. Particular attention in the review is given to microbial "shift reaction" bioreactors, which have potential for near-term practical applications.

2. INTRODUCTION

Hydrogen is an environmentally desirable fuel for the next century because burning of H_2 produces water, H_2 is a renewable form of energy, and it can be converted effectively to electricity. Certain organisms evolve H_2 either from reduced organic materials (i.e., biomass) or from water in reactions catalyzed by the nitrogenase or hydrogenase enzymes. Biological H_2 production has a number of advantages, and it could be considered for industrial applications if suitable low-cost bioreactors can be developed. This mini-review concentrates on the research and development of bioreactors for H_2 production with particular emphasis on the author's own data for the last eight years. Only a few reports are available on bioreactors for H_2 production. Some have been described in recent reviews (Benemann, 1994; Markov et al., 1995; Sasikala et al., 1993). More studies are available on the use of bioreactors in general (Asenjo and Merchuk, 1995; Pulz, 1994).

Bioreactors for H_2 productions can be divided into two separate groups based on the nature of H_2-evolving reactions:

- photobioreactors based on H_2 photoproduction,
- photobioreactors incorporating cyanobacteria, and
- photobioreactors incorporating purple bacteria, or
- "dark" bioreactors based on dark anaerobic H_2 production,
- "shift-reaction" bioreactors incorporating purple bacteria, and
- anaerobic fermentation bioreactors incorporating chemotrophic bacteria.

3. PHOTOBIOREACTORS

In cyanobacteria and purple bacteria, H_2 production is coupled to photosynthesis. Thus, it is possible to design photobioreactors in which solar energy can be used for the production of H_2.

3.1. Photobioreactors Incorporating Cyanobacteria

Compared to other photosynthetic H_2-producing organisms (purple bacteria, green algae, and higher plants), cyanobacteria are most attractive currently for H_2 photobioreactors because they use water as the electron source and solar energy as an energy source under aerobic conditions. In addition, they are fast-growing (doubling-times on the order of 20 h and less), with a flexible metabolism.

Three types of cyanobacterial photobioreactors are described here:

- glasstube photobioreactors,
- membrane photobioreactors, and
- spiral tubular photobioreactors.

3.1.1. Glasstube Photobioreactors. At M.V. Lomonosov Moscow State University, glasstube photobioreactors, which were originally constructed for biomass photoproduction by the Institute Biotehnika in Moscow (up to 10000 L volume), were used for H_2 production by the cyanobacterium *Spirulina* (Markov and Mecharikova, unpublished). The photobioreactor has a light-receiving glasstube part, mounted on vertical or horizontal bars. The photobioreactor also includes a pump for circulating the cyanobacterial suspension and a heat exchanger and gas exchanger for saturating the culture suspension with CO_2. The light-receiving glasstube part is equipped with a device for automatic cleaning of the inner surface of the glass pipes. The instrumentation includes automatic temperature and pH control systems.

Cyanobacterial H_2 production is a two-phase process. During the first light phase (20 W/m^2), cells take up nutrients and CO_2 and synthesize cell biomass. Biomass is harvested (collected) in a separator. In the second phase, cell biomass is used for dark H_2 production catalyzed by hydrogenase at rates of 1–2 mL/g cdw/h.

First phase (photosynthetic CO_2 uptake)

$$H_2O \longrightarrow PSII \longrightarrow PSI \longrightarrow [CHO]$$

with O_2 released from H_2O/PSII step and CO_2 taken up into [CHO] photosynthate.

Second phase (H_2 production)

$$2H^+ + 2e^- \rightarrow H_2$$

Hydrogen formation is measured amperometrically with a methyl viologen-dithionite system. This two-phase process resembles the two-step H_2 production system for marine cyanobacteria described by Mitsui et al. (1983). This cyanobacterial photobioreactor system is relatively simple; no fundamental obstacles can be seen and such a system is ready for commercialization if suitable cyanobacterial strains can be found. Application of heterocystous cyanobacteria for this type of photobioreactor, where H_2 production is catalyzed by a light-dependent nitrogenase system, will also require light in the second phase.

3.1.2 Membrane Photobioreactors. A membrane photobioreactor is a bioreactor within which membranes are used to separate cells from the H_2 stream. The application of membrane bioreactors could also permit the separation of gases such as H_2, O_2, and CO_2. Membranes can be obtained in hollow-fiber and flat or spiral sheet forms. Membranes are made from a variety of available materials, including cellulose acetate and nitrate, polyvinylidene difluoride, polysulfone, polypropylene, polytetrafluoroethylene, and polyacrylonitrile (Salmon and Robertson, 1995).

A novel hollow-fiber photobioreactor has been designed for continuous H_2 production using hollow-fiber immobilized cyanobacteria (Markov et al., 1992; Markov et al., 1996). Hydrophilic cuprammonium rayon hollow-fibers, 200 μm (Asahi Medical Co.Ltd, Tokyo, Japan), were cut into 110 mm lengths and sealed together at one end. Hollow fibers were packed into a water-jacketed glass column (length, 110 mm; volume, 15 mL). The cyanobacterial cells became immobilized on the outer surface of the fibers. The photobioreactor was designed so that the cyanobacterial growth medium passed from the outside of the fibers to the inner lumen space.

The hydrogen produced dissolved in the medium, went through the membrane to the lumen of the fibers, and from there was directed to a gas trap outside the column. The hydrogen in the gas phase (mixed with CO_2 and O_2) was measured in a gas chromatograph equipped with a thermal conductivity detector. The photobioreactor was illuminated continuously with white fluorescent lamps (5 W/m^2) and maintained at 25 °C.

A two-phase photobioreactor system for H_2 photoproduction by cyanobacteria has been selected for the practical demonstration (Markov et al., 1995)

First phase (photosynthetic CO_2 uptake)

Phase criteria: air conditions, CO_2 uptake, O_2 evolution, no H_2 production
Second phase (H_2 photoproduction)

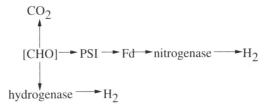

Phase criteria: partial vacuum, H_2 evolution, CO_2 evolution, no O_2 evolution

In the first phase, the cells take up CO_2 and N_2 from the air and synthesize the products which subsequently will be used for H_2 production. In the second phase, H_2 production occurs under partial vacuum with cells producing H_2 using the photosynthetic products.

It was possible to design such a two-stage photobioreactor system in which H_2 photoproduction is coupled with the uptake of CO_2 (Markov and Hall, 1994; Markov et al., 1995). The entire photobioreactor system was evacuated (250–300 torr) for maximum H_2 photoproduction. It was able to produce H_2 for 3–4 days, but then the rate gradually decreased to a lower steady state value. The decrease under partial vacuum was apparently due to the lack of CO_2 in the medium. Addition of CO_2 (air) is necessary to restore H_2 production.

The two-phase H_2 production system can be maintained continuously, but it is necessary to add CO_2 or air to the system every 7–10 days. The photobioreactor was run continuously for one year. Light energy conversion efficiencies (to H_2) in the hollow-fiber photobioreactor were calculated as follows:

Efficiency (%) = (H_2 production rate × H_2 energy content)/(incident light irradiance) × 100

The efficiency value obtained was 8% (Markov and Hall, 1994). This efficiency was obtained under optimal conditions: non-saturating light, partial vacuum, and 5% CO_2 in the gas phase, conditions which are difficult to maintain outdoors.

In conclusion, our studies of a hollow-fiber photobioreactor for H_2 production have demonstrated that there is potential for the use of cyanobacteria in photobioreactors for H_2 production. We found that cyanobacteria can evolve H_2 in a hollow-fiber photobioreactor at rates much higher (up to 20 mL/g cdw/h) than any reported so far for other cyanobacteria under other conditions. The cyanobacterial H_2 production photobioreactor can be operated continuously over the long term (more than one year) with good efficiencies (up to 8%).

3.1.3. Spiral PVC Tubular Photobioreactors. With the generation of *A. variabilis* mutants PK 84 and PK 17R (Mikheeva et al., 1995), without uptake hydrogenase activity, it was possible to build a one-stage photobioreactor where H_2 was produced at the same time as photosynthesis (Markov et al., 1997). The photobioreactor was 2 L (total volume) and 0.4 m high for *A. variabilis* PK84. The photobioreactor consisted of a transparent PVC (Nalgene, United States) tube, 7.9-mm internal diameter, which was wound helically on a vertical, transparent, cylindrical supporting structure. The bioreactor suspension was bubbled with a mixture of CO_2 (about 5%) and air through a needle/septum connection at the base of the photobioreactor to supply the cells with a carbon source and remove H_2. Hydrogen production rates up to 19 mL/m²/h were observed under light intensities of 3 W/m². The important new result was that H_2 could be produced from water under ambient conditions. Previous work employed an argon-sparging system (90% argon with CO_2 and N_2) or application of a partial vacuum to produce H_2 from a tubular (glass) photobioreactor or a hollow-fiber photobioreactor, inherently more complex systems.

3.2. Photobioreactors Incorporating Photosynthetic Bacteria

Most of the work in the photobioreactor field has been made with photobioreactors incorporating photosynthetic purple bacteria (Sasikala et al., 1993). These bacteria are able to photoproduce H_2 from simple organic compounds such as glucose or organic acids under anaerobic conditions, but H_2 photoproduction by photosynthetic bacteria has some commercial promise only if it uses organic wastes as a substrate. Different types of bioreactors were suggested for this type of H_2-producing activity: spiral tubular photobioreactor (Delachapelle et al., 1991), rectangular glass chamber with a porous glass sheet for bacteria immobilization (Tsygankov et al., 1994), floating plastic bag (Hydrogen Production 1995 Interim Evaluation Report, NEDO, Japan), parallel plate-type photobioreactor (Akano et al, 1996), cylindrical (Aric et al., 1996), plate loop-type photobioreactor (Tramm-Werner et al., 1996), stirred tank (D'Addario et al., 1996), and plate-type photobioreactor (Morimoto et al., 1996).

Although most work has been done with photobioreactors incorporating purple photosynthetic bacteria, so far these photobioreactors suffer from inefficiencies and difficulties in using them for waste treatment.

4. "DARK" BIOREACTORS

4.1. "Shift Reaction" Bioreactors

"Shift reaction" bioreactors are based on a unique type of H_2-producing activity originally found in a strain of photosynthetic bacteria by Uffen (1976); until now, this has been the most promising type of bioreactor for near-term commercial applications (Benemann, 1996). Fermentative cultures of Uffen's strain in complex media carry out a water-gas shift reaction in darkness to produce H_2 according to the equation:

$$CO + H_2O \rightarrow CO_2 + H_2$$

The great advantage of the microbial shift reaction is that it operates at room temperature and in one pass, in contrast to chemical catalysts that require high temperature or radiation and multiple stages. Numerous strains of photosynthetic bacteria, including *Rhodobacter* sp. CBS, have been isolated that utilize CO in light as well as in darkness and do not require complex organic substrates (Maness and Weaver, 1994). Additionally, these strains quantitatively shift the CO component of synthesis gas (e.g., from thermally gasified biomass) into H_2. However, mass transport of gaseous CO into an aqueous bacterial suspension is the rate-limiting step in the process and is the main challenge for bioreactor design. Two types of shift-reaction bioreactors are discussed here: membrane bioreactors and spiral tubular bioreactors.

4.1.1. Membrane Bioreactors. A simple method using hollow-fiber membrane technology to enhance mass transfer of CO has proven effective (Markov et al., 1996). A 250 mm × 33 mm AM-4OM-SD cartridge made of hydrophilic cuprammonium rayon hollow fibers (Asahi Medical CO. Ltd., Japan) was used as the bioreactor column. The total surface area of the 180 μm diameter hollow fibers was 0.8 m^2, and the column volume outside the fibers was 48 mL. The bioreactor was designed such that both a basal growth medium and CO (10% in N_2) passed from the inside of the fibers to the outside within the column. A photosynthetic inoculum readily adsorbed to the outer surface of the hollow fibers where bacteria were producing H_2, and the column was maintained at 26 °C in darkness. The hydrogen produced was measured using a gas chromatograph. Hydrogen production by *Rhodobacter* sp. CBS from CO at an average rate of 125 mL/g cdw/h (maximum rate of 700 mL/g cdw/h) was observed for more than nine months. This type of bioreactor can be scaled-up using industrial spiral membrane cartridges already available for filtration applications and produced by Amicon, Inc. (Beverly, Massachusetts).

4.1.2. Spiral PVC Tubular Bubble-Train Bioreactors. Spiral tubular bioreactors were also analyzed for this type of H_2-producing activity (Markov et al., 1997). A 0.5-L (total volume), 0.8-m high bioreactor for photosynthetic bacterial production of H_2 from CO (and H_2O) was built. The bioreactor used transparent PVC (Tygon, United States) 6.3-mm inner diameter tubing, wound helically on a vertical cylindrical supporting structure; a pump (Masterflex, Cole-Palmer Instrument Co., USA) for circulating the bacterial suspension (15 mL/min) and injecting CO; and a gas reservoir. The bioreactor was designed so that small bubbles containing CO were injected continuously through a needle/septum connection from the gas reservoir (20% CO). The bubbles rose (3.5 min transit time) with the pumped medium from the bottom of the bioreactor to the top. The high surface area of the bubble train promoted enhanced mass transport of gaseous CO into the aqueous bacterial suspension. In order to keep the pH from dropping due to CO_2 build-up, the gas phase of the bioreactor was degassed with N_2 once every day, and then 20% CO was reinjected into the system. The bioreactor was covered with a black cloth to prevent photosynthetic H_2 consumption from exposure to ambient light. Up to 140 mL/g cdw/h of H_2 production activity was calculated using this bioreactor.

4.2. Anaerobic Fermentation Bioreactors

Anaerobic bacteria can produce H_2 from organic substrates according to scheme:

$$[CHO] \rightarrow \text{Ferredoxin} \rightarrow \text{Hydrogenase} \rightarrow H_2$$
organic substrate

Hydrogen production by anaerobic fermentation has gained little attention compared to hydrogen evolution by photosynthetic microorganisms because the yield of such fermentation is relatively low, 1–2 mol H_2 per mol glucose. Hydrogen production by fermentation has, however, many advantages for industrial H_2 production. Among these advantages are high production rates (i.e., up to 400 mL H_2/g cdw/h for *Enterobacter aerogenes*) and high growth rate of bacteria (Tanisho, 1996). Additionally, it is not necessary to design novel bioreactors for this type of H_2 production; existing low-cost bioreactors, such as those used in ethanol fermentation, can be used for dark H_2 production by anaerobic fermentation (Tanisho, 1996).

5. CONCLUSION

In this mini-review, only some of the bioreactors for H_2 production were described and discussed to show general trends in bioreactor development and the main types of bioreactors (Table 1). As for photobioreactors, even from a very optimistic point of view,

Table 1. Hydrogen production by microorganisms in bioreactors

	Microorganism	Rate of H_2 production (mL/g cdw/h)	Reference
Photobioreactors	*Anabaena variabilis*	up to 20	Markov et al., 1994
	Rhodobacter sphaeroides	up to 62	D'Addario et al., 1996
"Dark" bioreactors	*Rhodobacter* sp. CBS	up to 700	Markov et al., 1996
	Enterobacter aerogenes	up to 400	Tanisho, 1996

enormous long-term research and development efforts are needed to introduce practical systems. Of course, it is still necessary to design different lab-scale photobioreactors to determine H_2 production properties and the potential of different photosynthetic microorganisms. In contrast "dark" bioreactors have near-term potential for commercialization. Hydrogen obtained in "shift reaction" bioreactors, containing up to 20% H_2, is sufficiently clean for direct injection into H_2 fuel cells.

ACKNOWLEDGMENTS

This review was made possible in part by a grant from the NREL-FIRST program. The author's research reported in this review has been supported by grants from the U.S. Department of Energy Hydrogen Program, RITE-NEDO (Japan), the British Council, King's College London, the International Science Foundation, and the Ministry of Science and Education of the Russian Federation. The research was carried out in part in the laboratories of Drs. D. O. Hall, M. Seibert, and P. F. Weaver.

REFERENCES

Akano, T., Miura, Y., Fukatsu, K., Miyasaka, H., Ikuta, Y., Matsumoto, H., Hamasaki, A., Shioji, N., Mizoguchi, T., Yagi, K., and Maeda, I., 1996, Hydrogen production by photosynthetic microorganisms, *Appl. Biochem. Biotechnol.*, 57/58:677–688.

Aric, T., Gunduz, U., Yucel, M., Turker, L., Sediroglu, V., and Eroglu, I., 1996, Photoproduction of hydrogen by *Rhodobacter sphaeroides* O.U. 001, in *Hydrogen Energy Progress XI*, Proceedings of the 11th World Hydrogen Energy Conference, Veziroglu, T.N. et al. (eds.), Stuttgart, Germany, June 23–28, 1996, pp. 2417–2426.

Arsenjo, J.A., and Merchuk, J.C. (eds.), 1995, *Bioreactor System Design*, Marcel Dekker, Inc., New York, 615 p.

Benemann, J., 1994, Feasibility analysis of photobiological hydrogen production, in *Hydrogen Energy Progress X*, Proceedings of the 10th World Hydrogen Energy Conference, Block, D.L., and Veziroglu, T.N. (eds.), Cocoa Beach, Florida, United States, June 20–24, 1994, pp. 931–940.

Benemann, J., 1996, Hydrogen biotechnology: progress and prospects, *Nature Biotechnology*, 14:1101–1103.

D'Addario, E., Fascetti, E., and Valdiserri, M., 1996, Hydrogen production from organic wastes by continuous culture of *Rhodobacter sphaeroides RV*, in *Hydrogen Energy Progress XI*, Proceedings of the 11th World Hydrogen Energy Conference, Veziroglu, T.N. et al. (eds.), Stuttgart, Germany, June 23–28, 1996, pp. 2577–2582.

Delachapelle, S., Renaud, M., and Vignais, P.M., 1991, Etude de la production d'hydrogene en bioreacteur par une bacterie photosynthetique *Rhodobacter capsulatus*. 1. Photobioreacteur et conditions optimales de production d'hydrogene, *Revue des sciences de l'eau*, 4:83–99.

Maness, P.C., and Weaver, P.F., 1994, Production of poly-3-hydroxyalkanoates from CO and H_2 by a novel photosynthetic bacterium, *Applied Biochemistry and Biotechnology*, 45/46:395–406.

Markov, S.A., Rao, K.K., and Hall, D.O., 1992, Hollow-fiber photobioreactors for continuous production of hydrogen by immobilized cyanobacteria under partial vacuum, in *Hydrogen Energy Progress IX*, Proceedings of the Ninth World hydrogen Energy Conference, Veziroglu, T.N. et al. (eds.), Paris, pp. 641–650.

Markov, S.A., and Hall, D.O., 1994, Photoproduction of hydrogen by cyanobacteria under partial vacuum in batch culture or in a photobioreactor, in *Hydrogen Energy Progress X*, Proceedings of the 10th World Hydrogen Energy Conference, Block, D.L., and Veziroglu, T.N. (eds.), Cocoa Beach, Florida, United States, June 20–24, 1994, pp. 941–949.

Markov, S.A., Bazin, M.J., and Hall, D.O., 1995, Potential of using cyanobacteria in photobioreactors for hydrogen production, in *Advances in Biochemical Engineering/Biotechnology*, vol. 52, Fiechter, A. (ed.), Springer-Verlag, Berlin, pp. 59–86.

Markov, S.A., Weaver, P., and Seibert, M., 1996, Hydrogen production using microorganisms in hollow-fiber bioreactors, in *Hydrogen Energy Progress XI*, Proceedings of the 11th World Hydrogen Energy Conference, Veziroglu, T. N. et al. (eds.), Stuttgart, Germany, June 23–28, 1996, pp. 2619–2624.

Markov, S.A., Weaver, P.F., and Seibert, M., 1997, Spiral tubular bioreactors for hydrogen production by photosynthetic microorganisms: design and operation, *Appl. Biochem. Biotechnol.*, 63–65:557–584.

Mikheeva, L.E., Schmitz, O., Shestakov, S.V., and Bothe, H., 1995, Mutants of the cyanobacterium *Anabaena variabilis* altered in hydrogenase activities, *Zeitschrift für Naturforschung*, 50:505–510.

Mitsui, A., Phlips, E.J., Kumazawa, S., Reddy, K.T., Ramachandran, S., Matsunaga, T., Haynes, L., and Ikemoto, H., 1983, Progress in research toward outdoor biological hydrogen production using solar energy, sea water, and marine photosynthetic microorganisms, *Ann. New York Acad. Sci.*, 413:514–530.

Morimoto, M., Kawasaki, S., El-Shishtawy, R.M.A., Ueno, Y., and Kunishi, N., 1996, Solar energy-induced and diffused photobioreactor, in *Hydrogen Energy Progress XI*, Proceedings of the 11th World Hydrogen Energy Conference, Veziroglu, T.N. et al. (eds.), Stuttgart, Germany, June 23–28, 1996, pp. 2761–2766.

Pulz, O., 1994, Laminar concept of closed photobioreactor designs for the production of microalgal biomass, *Russian J. Plant Physiology*, 41:256–261.

Salmon, P.M., and Robertson, C., 1995, Membrane reactors, in *Bioreactor System Design*, Arsenjo, J.A., and Merchuk, J.C., (eds.), Marcel Dekker, New York, pp. 305–338.

Tanisho, S., 1996, Feasibility study of biological hydrogen production from sugar cane by fermentation, in *Hydrogen Energy Progress XI*, Proceedings of the 11th World Hydrogen Energy Conference, Veziroglu, T.N. et al. (eds.), Stuttgart, Germany, June 23–28, 1996, pp. 2601–2606.

Tramm-Werner, S., Hackethal, M., Weng, M., and Hartmeier, W., 1996, Photobiological hydrogen production using a new plate loop reactor, in *Hydrogen Energy Progress XI*, Proceedings of the 11th World Hydrogen Energy Conference, Veziroglu, T.N. et al. (eds.), Stuttgart, GermanyJune 23–28, 1996, pp. 2407–2416.

Tsygankov, A., Hirata, Y., Miyake, M., Asada, Y., and Miyake, J., 1994, Photobioreactor with photosynthetic bacteria immobilized on porous glass for hydrogen production, *J. Ferment. Bioeng.*, 77:575–578.

Uffen, R.L., 1976, Anaerobic growth of a *Rhodopseudomonas* species in the dark with carbon monoxide as sole carbon and energy substrate, in *Proceedings of the National Academy of Sciences (USA)*, 73:3298–3302.

48

A TUBULAR INTEGRAL GAS EXCHANGE PHOTOBIOREACTOR FOR BIOLOGICAL HYDROGEN PRODUCTION

Preliminary Cost Analysis

Mario R. Tredici,[1] Graziella Chini Zittelli,[1] and John R. Benemann[2]

[1]Dipartimento di Scienze e Tecnologie Alimentari e Microbiologiche
dell' Universitaae di Firenze, Italy
[2]Department of Civil Engineering
University of California
Berkeley, California

1. SUMMARY

Very low-cost closed photobioreactors are required for photobiological production of H_2 from water and sunlight using microalgae as catalysts. Assuming a 10% solar energy conversion to H_2 fuel, such photobioreactors should cost at most about $50/m² for single-stage processes, or somewhat over $100/m² for two-stage systems. The basic requirements for such photobioreactors are:

- maximum light interception per unit area of reactor to minimize total costs;
- large unit size, over 100 m², to minimize capital and operating costs;
- low-cost, long-lasting, transparent materials to contain the culture and H_2;
- low-cost assembly, site preparation, and erection of the photobioreactor system;
- effective gas exchange to remove H_2 and O_2 or CO_2, as required;
- effective temperature control for the local climate and specific organisms used;
- minimum energy inputs for mixing, culture handling, and gas exchange;
- lack of fouling and/or ability to clean the photobioreactor, as required; and
- ease of operation, maintenance, and troubleshooting, for lowest possible costs.

Photobioreactors are of a few basic design concepts, such as tubular or flat plate with internal or external gas exchange, with myriad variations and elaborations described. However, no design has proved superior or commercially successful, even for high-value products. A tubular photobioreactor that can meet the above requirements has been demonstrated at the University of Florence. This near-horizontal tubular reactor (NHTR) has the following general attributes:

- tubes about 4 cm (range 3–5 cm) in diameter, about 40 m long (20–60 m);
- tubes placed on corrugated sheets to maintain straight lines and inclination;
- inclination of about 10% slope (6–12% likely range), with reactors set on earthworks;
- 40 (20–50) tubes connected by manifolds at the bottom and a degasser on top;
- internal gas exchange using compressed gas, with gas flow also providing mixing; and
- cooling with water sprays, with the cooling water collected and re-used.

In Florence, such NHTRs, ranging from 6–87 m, have been tested with Spirulina (*Arthrospira platensis*), with volumetric productivities of up to 1.5 g/L/day (for 4-cm diameter tubes). One important attribute of this reactor is little fouling, due to the scouring effect of the gas bubbles. Two photobioreactors, each with eight 20-m long tubes, were installed at the University of Hawaii in early 1997 under a collaborative project. A very preliminary cost analysis for a conceptual single-stage biophotolysis process is presented, based on many very favorable assumptions, such as 10% solar energy conversion efficiencies and a very low cost of capital. It suggests that such NHTRs might cost only $50/m^2, and projects H_2 costs of $15/GJ. This projection, however, does not consider many factors and a more detailed analysis is required.

2. INTRODUCTION

Solar biohydrogen production includes "biophotolysis," the production of H_2 (and O_2) from water and sunlight by microalgae (green algae and cyanobacteria), and "photofermentations," the conversion of organic substrates to H_2 and CO_2 by photosynthetic bacteria. Although solar biohydrogen production has been a subject of basic and applied research and development for some 25 years (Benemann, 1996), relatively little work has been carried out on process scale-up or cost analysis. One reason is that this technology is not yet very advanced in terms of light conversion efficiencies and process stabilities achieved, even in laboratory demonstrations, and much less so using sunlight in outdoor tests. The development of the microbial catalysts required for such processes must be the major focus of any such R&D effort in this field. However, biosolar H_2 production also requires very low-cost photobioreactors to maintain the microbes in a favorable environment while allowing collection of the H_2 produced.

Biophotolysis processes can be direct or indirect, that is, without or with CO_2 fixation as intermediate between the O_2 and H_2 evolution reactions. The indirect processes can be either single- or two-stage. For single-stage systems, the entire process would take place in closed photobioreactors, while in two-stage systems (actually multi-stage if dark fermentations and storage are included), large open ponds are followed by much smaller photobioreactors (Benemann, 1998).

At present, photosynthetic solar energy conversion into microalgae biomass is maximally 2–3%, and H_2 production efficiencies are lower still. A 10% efficiency goal for converting sunlight into H_2 fuel is generally assumed feasible in the development of microalgal biophotolysis catalysts (Bolton, 1996). One approach to this goal is to reduce the amount of light-gathering chlorophylls and other pigments in these photosynthetic microbes (Melis et al., 1998).

In addition to high efficiencies, the overall process must be of very low cost. Few economic analyses of biohydrogen production have been carried out (Benemann, 1994, 1998), and even fewer have considered the engineering and economic constraints of the photobioreactors required for such a process. Here we very briefly review alternative photobioreactor designs and the economics of biophotolysis, and then present a preliminary cost analysis for an NHTR described by Tredici et al. (1993). The objective is not so much to arrive at an actual current cost for such photobioreactors, but to understand how far we are from reaching the goal of economic H_2 production through biophotolysis.

3. PRIOR COST ANALYSIS OF BIOPHOTOLYSIS PROCESSES

Even if a highly efficient and stable algal biophotolysis catalyst is developed, the value of H_2 produced per square meter of area in very favorable climatic zones would only be about $10 per year or less, depending on the cost of competitive energy sources (Benemann, 1994). This sets a limit to allowable capital investment and operating costs. A preliminary cost analysis of a large-scale (3,600 GJ H^2/day), indirect, two-stage biophotolysis process was recently published (Benemann, 1998). In that concept, open ponds produce microalgae biomass with high contents of fermentable carbohydrates (starch, glycogen), followed by a dark fermenter and then a closed photobioreactor. The analysis was based on earlier cost estimates for large-scale open pond systems for algal biomass production (Benemann, 1993, and references therein), to which were added available cost estimates for H_2 handling (clean-up, compression, storage). The photobioreactor was assumed, without any further analysis or design specification, to cost $100/m^2 (plus 40% for engineering, contingencies, and other indirect costs). Total system capital costs were projected at about $40 million, including $10 million for 140 ha of open ponds and $20 million for 14 ha of photobioreactors, with the remainder for H_2 handling (clean-up, compression, storage) and ancillary systems. At a 25% annual capital charge (depreciation, insurance, maintenance, taxes, and cost of capital), capital costs were almost 90% of total costs, estimated at $10/GJ H_2.

It must be emphasized that the analysis was based on the assumption that highly efficient and stable microalgae catalysts would be developed and deployed at a very favorable site. Although highly preliminary, incomplete, and optimistic and based on prior studies and a simple assumption of photobioreactor costs, this cost estimate suggested that such two-stage photobiological H_2 production processes may have sufficient promise to justify further R&D and process analysis. By contrast, a single-stage process, in which the entire (140 ha, again assuming a 10% efficiency) area was covered with closed photobioreactors, would cost over fivefold as much (e.g., some $50/GJ), based on similar assumptions. In brief, single-stage processes, (e.g., direct biophotolysis, heterocystous cyanobacteria, and others) could only be considered if total photobioreactor costs were at most about $50/m^2.

The present analysis was undertaken to determine if, at least in principle, the NHTR with internal gas exchange developed in Florence could be of sufficiently low cost to allow its consideration for single-stage biophotolysis processes, and which would be less complex than two-stage processes and might be applicable to smaller, even roof-top, scale systems.

4. CLOSED PHOTOBIOREACTORS FOR BIOHYDROGEN PRODUCTION

Closed photobioreactors are algal culture systems that prevent direct exchange of gases (CO_2, O_2, H_2O, and even H_2) or contaminants (dust, microorganisms) between the culture and atmosphere. Many closed photobioreactor designs have been studied for some fifty years, as these are thought to have advantages over open ponds:

- They result in higher cell densities, reducing handling and harvesting costs.
- They allow better control over conditions (pH, pCO_2, pO_2, temperature, etc.).
- They prevent, or at least minimize, contamination by other algae, grazers, and so forth.

However, productivities have generally been only modestly higher than for open ponds. The reduced media handling, harvesting costs, and CO_2 use efficiencies are also not decisive advantages. It is their ability to cultivate algal strains that cannot be easily grown in open ponds that is their main advantage (Weissman et al., 1988).

Closed photobioreactors can be made of tubes of plastic or glass arranged in various ways—horizontal, serpentine, helical, and other permutations—and operated in series or parallel as single-phase or bubble columns (Chaumont, 1993; Tredici and Chin Zittelli, 1997; Tredici and Chini Zittelli, 1998). Mixing can be by air-lift or pumps. Flat plate systems can be shallow (<10 cm) tanks or made of commercially available "alveolar" reactors, two plastic sheets separated by cross-sections creating channels These, too, can be arranged in various configurations with internal or external gas exchange (Tredici et al., 1991; Tredici and Chini Zittelli, 1997; Pulz and Scheibenbogen, 1998). Internal gas exchange (bubble reactor) is preferable to external gas exchange, in particular for H_2 production, as large external gas exchangers would be required. Another advantage of internal gas exchange is that for tubular reactors, these allow larger unit scales, over 100 m^2. Internal gas exchange can also mix the reactors through an air-lift effect, with low power inputs. Assuming a gassing rate of <0.05 vol/vol reactor/min, energy inputs would be a small fraction of projected H_2 outputs. Closed photobioreactors also require temperature control. Thermophilic species have been studied for biohydrogen production, but some cooling will generally be required. Options are cooling ponds or water sprays, with the latter the method of choice except in humid climates.

The photobioreactor design that seems to best meet the requirements of biophotolysis is an NHTR with internal gas exchange and water spray cooling, a system first described in detail by Tredici et al. (1993) and demonstrated with *Arthrospira* (Spirulina) cultures by Chini Zittelli et al. (1996). Such a photobioreactor was recently installed in Hawaii to develop a two-stage biophotolysis process (Szyper et al., 1998). The design of this reactor is described next.

5. DESIGN AND OPERATION OF NHTR PHOTOBIOREACTORS

The NHTR design is for long, parallel tubes, with a small slope, bottom manifold for gassing and liquid recirculation, and top degasser. The general schematic of the NHTR design is shown in Figure 1. Essentially, this photobioreactor consists of a number of parallel tubes, with a slope of typically about 10% (although this could range from perhaps 6–12%). The tubes are connected on the bottom by a manifold tube that allows recirculation between tubes and contains a small gassing tube, and at the top by a larger degasser. Compressed gas is introduced into the bottom of most of the tubes, typically four out of five. The gas flows up the top of the tubes as relatively large gas bubbles, which displace some of the liquid into the degasser. This sets up an air-lift recirculation flow through the tubes that are not gassed, resulting in mixing of the entire culture. The key attributes of this reactor are the relatively long lengths and large numbers of tubes that can be ganged together by the bottom manifold and top degasser into a single operating unit.

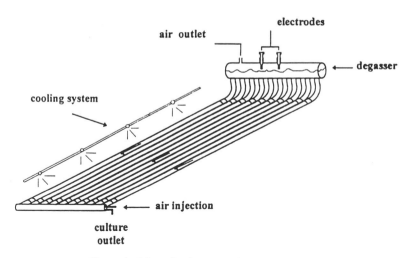

Figure 1. Schematic of a near-horizontal tubular reactor.

Thus far, systems of only one to eight tubes and 6–90 m long have been studied in Florence, and the reactors installed in Hawaii consist of eight tubes, 20 m long. Table 1 summarizes some initial results from the work in Florence with the cultivation of *Arthrospira*. Overall, the productivities obtained with this reactor were similar to those observed with other designs, such as alveolar panels and helical reactors. Thus, the fundamental advantages of this photobioreactor design are not any increases in productivity but in lower capital and operating costs.

The hydrodynamics, gas exchange, and other operating aspects must still be studied as functions of the various design parameters, including slope, length, tube diameter, gassing rate, and so forth. However, reasonable assumptions can be made and these allow at least an initial conceptual design and cost analysis of such a photobioreactor. One issue is slope, which affects bubble rise velocities, hold-up and gas transfer coefficients, as well as the earthwork required for the platform on which the reactors would be installed and the

Table 1. Initial results with near-horizontal tubular reactors[a]

Reactor design (no. of tubes, length, diameter)	Productivity (g/L/day)	Solar radiation (MJ/m^2/day)	Duration (avg. months [year])
1, 6 m, 4 cm	1.63	24.0	May–June (1996)
	1.40	22.2	July–August (1996)
	0.82	13.3	September–October (1996)
3, 6 m, 4 cm	0.90	22.2	July–August (1996)
1, 84 m, 4 cm	0.61	15.5	September (1996)
2, 84 m, 4 cm	0.52	15.5	September (1996)
3, 84 m, 4 cm	0.42	15.5	September (1996)
	0.33	13.1	October (1996)
1, 6 m, 5 cm	1.05	23.4	July (1995)
8, 6 m, 5 cm	0.67	15.3	September (1995)

[a]Based on Chini Zittelli et al., 1996; Chini Zittelli and Tredici, unpublished.

power required for gas compression. These parameters, in turn, determine the optimal length and numbers of tubes, which set the unit size of the reactors. For this analysis, the slope is assumed to be 8%, the length of tubes 50 m, with 40 tubes in a single unit, and a diameter of 4.2 cm, with tubes spaced 5 cm (0.8 cm apart, which is assumed to result in negligible light losses). The tubes would be connected at the lower end with a gassing manifold and on the top with a degasser, each 2 m wide, for nominally 100 m^2 arrays. Gas and liquid delivery and recovery pipes would be connected with Y-tubes that tie into the general supply-drainage lines, with a single shut-off valve for each connection. Thus, two 100-m^2 arrays would be operated in tandem, resulting in a single 200-m^2 operating unit.

Gassing has the objective of driving H_2 (and CO_2 or O_2) out of solution and must overcome liquid-side overpressures, which can be large and could inhibit the H_2-producing reactions. The gas used would be that generated by the process. For a high insolation location (21 MJ/m^2/day of solar energy) and a 10% solar conversion efficiency, an average H_2 production is 15 g/m^2/day, or 170 L H_2 produced, or some 425 L H_2/tube/day. To this would need to be added half a volume of either O_2 (direct biophotolysis) or CO_2 (indirect biophotolysis). Thus, the actual gas volume produced per tube is 630 L/day, or about 1 L/min/tube, or 0.014 vol/vol/min. This would also appear to be a reasonable estimate of the minimum gassing rate required for such photobioreactors. In practice, during periods of high gas production, gas recirculation could be reduced as much of the mixing would be effectuated by the internally generated gas. Indirect, two-stage biophotolysis systems are projected to produce up to ten times as much gas. This may require significant design changes, including possibly additional external gas transfer devices. This needs further study.

The gas recirculation system requires a compressor, with filter, water trap and regulator, valves, controls, and gas lines to the reactor arrays. Compression energy inputs depend on pressure drops in the distribution pipes at the orifice due to the hydrostatic head of the water. Assuming a constant 1 L/min/tube gassing rate, a 4000 tubes/ha module, and a hydrostatic head of 4 m (50-m long tubes, 8% slope), and allowing for compressor efficiencies (about 80%), this requires an overall power input equivalent to compression of the gas to 1 atm, about 44.4 KJ/m^2/day, compared to a gross H_2 output about 2100 KJ H^2/m^2/day. Thus, power inputs for gas compression and utilization at these recirculation rates are only slightly above 2% of H_2 outputs, not a limiting factor. Energy required for H_2 clean-up, compression, and storage are not further considered herein.

Several plastic materials are available that have a relatively low diffusivity to H_2 (Kjeldsen, 1995), but it is not clear if these will be suitable for this application. Cheap tubular glass is also available, but would require field assembly while plastic systems could be mostly pre-fabricated, greatly reducing installation costs. Installation requires a support, for which the lowest cost option would be an earthworks platform with a corrugated plastic support to contain the tubes, an important design feature. Provisions must also be made to allow for drainage (cooling, rainfall). A central supply system would provide compressed gas for mixing, media and catalyst (the starter cultures), power, cooling water, gas handling, drainage, and so forth. The above provides at least an initial basis for carrying out a preliminary cost analysis for such a near-horizontal tubular reactor, presented next.

6. PHOTOBIOREACTOR COST ESTIMATES

The total area required for such a photobioreactor system, including access between arrays, drainage, roads, etc., would be about twice that taken up by the tubular arrays themselves. Piping (diameters, wall thickness, materials, both trunk and laterals) must be specified for the liquid supply (water, media, catalyst) and drainage, the gas take-off and recirculation, and the cooling water. These set the sizes of the shut-off valves, both for 200-m^2 units and also for larger modules.

The largest single cost is the tubular material, either plastic or glass. Both appear to be available in quantity for about $1/m, or $20/$m^2$. The tubes would rest on a corrugated plastic sheet, which also keeps the tubes aligned and provides drainage. Installed costs are estimated at about $5/$m^2$. Installation is on top of a sloped, compacted earthwork. A 8% grade over 50 m gives a total 4-m rise, which, assuming initially flat land, requires 1 m^3 cut-and-fill per cubic meter array. Additional earthwork for backslopes, drainage, pipe routing, access and roads, etc., will increase the total cut-and-fill needed by about half. Precision earthwork with laser-guided earth movers is relatively inexpensive for larger systems of repetitive designs and are estimated at $3/$m^2$ of total reactor area. Additional site preparation costs for initial rough grading and final laser leveling is estimated at $1/$m^2$ for total site area for multi-hectare systems and assuming initially near-flat land. Drainage and flood protection would add roughly another $1/$m^2$. All these figures are preliminary estimates for large installations rather than based on specific cost quotes.

During operation, the top degasser is partially filled with liquid generated from the lift of the gas in the tubes, providing the head for the return flow in the tubes not gassed (typically 20% of the tubes are return tubes). The bottom manifold allows liquid circulation and contains an internal 5-cm gas supply line. We assume that the bottom manifold is made of 20-cm diameter PVC tubing. The top degasser is assumed to be 40 cm in diameter, which, at a bubble rise velocity of 30 cm/sec, restricts the gassing rate to under 0.025 vol/vol/min (allowing for some space in the degasser). This is almost twice the 0.014 vol/vol/min estimated above as sufficient for gas exchange. The manifold and degasser designs need not be further specified here. For the gas and liquid supply lines connecting to the arrays, 5-cm plastic pipes should suffice, with gas recovery (from the degasser) and liquid drainage specified at 10 cm. As noted above, two 100-m^2 panels are joined together with a Y-connector and attached to the supply line with a single shut-off valve. Four such connectors are required, one each for the liquid and gas supply and drainage lines.

Reservoirs for culture water storage (including cooling water), media preparation, etc. should be able to hold the entire reactor volumes, some 300,000 L/ha. Two 100-m^3 plastic-lined storage ponds are required, at $10,000 each, one below-grade for gravity draining and another above-grade, to allow media (or inoculum) supply, also by gravity. Pumps would fill and empty these reservoirs as needed. The cooling water would be supplied through an automatically controlled sprinkler system with one supply line for each 200-m^2 unit. Sprinkler coverage need not be complete and excess cooling water would be drained to a reservoir and recycled as appropriate. At 1 cm/day (including blow-down), a maximum supply rate of 200 L/min/ha would be required (about 4 m^3/m^2/year), with a 4-m lift and an additional 2-m pressure drop for the sprinklers and pipes, a modest energy input. Other support systems include a storm drainage system (part of the earthworks), electricity supply, waste treatment and disposal, overall process control systems, laboratory and other general facilities, and, perhaps most important, catalyst (culture) production. Labor and general support depend on scale. Ten catalyst changes a year appear to reasonable, with the catalyst production carried out in photobioreactors of various designs. Ten percent is added to total capital and operating costs for catalyst production. Many problems were ignored above, such as fouling and cleaning, contamination, and other process upsets. We assume that R&D will solve these within acceptable limits.

From the above, rather sketchy general discussion, some initial capital and operating costs can be estimated. Site preparation costs, including grading, earthworks, drainage, and site-related support systems, were estimated at $10/m^2 and the tubes (glass or plastic) at $20/m^2. Other components include, per hectare of reactor:

- bottom manifolds and top degassers (2 m each, 20- and 40-cm diameters, respectively, end plates, and 40 × 4.2 cm nipples to attach the tubing);
- cooling water supply (250 m × 10 cm main and 2500 m of 5-cm lateral PVC lines, with 1,000 sprinklers, valves, a water reservoir, and pump);
- water, media, and culture inoculum supply system, with 250 m × 10 cm PVC line, 100-m^3 reservoirs, pumps, valves, fertilizer storage, mixer, etc.;
- drain for spent culture (catalyst), 250 m × 10 cm PVC line, to a below-grade reservoir and pump to an anaerobic digester (not included herein);
- gas collection system, consisting of 250 m of 10-cm PVC line, from the degasser to a central gas processing station (not further described herein);
- gas recirculation line, including blower, 250 m of 5-cm distribution line, valves, in-line check valves, and related control and safety equipment;
- data collection and automation system, including four sets of sensors in the systems (pH, temperature, gas analysis, turbidity, etc.) and four sets of automatic control valves (sprinklers, gas recirculation, etc.);
- other support services, such as power, water supply and treatment, waste treatment and brine disposal, maintenance and office facilities, etc. Pipe diameters are not based on detailed analysis but on rough estimates of likely requirements for such a system. Pipe lengths are derived from simple geometry.

7. BIOHYDROGEN PRODUCTION COSTS

From the above, a preliminary cost for biohydrogen production can be estimated. We assume that the photobioreactor is prefabricated and field installations are only required for the piping, pumps, etc., at a cost equal to that of the materials. Engineering and contingency costs are ignored, a reasonable assumption for a mature and modular technology. Cost of capital (loans and investment) is assumed at only 5%, reflecting real, riskless, long-term returns on capital. Depreciation and maintenance of 10% a year account for replacement of the tubes and other materials. Taxes and insurance are set at 1% a year of total capital costs.

The operating or variable costs include power, water, labor, waste disposal, general supplies, indirect costs, and income taxes. Power use is, as discussed above, rather modest, equivalent to 3% of total H_2 energy output. Some of this power may be produced from the digestion of the wasted biomass, but that is not considered here. Water consumption, at some 4 m^3/m^2/year, is based on a cost of some $50 per 1000 m^3 delivered to the site, or $2000/ha. Labor costs, including monitoring, operations, culture replacement, cleaning, supervision, laboratory, etc. (but excluding maintenance, included under capital costs), is assumed at four times those of open pond systems (Benemann, 1993), or $16,000/ha, including overhead. This assumes relatively large systems for economies of scale.

General supplies include the nutrients, N, P, etc., and make-up CO_2 for indirect processes. Assuming a total catalyst (dry weight microalgal biomass) production (and, thus, wastage) of 1 kg/m^2/year, or 10 tons/ha per year, this would amount to only 50 g of N and 5 g of P per square meter a year and smaller amounts of other nutrients. This is a cost of less than $400/ha/year. A 1% loss of CO_2 per cycle represents 12 kg CO_2/m^2/year, or, at $100/ton CO_2, $1200/m²/year, a significant cost, though lower cost sources of CO_2 may likely be available, allowing for lower costs or higher losses. A sum of $2000/ha/year is budgeted for nutrients, CO_2, and general supplies. Overhead, licensing and other general indirect costs, income taxes, etc., are ignored, as these would vary greatly from case to case and could be rather low in some circumstances.

From the above, total system and resulting H_2 production costs are summarized in Table 2. If the process is stable and can be automated to function with minimal supervision, and if support systems (gas handling) can be miniaturized, then smaller, dispersed (e.g., roof-top) systems may be possible. However, an analysis of such an option is beyond the present effort. In conclusion, this analysis projects biohydrogen costs of some $15/GJ, exclusive of any gas purification, handling, or storage costs. Of course, this is based on very preliminary estimates and favorable assumptions.

8. DISCUSSION AND CONCLUSIONS

The above cost analysis is only an initial, highly conceptual, and preliminary estimate of costs of such photobioreactors and biohydrogen production. The main objective was to determine if such designs could, in principle at least, meet the stringent cost limitations for biophotolysis processes, in particular single-stage systems. As discussed earlier, two-stage processes could allow somewhat larger costs per unit area, but their very high gas production rates would likely require additional gas exchange. Much more work is required to arrive at more realistic and complete cost estimates. Still, this preliminary analysis suggests that it may indeed be possible to produce H_2 at reasonable costs with photobioreactors. Of course, this assumes the development of very efficient and long-lasting catalysts, operating at very favorable sites.

One major limitation is that this analysis did not include costs for gas handling (drying, compression, and separation of CO_2 or O_2). Another major issue is the cost routines used, such as a total capital charge (cost of capital, depreciation, taxes, etc.) of only 17%. Even then, capital (fixed) costs are some 80% of total costs. And this estimate did not include allowances for engineering, construction, contingency, or working capital. More importantly, this analysis is based on only very preliminary and superficial estimates for materials, piping, valves, holding ponds, etc. A much more detailed design and cost analysis is required. The availability of suitable, long-lasting, and H_2-impervious tubular materials, at about $20/m², must be verified. Somewhat surprising, the piping, valves, and other materials and their assembly were not cost-prohibitive, that is, within the assumptions made, and neither were nutrient supplies, energy inputs, or other operating costs. These too need to be verified.

Perhaps the major conclusion from this exercise is not the final $/GJ H_2 projected, but that it suggests that sufficiently low-cost photobioreactors are plausible that could possibly allow even single-stage biophotolysis processes. It remains for future analysis and R&D to determine if such reactors can indeed be developed in practice and whether the cost estimates and routines used above could be realistic in a commercial setting. Indeed, the many assumptions made and required could as well argue for the contrary position. Indeed, the commercial feasibility of any photobioreactor, regardless of the value of the product, is still unproven. Although extensively studied (Tredici et al., 1992; Pulz and Scheibenbogen, 1998), no commercial venture using such reactors has yet succeeded, and several have failed, even for high-value products (Tredici and Chini Zittelli, 1995). What-

Table 2. Capital and operating costs. Capital costs $/ha of photobioreactor arrays for a large-scale (>100 ha) system. Assumes average insolation of 21 MJ/m^2/day and 10% solar conversion efficiency

No.	Item	Unit	$/unit	#/ha	$/ha
A. Capital costs: site and materials					
1.	Site preparation	m^2	1	10000	10000
2.	Cut/fill, drains	m^3	2	15000	30000
3.	Corrugated support	m^2	5	10000	50000
4.	Liners	m^2	3.3	3300	10000
5.	Tube materials	m	1	200000	200000
6.	Manifold, 15 cm	m	50	200	10000
7.	Degasser, 30 cm	m	100	200	20000
8.	Gas supply, 5/10 cm	m	1/3	250/50	400
9.	Gas removal, 10/20 cm	m	3/15	250/50	1500
10.	Blower, 4 m^3/min	each	5000	1	5000
11.	Cool. main, 10/20 cm	m	3/15	250/50	1500
12.	Cool. lateral, 5 cm	m	1	2500	2500
13.	Sprinklers	each	1	1000	1000
14.	Media/water, 10/20 cm	m	3/15	250/50	1500
15.	Drain, 10/20 cm	m	3/15	250/50	1500
16.	Valves, 8, 9, 12, 14, 15	each	10/20	100/150	4000
Subtotal					350000

No.	Item	$/ha/year
B. Capital costs: installation, support, general		
17.	Installation (= #6–16)	50000
18.	Automation system	15000
19.	Waste holding pond	10000
20.	Waste disposal	Not included
21.	Water supply	10000
22.	Media preparation	10000
23.	Power supply	5000
24.	Other support facilities	5000
25.	Catalyst production 10% of above	45000
26.	Product gas handling	Not included
27.	Land costs	Not included
28.	Engineering, construction contingencies	None charged
29.	Interest, working capital	None charged
Subtotal		150000
Total capital costs		500000
C. Annualized costs: capital charges and operations		
30.	Depreciation, maintenance, 10%	50000
32.	Insurance, taxes, 2%	10000
33.	Capital cost (interest, ROI), 5%	25000
34.	Labor	16000
35.	Water, nutrients, supplies	4000
36.	Catalyst production, 10% of 34–36	2000
37.	Power, increase final cost by dividing by 0.97	
Total annual costs		110000
Costs per GJ H$_2$		$15/GJ

ever the reason for these failures, the above exercise does suggest that photobioreactors need not necessarily be prohibitively expensive, perhaps even for biohydrogen production. However, much more analysis, research, and development of photobioreactors are required and should proceed alongside development of the biophotolysis catalysts.

REFERENCES

Benemann, J.R., 1993, Utilization of carbon dioxide from fossil fuel-burning power plants with biological systems, *Energy Cons. Mgmt.*, 34:999–1004.
Benemann, J.R., 1996, Hydrogen biotechnology: progress and prospects, *Nature Biotechnology*, 14:1101–1103.
Benemann, J.R., 1998, Processes analysis and economics of biophotolysis: a preliminary assessment. Report to the International Energy Agency, Subtask B, Annex 10, Photoproduction of Hydrogen Program.
Benemann, J.R., Feasibility analysis of photobiological hydrogen production, in *Hydrogen Energy Progress X*, Proceedings of the 10th World Hydrogen Energy Conference, Block, D.L., and Veziroglu, T.N. (eds.), Cocoa Beach, Florida, United States, June 20–24, 1994, pp. 931–940.
Bolton, J.R., 1996, Solar photoproduction of hydrogen, Report to the International Energy Agency, Agreement on the Production and Utilization of Hydrogen IEA/H2/TR-96.
Chaumont, D., 1993, Biotechnology of algal biomass production: a review of systems for outdoor mass culture, *J. Appl. Phycol.*, 5:593–604.
Chini Zittelli, G., Tomasello, V., Pinzani, E., and Tredici, M., 1996, Outdoor cultivation of *Arthrospira platensis* during autumn and winter in temperate climates, *J. Appl. Phycol.*, 8: 293–301.
Kjeldsen, P., 1993, Evaluation of gas diffusion through plastic maerials used in experimental and sampling equipment, *Wat. Res.*, 27(1):121–131.
Melis, A., Neidhardt, J., Baroli, I., and Benemann, J.R., 1998, Maximizing photosynthetic productivity and light utilization in microalgae by minimizing the light-harvesting chlorophyll antenna size of the photosystems, in *Proceedings of BioHydrogen 97*, Plenum Publishing, New York, in press.
Pulz, O., and Scheibenbogen, K., 1998, Photobioreactors: design and performance with respect to light energy input. *Advances in Biochemical Engineering*, 59:123–152.
Szyper, J. et al., 1998, in *Proceedings of BioHydrogen 97*, Plenum Publishing, New York, in press.
Tredici, M.R., Carlozzi, P., Chini Zittelli, G., and Materassi, R., 1991, A vertical alveolar panel (VAP) for outdoor mass cultivation of microalgae and cyanobacteria, *Bioresource Technology*, 38:153–159.
Tredici, M.R., and Materassi, R., 1992, From open ponds to vertical alveolar panels: the Italian experience in the development of reactors for the mass cultivation of phototrophic microorganisms, *J. of App. Phycol.*, 4:221–231.
Tredici, M.R., and Chini Zittelli, G., 1995, Scale-up of photobioreactors to commercial size, in *Proceedings of the Second European Workshop on Biotechnology of Microalgae*, September.
Tredici, M.R., and Chini Zittelli, G., 1997, Cultivation of *Spirulina (Arthrospira) platensis* in flat plate reactors, in Spirulina platensis *(*Arthrospira*): Physiology, Cell Biology and Biotechnology*, Vonshak, A. (ed.), Taylor & Francis.
Tredici, M.R., Chini Zittelli, G., Biagiolini, S., Carobbi, R., Favilli, F., Mannelli, D., and Pinzani, E., 1993, Impianto a fotobioreattori tubolari per la coltura industriale di microrgansimi fotosintetici, Italian patent.
Tredici, M. R., and Chini Zittelli, G., 1998, Efficiency of sunlight utilization: tubular vs. flat photobioreactors, *Biotechnol. Bioeng.*, 57:187–197.
Weissman, J.C., Goebel, R.P., and Benemann, J.R., 1988, Photobioreactor design: comparison of open ponds and tubular reactors, *Bioeng. Biotech.*, 31:336–344.

49

A TUBULAR RECYCLE PHOTOBIOREACTOR FOR MACROALGAL SUSPENSION CULTURES

Ronald K. Mullikin and Gregory L. Rorrer

Department of Chemical Engineering
Oregon State University
Corvallis, Oregon 97331

Key Words

macroalgae, cell culture, photobioreactor, tubular

1. SUMMARY

Macrophytic marine algae are a rich source of pharmacologically active compounds. Cell and tissue suspension cultures established from macrophytic marine algae have the potential to biosynthesize these compounds in a controlled environment within bioreactor systems at a scale required for further product development or commercial production. Toward this end, we recently developed a tubular recycle photobioreactor that integrates the best features of tubular, stirred-tank, and air-lift bioreactor configurations into one system. The tubular recycle photobioreactor consisted of a tubular section, stirred aeration tank, and recycle line with a peristaltic pump to move the culture between the tubular section and the aeration tank. In the tubular section, 39 m of silicone tubing of 0.8-cm inner diameter (1.9-L culture volume) was wrapped around a 30.5-cm diameter cylindrical frame. The tubular section was radially illuminated by two 20-W fluorescent lamps positioned at the center of the coil to provide an incident light intensity of 18 $\mu E\ m^{-2}\ s^{-1}$. The culture was pumped through the tubular section at a rate of 125 mL min^{-1} to provide residence times of 15 min in the tubular section and 6 min in the aeration tank. An air flow rate of 1.5 vvm in the aeration tank provided all the CO_2/O_2 gas exchange requirements for the culture. Air was also injected directly into the tubular section at 2 L min^{-1} for 90 sec every 90 min to remove cell mass attached to the tubing wall. Under these conditions, gametophyte cell suspension cultures of the macrophytic brown alga *Laminaria saccharina* achieved a specific growth rate of 0.1 day^{-1} and a maximum cell density exceeding 1000 mg DCW L^{-1} over a 30-day cultivation period. The efficient delivery of light to the suspension culture

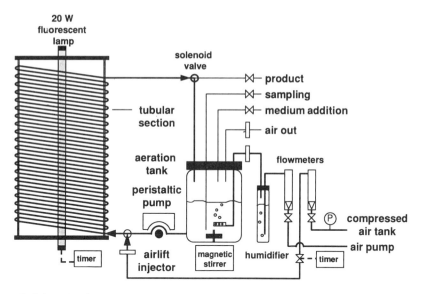

Figure 1. Schematic of tubular recycle photobioreactor for cultivation of macroalgal suspension cultures.

makes this system an attractive alternative to externally illuminated stirred-tank or air-lift bioreactors for the production of photosynthetic biomass.

2. INTRODUCTION

Macrophytic marine algae, commonly known as seaweeds, are a rich source of unique and high-value natural products, including pharmaceutically active compounds (Carte, 1996) and specialty polysaccharides (Radmer, 1996). Many seaweeds that produce these compounds are delicate, anatomically complex plants sparsely distributed in fragile marine environments such as tide pools or coral reefs, and hence are difficult to mariculture. However, cell and tissue suspension cultures established from marine seaweeds have the potential to biosynthesize these compounds in a controlled environment within bioreactor systems at a scale required for further product development or commercial production. Toward this end, we have developed several unique phototrophic cell and tissue suspension culture systems representing all three major divisions of marine macroalgae, including brown macroalgae (Qi and Rorrer, 1995), green macroalgae (Rorrer et al., 1996), and red macroalgae (Rorrer et al., 1997a), and demonstrated their cultivation in externally illuminated stirred-tank and bubble-column bioreactors.

An important issue in the bioreactor scale-up of phototrophic algal cultures is the efficient delivery and utilization of light (Richmond, 1996). Although the growth of macroalgal suspension cultures saturates at fairly low light intensities below 100 $\mu E\ m^{-2}\ s^{-1}$ (Qi and Rorrer, 1995; Rorrer et al., 1996), insufficient light transmission through the culture ultimately reduces biomass productivity upon scale-up. To address this problem, we considered the tubular recycle photobioreactor for the cultivation of macroalgal suspension cultures (Figure 1). This bioreactor combines the efficient gas-liquid mass transfer of stirred-tank bioreactors with the efficient light transmission of tubular photobioreactors, and can be operated in batch, semi-continuous, or continuous-recycle modes. Although this photobioreactor configuration is largely unstudied, similar helical coil photobioreactors have recently been described for the cultivation of microalgae (Lee and Bazin, 1990; Chrismadha and Borowitzka, 1994; Watanabe et al., 1995; Watanabe and Hall, 1996).

The model culture system chosen for study is a gametophyte cell suspension culture established from the macrophytic brown alga *Laminaria saccharina*. This culture produces 15-HETE, a potent pro-inflammatory agent derived from ω-σ lipoxygenase metabolism of arachidonic acid (Rorrer et al., 1997b). The cultivation of *L. saccharina* gametophyte cell suspensions has been studied in both externally illuminated stirred-tank (Qi and Rorrer, 1995) and bubble-column (Zhi and Rorrer, 1996) bioreactors. In this paper, we demonstrate the cultivation of *L. saccharina* gametophyte cell suspension in a bench-scale tubular recycle photobioreactor and assess the significance of light attenuation in the process. We also describe the unique features of this photobioreactor system and suggest its applicability to the cultivation of phototrophic algae that produce volatile and sparingly soluble metabolites such as hydrogen gas.

3. TUBULAR RECYCLE PHOTOBIOREACTOR DESIGN

A schematic of the bench-scale tubular recycle photobioreactor is provided in Figure 1. Relevant dimensions and operating conditions of the photobioreactor are provided in Table 1. The photobioreactor has five principal elements: 2-L tubular section, illumination stage, peristaltic pump, air-lift injector, and 2-L aeration tank. The tubular section consists of autoclavable silicone tubing (Cole-Parmer Masterflex 96410-18, 1.11-cm outer diameter, 0.79-cm inner diameter, 0.16-cm wall) helically wrapped around a cylindrical stainless steel frame. The tubular section is radially illuminated by two 20-W fluorescent

Table 1. Process parameters for bench-scale tubular recycle photobioreactor

Process parameter	Variable	Value and units
Tubular section		
Inner tubing diameter	d	0.79 cm
Outer tubing diameter	d_o	1.11 cm
Tubing length	L	39.0 m
Total tubing volume	V_t	1.9 L
Coil frame diameter	D	30.5 cm
Flow system		
Culture flow rate	v_m	125 mL min^{-1}
Residence time	τ_t	15 min
Reynolds number	Re	330
Recycle rate	r	0 mL min^{-1}
Illumination system		
Incident light intensity	I_o	18 µE m^{-2} sec^{-1}
Photoperiod	f_t	21.5 hr ON/2.5 hr OFF
Airlift injector		
Timer program		1.5 min ON/90.0 min OFF
Air flow rate	v_a	2000 mL min^{-1}
Stirred aeration tank		
Culture volume	V_s	700 mL
Air flow rate	v_a/V_s	1.5 vvm
Culture residence time	τ_s	5.6 min

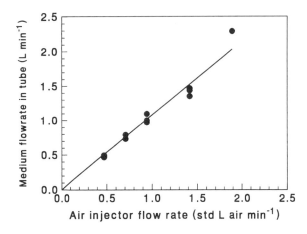

Figure 2. Effect of air injector flow rate on medium flow rate in the tubular section. The slope of least-squares line is 1.08 ± 0.03 (1s, $r^2 = 0.961$, n = 12).

lamps (61-cm length) positioned axially at the center of the coiled tubular section. A peristaltic pump (Cole-Parmer Masterflex no. 7518-10, flow-rate range 20–350 mL min^{-1}) moves the culture from the aeration tank to the coiled tubular section, and then up the coiled section and back to the tank. There is no free gas space within the tubular section in order to maximize the use of reactor volume. The air-lift injector does not continuously aerate the culture within the tubular section. Instead, at periodic intervals, the direct injection of air into the inlet of the tubular section for a short time forces the liquid culture through the tube at a high velocity sufficient to shake loose biomass attached to the tube walls. The liquid flow rate through the tubular section is directly proportional to the air injection flow rate (Figure 2). A 2-L Virtis Omni-Culture fermenter (Virtis, Inc., 15.2-cm inner diameter, 22.5-cm height, 3-L total volume, 2-L working volume) serves as the aeration tank for gas-liquid mass transfer (CO_2 absorption and dissolved O_2 stripping). The air sparger in the aeration tank has five 1-mm diameter holes. The tank is not directly illuminated. The aeration tank also serves as a surge tank for medium addition, culture inoculation, and rapid evacuation of culture from the tubular section during operation of the air-lift injector. All tubing materials were autoclavable silicone or stainless steel. The entire bench-scale system is contained within an illuminated low-temperature incubator.

The tubular recycle photobioreactor can be operated in three modes: batch, semi-continuous, and continuous-recycle. A three-way solenoid valve at the tubular section outlet sets the mode of cultivation. In batch operation, the solenoid valve is closed, then the culture is returned to the aeration tank. In semi-continuous or continuous recycle operation, the solenoid valve is opened and closed periodically, and a portion of the culture is drawn off as product while the remainder is returned to the tank. Fresh medium is concurrently added to the tank while product is being removed. The recycle ratio, defined as the ratio of returned product to removed product, is proportional to the ratio of on/off cycle times for the solenoid.

4. EXPERIMENTAL

4.1. Culture Maintenance

The development, characteristics, and maintenance of *L. saccharina* gametophyte cell suspension cultures are described in our previous work (Qi and Rorrer, 1995; Zhi and Rorrer,

1996). Cultures were maintained without agitation in 250-mL flasks (100 mL culture per flask) in GP2 artificial seawater medium (pH 8.5 buffered with sodium bicarbonate) under 20 $\mu E\ m^{-2}\ s^{-1}$ cool white fluorescent light (16:8 LD photoperiod) within a low-temperature incubator at 12 °C. The details of GP2 medium composition and preparation are detailed by Zhi and Rorrer (1996). Selected flasks were subcultured every four weeks at 25% v/v. Prior to inoculation, cultures were blended for 5 sec at "grind" speed in an Osterizer blender (Sunbeam, Inc.) to disperse the clumped cell mass, and then sterile filtered on 100-µm nylon mesh. The filtered cells were washed with and re-suspended in fresh GP2 artificial seawater medium. All procedures were carried out using sterile techniques in a laminar flow hood.

4.2. Start-Up and Sampling of Tubular Recycle Photobioreactor

The start-up and sampling procedures for the tubular recycle photobioreactor are provided below. The tubing in the coiled tubular section was flushed with tap water for 2 min. The inner surface of the tubing in the coiled tubular section was cleaned by forcing a compressible foam plug (1.5-cm diameter, 1.5-cm length) through the tube under water pressure from a laboratory faucet over the entire 39-m tubing length. The tubing section was flushed thoroughly with tap water again. The coiled tubular section and aeration tank assembly were autoclaved at 121 °C, 205 kPa for 20 min. To prepare the inoculum, 500 mL of *L. saccharina* culture from five flasks were pooled and blended for 5 sec, sterile-filtered on 100-µm nylon mesh, and then re-suspended in 300 mL of GP2 medium. The 300-mL inoculum was poured into the autoclaved aeration tank, directly followed by 2500 mL of sterilized, cooled (12 °C), GP2 medium containing 4X macro- and micro-nutrients (includes 254 mg $NaNO_3\ L^{-1}$, 26 mg $NaH_2PO_4\ L^{-1}$) and 2X bicarbonate (680 mg $NaHCO_3\ L^{-1}$). The initial cell density was typically 240 mg DCW L^{-1}. All culture transfers were performed using sterile techniques in a laminar flow hood. The aeration tank and tubular section were transferred to the low-temperature incubator and all connections were finalized. The culture pump and air flow were turned on and adjusted to the desired values. At a steady-state, there were initially 1900 mL of culture volume in the tube and 900 mL of culture volume in the aeration tank. The air injector timer program was set, typically for 90 sec at 2 L min^{-1} air flow every 90 min. At this air injection flow rate, all the culture was transferred from the tubular section to the aeration tank. The culture was sampled only at this condition to ensure a uniform suspension. The culture was typically sampled at 2–5 day intervals by removing 40 mL of sample through a 0.457-cm inner diameter tube under sterile suction. At least 400 mL of the total culture volume was removed by sampling.

4.3. Analytical Techniques

Dry cell density was determined according to Zhi and Rorrer (1996). The pH of the culture samples was measured with a pH electrode and meter immediately after removal from the bioreactor. Irradiance was measured in units of $\mu E\ m^{-2}\ s^{-1}$ with a LI-COR 190SA PAR quantum sensor and LI-COR 189 quantum radiometer.

4.4. Light Attenuation Through Culture and Tubing

The apparent light attenuation constant k' is an aggregate measure of light absorption, light diffusion, and light diffraction in the culture and must be empirically determined. The apparent light attenuation constant though the culture at a given cell density was estimated according to the following procedure. Freshly blended *L. saccharina* gametophyte cell suspen-

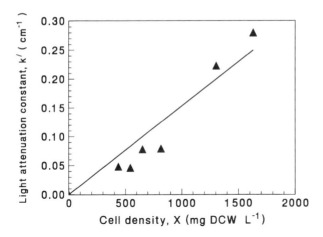

Figure 3. Effect of dry cell density on apparent light attenuation constant k' for gametophyte cell suspension cultures of the macrophytic brown alga *Laminaria saccharina*. From the least-squares line, $k_c = 0.153 \pm 0.014$ cm^2 mg^{-1} DCW (1s, r^2 = 0.889, n = 6).

sions were re-suspended in GP2 medium to dry cell densities ranging from 435–1630 mg DCW L^{-1}. The suspension was poured into a glass culture dish (9.6 cm wide by 5 cm deep) to a given depth. Two 9-W Dulux cool-white fluorescent lamps were placed directly above the culture dish, and a LI-COR 190 SA PAR quantum sensor was positioned underneath the dish facing the lamp. The incident light intensity to the culture surface was set to 50 µE m^{-2} s^{-1} by adjusting the lamp distance. The light intensity through the culture suspension was measured at culture depths in the dish ranging from 0.5–3.0 cm. The apparent attenuation constant (k') at each cell density was estimated using Beer's Law of the form

$$I_z = I_o e^{-k'z} = I_o e^{-k_c X z} \tag{1}$$

from the least-squares slope of the measured light intensity I_z vs. culture depth z at fixed incident light intensity I_o. A plot of the apparent light attenuation constant k' with cell density X is provided in Figure 3. The least-square slope forced through the origin is k_c, equal to $k_c = 0.153 \pm 0.014$ cm^2 mg^{-1} DCW (1s, r^2 = 0.889, n = 6).

The light attenuation constant through the silicone tubing material was also determined. Specifically, the tubing was cut open and flattened to form a 3.5-cm square slab of 0.16 cm thickness (l). The tubing section was placed over the LI-COR 190SA PAR quantum sensor, and the irradiance exiting the tubing material ($I_{t,o}$) was measured at different incident light intensities (I_o) ranging from 10–100 µE m^{-2} s^{-1}. The apparent attenuation constant for the tubing material (k_s) was estimated from $I_{t,o}$ vs. I_o data using Beer's Law of the form

$$I_{t,o} = I_o e^{-k_s l} \tag{2}$$

The least squares value of k_s was 0.674 ± 0.088 cm^{-1} (1s, n = 8) for l equal to 0.169 cm.

5. RESULTS AND CONCLUSIONS

5.1. Growth Curves

Two representative growth curves for cultivation of the *L. saccharina* gametophyte cell suspension in the tubular recycle photobioreactor are provided in Figure 4 at the operating

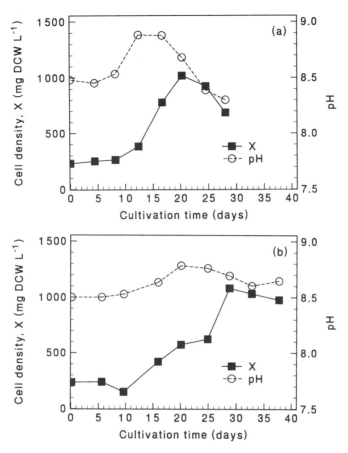

Figure 4. Dry cell density and pH vs. time for batch cultivation of *Laminaria saccharina* gametophyte cell suspension in the tubular recycle photobioreactor at 12 °C. Other process conditions are given in Table 1. (a) Run 1; (b) Run 2. Specific growth rates are given in Table 3.

conditions detailed in Table 1. The bioreactor was operated in batch mode with complete recycling of tubular section culture effluent back to the aeration tank. Peak cell densities exceeding 1000 mg DCW L^{-1} were attained within a 30-day cultivation time under unoptimized process conditions at 12 °C. This biomass productivity was comparable to cultivation of *L. saccharina* gametophyte cell suspensions in stirred tank and bubble-column bioreactors under optimal conditions (Qi and Rorrer, 1995; Zhi and Rorrer, 1996). However, relative to these previous studies, cultivation of *L. saccharina* in the tubular recycle photobioreactor had a relatively long lag phase of 10 days. Furthermore, no stationary phase of growth with a constant final cell density was observed; the culture went from late exponential phase directly to a post-stationary phase where cell biomass decreased with time.

The specific growth rate (μ) for the exponential phase of growth and the specific death rate for the post-stationary phase (k_d) were estimated from the least-square slope of cell density (X) vs. cultivation time on a semi-log plot. Values for μ and k_d the cultivation times associated with each estimate are compared in Table 2. Specific growth rates were comparable at 0.1 day^{-1}, but the specific death rates were more erratic and varied from 0.01–0.05 day^{-1}. Microscopic observation of the *L. saccharina* cell suspension in the tubu-

Table 2. Specific growth rate during exponential phase and specific death rate during post-stationary phase for batch cultivation of *L. saccharina* gametophyte cell suspension in the tubular recycle photobioreactor (see Figure 3)

Run	Specific growth rate μ (day^{-1})	Specific death rate k_d (day^{-1})
Run #1	0.118 ± 0.012 (1s)	−0.050 ± 0.018 (1s)
	(8.2–20.3 days)	(20.3–28.1 days)
Run #2	0.093 ± 0.016 (1s)	−0.011 ± 0.001 (1s)
	(9.6–28.8 days)	(28.8–37.7 days)

lar recycle photobioreactor revealed that during the exponential phase of growth, the filamentous cells were healthy and formed uniform clumps about 500 μm in diameter. However, in the post-stationary phase, the cell filament clumps dispersed and many of filaments were dead, as evidenced by Evan's blue stain. A photomicrograph of the cell mass in the post-stationary phase of growth is presented in Figure 5. Since this phenomenon was not observed in stirred-tank or bubble-column bioreactor cultivation, it is possible that shear forces exerted by the pump on the more fragile late-exponential phase cells caused cell damage leading to attrition of cellular biomass.

5.2. Light Transfer in Tubular Recycle Photobioreactor

An important advantage of the tubular recycle photobioreactor for cultivation of photolithotrophic macroalgal suspension cultures is that the light path for the transfer of

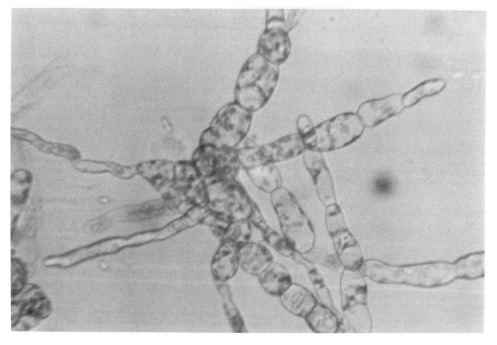

Figure 5. Photomicrograph (430X) of *Laminaria saccharina* gametophyte cells in post stationary phase of culture.

Table 3. Attenuation of light through culture and tubing for *Laminaria saccharina* gametophyte cell suspensions in the tubular recycle photobioreactor

Resistance	Light attenuation constant		Light path		Light attenuation	
	Variable	Value (cm^{-1})	Variable	Value (cm)	Variable	Value
Tubing wall	k_s	0.674	l	0.16	$I_{t,o}/I_o$	0.90
Culture (1000 mg DCW/L)	$k_c X$	0.153	d	0.79	$I_m/I_{t,o}$	0.94
Culture and tubing (1000 mg DCW/L)	–	–	–	–	I_m'/I_o	0.85

light through the culture is contained within the tubing diameter. The mean light intensity I_m is a useful parameter for assessing light limitations in phototrophic algal suspension cultures (Rabe and Benoit, 1962). The mean light intensity is the mean integral of the light distribution function over the light path length of the culture. For example, if it is assumed that (1) the culture is illuminated only from one side, (2) the light flux through the culture is one-dimensional along z over path length d, and (3) the light attenuation through the culture is defined by Beer's Law via equation 1, then

$$I_m = \frac{1}{d}\int_0^d I_z dz = \frac{I_{t,o}}{k_c X d}\left[1 - e^{-k_c X d}\right] \qquad (3)$$

In the illuminated tubular section of the photobioreactor, the inner tubing diameter d approximates the maximum light path length through the culture, and the vertically aligned fluorescent lamp provides a uniform, symmetric, one-dimensional light source. The irradiance delivered to the surface of the tubing is attenuated by the tubing wall itself and by the culture. Therefore, when the light attenuation though the tubing described by equation 2 is inserted into equation 3, the adjusted mean intensity I_m' can be related to the incident light intensity to the outer surface of the tube (I_o) by

$$I_m' = \frac{I_o}{k_c X d}\left[e^{-k_s l} - e^{-(k_s l + k_c X d)}\right] \qquad (4)$$

Input values for equation 4 are given in Table 3, along with estimates of $I_m/I_{t,o}$ and I_m'/I_o in the bench-scale photobioreactor. The adjusted mean light intensity is 85% of the incident light intensity at the peak cell density of 1000 mg DCW L^{-1}, suggesting that light attenuation is not significant in the bench-scale photobioreactor. The silicone tubing, though translucent, reduces the incident light intensity available to the culture by 10% ($I_{t,o}/I_o = 0.9$). If the tubing diameter is increased to 3 cm and the tubing wall thickness is increased to 0.5 cm, which are the envisioned tubing dimensions for a process-scale unit, then the adjusted mean light intensity is still 57% relative to the incident light intensity at 1000 mg DCW L^{-1} cell density.

The overall light delivery to the culture is a function of the culture residence times of the tubular section and the aeration tank. Although the tubular section is illuminated, the aeration tank does not receive direct illumination. Therefore, the net photoperiod of the culture is determined by

$$f_o = \frac{\tau_t}{\tau_t + \tau_s} f_t \qquad (5)$$

where f_o is the overall fraction of time within a 24-h photoperiod that the culture is illuminated, f_t is the fraction of time within a 24-h photoperiod that the tubular section is illuminated, τ_s is culture residence time in the aeration tank, and τ_t is the culture residence time in the tubular section. For example, for the values given in Table 1, if the tubular section is illuminated on a 21.5:2.5 LD photoperiod (f_t of 0.896), then f_o is equal to 0.67, which is equivalent to a 16:8 LD photoperiod.

5.3 Summary of Features and Advantages of the Tubular Recycle Photobioreactor for Cultivation of Macroalgal Cell Suspensions

The tubular recycle photobioreactor shown in Figure 1 is a viable new bioreactor configuration for photolithotrophic cultivation of macroalgal cell or tissue suspensions. It combines the best features of tubular, stirred-tank, and air-lift bioreactors into one system. Light is efficiently delivered to the coiled tubular section with only modest attenuation of light. Carbon dioxide and oxygen gas exchange is provided in the stirred aeration tank. The air-lift injector augments gas exchange and removes attached biomass from the tubing by rapidly pushing the culture through the tubing for short times under low shear. The bioreactor can be operated in batch, continuous-recycle, or semi-continuous modes. Scale-up is accomplished by increasing the length of tubing and, if necessary, the diameter of the tubing while keeping the residence time constant. An apparent disadvantage is that continuous pumping of the culture though the tubular section by the peristaltic pump can cause cell damage and biomass attrition, especially in the stationary growth phase. However, this problem can potentially be avoided or minimized by choosing an appropriate low-shear pump.

5.4. Possible Applications of the Tubular Recycle Photobioreactor to Biohydrogen Production

Many photobioreactor designs have been developed and tested for the biological production of hydrogen gas from microalgae (Akano et al., 1996). However, there are no reported studies that utilize a tubular recycle photobioreactor or closely related bioreactor configuration for the biological production of hydrogen gas. Therefore, the tubular recycle photobioreactor has potential advantages for biological hydrogen production in addition to the generic advantages mentioned above. For example, evolved hydrogen gas can be collected at the tubular section effluent point before the culture is returned to the aeration tank to facilitate the separation of hydrogen product gas from the aeration gas (e.g., air) in the aeration tank. Also, tailored gas mixtures could be directly injected into the tubular section to promote additional nitrogenase or hydrogenase activity while providing an additional mass transfer sink for evolved oxygen and hydrogen. These may be worthy topics for future research.

6. NOMENCLATURE

d = inner diameter of silicone tubing (cm)
f_o = net fractional photoperiod of culture in tubular section and aeration tank (h light ON h^{-1} of photoperiod)

f_t = fractional photoperiod of light delivered to tubular section (h light ON h^{-1} of photoperiod)
I_m = mean light intensity of culture (μE m^{-2} s^{-1})
I_m' = adjusted mean light intensity accounting for light attenuation through tubing (μE m^{-2} s^{-1})
I_o = incident light intensity (μE m^{-2} s^{-1})
$I_{t,o}$ = incident light intensity to culture (μE m^{-2} s^{-1})
I_z = light intensity at position z (μE m^{-2} s^{-1})
k_c = light attenuation constant through culture (L mg^{-1} DCW cm^{-1})
k_d = specific culture death rate in stationary phase (day^{-1})
k_s = light attenuation constant through silcone tubing (cm^{-1})
l = wall thickness of silicone tubing (cm)
v_a = volumetric flow rate of air delivered to air-lift injection system (mL min^{-1})
v_m = volumetric flow rate of culture through tubular section (mL min^{-1})
X = dry cell density (mg DCW L^{-1})
μ = specific growth rate during exponential phase (day^{-1})
τ_s = culture residence time in aeration tank (min)
τ_t = culture residence time in tubular section (min)

ACKNOWLEDGMENTS

This work was supported by grant no. NA36RG0451 (project no. R/BT-8) from the National Oceanic and Atmospheric Administration to the Oregon State University Sea Grant College Program under the National Sea Grant Program Marine Biotechnology Initiative.

REFERENCES

Akano, T., Miura, Y., Fukatsu, K, Miyasaka, H., Ikuta, Y., Matsumoto, H., Hamasaki, A., Shioji, N., Mizoguchi, T., Yagi, K., and Maeda, I., 1996, Hydrogen production by photosynthetic microorganisms, *Appl. Biochem. Biotech.*, 57/58:677–688.

Carte, B.K., 1996, Biomedical potential of marine natural products, *BioSci.*, 116:271–286.

Chrismadha, T., and Borowitzka, M.A., 1994, Effect of cell density and irradiance on growth, proximate composition, and eicosapentaenenoic acid production of *Phyaedactylum tricornutum* grown in a tubular photobioreactor, *J. Appl. Phycol.*, 6:67–74.

Lee, E.T.Y., and Bazin, M., 1990, A laboratory scale air-lift helical photobioreactor to increase biomass output rate of photosynthetic algal cultures, *New. Phytol.*, 116:331–335.

Qi, H., and Rorrer, G.L., 1995, Photolithotrophic cultivation of *Laminaria saccharina* gametophyte cells in a stirred-tank bioreactor, *Biotech. Bioeng.*, 45:251–260.

Rabe, A.E., and Benoit, R.J., 1962, Mean light intensity—a useful concept in correlating growth rates of dense cultures of microalgae, *Biotech. Bioeng.*, 4:377–390.

Radmer, R.J., 1996, Algal diversity and commercial algal products, *BioSci.*, 46:263–270.

Richmond, A., 1996, Efficient utilization of high irradiance for production of photoautotrophic cell mass: a survey, *J. Appl. Phycol.*, 8:381–387.

Rorrer, G.L., Polne-Fuller, M., and Zhi, C., 1996, Development and bioreactor cultivation of a novel semi-differentiated tissue suspension derived from the marine plant *Acrosiphonia coalita*, *Biotech. Bioeng.*, 49:559–567.

Rorrer, G.L., Yoo, H.D., Huang, B., Hayden, C., and Gerwick, W.H., 1997a, Production of hydroxy fatty acids by cell suspension cultures of the marine brown alga *Laminaria saccharina*, *Phytochem.*, 46:871–877.

Rorrer, G.L., Mullikin, R.K., Huang, B., Gerwick, W.H., Maliakal, S., and Cheney, D.P., 1997b, Production of bioactive metabolites by cell and tissue cultures of marine macroalgae in bioreactor systems, in *Plant Cell and Tissue Culture for Food Ingredient Production*, Fu, T.J. et al. (eds.), Plenum Press, New York, in press.

Watanabe, Y., de la Noüe, J., and Hall, D.O., 1995, Photosynthetic performance of a helical tubular photobioreactor incorporating the cyanobacterium *Spirulina platensis*, *Biotech. Bioeng.*, 47:261–269.

Watanabe, Y., and Hall, D.O., 1996, Photosynthetic production of the filamentous cyanobacterium *Spirulina platensis* in a cone-shaped helical tubular photobioreactor, *Microbiol. Biotechnol.*, 44:693–698.

Zhi, C., and Rorrer, G.L., 1996, Photolithotrophic cultivation of *Laminaria saccharina* gametophyte cells in a bubble-column bioreactor, *Enz. Microb. Tech.*, 18:291–299.

50

TECHNOECONOMIC ANALYSIS OF ALGAL AND BACTERIAL HYDROGEN PRODUCTION SYSTEMS

Methodologies and Issues

M. K. Mann and J. S. Ivy

National Renewable Energy Laboratory
1617 Cole Boulevard
Golden, Colorado 80401

Key Words

technoeconomic analysis, feasibility issues, hydrogenase-based algal hydrogen, bacterial water-gas shift

1. SUMMARY

The goal of this work is to provide direction, focus, and support to the development and introduction of renewable hydrogen through the evaluation of the technical, economic, and environmental aspects of hydrogen production technologies. The advantages of performing analyses of this type within a research environment are severalfold. First, the economic competitiveness of a project can be assessed by evaluating the costs of a given process compared to the current technology. These analyses can therefore be useful in determining which projects have the highest potential for near-, mid-, and long-term success. Second, the results of a technoeconomic analysis are useful in directing research toward areas in which improvements will result in the largest cost reductions. Finally, as the economics of a process are evaluated throughout the life of the project, advancement toward the final goal of commercialization can be measured. The production of hydrogen by biological processes has been analyzed to identify key issues relating to the conceptual design and costing of algal- and bacterial-based systems. Possible productivity advances and gas separation were the two most important considerations in a system that uses bacteria to produce hydrogen from biomass synthesis gas. Pond-based algal system studies have found gas separation, auxiliary equipment requirements, and pond designs to be key issues. The methodology for analyzing these systems and a discussion of key issues will be presented.

2. INTRODUCTION

Biological processes have the potential to be an important source of hydrogen in the 21st century. Many aspects of algal- and bacterial-based systems make them uniquely suited to meeting future energy demands. Their inherent renewability and flexibility, for example, may greatly reduce our use of fossil fuels and make local energy production possible. Drawbacks exist, however, and must be faced early in our research programs. Specifically, high cost and low efficiency make the prospects for biological hydrogen production in the near-term unlikely.

Performing technoeconomic analyses is an effective way to determine the likelihood of a process ever being commercially competitive. As its name implies, both the technical and economic feasibility of a system are examined, with the main goal being to identify necessary research improvements and goals. That is to say, rather than using technoeconomic analysis to "kill" projects that don't measure up to current technologies, it is best used as a tool to guide the research toward areas that will result in the greatest reduction in system cost. However, the results of such studies may also lead the researchers and program managers to believe that the technology will never compete no matter how many research breakthroughs are made. However, such decisions are hardly ever made without the use of other tools such as program portfolio planning, market analysis, and although we hate to admit it, politics.

The work reported here is sponsored by the U.S. Department of Energy's Hydrogen Program. The discussion is limited to analysis work being conducted on biological hydrogen projects at the National Renewable Energy Laboratory (NREL), and so should not be seen as an overall assessment of the field. The two technologies studied here are bacterial water-gas shift (studied at NREL by P. Weaver and P. Maness) and hydrogenase-based algal hydrogen production (studied at NREL by M. Ghirardi and M. Seibert).

The level of technoeconomic analysis appropriate for a project is dictated by the maturity of the technology and the amount of data that has been gathered, as well as the purpose of the evaluation. A new initiative may require only a literature search in order to determine if process costs can be estimated from related systems or if laboratory data are needed first. Alternatively, a cost boundary analysis is appropriate for a project that has been conceptualized but has not generated a significant amount of data. This type of analysis determines the minimum yields, maximum manufacturing costs, or necessary market conditions for the technology to be economically competitive. Finally, a highly detailed analysis can be performed on processes that have been more or less fully conceptualized, and laboratory work has provided yield and operating data. This type of analysis includes a process flowsheet, material and energy balances, equipment cost estimation, and the determination of the cost of the final product over the expected life of an industrial-scale plant.

The amount of data currently available for the two biological hydrogen processes of interest precludes the possibility of performing a detailed analysis. Assumptions on hydrogen yield and oxygen tolerance for the algal-based system are speculative based on achievements in other biological processes. Maximum yield and appropriate reactor design for the bacterial water-gas shift process are better defined but still contribute considerable uncertainty to our understanding of what a commercial system will look like.

A preliminary economic analysis of the two NREL biological hydrogen production processes has previously been assessed and reported (Mann, 1996) and will not be reiterated here. Rather, this work serves to identify some of the key issues and difficulties these processes face. Specifically, the relationship between productivity, gas separation, and bioreactor operating conditions should be considered in designing the bacterial water-gas shift system. Similarly, the importance of gas separation, auxiliary equipment requirements, and pond design are generally not the focus of the research taking place but were found to be troubling hurdles.

3. EXPERIMENTAL AND RESULTS

3.1. The Bacterial Water-Gas Shift Process

Biomass, in the form of trees and grasses, can be converted through thermochemical processes to electricity, liquid transportation fuels, and hydrogen. Basic block flow diagrams for the major processes for producing hydrogen from biomass are shown in Figure 1. Gasification followed by steam reforming is the technology most likely to make it to the market first. This is largely because gasification reactors have been successfully demonstrated for the purpose of producing electricity in combined-cycle power plants, and conventional natural gas reformers need only slight modification to be able to use treated biomass-derived product gas. Hydrogen production from biomass through pyrolysis followed by steam reforming is an alternative to gasification, and allows for the possibility of using cheaper waste feedstocks to produce an oil that can be more easily transported than biomass. Additionally, depending on the reforming technology, a co-product can be made from a fraction of the oil. Since the majority of the research in the areas of energy and fuels has been on thermochemical processes, these are generally considered to have economic promise barring expensive feedstocks.

Researchers at NREL have been working on a unique type of hydrogen-producing activity in a strain of photosynthetic bacteria that function only in darkness to shift CO (and H_2O) into H_2 (and CO_2). A discussion of this effort can be found in Weaver, et al. (1996, 1997). Mass transfer of the CO to the active sites in the bacteria currently limit higher productivities, although research on this front is progressing well. Looking at this process from a systems standpoint has identified trade-offs between productivity and lower capital cost depending on the operating conditions chosen.

Higher operating temperatures in the bioreactor may increase productivity because of higher reaction rates. However, our analysis has identified this as a possible problem for the feasibility of this system. The solubilities of the CO feed and CO_2 product are decreased at increased temperatures, possibly making the system less efficient and more expensive. A decreased amount of CO dissolved in the water phase will further reduce mass transfer to the active sites. Furthermore, if less product CO_2 dissolves in the water flowing through the bioreactor, an expensive separation step to produce a hydrogen-rich product will be necessary.

Figure 2 shows the possible scenarios for producing hydrogen from biomass using bacteria to effect the water-gas shift reaction. Both the gasification and the bacterial process can occur at either high or low pressures. The low-pressure gasifier would likely be an indirectly heated dual-bed system like the one developed by Battelle and now licensed by FERCO. It's called indirectly heated because steam rather than air or oxygen is fed to the gasifier, and the heat necessary for the endothermic gasification reactions is supplied by

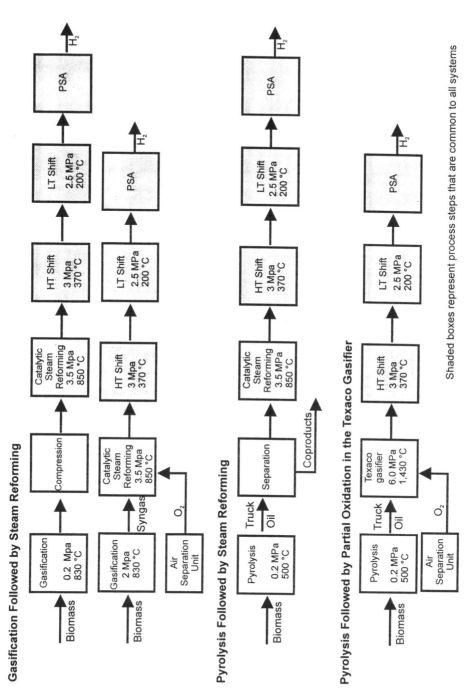

Figure 1. Major thermochemical options for the conversion of biomass into hydrogen.

Technoeconomic Analysis of Algal and Bacterial Hydrogen Production Systems 419

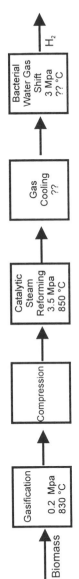
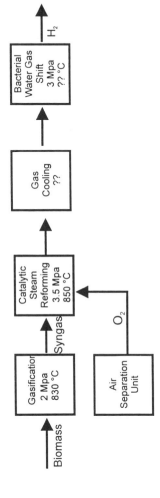
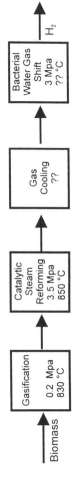

Figure 2. Options for the bacterial water-gas shift process.

hot sand circulating between the char combustor and gasifier. The high-pressure gasifier could be one like that developed by the Institute of Gas Technology. For more information on the configuration of these gasifiers, see Craig and Mann (1996).

The first option shown in Figure 2, low-pressure gasification followed by a high-pressure bioreactor, has two advantages. First, an air separation unit to provide oxygen for the gasifier is not needed. Second, at higher pressures, the solubility of CO_2 in the water flowing through the bioreactor will be higher, thus providing a higher-purity hydrogen product stream without gas separation. However, higher operating pressures mean a higher capital cost for the bioreactor and auxiliary equipment.

Figure 3 shows the results of preliminary mass balance calculations for the low-pressure gasifier system. Because the bacteria are only capable of converting CO to hydrogen, the scrubbed gasifier product gas must first pass through a steam reformer to break down the hydrocarbons. Those remaining in the syngas pass through the bioreactor and end up in the product gas. Depending on the species of hydrocarbons present and the purity requirements for the hydrogen, no further gas cleaning may be necessary. Generally, catalytic steam-reforming efficiently converts hydrocarbons into CO and CH_4 (plus H_2O, CO_2, and H_2). If the hydrogen is to be used as a fuel in an internal combustion engine, the lighter hydrocarbons shouldn't pose any problems. Some fuel cells may also be able to handle some methane, although they would suffer from decreased efficiency. Additionally, the product gas would contain the unreacted methane, although at levels far below its flash point.

3.2. Hydrogen from Hydrogenase-Based Algae Systems

Systems that use hydrogenase-catalyzed algae to produce hydrogen are being developed at NREL and other laboratories. Though potentially very efficient, these systems currently suffer from low light saturation and low oxygen tolerance. The first problem, which hinders the algae from using all of the energy available to it, is being worked on by researchers at Oak Ridge National Laboratory (Greenbaum, et al., 1997). The second problem is being studied at NREL and is part of what will be discussed here. For more information on the system, consult Ghirardi et al. (1996, 1997) and Greenbaum et al. (1996, 1997).

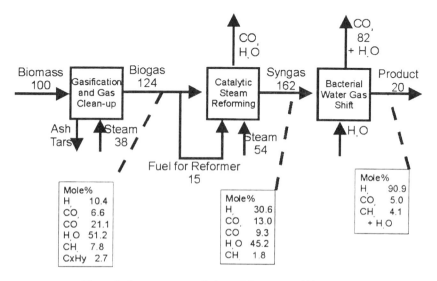

Figure 3. Stream summary for bacterial water-gas shift process.

Previous analyses of algal hydrogen production have been conducted by NREL and others. They have shown that these systems have the *potential* to be economically feasible, but failed to identify the key cost-drivers. Therefore, for this analysis, algae systems were studied from a process standpoint, with the purpose of aiding the researchers in design. At this time, too many unknowns exist for a full cost analysis. The final mutant has not been selected, likely yield and the degree to which the algae can tolerate oxygen have not been determined, and full sunlight effects are untested. Without this information, a broader process study has been undertaken. The feasibility of the one-pond versus two-pond configuration and gas separation issues have been considered.

Because the area required for daytime solar collection is larger than that needed for nighttime hydrogen production, a single covered pond will require a larger cover than one where the two activities take place in separate ponds. Generally it's assumed that because of higher cover costs, a two-pond system will be more economical than a one-pond system. However, because of the extra auxiliary equipment and operating costs associated with a two-pond system, a single pond will be cheaper and easier to operate. This is easiest to understand when it's observed that the entire volume of the growth pond has to be transferred into the fermentation pond in as short a time as possible every day as sunlight levels decrease. In the morning, the contents have to be transferred back again. This transferring results in very high pumping and piping costs. Furthermore, because the hydrogen production pond is smaller than the growth pond, a settling tank and water storage tank must also be added. These extra costs are greater than the likely cost of extra covers, and add complexity to the operation of the system.

Because hydrogen and oxygen are simultaneously produced by these systems, gas composition and separation are important issues. The concentration of hydrogen that will result in an explosion is 18% by volume in air, while the lower flammability limit is only 4.1% in air (Lewis and von Elbe, 1961). These limits are lower in an oxygen-rich environment. Therefore, to maintain a safe system, the concentration of hydrogen inside the pond cover and process piping must be kept at levels substantially lower than the two-thirds hydrogen, one-third oxygen ratio at which it's produced. Adding to the complexity of the problem, the algae are not currently able to tolerate even modest levels of oxygen. A solution might be the use of a recycling purge gas, preferably product hydrogen or air instead of pure nitrogen.

If the hydrogen product is used as the recycle gas, only one separations step will be required. Figure 4 shows the amount of hydrogen product recycle as a function of the hydrogen concentration. For safe operating in oxygen-rich conditions, hydrogen must be below 4% or above 96% by volume (Coward and Jones, 1952, Payman and Titman, 1936). This figure shows that more than 10 mol of hydrogen need to be recycled through the system for each mole produced in order to obtain the upper limit. This increased flow-rate will make the piping, blower, and separations equipment much more expensive and will reduce the overall amount of hydrogen recovered. Furthermore, leaks from the system will be more dangerous and expensive.

An alternative to using product hydrogen would be to use air or nitrogen and air. Figure 5 shows the amount of pure nitrogen that must be added to the air recycle versus oxygen tolerance for different air flow-rates. Higher oxygen tolerances are required when lower amounts of nitrogen are added to the air recycle. The lines cross at the concentration of oxygen in air. Figure 6 more plainly illustrates this transition. When the oxygen tolerance is below 21%, more nitrogen is required as more air is added to the system; above 21%, the required nitrogen decreases with increased air flow-rates. Thus, it is imperative that the algae be able to tolerate oxygen at concentrations greater than atmospheric. Achieving this will substantially reduce the amount of recycle gas needed to maintain the hydrogen con-

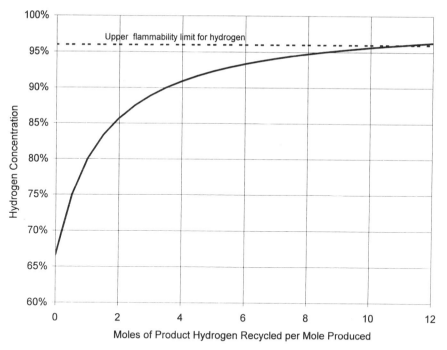

Figure 4. Product recycle rate necessary to obtain a safe hydrogen concentration.

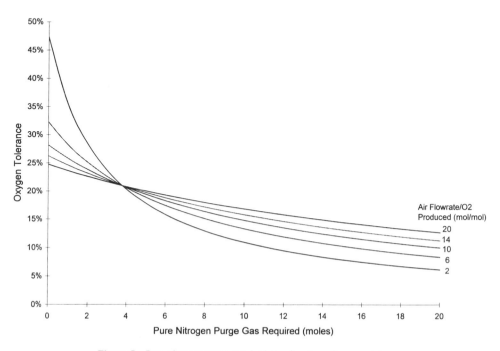

Figure 5. Pure nitrogen purge required as a function of oxygen tolerance.

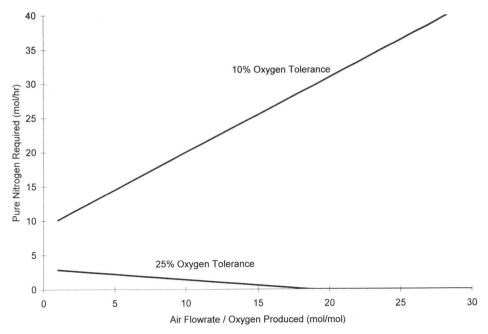

Figure 6. Pure nitrogen requirements for oxygen tolerances above and below atmospheric conditions.

centration at safe limits. However, because the concentration of the hydrogen in the product gas will be low, separations equipment will be expensive and hydrogen recovery very poor.

4. CONCLUSIONS

Biological hydrogen production has been shown to have potential for meeting future energy needs. However, in addition to the problems being worked on in the laboratory, technoeconomic analysis has identified process concerns that will also need to be addressed.

In the production of hydrogen from bacteria effecting the water-gas shift reaction, choosing the right operating conditions is vital to obtaining a balance between high productivities and low capital cost. Higher operating temperatures in the bioreactor will increase reaction rates but decrease the solubility of the CO substrate and CO_2 product. This decrease in solubility will result in poorer mass transfer of the CO to the active sites and increased separations costs. Operating pressure also has a significant effect on the feasibility of the system. The gasifier should be at low pressure to reduce the need for pure oxygen feed, while operating the bioreactor at high pressure will decrease product separations costs by increasing the solubility of CO_2. However, higher operating pressures will mean higher bioreactor capital and operating costs. Future work will seek to quantify the appropriate balance between high and low temperatures and pressures.

Operational problems also exist for the production of hydrogen by photosynthetic algae in ponds. Single-pond systems will be cheaper and easier to operate than dual-pond systems because of the reduction in equipment needed to transfer the entire volumes between ponds twice each day. Of greater concern than this issue, however, is the inherent danger of simultaneously producing hydrogen and oxygen in these systems. To maintain hydrogen levels either above or below safe concentrations, a recycle purge gas will be necessary. Using product hydrogen will result in extremely large volumes of gas needing to be transferred and purified, and will make leaks from the system more dangerous. Using air or air mixed with pure nitrogen will be more feasible but still means extra gas to circulate and purify. Improving the oxygen tolerance of the algae to levels above atmospheric concentrations is critical. Future work will incorporate the different recycle gas and purification options into an existing economic analysis.

ACKNOWLEDGMENTS

We would like to thank the U.S. Department of Energy for sponsoring this work. Additionally, many thanks to Dr. Maria Ghirardi, Dr. Mike Seibert, Dr. Paul Weaver, and Dr. Pin-Ching Maness for providing lessons in biological hydrogen production and for their careful review of our work.

REFERENCES

Craig, K.R., and Mann, M.K., 1996, Cost and performance analysis of biomass-based integrated gasification combined-cycle (BIGCC) power systems, National Renewable Energy Laboratory Report No. NREL/TP-430–21657, Golden, Colorado.

Coward, H.F., and Jones, G.W., 1952, Limits of flammability of gases and vapors, *Bureau of Mines Bulletin 503, N70–74476*, and *AD-701575*.

Ghirardi, M., Markov, S., and Seibert, M., 1996, Development of an efficient algal H_2-producing system, in *Proceedings of the 1996 U.S. Department of Energy Hydrogen Program Review*, vol. 1, National Renewable Energy Laboratory Report No. NREL/CP-430–21968, Golden, Colorado.

Ghirardi, M., Flynn, T., Markov, S., and Seibert, M., 1997, Development of an efficient algal H_2-producing system, in *Proceedings of the 1997 U.S. Department of Energy Hydrogen Program Review*, National Renewable Energy Laboratory Report No. NREL/CP-430–21968, Golden, Colorado.

Greenbaum, E., Lee, J., and Tevault, C., 1996, Hydrogen production by photosynthetic water splitting, in *Proceedings of the 1996 U.S. Department of Energy Hydrogen Program Review*, vol. 1, National Renewable Energy Laboratory Report No. NREL/CP-430–21968, Golden, Colorado.

Greenbaum, E., Lee, J., Blankinship, S., and Tevault, C., 1997, Hydrogen and oxygen production in mutant FUD26 of *Chlamydomonas Reinhardtii*, in *Proceedings of the 1997 U.S. Department of Energy Hydrogen Program Review*, National Renewable Energy Laboratory Report No. NREL/CP-430–21968, Golden, Colorado.

Lewis, B., and von Elbe, G., 1961, *Combustion, Flames and Explosions of Gases*, 2nd ed., Academy Press, New York.

Mann, M., Spath, P., and Kadam, K., 1996, Technoeconomic analysis of renewable hydrogen production, storage, and detection systems, in *Proceedings of the 1996 U.S. Department of Energy Hydrogen Program Review*, vol. 1, National Renewable Energy Laboratory Report No. NREL/CP-430–21968, Golden, Colorado.

Payman, W., and Titman, H., 1936, Limits of inflammability of hydrogen and deuterium in oxygen and air, *Nature*, 137:160–168.

Weaver, P., Maness, P.C., Markov, S., and Martin, S., 1996, Biological H_2 from syngas and from H_2O, in *Proceedings of the 1996 U.S. Department of Energy Hydrogen Program Review*, vol. 1, National Renewable Energy Laboratory Report No. NREL/CP-430–21968, Golden, Colorado.

Weaver, P., Maness, P.C., Markov, S., and Martin, S., 1997, Biological H_2 from syngas and from H_2O, in *Proceedings of the 1997 U.S. Department of Energy Hydrogen Program Review*, National Renewable Energy Laboratory Report no. NREL/CP-430 21968, Golden, Colorado.

51

ENVIRONMENTAL ASPECTS OF LARGE-SCALE MICROALGAE CULTIVATION

Implications for Biological Hydrogen Production

Roger Babcock, Jr.

Civil Engineering Department
University of Hawaii at Manoa
2540 Dole Street, Holmes Hall 383
Honolulu, Hawaii 96822

Key Words

microalgae, mass cultivation, environmental

1. OUTLINE

The environmental issues involved with large-scale production of microalgae or bacteria are closely associated with production efficiency. In microalgae production, it is necessary to maximize the cell production rate (i.e., the cell yield that is equivalent to the grams of cells produced per gram of substrate) when the cells or a cellular component are the desired product. In other cell cultivation systems where the desired product is not the cell mass, or even contained within the cell mass, such as biohydrogen production, it is necessary to minimize the cell yield and maximize the production of the desired product (i.e., the grams of H_2 per gram of substrate). In each case, it is necessary to minimize substrate and energy inputs. These are production issues, but are also directly related to waste issues. Generally, as the mass production rate of the desired product increases, so does the rate of by-product formation. The undesired by-products must be separated and disposed of as waste. In addition, any required purification of the desired products may result in additional waste product streams. These waste by-product streams are a potential source of environmental pollution. The type and quantity of waste products depend on the type of growth reactor, type of microbial culture, type of substrate, conversion efficiency, and growth environment. In addition to solid waste products from cell wasting, other waste products include liquid waste (unused substrate, converted or residual nutrients, salts, etc.) and off-gases (odor, CO_2, CH_4, etc.). Requirements for further processing (treatment) of

BioHydrogen, edited by Zaborsky *et al.*
Plenum Press, New York, 1998

wastes depend on local conditions, various regulations, and available disposal methods. For the wasted cell solids, dewatering alone may be suitable prior to disposal. For the spent liquid stream, the water quality (concentrations of organic material, metals, trace organics, salts, nutrients, etc.) will dictate what degree of additional treatment is required for any particular disposal method. Other important environmental issues are related to agitation and aeration (transfer efficiency, substrate contact, power requirements), processing (solid-liquid separation, recovery [concentration], purification), environmental requirements (C, N, P, trace minerals, vitamins, energy, temperature, mixing, pH, etc.), and maintenance of strain purity. This poster gives an overview of the environmental issues, problems, and needs associated with large-scale outdoor cell cultivation systems.

2. ALGAL MASS PRODUCTION SYSTEMS

Mass production of microalgae has been suggested for production of specialty chemicals, such as carotenoids, pigments, compounds with medical value including anti-cancer agents, food supplements for humans and animals, wastewater treatment, aquaculture, and even energy production since the 1950s (Dugan, 1980). Microalgae are attractive because as photoautotrophs their nutritional requirements are simple, requiring only inorganic nutrients and sunlight, both of which are abundant and "free." Currently, wastewater treatment is likely to be the largest intentional use of algae on a worldwide basis (in terms of mass of algae used and the number of facilities); however, in this application, there is little interest in maintaining any specific species or in harvesting the algae. The currently most economically successful mass-culturing of a species of algae for the generation of a saleable product is likely to be the freshwater blue-green algae Spirulina, which has a high protein content ($\approx 60\%$). Laboratory-scale algal growth systems (photobioreactors) used to study growth and other physiological characteristics can have many forms, from simple plastic bags and shake-flasks to chemostats and complicated bioreactors that rotate about a light source. The bench-scale units have light-energy requirements that can be easily obtained using artificial sources in the laboratory. In contrast, algal mass-culture systems are interesting only to the extent that they are able to capture the "free" supply of incident solar radiation. For economic reasons, this generally means that large, outdoor, shallow ponds are required to maximize the surface area exposed to the sunlight. Deep, well-mixed photobioreactors have been studied in the past, but it is now generally accepted that the energy input required for mixing such reactors is well in excess of the benefit gained by reduced land area. Even for shallow (<1.5 feet) photobioreactors, there is a trade-off between energy input and the beneficial effects of reactor mixing on net algae growth rate, which generally suggests that only a moderate degree of mixing is cost-effective.

3. ALGAL MASS PRODUCTION ISSUES

Algal mass-culturing problems include culture species control, culture mixing, nutrient availability and addition, carbon supply source, water supply, evaporation, separation and harvesting of algae, and wastewater disposal.

The factors that would be essentially uncontrollable in a large-scale mass-culturing system include sunlight (intensity and duration) and temperature. The factors that are more or less controllable include nutrient additions (which should be in excess), algae concentration (related to harvest rate), mixing, CO_2 availability, pH, fluid depth, dilution rate (retention time), and predator species.

There are several potential environmental impacts as a result of algae mass-culturing. First, there would be impacts on land resources (land area for gathering sunlight), which can be significant because shallow ponds are generally the most effective. Probably, such systems should not use prime farmland. Second, there could be impacts on water resources (water supply) due to the need for regular dilution. The water supply impacts, if any, will depend on local process site conditions; in water-scarce locations, the existing supply may be needed for "higher" uses such as drinking water. In the general case, water utilized will not be available for other potential uses. Third, there would be energy demands, including direct energy demands for electricity (and possibly petroleum) for various powered equipment, as well as indirect energy demands such as that required to produce fertilizer (nutrients), feeds, additives, and pesticides.

Fourth, there could be impacts of emissions on degradation of air quality in the vicinity of the production facility. Air quality could be degraded due to emission of odors, greenhouse gases, and combustion emissions if used for energy recovery. Fifth, there would be impacts on water quality from wastewater emissions. Water quality could be impacted by intentional point sources such as flow-through reactor effluent and separated water (centrate) from solids dewatering processes. If intentionally disposed directly to surface waters, then issues such as erosion, sediment load, nutrient loads, organic load, pesticides load, and algae blooms would be of primary concern. Water quality could also be impacted by unintentional non-point sources, such as rainfall runoff from the site and overflow of ponds during heavy rains. In addition, unlined ponds could contaminate the underlying soil and groundwater aquifers with nutrients, pesticides, salts, and/or organic material. Lastly, there could be impacts due to solid waste generation and disposal requirements. Solid waste could be generated due to harvesting, dewatering, and/or extraction inefficiencies. Unless a use can be found for the waste solids, such as for a soil amendment, they would have to be disposed of at a landfill or other suitable facility.

Culture species control is dependent upon virtually all of the factors which are important for growth optimization including: temperature, dilution rate and media replacement cycle, pH, nutrient concentrations, bioreactor cleanliness, and biomass concentration. The most general control method is to maintain relatively constant environmental conditions in the production unit. Other control methods such as pH, temperature, or nutrient concentration manipulation may be successful for control of specific predator species in some cases (Tseng et. al., 1991; Laws et. al., 1983).

4. WASTE PRODUCTION AND TREATMENT ISSUES

Algae are known to excrete waste products (metabolites) that are toxic to other species as well as to the source species (Hutchinson, 1967). During mass production of algae, these waste products must be removed at a proper rate or interval to ensure high growth rates. Generally, the waste metabolites are removed by harvesting a portion of the bioreactor contents (not just the algae), followed by dilution with fresh medium. The harvested portion consists of the desired algal biomass and the undesired waste (spent) medium. One or more separation process is then employed to remove the algae and produce a wastewater. Most of the algae mass production literature does not discuss wastewater production and disposal issues. Often, the quantity and quality of wastewater generated are not reported, and sufficient information is not provided to allow determination via calculation.

There is evidence (Laws et al., 1986) that it is best to not grow algae in a strictly continuous-flow mode in which dilution water is added continuously and the bioreactor is harvested at the same constant rate. Daily or other longer cycle dilutions of a fixed percentage of the bioreactor could be considered a semi-batch operation. Laws et al. found that optimal algal production rates were achieved when 86% of their bioreactor contents were replaced with fresh medium on every third day. This is a semi-batch operation because a true dilution rate cannot be calculated and the dilution process must be described. An average dilution rate could be calculated as 0.86 divided by 3, which equals 0.287 d^{-1}; however, this is not extremely useful in terms of process operation since if one were to waste 28.7% of the bioreactor contents each day (as implied by the dilution rate), one would not expect to obtain the same results reported by Laws et al. However, this "average" dilution rate is useful for determination of the quantity of wastewater that is generated and which must be disposed following any necessary treatment.

Several investigators have indicated that dilution rates between 0.5 and 1.0 d^{-1} could work well and that approximately 0.7 d^{-1} may be optimal for the general case (Goldman, 1979). A dilution rate of 0.7 d^{-1} will result in a large quantity of wastewater which must be disposed of (70% of the bioreactor contents each day).

The spent liquid stream water quality (concentrations of dissolved and suspended organic material, metals, trace organics, salts, nutrients, etc.) will dictate what degree of additional treatment is required for any particular disposal method. Potential disposal methods might include surface water discharge, injection wells, and ocean disposal.

5. BIOHYDROGEN IMPLICATIONS

Many of the potential biohydrogen organisms are photosynthetic (anaerobic purple-sulfur and non-sulfur bacteria, and aerobic filamentous cyanobacteria), and therefore could be grown in large, well-mixed, slurry-type photobioreactors like those used for microalgae. However, these photobioreactors would have to be covered in order to contain the desired gaseous product. The photobioreactor might resemble a sealed greenhouse or a system similar to some anaerobic digesters, which are essentially shallow ponds covered with plastic or simply large, sealed bags. For a photosynthetic system, the cover or bag would need to be clear plastic (or other material) to allow light transmission. Brouers and Hall (1985) conducted experiments using cyanobacteria and found that immobilized cells had higher (by a factor of two) H_2 production rates than free-living cells. This means that an attached-growth type system similar to a rotating biological contactor with a clear cover might be appropriate.

Many of the environmental issues discussed above for microalgae mass-culture systems will also be applicable to large-scale biohydrogen production systems. Similarities will likely occur with respect to wastewater quantity and quality issues, wastewater disposal methods, water supply and evaporation, and solid waste generation, dewatering, and disposal issues.

Differences will occur due to the potential for anaerobic waste stream odors because all of the biomass is a by-product requiring disposal, and because gas purification by-products (off-gases) may be odorous or have pollution potential.

REFERENCES

Brouers, M., and Hall, D.O., 1985, Hydrogen production, ammonia production, and nitrogen fixation by free and immobilized cyanobacteria, in *Proceedings of the Third International Conference on Energy from Biomass*, Venice, Italy, March 25–29, 1985, pp. 387–392.

Dugan, G.L., 1980, *Algal Mass Culture: Principles, Procedures, and Prospects*, Hawaii Natural Energy Institute, University of Hawaii, Honolulu, Hawaii.

Goldman, J.C., 1979, Outdoor algal mass cultures-II: photosynthetic yield limitations, *Water Research*, 13:119–136.

Laws, E.A., Terry, K.L., Wickman, J., and Chalup, M.S., 1983, A simple algal production system designed to utilize the flashing light effect, *Biotechnology and Bioengineering*, 25:2319–2335.

Laws, E.A., Taguchi, S., Hirata, J., and Pang, L., 1986, High algal production rates achieved in a shallow outdoor flume, *Biotechnology and Bioengineering*, 28:191–197.

Tseng, K.F., Huang, J.S., and Liao, I.C., 1991, Species control of microalgae in an aquaculture pond, *Water Research*, 25(11):1431–1437.

AN AUTOMATED HELICAL PHOTOBIOREACTOR INCORPORATING CYANOBACTERIA FOR CONTINUOUS HYDROGEN PRODUCTION[*]

Anatoly A. Tsygankov,[1] David O. Hall,[2] Jian-guo Liu,[3] and K. Krishna Rao[2]

[1]Institute of Soil Science and Photosynthesis
Russian Academy of Sciences
Pushchino, Moscow Region, 142292, Russia
[2]Division of Life Sciences
King's College London
Campden Hill Road, London W8 7AH
[3]Institute of Oceanology
Chinese Academy of Sciences
Qingdao 266071, China

Key Words

Anabaena, cyanobacteria, hydrogen production, nitrogenase, photobioreactor

1. SUMMARY

A laboratory-scale, helical tubular photobioreactor, made of polyvinyl chloride tubing, was constructed to study photobiological H_2 production by cyanobacteria catalyzed by cellular nitrogenase. The photobioreactor was connected to a computer, with specially written software and hardware, for automated control of the biochemical and environmental processes regulating H_2 evolution. A basic photobioreactor unit with a photostage volume of 4.35 L and containing *Anabaena azollae* grown in 98% air + 2% CO_2 gas mixture produced H_2 at a rate of 13 mL $H_2 \cdot h^{-1} \cdot L^{-1}$ culture suspension when the gas phase was

---

[*] Abbreviations: OD, optical density; PhBR, photobioreactor; PVC, polyvinylchloride.

BioHydrogen, edited by Zaborsky *et al.*
Plenum Press, New York, 1998

changed to Ar. The H_2 production was sustained for 6 h. No H_2 evolution was observed in the presence of photosynthetic O_2.

Techniques were introduced for keeping the cyanobacterial suspension in a turbulent state and for scrubbing the inside walls of the PVC tubing during the operation of the photobioreactor. A method for scale-up of the bioreactor is proposed.

2. INTRODUCTION

Photobiological H_2 production processes make use of the photosynthetic and enzymatic assemblies in eukaryotic green algae or prokaryotic cyanobacteria and photosynthetic bacteria. While green algal H_2 production is catalyzed by an inducible hydrogenase, the H_2 evolution from photosynthetic prokaryotes is catalyzed mainly by nitrogenase. Cyanobacteria are ideal microorganisms for solar energy conversion to the chemical energy of H_2 because water provides the protons and electrons for H_2 formation (Bothe and Kentemich, 1990; Markov et al., 1995; Hall et al., 1995). H_2 photoproduction studies in our laboratory have been carried out using the aerobic, filamentous, heterocystous, nitrogen-fixers *Anabaena* and *Nostoc* spp. H_2 evolution in these organisms is catalyzed by the nitrogenase located in the heterocysts. H_2 is formed as a co-product along with NH_3 during the ATP-dependent reduction of N_2.

$$N_2 + 8H^+ + 8e^- + 16\,ATP \rightarrow 2NH_3 + H_2 + 16ADP + 16Pi$$

Although many advances have been reported in the last few years on the technology and yield of H_2 production, the overall conversion efficiencies of light to chemical energy are still much lower than the theoretical (Rao and Hall, 1996). The factors that affect sustained H_2 production include oxygen sensitivity of the nitrogenase and the photosynthetic apparatus, photoinhibition of photosynthesis, oxidation of the H_2 formed by an uptake hydrogenase located in the heterocysts, etc. The cyanobacterial strain selected and the age of the culture are also important for better yields of H_2. This paper describes the design and assembly of a bench-top photobioreactor (PhBR) for the continuous production of H_2 from water, catalyzed by cyanobacteria.

3. EXPERIMENTAL

3.1. Organism and Growth

Batch and continuous culturing of *Anabaena variabilis* ATCC 29413, mutant PK84 (Mikheeva et al., 1995), and *A. azollae* Strassburg were performed in Allen and Arnon (1955) medium (30 °C, pH 7.0). The medium contained 1.0 µM Na_3VO_4 and no added Mo. In the batch culture mode, the photobioreactor was flushed with a mixture of 98% air and 2% CO_2 (250–300 mL/min). The pH was adjusted by automated addition of 0.1 N NaOH to the medium.

3.2. Activity Assays

The rates of H_2 photoproduction and C_2H_2 reduction (equivalent to N_2 fixation) by cell suspensions were assayed by gas chromatography in a Hewlett Packard 5890A. A 2-

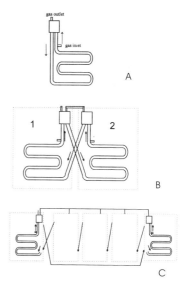

Figure 1. Tubular helical PhBR made of PVC tube: (A) main unit; (B) two main units connected in series; (C) proposed method for connection of any number of units in series.

mL cell suspension was incubated in 14-mL glass vials closed with Suba Seal stoppers in a thermostated water bath (30 °C) mounted over a magnetic stirrer and illuminated laterally with daylight fluorescent lamps (140 µE/m²/s). When assaying H_2 photoproduction, the air in the gas phase was replaced by Ar, whereas in the C_2H_2 reduction assays, the gas phase inside the vials contained Ar with 20% C_2H_2. The time of assays was 30–45 min with 5–10 min of measurement intervals.

3.3. The Photobioreactor

A helical tubular photobioreactor made of PVC tubing with a 10-mm inner diameter was used in this study (Watanabe et al., 1995) (Figure 1a). This PhBR has a very good surface/volume ratio (200 m^{-1} without gas exchanger unit) and can provide uniform illumination to the cultures. The main disadvantage of this type of PhBR lies in the separation of the process of photosynthesis and gas exchange. During cyanobacterial photosynthesis, O_2 accumulates and CO_2 decreases inside the photosynthetic section of the PhBR. Hence, the scale-up of this PhBR by enlargement of the photosynthetic section (the length of PVC tube) has limitations; accumulation of O_2 and/or decrease of CO_2 (and pH shift due to CO_2 changes) along the tube will inhibit growth and product (H_2) formation by the cells when the length of the tube is too great.

The maximal length of the tube (L) that would cause no inhibition or limitation of cultures can be calculated on the basis of the linear flow rate of the culture (V) and maximal non-inhibiting concentration of O_2 (minimal non-limiting concentration of CO_2) C_{out}: L = Vt, where t is time of culture flow along the tube with accumulation of a non-inhibiting concentration of O_2·t = $|C_{out}-C_{in}|$/XP where C_{in} is the concentration of O_2 (CO_2) at the gas exchanger, X is the biomass concentration, and P the specific rate of photosynthesis by the culture (rate of oxygen evolution or CO_2 fixation per unit of biomass per minute).

Thus, for any culture concentration with a known rate of photosynthesis, a maximal length of photosynthetic path exists (Figure 1a). The scale-up procedure must use a modified type of PhBR for enlargement. The PhBR illustrated in Figure 1A was used as the main unit for scaling-up. Two of these units were connected as shown in Figure 1B. Using this method of unit addition in series, it is possible to construct a PhBR of any volume without causing a decrease in the specific rate of the process (Figure 1C). All the photosynthetic parts of these units can be coiled together into a large, hollow cylinder. Thus, the large-scale PhBR is similar to a single, helical, coiled PVC tube with one difference: it has as many separate gas exchangers as there are units connected in series. This PhBR has only one technical disadvantage: it has many inlets for gas mixtures and many gas exchangers.

3.4. Prevention of Biofilm Formation on the Walls

The culture suspension flow in this PhBR is laminar and, thus, with time biofilm formation occurs on the walls that prevents light penetration inside the tube. Also, with the biofilm deposit, it is difficult to control the parameters of the process due to differences in environmental factors inside the biofilm and in the cell suspension. To prevent the deposition of culture on the inside walls, balls made of polyurethane foam were used. The balls were inserted into the PhBR made of two units (Figure 1B) and they circulated (at the rate 16–20 cm/sec) continuously through the reactor, thereby scrubbing the walls. It is essential to keep the inner space of the PhBR uniform and to make the balls with precise diameter to ensure continuous circulation of the balls inside the PhBR.

3.5. Installation of Sensors

To control the cultivation parameters, it is necessary to introduce sensors inside the PhBR. The most important place for pO_2 control is at the end of the photosynthetic section of the PhBR because oxygen accumulates there. The pO_2 sensor was introduced in the loop incorporating the cell with an inner diameter greater than the inner diameter of the tube. The circulation of the balls was not stopped in the cell and due to pulsed movements even cleaned the surface of the sensor. However, sensors for temperature, pH, and optical density required more active anti-sedimentation treatment because they prevented circulation of the balls. To overcome this, the sensors were introduced in a special bypass between neighboring loops. The linear rate of suspension flow inside the bypass was equal to the rate inside the loop connected with the bypass in parallel. To prevent the balls passing into the bypass, a grid was added at the entrance of the bypass. To prevent the sedimentation of culture on the sensors, a special type of shaker was installed in series with the chamber of sensors (Figure 2). Pulses of air pressure through the shaker activated pulses of suspension flow through a silicone tube inside the shaker and prevented cyanobacterial sedimentation on the surfaces of the sensors.

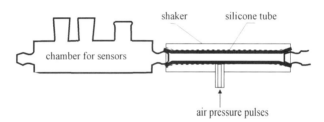

Figure 2. Shaker connected with chamber for sensors (temperature, optical density, pH) in series.

3.6. Continuous Culturing

For continuous cultivation, a constant volume of suspension inside the PhBR is essential. Inlets for media and titrants were installed near the gas inlet of one of the units for better mixing of fresh medium and titrants with the suspension. On addition of the medium, the level of suspension in the gas exchanger increased. The gas outlet was through a tube introduced from the top of the gas exchanger (Figure 1A). When the liquid level reached the lower end of this tube, excess liquid together with the gas mixture was pushed out into a culture collector.

3.7. Computer Control

Using modern IBM-compatible computers, it is now possible to create very user-friendly hardware and software for computer control of the culturing of microorganisms. Markov et al. (1997) were the first to report the design of a computer-controlled photobioreactor for H_2 production from *Anabaena variabilis*. The system we have now set up is much more advanced and versatile and can be used for the control of PhBRs to produce H_2 or for the cultivation of any phototroph.

3.8. Hardware

In the present system, computer control was realized at two levels: low-level and high-level. Low-level control was achieved through a device for automation of photobioreactors comprising a digital measurement and control system based on a standard AT286 motherboard. This control system combines preamplifiers of the signals from the sensors and digital processing of data. It uses its own software and works independently of the high-level computer. It is programmed for: (a) measurement of signals from sensors

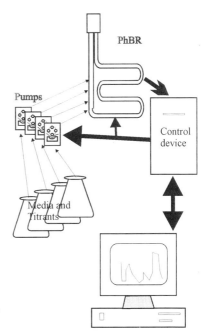

Figure 3. An automated helical photobioreactor incorporating cyanobacteria for continuous H_2 production.

and transformation of data into digital form; (b) storage of data as mean, minimal, and maximal values for every sensor for each hour; (c) realization of several modes of cultivation: uncontrolled regime, batch culture, chemostat, and turbidostat continuous culture (the regime is predefined at the start of the experiment and can be changed by the high-level computer), (d) control of pH, temperature, and optical density (in turbidostat regime) at the set level by switching on and off the pumps to control mode and intensity of illumination; and (e) transfer of the information about the process to the high-level computer.

The high-level computer is programmed for: (a) calibration of sensors and pumps, (b) altering the characteristics of the operation (settings, modes of cultivation, mode of illumination, tuning of sensors, etc.), (c) continuous visual presentation of data of the process as a graph, and (d) quick storage of data as a table of digits or as a graph. The high-level computer is connected with the control device via a RS232 port.

3.9. Software

The software has two closely connected parts: low-level software (for the control system) and high-level software for the high-level computer. Low-level software, written in PASCAL, allows the measurement and control of the cultivation parameters to store the data about the process of cultivation and to connect (if necessary) with the high-level computer. High-level software for the computer with Windows 3.1 or more advanced versions is written in DELPHI and enables the investigator to recalibrate the sensors, change any parameter (or regime of cultivation), monitor the operation by visual display unit as a graph or a table, store quick data (with discretion between the points from 2 sec), and file as a table or as an image. The main program has less then 400 KB volume, is friendly, and easy to use. It is possible to use the computer to operate the photobioreactor simultaneously for any other routine application.

3.10. Control of the Process

The control device has outputs of 220–240 V and controls the pumps by switching on and off the power supply. It is possible to use any kind of pump, including the simplest type without facilities for flow-rate regulation. In addition to pumps, it controls the valve for cooling water (220–240 V), light (220–240 V) and the additional device inserted specifically for air pulses (24 V, 0.5 A).

4. RESULTS AND CONCLUSIONS

4.1. Batch Culture of Cyanobacteria

A PhBR was constructed of two basic units with 20 m PVC tubing in each unit with gas and heat exchangers of 0.5 L volume each. Tubes of both units were coiled into one cylinder with a diameter of approximately 50 cm. The overall volume of the PhBR with connecting tubes and accessories was 4.35 L. Inside the tubular cylinder, four 55-W and two 32-W fluorescent lamps were introduced. By switching the lamps on in different combinations, the light intensity incident on the surface of the tubes could be changed from 0 to 230 $\mu E/m^2/sec$. At a gas flow rate of 0.3 and 0.2 L/min in unit 1 and unit 2, respectively, (Figure 1b), the linear flow rate of the suspension inside the tube was 22–24 cm/sec. The time for an overall cycle inside the PhBR under these conditions was 6.5–7.0 min.

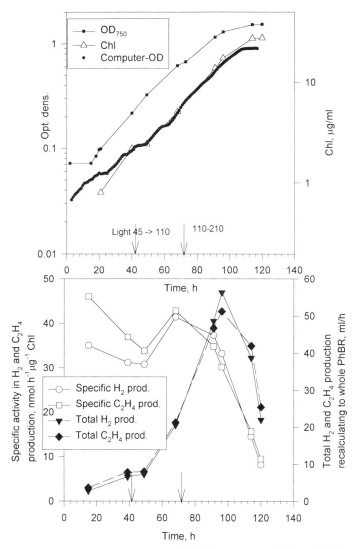

Figure 4. Batch culture of *A. azollae* grown in automated PhBR. Down arrows indicate the times when light intensities (µE/m²/sec) were changed.

A typical growth curve of the vanadium-culture of *A. azollae* in the PhBR is shown in Figure 4. In preliminary experiments, it was found that strong illumination at the start of cultivation inhibited the growth of cyanobacteria; thus, a higher light intensity was applied only later to more dense cultures by switching on additional lamps. Exponential growth was observed up to 15 µg chlorophyll/mL. Kinetics of the OD (at 670 nm) measured on-line by computer was well-correlated with the OD (at 750 nm) of samples simultaneously removed and measured with a spectrophotometer and with the chlorophyll a. concentration which was determined separately. We conclude that computerized measurement of OD is a very useful tool for the control of culture density as it can be recalibrated in terms of chlorophyll a. concentration.

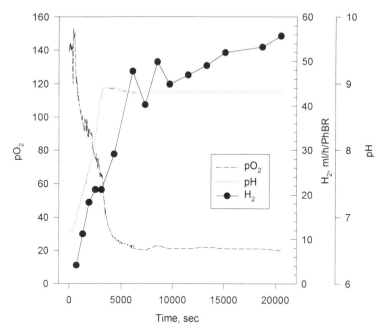

Figure 5. H_2 production by automated PhBR incorporating *A. azollae* grown in turbidostat culture. At the start (0 sec) the gas phase was changed from 98% air + 2% CO_2 to 100% Ar.

4.2. Enzyme Activity

Specific nitrogenase activity measured at two intervals did not change significantly during exponential growth (Figure 4), but decreased dramatically after the exponential phase. This differs from the data obtained with batch cultures growing under constant illumination where specific nitrogenase activity had a sharp maximum at the start of growth (Tsygankov et al., 1997). Thus, addition of saturating light to the culture prolonged the maximal nitrogenase activity up to the end of the exponential phase (Figure 4). As a result of high specific nitrogenase activity, the potential activity of whole PhBR in H_2 production reached 57 mL/h/PhBR (13 mL/h/L suspension). Compared with other data available (not higher than 5 mL/h/L suspension) in the literature (Kentemich et al., 1991; Kentemich et al., 1990; Ni et al., 1990; Yakunin et al., 1991) *A. azollae* grown in the PhBR showed rather high specific and nitrogenase activity for H_2 production.

4.3. Hydrogen Production by a Photobioreactor

To investigate the possibility of using the PhBR for sustained H_2 production, *A. azollae* was cultivated in a turbidostat with an OD of 0.7 (computer-monitored) and maximal H_2 evolution (Figure 4). The culture showed stable growth with a rate of 0.03 h^{-1} (data not shown). Nitrogenase activity measured in the vials was essentially the same as in the batch culture. Thus, the turbidostat regime allowed the culture to be maintained with a high nitrogenase activity for a long period.

After a change of gas phase from 98% air + 2% CO_2 to 100% Ar, the oxygen partial pressure decreased (Figure 5) and the pH increased, probably due to CO_2 loss by the Ar flush and by consumption by the culture. The presence of H_2 in the outlet gas was observed 10 min after the switch to Ar. After 1.5 h of exposure of culture under Ar, the pH stabilized at 8.9, and the pO_2 stabilized at a low level (showing a lack of photosynthesis due to the absence of CO_2), and H_2 production reached 45 mL/h/PhBR. The rate of H_2 production further increased and reached 56 mL/h/PhBR 6 h after the gas phase change, which is essentially the same rate as that measured in small vials (57 mL/h/PhBR). Thus, *A. azollae* produced H_2 under Ar in the PhBR at the same rate as in batch experiments in small vials, indicating that the method of scaling-up from 2 mL to 4.35 L did not cause any decrease in specific or volumetric activity.

After H_2 production under the Ar for 6 h, the culture was bubbled with a mixture of 98% air + 2% CO_2 (Figure 5). The culture then exhibited stable growth (data not shown). After 20 h of turbidostat growth, the gas phase was replaced again with pure Ar and the culture showed increased H_2 production at approximately the same rate as before (Figure 5).

It should be pointed out that *A. azollae* did not produce H_2 after replacement of air + CO_2 with Ar + CO_2 (data not shown). In the presence of CO_2, the partial oxygen pressure was not lower than 60% of air saturation (100% of air saturation equals 6.3 µL O_2/mL suspension). Taking into account the high nitrogenase activity of the cells, it is possible to assume that H_2 was evolved under these conditions but was oxidized by the uptake hydrogenase present in the cells.

5. CONCLUSIONS

(1) A method for scaling-up the PhBR without a decrease in specific rate of H_2 production is proposed. (2) Connection of the tubular helical PhBR to a computer with specially written soft and hardware programs for automated control of the process is described. (3) Model experiments with this scaleable PhBR showed the possibility of high rates of H_2 production by the PhBR (up to 13 mL H_2/h/L suspension).

Future investigations should be directed to research on: (a) selection of cyanobacterial strain with the highest rate of H_2 production, (b) optimal interval between time of H_2 evolution and time of culture regeneration, (c) optimal pH, temperature and light intensity during regeneration of culture and during H_2 production, (d) adoption of the process for natural light conditions (high light intensity in the day and darkness at night), and (e) studying the possibility of scaling-up this PhBR to hundreds of liters.

ACKNOWLEDGMENTS

Financial support for this research from RITE (Japan) and INTAS (Brussels) is gratefully acknowledged. Thanks are also due to D. Sveshnikov for discussions and suggestions on the assembly of the PhBR.

REFERENCES

Allen, M.B., and Arnon, D.I., 1955, Studies on nitrogen-fixing blue-green algae. 1. growth and nitrogen-fixation by *Anabaena cylindrica* lemm, *Plant Physiol.*, 30:366–372.
Bothe, H., and Kentemich, T., 1990, Potentialities of H_2 production by cyanobacteria for solar energy conversion programmes, in *Hydrogen Energy Progress VIII*, Veziroglu, T.N., and Takahashi, P.K. (eds.), Pergamon Press, New York, pp 729–734.

Hall, D.O., Markov, S.A., Watanabe, Y., and Rao, K.K., 1995, The potential applications of cyanobacterial photosynthesis for clean technologies, *Photosynthesis Research*, 46:159–167.
Kentemich, T., Haverkamp, G., and Bothe, H., 1990, The generation of molecular hydrogen by cyanobacteria, *Naturwissenschaften*, 77:12–18.
Kentemich, T., Haverkamp, G., and Bothe, H., 1991, The expression of a third nitrogenase in the cyanobacterium *Anabaena variabilis*, *Z. Naturforsch*, 46c:217–222.
Markov, S.A., Bazin, M.J., and Hall, D.O., 1995, The potential of using cyanobacteria in photobioreactors for hydrogen production, *Adv. Biochem. Eng. Biotechnol.*, 52:59–86.
Markov, S.A., Thomas, A.D., Bazin, M.J., and Hall, D.O., 1997, Photoproduction of hydrogen by cyanobacteria under partial vacuum in batch culture or in a photobioreactor, *Int. J. Hydrogen Energy*, 22:521–524.
Mikheeva, L.E., Schmitz, O., Shestakov, S., and Bothe, H., 1995, Mutants of the cyanobacterium *Anabaena Variabilis* altered in hydrogenase activities, *Z. Naturforsch*, 50c:505–510.
Ni, C.V., Yakunin, A.F., and Gogotov, I.N., 1990, Influence of molybdenum, vanadium, and tungsten on growth and nitrogenase synthesis of the free-living cyanobacterium *Anabaena azollae*, *Microbiology*, 59:395–398.
Rao, K.K., and Hall, D.O., 1996, Hydrogen production by cyanobacteria: potential, prospects, *J. Mar. Biotechnol.*, 4:10–15.
Tsygankov, A.A., Serebryakova, L.T., Sveshnikov, D.A., Rao, K.K., Gogtov, I.N., and Hall, D.O., 1997, Hydrogen photoproduction by three different nitrogenases in whole cells of *Anabaena variabilis* and the dependence on pH, *Int. J. Hydrogen Energy*, in press.
Watanabe, Y., de la Noue, J., and Hall, D.O., 1995, Photosynthetic performance of a helical tubular photobioreactor incorporating the cyanobacterium *Spirulinaplatensis*, *Biotechnol. Bioeng.*, 47:261–269.
Yakunin, A.F., Ni, C.V., and Gogotov, I.N., 1991, Growth and synthesis of different nitrogenases in *Anabaena variabilis* in dependence of presence of Mo, V and W in medium, *Mikrobiologiya* (Russian), 60:71–76.

53

INTERNAL GAS EXCHANGE PHOTOBIOREACTOR

Development and Testing in Hawaii

James P. Szyper,[1] Brandon A. Yoza,[1] John R. Benemann,[2] Mario R. Tredici,[3] and Oskar R. Zaborsky[1]

[1]Hawaii Natural Energy Institute
School of Ocean and Earth Science and Technology
University of Hawaii at Manoa
2540 Dole Street, Holmes Hall 246
Honolulu, Hawaii 96822
[2]Department of Civil Engineering
University of California
Berkeley, California
[3]Dipartimento di Scienze e Tecnologie Alimentari e Microbiogiche
University of Florence, Italy

Key Words

cyanobacteria, *Arthrospira*, hydrogen, tubular photobioreactor

1. SUMMARY

Two eight-tube internal gas exchange photobioreactors were deployed on an inclined platform at an ocean-front research facility in Honolulu, Hawaii. The photobioreactors consisted of 3.8-cm inside diameter, 20-m long tubes joined at the bottom to a manifold aerator and at the top to a degassing reservoir, with a total volume of 230 L and a nominal culture volume of 200 L. The eight-tube bioreactors were operated with the two outside tubes unaerated, thus establishing a recirculating flow through the bioreactors. *Arthrospira* (*Spirulina*) sp. was cultured using Zarrouk's medium. Cultures were monitored daily by analyzing samples for dry weights and optical cell densities. An automated system recorded incident light, culture temperature, pH, and dissolved oxygen concentrations.

Carbon dioxide was supplied once a day by bubbling pure CO_2 through the culture. As needed, cultures were cooled by a sprinkler system. Cultures were inoculated from laboratory stocks into a single-tube outdoor bioreactor at a density of less than 0.2 g/L, and maintained under 80% shade cloth to prevent photooxidation. Initial instantaneous growth rates ranged from 0.3 to 0.5 per day with final densities up to 2.5 g/L. An initial trial with a dilution of 30% per day under a 50% shade cloth demonstrated a productivity of about 1 g/L/day over a six-day period and an average of 0.7 g/L/day for a 19-day period. The utility of tubular photobioreactors for the development of a biophotolysis process is being explored further.

2. INTRODUCTION

Biological hydrogen production could provide another source of renewable fuel, required for a sustainable society of the future. Biophotolysis of water, using microalgae cultures, is one of the long-term goals of the U.S. Department of Energy's biohydrogen research and development program, as it would not be limited by the need for some type of waste or very low-cost fermentable biomass substrate. The various mechanisms of biohydrogen and the potential of this field have been reviewed recently (Benemann and Zaborsky, 1996; Benemann, 1996).

It is not certain yet which, if any, of the different modes of biophotolysis will be able to be developed to a practical scale. However, all modes require a photobioreactor (PBR) for the containment of the reaction and the capture of hydrogen. For biophotolysis processes, the photobioreactors must be of very low capital and operating cost, in particular for energy inputs, and must be able to have a relatively large unit scale, preferably >100 m^2. There are many types of photobioreactors—tubular, flat plate, bag-type—arranged in many configurations. From a gas exchange perspective, bioreactors can be classified as either internal (two-phase) or external (single-phase). Gas exchange, CO_2 and O_2, is the limiting factor in conventional algal photobioreactors (Weissman et al., 1988). This will be even more the case for biophotolysis processes where rates of hydrogen and oxygen production would be even higher, due to the much higher projected productivities. Thus, internal gas exchange bioreactors must be considered. These are well-known in the art but the scale-up of internal gas exchange PBRs is rather recent.

One of the most widely known types is the Biocoil reactors (Biotechna Ltd.), which consist of long tubes wound around a central cylinder, and can be as high as 3 m and 3 m in diameter. However, such geometry creates both undue pressure losses and significantly reduces the fraction of area exposed to sunlight, both of which are not acceptable. Bubbled flat plate and vertical tubular PBRs have also been described, but suffer from relatively small unit size (<10 m^2) (Benemann, 1997). For the Hawaii biophotolysis project, the internal gas exchange PBR developed by Tredici and colleagues (Tredici et al., 1994; Chini Zittelli et al., 1996; Tredici and Chini Zittelli, 1997) has been selected, as it is the only design that in principle could meet the minimal requirements of the envisioned Hawaii bioprocess involving hydrogenase-induced cyanobacterial cultures.

Controlled, sustainable biophotolysis is the overall goal of the BioHydrogen Program at the University of Hawaii, based in the Marine Biotechnology Center of the Hawaii Natural Energy Institute. Specific research objectives include design, construction, testing, and validation of a sustainable and efficient PBR system that takes advantage of Hawaii's low-latitude, high-insolation climate. Testing and validation include development of protocols for cultures of potentially hydrogen-producing cyanobacteria and mi-

Internal Gas Exchange Photobioreactor

Figure 1. The Tredici-design tubular photobioreactor located at Kewalo Basin at a waterfront research facility of the University of Hawaii.

croalgae at high biomass densities and turnover rates. Such protocols may well be specific to local conditions, which include daily maximum quantum flux values in excess of 2000 μmol m^2/s and daily totals exceeding 60 mol m^2/day during the spring season.

In this paper, we report the construction and development of a photobioreactor of Tredici design at our facility and early results of culture trials with a locally available strain of the filamentous cyanobacterium *Arthrospira platensis* (formerly *Spirulina*) (Tomaselli et al., 1996).

3. EXPERIMENTAL

The PBR platform, approximately 5 m wide and 20 m long, was assembled of standard construction scaffold topped with wooden walkways on each side and the middle. The platform had a slope of 6.5%, with an elevation of 1.4 m at the lower front and 2.7 m at the higher back end. Corrugated sheet metal was placed between the walkways to provide straight channels for the plastic tubes (Figure 1). The eight tubes for each reactor were attached to the manifolds on the bottom and top. The bottom manifold was equipped with an aeration port, which permitted the tubes to be aerated evenly. The aeration to the two outside tubes in each reactor was shut-off to allow these to serve as return channels for liquid, thus creating a circulation pattern throughout the reactor. The top manifold had a volume of about 30 L, to hold liquid displaced by the aeration. Air was supplied from an air compressor.

A strain of *Arthrospira* sp. was cultivated in the PBR using a high alkalinity (16 g/L bicarbonate) medium (made with tap water), as described by Zarrouk (1966). The cultures

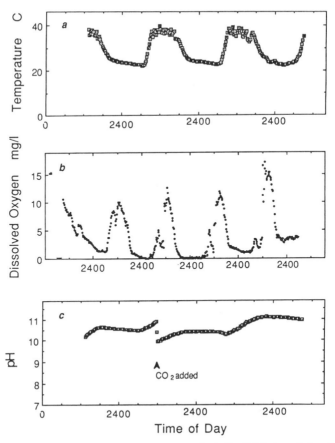

Figure 2. Representative diel cycles of monitored parameters in the Tredici design tubular photobioreactors. Data points were taken at 15-min intervals.

were initiated in the laboratory and then transferred to a single-tube PBR which was used to produce inoculum for the eight-tube PBRs to an initial density between 0.2–0.5 g/L. Culture growth and system performance were monitored by daily manual sampling for biomass determination as filter-collected (Whatman GFC) dry weight and in vivo absorbance at 665 nm, and by automated recording at 15-min intervals of incident light (PAR quantum flux) and culture temperature, dissolved oxygen concentration, and pH with a system of off-the-shelf hardware described by Szyper and Ebeling (1993). The cultures were provided manually with CO_2 on a daily basis. Cooling was provided by a sprinkler, also operated manually, when the temperature in the reactor exceeded 37 °C.

4. RESULTS AND CONCLUSIONS

Figure 2 provides a typical readout from the monitoring system over several days. Dissolved oxygen tension, measured in the top manifold (degasser), increased above saturation during the day, but was quite low at night, suggesting depletion of oxygen in the culture. It can be seen from Figure 2 that oxygen tension did not exceed 15 mg/L, that is,

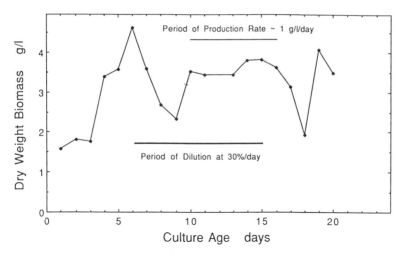

Figure 3. Time course of *Arthrospira* biomass during the growth trial with a period of constant daily medium replacement.

it never reached inhibiting values. The pH went to 11 when CO_2 was not added daily, but once-a-day addition of this nutrient appears to have been sufficient for this high bicarbonate medium.

Culture grew well in the single-tube PBR used to produce inoculum, from less than 0.2 g/L after inoculation to a high of 3.9 g/L after completion of the batch growth. Instantaneous growth rates of 0.4 per day were sustained for six days under 80 or 50% shade cloth, with a maximum of 0.7 per day. Production in this reactor was in excess of 0.5 g/L/day under 50% shade cloth. The culture was not homogenously dispersed through the reactor, however.

The cultures had to be kept under shade cloth for most of the time. When exposed to full sunlight, they continued to grow but lost pigments, as seen from a change in color from dark blue-green to paler yellow-green, with this effect not being reversed in a few days under shade cloth. A five-day comparison between cultures exposed to full sunlight and shade cloth revealed almost the same productivity in both cultures. Both observations are due to the fact that optimal cell density had not been obtained before exposing the cultures to full sunlight.

A six-day replication trial in which both eight-tube PBRs were managed identically was carried out. Actually, the two bioreactors started out with a 22% difference in biomass densities, but these converged and over the next five days differences were only between 1 and 11%, while densities increased from 1.6 to 4.6 g/L (with 80% shade cloth reduced to 50% on day 3). Maximum daily productivities were 1.3 and 1.6 g/L/day for the two reactors. A constant dilution regime of 30% per day under 50% shade cloth resulted in a sustained production of about 1 g/L/day, at a biomass density of about 3.5 g/L (Figure 3). The reason for the fluctuations before and after this period were not clear.

These results are only initial tests of the system that still is under development. For example, automatic CO_2 addition and temperature control are required. Also, a better inoculum production system is required, to allow start-up of the cultures rapidly. Further, the PBRs need to be operated at full sunlight intensities. However, present results indicate that the tubular PBRs are satisfactory in terms of performance and ability to replicate results.

The near-term objectives of this project are to determine the parameter space for obtaining maximum productivity with this culture under full sunlight. It should be noted that even with shade cloth, these cultures are already outperforming cultures observed in Florence, Italy. This is likely due to the more favorable climatic conditions, especially temperature and insolation, found in Hawaii.

The longer-term objectives are to demonstrate high productivity, high carbohydrate content, and N-limited cultures to provide the fermentable substrate for the subsequent anaerobic hydrogen production stage. Also, the reactors will be used to test algal cultures identified as potential superior species from laboratory screening protocols. Hydrogen leakage will be tested and, if necessary, an all-glass bioreactor will be constructed in the future for the light-driven hydrogen production stage.

ACKNOWLEDGMENTS

This material is based on work supported by the U.S. Department of Energy, Hydrogen Program, under award No. DE-FC36–97GO10202, O.R. Zaborsky, principal investigator. Any opinions, findings, and conclusions or recommendations expressed in this publication are those of the authors and do not necessarily reflect the views of the Department of Energy. We thank Dulal Borthakur, University of Hawaii, and Gerald Cysewski, Cyanotech Corporation, for starter cultures, and Karynne Morgan and the support staffs of HNEI and Look Laboratory of Ocean Engineering for other valued assistance.

REFERENCES

Benemann, J.R., 1996, Hydrogen biotechnology: progress and prospects, *Nature Biotechnology*, 14:1101–1103.

Benemann, J.R., and Zaborsky, O.R., 1996, Biohydrogen: market potential, in *Seventh Annual U.S. Hydrogen Meeting Proceedings: Planning for a Hydrogen Future*, National Hydrogen Association, Washington, D.C., pp. 369–380.

Chini Zittelli, G., Tomasello, V., Pinzani, E., and Tredici, M.R., 1996, Outdoor cultivation of *Arthrospira platensis* during autumn and winter in temperate climates, *Journal of Applied Phycology*, 8:293–301.

Tomaselli, L., Palandri, M.R., and Tredici, M.R., 1966, On the correct use of the Spirulina designation, *Arch. Hydrobiol.* (Algal Studies), 83(Suppl. 117):539–548.

Tredici, M., Biagiolini, S., Chini Zittelli, G., Montaini, E., Favili, F., Mannelli, D., and Materassi, R., 1994, Fully-controllable, high surface-to-volume ratio photobioreactors for the production of specialty chemicals from oxygenic photoautotrophs, in *ECB6: Proceedings of the Sixth European Congress on Biotechnology*, Alberghina, L., Frontali, L., and Sensi, P. (eds.), pp. 1011–1016.

Tredici, M.R., and Chini Zittelli, G., 1997, Cultivation of *Spirulina* (*Arthrospira*) *platensis* in flat plate reactors, in Spirulina platensis *(Arthrospira), Physiology, Cell-Biology, and Biotechnology*, Vonshak, A. (ed.), Taylor and Francis, pp. 117–130.

Szyper, J.P., and Ebeling, J.M., 1993, Photosynthesis and community respiration at three depths during a period of stable phytoplankton stock in a eutrophic brackish water culture pond, *Marine Ecology Progress Series*, 94:229–238.

Weissman, J.C., Goebel, R.P., and Benemann, J.R., 1988, Photobioreactor design: mixing, carbon utilization, and oxygen accumulation, 31:336–344.

Zarrouk, C., 1966, Contribution a L'etude d'une cyanophycee. Influence de divers facteurs physiques et chimiqus sur la croissance et la photosynthese de *Spirulina maxima*, Ph.D. thesis, University of Paris, France.

54

INTERNATIONAL COLLABORATION IN BIOHYDROGEN

"An Opportunity"

Neil P. Rossmeissl

Office of Utility Technologies
U.S. Department of Energy
Washington, D.C. 20585

Over the last two decades, federal, state, and local institutions have been actively involved in developing technologies to reduce or eliminate air pollution. Hydrogen has great promise as an energy carrier and fuel in reducing greenhouse gas emissions and other toxic emissions from end-use technologies. However, hydrogen is not found in the free state in nature but must be produced from primary energy sources. Producing hydrogen from renewable energy offers the potential for cooperative pre-proprietary research and development projects due to their long-term nature and high risk associated with these technologies.

Hydrogen production using biological processes is new, innovative, and potentially more efficient in the direct conversion of solar energy and biomass to hydrogen. Such processes use, adapt, or genetically engineer the biochemical mechanisms present in microalgae or bacteria for the production of hydrogen. In order to achieve the goals of practical biohydrogen processes, advanced low cost bioreactors, and systems with oxygen-tolerant hydrogenate and high efficiencies (greater than 10% total solar conversion), need to be developed and engineered. The work is challenging, but provides an excellent opportunity for the biological community to work across political and ideology barriers to develop collaborations.

This paper presents some of the opportunities to be developed under programs such as the Climate Technology Initiative.

I am honored to be here with you today. I want to speak of the context in which this group's agenda must be carried out. My theme today is creation.

On Day One, God created the heavens and the Earth.
On Day Two, God created light, and separated light from darkness.
On Day Three, God created land and pooled the waters separate from the land.
On Day Four, God created vegetation and day and night.

On Day Five, God created living creatures in the waters.

On Day Six, God created the beasts including man. God said to man, "Be fruitful, multiply, replenish the Earth, subdue the Earth, and assume dominion over the Earth." God gave man the first power, the power to think and name, directing that man should name the animals and the life that was created.

Think about it for a minute. Man has the power to name, effectively drawing lines around it, defining it and only sometimes rightly. Think of the effect of names such as long-term, incremental, costly, and inefficient, all with a negative connotation. Well, I am delighted to be here, not only to visit with you, but also this gives me the opportunity to talk about an issue that has been very much on my mind since I joined the Hydrogen Program. The issue is the very survival of our efforts, all of ours, to promote the use of biological systems to produce energy and other commodity chemicals and compete with existing fossil based processes.

There is an active debate in Washington about whether the United States has an energy problem and if the federal government has any role to play in the energy marketplace. The debate not only focuses on the use of natural gas, but also if renewable energy technologies deserve the support of federal investment.

I believe with absolute conviction that energy efficiency, energy technologies, and biological processes are critical to a prosperous economic future for our nation and the rest of the world. I believe with the same conviction the people of this and other energy-consuming countries want their governments to play a role in developing and deploying technology.

I believe that you already agree with me, but I hope that you will be inspired, at the end of this talk, to work even harder to share these beliefs and work to develop international collaborations that will energize others.

What is the debate and how does it affect your work?

As you know, biological hydrogen production has been around for a long time. According to Dr. John Benemann, the first indication of an organism producing hydrogen was discovered over 100 years ago by Jackson and Elms using Anabaena. The cyanobacterium produced gas for several days and was almost pure hydrogen. The scientific credit for discovering algal hydrogen evolution was Gaffron and Rubin in 1942, who studied the process quantitatively and initiated the field of photosynthetic hydrogen metabolism.

A meeting sponsored by the National Science Foundation in 1973 summarized the state of knowledge of biohydrogen as low rates of production, lack of involvement of water-splitting, and re-absorption of hydrogen, all limiting factors. A genetic approach was recommended.

- The potential of hydrogen production from biological systems is beyond doubt. The maximum efficiency for plant photosynthesis is 10%.
- Processes proposed require modification of the organism (algae) such that hydrogen is the sole product.
- Lack of knowledge of cellular processes and how to "rearrange and stabilize" biological systems make any rational assessment of technical and economic feasibility unwarranted.

There are considerable technical challenges to producing a biological system to produce hydrogen.

- Nitrogenase versus hydrogenase: The high metabolic energy requirement for nitrogenase systems vs. the redox potential of the hydrogenase reductant supply. In

the U.S. Hydrogen Program, the experts' consensus view is to concentrate on hydrogenase-based systems to reach the photosynthetic efficiencies required. The Japanese RITE program is conducting research on nitrogenase-based systems using photosynthetic bacteria.
- Oxygen sensitivity of hydrogen-evolving enzymes: Oxygen inhibition is overcome in cyanobacteria by spatial or temporal separation of hydrogen and oxygen. While there are known oxygen-stable hydrogenases, the issues of enzyme induction, repression, inhibition, and inactivation are priority research areas.
- Direct and indirect biophotolysis systems: Two-stage systems use carbon dioxide as an intermediate electron carrier between the oxygen and hydrogen evolution. Single-stage systems produce the oxygen and hydrogen simultaneously. The hydrogen production is coupled to water splitting with intermediate carbon dioxide fixation. The "Osaka" process in Japan uses algal biomass grown in open ponds to generate, by fermentation, substrates used by photosynthetic bacteria to produce hydrogen. In the U.S. Hydrogen Program, a single-stage system is under development by the National Renewable Energy Laboratory and the Oak Ridge National Laboratory, while a two-stage process is under development at the University of Hawaii.
- Open pond versus bioreactors: Large commercial pond production systems are already in operation; however, our analyses have shown they are more costly than bioreactors. Many types of bioreactors have been described, but hydrogen production presents some very unique challenges. The control of gas transfer, temperature, fouling, and purification are just some of the engineering developments required.
- Molecular genetics to increase system efficiency: The use of molecular genetics to transform photosynthetic bacteria to make them more light-tolerant, the introduction of an oxygen-tolerant hydrogenase into a different organism, the elucidation of the photosystem II-type photosynthesis.
- Dark fermentation and other biologically catalyzed processes: The use of waste as substrates, such as sewer sludge, could use conventional digestion reactors if the metabolic pathways can be shifted to hydrogen production from other products, such as methane and contamination, can be avoided. The low temperature water gas shift reaction using photosynthetic bacteria can carry out the shift reaction in the dark and produce final gas streams that have less than 1 part per million carbon monoxide.

These research challenges are achievable; however, some groups are debating the issue. Rather than focus the resources to conduct the appropriate research, they would rather eliminate the work completely. They point to the German program on biohydrogen and indicate such technologies are not cost-effective, but are wasteful. Their program did not achieve the objectives defined by their funding agency, so the program was eliminated. As the NSF study pointed out, the rhetoric is factually wrong. The *names* that are used (longer-term, incremental, costly, inefficient) are prematurely negative and force research organizations to align themselves to preserve funding instead of collaborating.

Sustainable energy development and hydrogen are not partisan nor ideological issues. They are part of a commonly held vision of our future, part of common expectations and dreams. These so-called experts should be aware of the accomplishments over the last five years before advising us of the absurd nature of our dreams.

In recent years, the issue of climate change has become a very real concern. More than 2,400 scientists urged the President to take actions that will reduce man-made pollution that is warming the Earth. These scientists want the administration to adopt a plan that will lead to at least a 10% reduction in carbon dioxide emissions by 2010, compared with the 1990 levels. European countries want a 15% reduction by industrial nations. As a result, an initiative was developed through the International Energy Agency and United Nations referred to as the "Climate Technology Initiative." This program is a joint collaboration between Japan, the United States, Netherlands, and Germany to fund projects in the separation and removal of greenhouse gas emissions. Hydrogen coupled with carbon dioxide sequestration is viewed in Europe as one method to limit global emissions.

In the United States, the consumption of fossil fuels in the transportation sector in urban areas produces 80% of the pollution. In addition, we are importing record amounts of petroleum more than we produce and over 60% of our petroleum is used for transportation. Not many people realize that the national energy bill is almost as much as the federal income tax bill. Americans spend an estimated $490 billion for energy. We spend a lot of time worrying about income tax, but we virtually ignore the economic and environmental impacts of inefficient energy production and consumption.

This brings up an important question, Where does hydrogen fit in?

Hydrogen and fuel cells are two technologies that will have a major effect on the transportation sector. In the last three years, developments relating to the proton exchange membrane fuel cell have been dramatic. Power densities have been increased while platinum levels have been reduced. Manufacturing techniques such as the fabrication of the membrane electrode assemblies have improved such that a plausible case can be made that the PEM fuel cell vehicle will cost only marginally more than a conventional internal combustion engine vehicle, with a two- to threefold increase in fuel efficiency and near-zero pollutant emissions with no need for pollution control.

Hydrogen produced from renewable energy sources, rather than natural gas, provides a greater impact on emissions since there will be a larger mitigation of carbon dioxide. The successful development of PEM fuel cells would create a strong market pull for hydrogen as a fuel and potentially result in large environmental benefits.

Today we encourage energy efficiency. With hydrogen and other sustainable energy technologies, we will have better levels of comfort, prosperity, health, and safety with less consumption and waste.

What does this mean? It means we, as a community of scientists, need to work together on the research challenges. Biologically produced hydrogen is not proprietary. Most of the work is pre-proprietary and government, not industry, led. The cost to develop these technologies is significant and beyond any one program's ability to fund, and the development time is beyond the "near-term" nature of commercial processes.

This is an opportunity for us, as concerned scientists and engineers, to address. There are a number of programs within the United States, European communities, and Japan, all intent on evaluating and conducting research to "solve" the greenhouse gas problem. We can either all work independently on the same research, or we can develop collaborations at world class facilities to accelerate the development and integration of biological systems.

The U.S. Hydrogen Program is interested in developing collaborations either through the "Climate Technology Initiative," Greenhouse Gas, or the Hydrogen Implementing Agreements of the International Energy Agency, or directly through the organizations already funded.

The scientists in this room are leaders in their field. The challenge is to find a way to collaborate through the political barriers, to work together to accelerate the introduction of biological hydrogen processes for the cleanup of the environment, and the improvement of our economy.

Thank you for your attention, and the honor of speaking with you today.

55

PRINCIPLES OF BIOINFORMATICS AS APPLIED TO HYDROGEN-PRODUCING MICROORGANISMS

Lois D. Blaine[1] and Oskar R. Zaborsky[2]

[1]American Type Culture Collection
10801 University Boulevard
Manassas, Virginia 20110-2209
[2]University of Hawaii at Manoa
2540 Dole Street, Holmes 246
Honolulu, Hawaii 96822

Key Words

bioinformatics, biohydrogen information system, biohydrogen database

1. SUMMARY

Access to computer technology is so widespread that few biologists do not make use of automated systems for managing laboratory data. The growth of the Internet and the development of software tools that facilitate management and dissemination of data in a networked environment have resulted in the proliferation of biological data resources that are now available to researchers throughout the world. The biohydrogen community is no exception, in that data relevant to hydrogen-producing microorganisms are available through numerous resources, many of which may be accessed via the World Wide Web. From genomics and sequence databases to culture collections and specialized data, the hydrogen producers are well-represented. The challenge is to develop a platform on which these data can be merged into a working tool that will meet the various requirements of a broad range of potential users. Environmental conditions, media/nutrient requirements, biochemical characteristics, metabolic pathways, genomics, and industrial process modeling are all aspects of a knowledge base that will be required to track the growing body of information about these organisms and the distinguishing characteristics that may result in the production of energy sources to improve the environment. Development of a working data management tool to serve the biohydrogen community will require adherence to accepted technical and semantic standards. The interdependence of information in disparate databases must be recognized and support for integrated queries involving multiple databases must be maintained. The disparity in data resources relevant to hydrogen producers may be more acute than for other microorganisms because several distinct taxa are represented and will be studied from a functional, rather than a taxonomic, viewpoint. This requires an interdisciplinary approach to data management problems and will require cooperation from a social and technical standpoint. A data model for the biohydrogen community, based on existing software used by culture collections that includes links to networked data resources, is presented. Such a system will serve not only individual laboratories but, importantly, the global biohydrogen community.

2. INTRODUCTION

The great potential of biologically produced hydrogen to serve as a clean, efficient, and renewable energy source has attracted a substantial number of researchers throughout the world to study the intriguing problems related to fundamental and applied aspects, with the hope of achieving a sustainable production system in the not-too-distant future. Precisely because of the disparate nature of the research and the numerous pathways through which biohydrogen may be produced, there appears to be a lack of cohesion and close collaboration among investigators who are exploring the ability of various microorganisms, enzymes, environmental conditions, substrates, and collection systems to achieve the desired result. Large-scale funding for science usually involves the organization of a significant segment of the research community around a central theme with specified goals, for example, the human genome project. While it may be unrealistic to suppose that the biohydrogen community can generate the political and economic attention commanded by the genome community, there are concrete steps that can be taken to unify biohydrogen researchers and enhance collaboration. A key factor in the collaboration is the ability to compare and evaluate the data being generated throughout the world by these investigators.

One obvious step toward unification of the community would be the development of an international biohydrogen information system, that would enable capture, management, and dissemination of data according to prescribed standards and principles of bioinformatics.

Bioinformatics, the use of computational tools for the analysis and management of biological data, empowers us to examine data on a large scale, interdisciplinary basis from a perspective that provides new insights to fundamental biological processes and complex biological interactions. The biohydrogen community is generating diverse information that requires conceptual models for correct organization and interpretation. These data may be partially complete, come from the disciplines of microbiology, biochemistry, physics, or engineering, and may describe objects from different points of view and to various degrees of detail. The biohydrogen information system should be a dynamic federation of linked resources supported by a central core of fundamental data designed to express structural and functional relationships among the entities described in the database. It should enable data comparisons and analyses that lead to formulation and testing of hypotheses.

The advent of the World Wide Web and the browsers that enable linking of data resources throughout the world provides an ideal platform on which to build a global information resource designed for a special interest group such as the biohydrogen community. However, gathering resources via the Web is not, in itself, an adequate means on which to build a structured platform with appropriate standards for content and format. Development of a working data management tool to serve the biohydrogen researchers' needs will require adherence to technical and semantic standards accepted by the community. The interdependence of information in disparate databases must be recognized and support for integrated queries involving multiple databases must be maintained.

3. PROPOSED STRUCTURE FOR A BIOHYDROGEN INFORMATION SYSTEM

Biologists have traditionally relied on the relational database model on which to build their databases. Many powerful relational database management systems (RDBMS) are available that run on desktop computers as well as network servers. Data are managed and accessed through structured query language (SQL) that provides a cross-platform standard. Thus, data stored in SQL compliant RDBMSs may be extracted and exchanged among laboratories, provided the data content is either standardized or can be mapped across databases.

Recently, it has been recognized that the object oriented data model may be more amenable to managing biological data because sets of attributes may be stored as an "object" in the database, thus simplifying the relationships among objects and facilitating hierarchical relationships and dependencies of data entities.

An example of such an object-oriented database management system that could serve as a model for a BioHydrogen database is the Microbiological Information System (MICRO-IS), (Krichevsky 1997). MICRO-IS allows user defined multiple hierarchical tree views of the data. The database can contain multiple trees of overlaying data, such as phylogenetic and phenotypic taxonomies (Kingdom, Phylum, Class, etc.); field studies (site, environmental conditions, energy source, time, geographic parameters, etc.); biochemical parameters (substrate classifications, enzyme classifications, etc.); or genetic parameters (gene names, genetic loci, genotypes, etc.).

A few examples of top levels of organism, process, and product trees are illustrated below. Sublevels of each of these "nodes" on the tree would be added as appropriate, for example, Bacteria, Eubacteria, Cyanobacteria, Chroococcales, and Synochocystis. The user of the database would have multiple entry points into the database, depending on the nature of the inquiry.

MICRO-IS accepts binary, numeric, and textual data as well as graphics/images and bar-coded data. HTML codes may be embedded in textual data to enable links to external resources on the Web. The programs are designed to be interoperable with statistical analysis programs, for incidence associations, diversity calculations, sequence analysis, etc.

A method of standardizing terminology and content syntax is mandatory in creating a community database. A set of standardized descriptors forms the basis for storage and retrieval of data in MICRO-IS. The current set, called RKC Codes (Rogosa, 1986), han-

Table 1. Examples of organism, process, and product trees

Organism	Process	Product	Co-product
Algae	aerobic	H_2	O_2
Bacteria	anaerobic	O_2	ALA
Protist	hybrid	ALA	CO_2
		CO_2	

dles the morphological and biochemical characteristics of microorganisms. A few of the existing codes that apply to biohydrogen producing organisms are:

016069	Strain is photosynthetic
020012	Cells contain green photosynthetic pigments
020017	Cells contain chlorophyll a
020084	Cells contain chlorophyll b
024168	Molecular hydrogen is produced
016071	Growth occurs anaerobically in the light

Expansion and customization of the codes to include experimental protocols, environmental conditions, genetic parameters, process design, photo bioreactor design, metabolic engineering protocols, and other aspects of biohydrogen production would be quite possible. Initial recommendations have been made by an ad hoc panel on standardization regarding the reporting of H_2 production rates and efficiencies. Parameters such as quantitation of radiation, surface, volume of liquid, mass, chlorophyll type, time, and units of measurement for the product were considered during a BioHydrogen '97 standardization workshop. Conversion factors for the energy of light, light intensity, and energy were distributed in the form of a memo from Anthony San Pietro of Indiana University. It was suggested that a BioHydrogen Website might offer algorithms for automatic application of these conversion factors. These and other recommended standards would form the core set of guidelines for entry and retrieval of data from the BioHydrogen database.

The first step in construction of a database would be the development of a comprehensive set of descriptors to cover all aspects of biohydrogen production. Authorities would be sought in specific areas, such as for gene names, illustrated below. In this illustration, the sll, slr, and ssl numbers correspond to sequences of Synechocystis hydrogenase gene units and subunits with the gene names in parentheses (Kaneko, 1996). There is some evidence that there are sequence homologies to hydrogenase genes of other genera (Schmitz, 1995; Boison, 1996), but the naming conventions across genera are not consistent. For example, Synechocystis hoxF is homologous to hydrogenase subunit hndC of Desulfovibrio fructosovorans and Synechocystis hoxU is homologous to hydrogenase subunit hndD of Desulfovibrio fructosovorans. Therefore, homology determinations and mapping of the gene names would have to be part of the database in order to make cross-species comparisons for the genetic expression and physiological role of this enzyme. It is impractical to expect that genes coding for identical proteins or subunits of proteins are going to be named consistently across all organism types. However, an appropriately designed database with links to the sequence databanks can help to alleviate this problem.

Hydrogenase Genes of Synechocystis

sll1221:	hydrogenase subunit (hoxF)
sll1226:	hydrogenase large subunit (hoxH)
sll1223:	hydrogenase subunit (hoxU)
sll1224:	hydrogenase small subunit (hoxY)
slr2135:	hydrogenase accessory protein (hupE)
slr1675:	hydrogenase expression/formation protein HypA (hypA)
sll1079:	hydrogenase expression/formation protein HypB (hypB)
sll1432:	hydrogenase isoenzymes formation protein HypB (hypB)
ssl3580:	hydrogenase expression/formation protein HypC (hypC)
slr1498:	hydrogenase isoenzymes formation protein HypD (hypD)

sll1462:   hydrogenase expression/formation protein HypE (hypE)
sll0322:   transcriptional regulatory protein HypF (hypF)
sll1559:   soluble hydrogenase 45-kD subunit
sll1220:   potential NAD-reducing hydrogenase subunit
ssl3044:   hydrogenase component

Another important factor to consider in selection of a database management system for the BioHydrogen database would be cross-platform portability. The ideal system would consist of a Web-based server version of the database with corresponding desktop computer programs available for collection and maintenance of laboratory-based data, subsets of which could be imported into the larger community database on the Web. The object-oriented database management system running MICRO-IS provides such cross-platform capability. It is available for the Macintosh, the PC running Windows 95, and also for UNIX servers. Figure 1 illustrates the proposed model of a core database, curated by a database manager and subject matter experts, that would receive the data generated by the labs in a prescribed format and adhering to preestablished guidelines for content and syntax.

Software used to manage this database would be available to the research community to enable data entry, management, and retrieval at the desktop level and provide for direct submission of data to the centrally curated database. The BioHydrogen database would be available via the World Wide Web and would have links to related information resources of interest to the BioHydrogen community, but not necessarily fitting into the structured data format of the central database. There are numerous information resources on the web that are directly relevant to BioHydrogen research. Several of these resources are listed Appendix A. Obvious links that would be made, such as the sequence data banks, culture collection data banks, bibliographic reference databases, and so forth, are not included in this table.

A global community information resource is not a trivial undertaking and cannot be accomplished by volunteerism. Financial support for a technical staff must be available and a site with a robust connection to the Internet must be selected to disseminate the data. However, a good requirements analysis and preliminary design of the system can be carried out by willing members of the community. A steering committee composed of subject

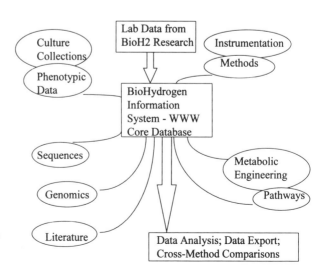

Figure 1. Schematic representation of BioHydrogen Information System.

matter specialists and bioinformaticians with representatives from major geographic regions having an interest in the project can develop a plan and proposals for funding from appropriate government agencies and private sources. This model has worked successfully for similar projects and should probably be initiated with a workshop during which the steering committee members can set the parameters for the information system and enumerate the issues that must be solved in order to develop a sound proposal for funding. A thorough study of related information systems and available software will identify tools and methodologies that have been successful in other areas of biology.

The need for an effective biohydrogen database has been recognized and initial efforts have begun. In particular, the information being assembled by Yasuo Asada of the National Institute of Bioscience and Human Technology, through the auspices of the International Energy Agency (IEA), on organisms known to produce hydrogen will be an immense aid to other researchers. Specifically, information available on a organism basis includes method of measurement, time period, light and other conditions, hydrogen production rate (given in µmoles of H_2 produced/hr/mg dry weight/mg chlorophyll a/mL cells), and associated references.

The growth of the Internet and the development of software tools that facilitate management and dissemination of data in a networked environment will enable rapid and economical development of a global "BioHydrogen Information Resource." The key ingredients for success are a strong commitment on the part of the community and strong leadership to make the vision a reality.

REFERENCES

Boison, G., Schmitz, O., Mikheeva, L., Shestakov, S., and Bothe, H., 1996, Cloning, molecular analysis and insertional mutagenesis of the bidirectional hydrogenase genes from the cyanobacterium *Anacystis nidulans*, *FEBS Lett.*, 394(2):153–158.

Kaneko, T., Sato, S., Kotani, H., Tanaka, A., Asamizu, E., Nakamura, Y., Miyajima, N., Hirosawa, M., Sugiura, M., Sasamoto, S., Kimura, T., Hosouchi, T., Matsuno, A., Muraki, A., Nakazaki, N., Naruo, K., Okumura, S., Shimpo, S., Takeuchi, C., Wada, T., Watanabe, A., Yamada, M., and Yasuda, M., 1996, Sequence analysis of the genome of the unicellular cyanobacterium *Synechocystis* sp. strain PCC6803 II; sequence determination of the entire genome and assignment of potential protein-coding regions, *DNA Res.*, 3(3):109–136.

Kazic, T., 1992, Metabolic pathways databases: challenges and ppportunities, in *CODATA Bulletin: Scientific Program and Abstracts*, Thirteenth International CODATA Conference, October 1992, Beijing, China, 24(2).

Krichevsky, M., 1997, The microbiological information system, in *Proceedings of the American Society for Microbiology Annual Meeting*, May 1997, Miami, Florida.

Schmitz, O., Boison, G., Hilscher, R., Hundeshagen, B., Zimmer, W., Lottspeich, F., and Bothe, H., 1995, Molecular biological analysis of a bidirectional hydrogenase from cynaobacteria, *Eur. J. Biochem.*, 233(1):266–276.

Rogosa, M., Krichevsky, M., and Colwell, R., 1986, *Coding Microbiological Data for Computers*, Springer-Verlag, New York, pg. 299.

U.S. Department of Energy, 1993, Meeting Report: DOE Informatics Summit, April 26–27, Baltimore, Maryland, available at bioinfo@oerv01.er.doe.gov

Appendix A. BioHydrogen Relevant Information Resources

Name of resource	URL of resource
Hawaii Natural Energy Institute, University of Hawaii	http://www.soest.hawaii.edu/HNEI/hnei_hydrogen.html
National Institute of Bioscience and Human-Technology, Agency of Industrial Science and Technology, MITI	http://www.aist.go.jp/NIBH/ourpages/iea/index.html

Cyanosite, A New Webserver for Cyanobacterial Research	http://www-cyanosite.bio.purdue.edu/
The AAA Protein Superfamily: ATPases associated with various cellular activities	http://yeamob.pci.chemie.uni-tuebingen.de/AAA/Description.html
Green Algae: Chlorophyta Database	http://www.sonoma.edu/biology/algae/Green.html
The American Hydrogen Association	http://www.getnet.com/charity/aha/ahafaq.html
PUMA: Metabolic pathways, hydrogen dehydrogenase	http://www.mcs.anl.gov/home/compbio/PUMA/Development/EnzymeObjects/1.12.1.2.html
Hydrogen Research at NREL	http://syssrv9.nrel.gov/lab/pao/hydrogen.html
Solstice: Center for Renewable Energy and Sustainable Technology (CREST)	http://solstice.crest.org/
HyWeb - latest news and detailed information from Germany on hydrogen	http://www.HyWeb.de/index-e.html
U.S. Department of Energy Hydrogen InfoNet	http://www.eren.doe.gov/hydrogen/
National Renewable Energy Laboratory	http://www.nrel.gov/lab/pao/hydrogen.html
The Global Energy Marketplace	http://gem.crest.org/search.html
National Hydrogen Association	http://www.ttcorp.com/nha/
Cyanolab!	http://131.252.86.36/
The Schatz Energy Research Center (SERC)	http://www.humboldt.edu/~serc/

56

PRODUCTION OF SULFOLIPIDS FROM CYANOBACTERIA IN PHOTOBIOREACTORS

Seher Dagdeviren, Karen A. McDonald, and Alan P. Jackman

Department of Chemical Engineering and Materials Science
University of California
Davis, California 95616

Key Words:

sulfolipid, cyanobacteria, *Synechocystis*, photobioreactor, fed-batch culture

1. ABSTRACT

Sulfolipids are compounds found in association with the photosynthetic apparatus in most photoautotrophic organisms, although there have been no quantitative studies on how environmental and biological factors influence the production of sulfolipids. They are of interest due to their possible pharmaceutical use (anti-tumor and anti-HIV) and also as an interesting new class of lipids. In this study, we are investigating the kinetics of growth and sulfolipid production using the cyanobacterium *Synechocystis PCC 6803*. The cyanobacteria are grown in a 2-L, fed-batch photobioreactor under well-defined conditions. We obtained a final biomass concentration of 5 g/L after 20 days and an overall sulfolipid productivity of 0.24 mg/(L-hr) at a light intensity of 216 $\mu E/m^2/s$.

2. INTRODUCTION

Sulfoquinovosyl diacylglycerol (SQDG), commonly referred to as "the plant sulfolipid," was first discovered by Benson et al. (1959) and has been reported to occur in a variety of organisms, including cyanobacteria, green, red, and brown algae (Barber and Gounaris, 1986), protozoa, fungi, and higher plants (Harwood, 1980).

The sulfolipids have a wide variety of structures in which the sulfur occurs as sulfate ester (RCH_2OSO_3) and a sulfonate (RCH_2SO_3). In the case of SQDG, this structure is in the form of a sulfonic acid group in which carbon is directly bonded to the sulphur atom (Haines, 1973). The two major fatty acids found in SQDG in most plants examined are li-

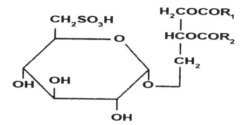

Figure 1. The structure of the plant sulfolipid SQDG.

nolenic (C18:3) and palmitic (C16:0) acids while linoleic (C18:2) and oleic (C18:1) acids occur in lesser amounts (Barber and Gounaris, 1986).

Most photosynthetic cells contain four major glycerol lipids in their chloroplasts (monogalactosyl diacylglycerol, digalactosyl diacylglycerol, sulfoquinovosyl diacylglycerol, and phosphatidylglycerol). Neither the function of SQDG nor the biosynthesis has yet been fully elucidated. However, all photosynthetic microorganisms and higher plants investigated so far contain the sulfolipid. As a result, sulfolipids are thought to play a role in photosynthesis (Haines, 1980).

Light is an important factor for the lipid composition of photosynthetic organisms. It has been observed that light-induced chloroplast development is accompanied by the increased production of sulfolipids (Rosenberg and Pecker, 1964). Benson also observed its formation at a rapid rate in *Chlorella*, or *Euglena*, than observed in the dark. These observations supported the direct correlation between the appearance of chlorophyll and SQDG. Furthermore it has been suggested that it may be involved in the orientation and functions of chlorophyll molecules (Barber, 1986).

Sulfolipids are worthy of further study due to their anti HIV-1 and anti-tumor activity. In in vitro studies, the sulfonic acid containing glycolipids have shown impressive anti-viral activity on human immunodeficiency virus (HIV-1) (Gustafson et al., 1989). The pure compound extracted from cultured cyanobacteria protected human lymphoblastoid T cells from the cytopathic effect of HIV-1 infection.

This work studies the kinetics of growth and sulfolipid production at various light intensities using the cyanobacterium *Synechocystis PCC 6803* in a fed-batch photobioreactor to assess the potential of such systems for sulfolipid production, as well as to provide additional sulfolipids for further characterization. Previous work has been done with *Anabaena* 7120, a filamentous cyanobacterium (Archer et al., 1997). S*ynechocystis PCC 6803*, a unicellular cyanobacterium, may be more amenable to large-scale culturing and genetic manipulation for increased sulfolipid productivity.

3. MATERIALS AND METHODS

3.1. Organism and Culture

The unicellular cyanobacterium *Synechocystis PCC 6803* was obtained from the Department of Microbiology at the University of California, Davis. The culture media employed in this work is BG-11 medium enriched with sodium nitrate and sodium N-tris (hydroxymethyl) methyl-2-aminoethanesulfonic acid (TES) as the nitrogen source. The cultures are subcultured every 20 days and maintained in 250-mL shake flasks containing

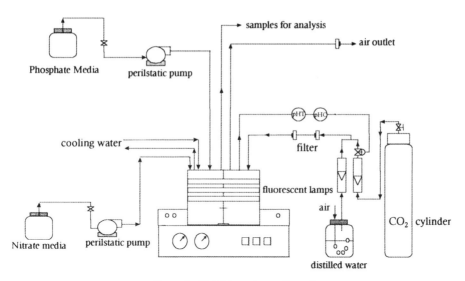

Figure 2. Photobioreactor configuration.

100 mL of medium on an orbital shaker illuminated by cool-white fluorescent lamps with an average light intensity of 40 µE/m^2/s.

3.2. Operation of Fed-Batch Photobioreactor

Synechocystis PCC 6803 grown in shake flasks is inoculated in a 2-L New Brunswick Multigen bioreactor under sterile conditions. Experiments are conducted under controlled conditions of external light intensity, temperature, agitation rate, and pH, in a fed-batch mode to prevent nutrient deprivation. The temperature is kept at 28 °C by cool water circulation in an indwelling heat exchange tube. The pH is measured by an indwelling pH probe (Ingold, type 465-35-90-K9) and regulated at 7.6 ± 0.1 by a pH controller (Cole-Parmer, Model 01970–00) connected to a solenoid valve manipulating CO_2 flow in the sparging gas. Humidified and filter-sterilized air is supplied at a rate of 450 mL/min while CO_2 is fed sporadically to the system at a rate of 30 mL/min. The working volume is maintained at approximately 1.5 L and agitated at 200 rpm with two flat-blade turbine impellers. Light is provided by two 32-W circular fluorescence lamps around the reactor surrounded by a cylindrical aluminum reflector. The experimental set-up is shown in Figure 2.

3.3. Nutrient Uptake

Light-limited experiments ensure that the only limiting factor is light and all other major nutrients such as nitrogen and phosphorus are in excess for normal growth and cell synthesis of algae. The concentrations of these components in the culture are maintained above limiting levels by addition of concentrated fresh medium. BG 11 medium (without phosphate) and dibasic potassium phosphate solutions are separated due to the tendency of phosphate to precipitate. The 30-mL samples are removed from the reactor daily and the concentrations of phosphate and nitrate are used as indicators of how much fresh medium to add. The concentrations of NO_3^- and PO_4^{-3} are maintained above 12.5 mM and 40 mg/L, respectively, in the culture. The working volume is maintained around 1.5 L by the

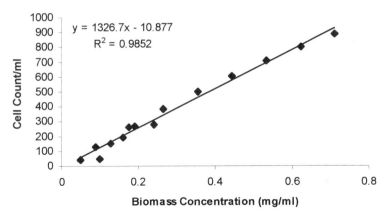

Figure 3. Calibration curve for biomass dry-weight concentration vs. cell count/volume.

addition of fresh medium after sampling and feeding medium. Biomass concentrations are corrected to take into account this dilution effect.

3.4. Analytical Methods

Experiments are run for at least three weeks and samples are taken from the reactor at the same time every day. The 30-mL samples are used for biomass, nitrate, phosphate, chlorophyll, and sulfolipid analyses. The biomass concentration is determined by filtering a known volume of three samples onto predried filter paper (0.2 μm, Millipore). To minimize analytical errors, the algal cells are washed with distilled water to remove salts and other contaminants (Becker, 1994). The biomass concentration is calculated by taking the average weight of these three samples after they are dried at 55 °C until their weight no longer changes.

Syenochocystis PCC 6803 cell concentration can be easily monitored by Coulter Counter and correlated to dry weight concentration. A linear correlation between the dry weight concentration of this strain and cell counts is given in Figure 3.

The concentrations of NO_3^- and PO_4^{-3} in the samples are measured by a nitrate electrode (Orion Model 93–07) and molybdate test for phosphate. The extraction of chlorophyll is performed by suspending and homogenizing the separated algal cells by tissue tearer in 90% methanol and the chlorophyll concentrations are calculated by reading the absorption of the pigment at 665 nm.

3.5. Sulfolipid Analysis

Folch et al. (1957) and Bligh and Dyer (1959) showed that a mixture of a nonpolar and polar solvent was most effective in extracting lipids from biological materials. We have used a modified method of the Folch procedure, in which cyanobacterial cells are homogenized with a methanol:chloroform (1:1) mixture. To minimize the risk of auto-oxidation of polyunsaturated fatty acids or hydrolysis of lipids, an antioxidant, butylated hydroxy toluene (BHT), at a concentration of around 0.05% (w/v), is added to the solvent mixture. The homogenized sample is filtered through a solvent-resistant syringe filter having a glass fiber prefilter layer (0.45 μm, Gelman), and 1% KCl is added to form a biphasic system. The upper phase containing water-soluble impurities is removed and the organic phase is concentrated by evaporation under a nitrogen gas stream.

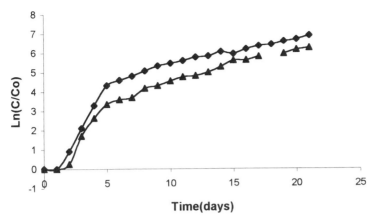

Figure 4. Dimensionless biomass concentration vs. time at various light intensities. [C-biomass concentration (g dw/ mL), C_o-initial biomass concentration (g dw/mL)]. ♦ = external light intensity 216 $\mu E/m^2/s$ with an initial biomass concentration 2×10^{-2} g dry weight/L, Δ = external light intensity 170 $\mu E/m^2/s$ with an initial biomass concentration 5×10^{-2} g dry weight/L.

Sulfolipids are quantified using thin layer chromatography (TLC) and scanning photodensitometry. TLC offers a number of advantages over other separation techniques. It is rapid, sensitive, applicable to small amounts of lipids, and requires very little equipment (Christie, 1973). Although staining intensity can be quite variable between the plates, requiring known concentrations of standards to be run on each plate, it gives a high degree of resolution and separated components can be recovered from the plates in amounts sufficient for further studies (Kates, 1986).

The crude lipid extracts obtained from the bioreactor samples and various amounts of purified SQDG standards are spotted on commercially precoated TLC plates (Kieselgel, HPTLC). The stationary phase is a silica gel layer and the mobile phase used is chloroform:methanol:acetic acid:water (85:15:10:3, v/v) system. The plate is inserted in a vertical position in a glass chamber containing the polar mobile phase. When the solvent front ascends the preset mark, the plate is removed from the chamber and the volatile solvent is allowed to evaporate off at room temperature. For the quantitative determination of lipids, cupric sulfate in phosphoric acid is used as a charring agent. The charred TLC plate is scanned with an AGFA Arcus Plus scanner connected to a Macintosh computer (Centris 650) and using the software program Adobe Photoshop. The areas of the peaks obtained by photodensitometric scanning are proportional to the amount of lipid originally present. The concentrations of sulfolipid in the samples are calculated by a standard curve of band mean density vs. sulfolipid amount. In order to find the intrinsic sulfolipid concentration (mg/g dry weight cells), the sulfolipid concentration in the sample is divided by the biomass concentration in the sample.

4. RESULTS AND DISCUSSION

Synechocystis PCC 6803 is grown at two different light intensities, 216 and approximately 170 $\mu E/m^2/s$. The growth curves are shown in Figure 4. The biomass data is plotted

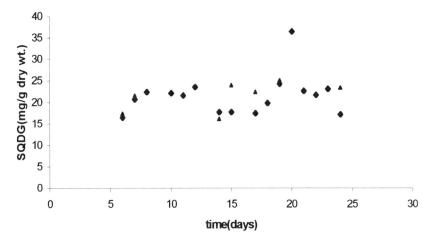

Figure 5. Sulfolipid concentration vs. time at light intensity of 216 $\mu E/m^2/s$.

as dimensionless biomass concentration, that is, biomass concentrations are divided by the initial inoculum biomass concentration for the data shown in Figure 3.

Figure 3 shows the characteristic stages of fed-batch cyanobacterial growth such as the lag phase (about two days) and two exponential growth phases, one in which light irradiance is fairly uniform throughout the reactor followed by the light-limited phase as the culture becomes very dense. It is also observed that there is not a true stationary phase in this system. This is due to the fact that nutrients are added to the culture and apparently there is no accumulation of by-products that affects growth rate over the duration of the experiment.

Sulfolipid production at a light intensity of 216 $\mu E/m^2/s$ is also shown in Figure 5. The sulfolipid concentrations are determined based on two different lipid extracts obtained from the same day to minimize the error accounting for extraction technique and oxidation of lipid samples. Although the data show an increase in sulfolipid production just after the exponential growth phase (days 5–8) for both extracts, there appears to be a difference in sulfolipid amounts in the samples between days 15 through 18. We believe that this discrepancy is due to the extraction error or oxidation for one set of the duplicate samples. Also, the SQDG content for one of the samples was unusually high, perhaps due to error

Table 1. Specific growth rate and mean doubling time of *Anabeana 7120* (Archer, 1996) and *Synechocystis PCC 6803*

Strain	Specific growth rate (day^{-1})	Doubling time (hrs)
Anabeana 7120 (first exponential phase)	1.37	12
Anabeana 7120 (light-limited phase)	0.11	150
Synechocystis PCC 6803 (first exponential phase)	1.1	15
Synechocystis PCC 6803 (light-limited phase)	0.2	83

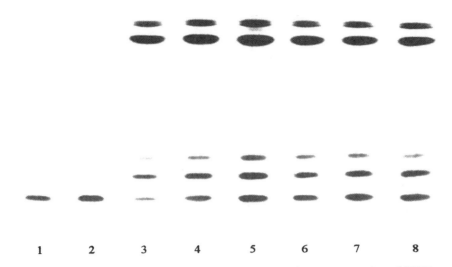

Figure 6. Thin layer chromatogram on a silica gel plate. Lanes 1–2 are known concentrations of SQDG standards obtained from NCI; (6, 9 µg), Lanes 3–8: lipid extracts from bioreactor.

in the concentration step, and believed to be an outlier. The final dry weight concentration of *Synechocystis* cells grown at 216 µE/m^2/s is 5 g dw/L and the average sulfolipid concentration is 21.61 mg/g dry weight after 20 days. This corresponds to an overall sulfolipid productivity of 0.24 mg/(L-hr) in our system. We expect that the sulfolipid productivity would increase for longer times since previous work for the *Anabaena* strain shows an increase in the intrinsic sulfolipid concentration (as well as biomass concentration) during the light-limited growth phase (Archer et al., 1997). Acher et al. reported a final sulfolipid level of 11.1 mg/g dry weight corresponding to an overall sulfolipid production rate of 0.06 mg/L/h after 20 days.

The specific growth rate (μ), the slope of the exponential growth phases, and the corresponding mean doubling time are given in Table 1 and compared with results obtained by Archer et al. (1997) for *Anabeana* 7120 grown in a similar system under similar conditions. The intitial growth rates for *Synechocystis* and *Anabeana* are similar, although the *Synechocystis* had a slightly higher growth rate during the light-limited phase. The final dry weight concentration of the *Synechocystis* cells is 4.6 g/L after 15 days. The maximum intrinsic chlorophyll concentration attained is 15 mg/g cell.

Figure 6 shows the TLC lipid profile for extracts of *Synechocystis PCC 6803* taken from the bioreactor on different days. For comparison, a sample from *Synechococcus PCC 7942* grown in shake flasks is also included. The intrinsic sulfolipid level of *Synechococcus PCC 7942* obtained from a single phase lipid extract was 9.7 mg/g dw.

5. CONCLUSION

Obtainable biomass and sulfolipid concentrations for *Synechocystis 6803* are much higher than *Anabeana 7120*. After 20 days, final dry weight concentrations of the *Synechocystis 6803* and *Anabaena 7120* at approximately the same light intensity are 5 g/L and 2.5 g/L, respectively. Intrinsic sulfolipid range of *Synechocystis 6803* is 20–30 mg/g cell at a light intensity of 216 and the overall sulfolipid production rate (dry cell weight/reactor volume·mass sulfolipid/gram dry weight) is between 0.2 and 0.315 (mg/L-hr) in our system. Biomass concentrations can be monitored by a Coulter Counter for this unicellular cyanobacterium.

Future work will include the study of the effect of various light intensities on the growth kinetic and sulfolipid production in fed-batch cultures and continuous cultures. We are currently developing a mathematical model for light-limited growth of *Synechocystis PCC 6803* and sulfolipid production by examining the relationship between sulfolipid production and light intensity in the culture.

ACKNOWLEDGMENT

This work was supported by the University of California Systemwide Biotechnology Research and Education Program. We would like to thank Professor Bruce German and Thomas Hanson for help in lipid analysis and Professor Jack Meeks and Eunice Tan for assistance in culturing cyanobacteria.

REFERENCES

Archer, D.S., McDonald, K.A., and Jackman, P.A., 1997, Effect of light irradiance on the production of sulfolipids from *Anabeana* 7120 in a fed-batch photobioreactor, *Applied Biochemistry and Biotechnology*, 67:15–28.

Archer, D.S., 1996, Effect of light on the production of sulfolipids from cyanobacteria, Ph.D. thesis, University of California, Davis.

Barber, J., and Gounaris, K., 1986, What role does sulfolipid play within the thylakoid membrane? *Photosynthetic Res.*, 9:239–249.

Becker, E.W., 1994, Applications of algae, in *Microalgae Biotechnology and Microbiology*, Cambridge University Press, Cambridge, pg. 253.

Benson, A.A., Daniel, H., and Wiser, R., 1959, in *Proceedings of the National Academy of Sciences (USA)*, 45:1582–1587

Benson, A. A., 1963, The plant sulfolipid, *Adv. Lipid Res.*, 1:387–394.

Bligh, E.G., and Dyer, W.J., 1959, A rapid method of total lipid extraction and purification, *Can. J. Biochem. Physiol.*, 37:911.

Christie, W.W., 1973, Chromatographic and spectroscopic analysis of lipids: general principles, in *Lipid Analysis*, Pergamon Press, New York, pp. 56–64.

Folch, J., Lees, M., and Stanley, G.H.S., 1957, A simple method for the isolation and purification of total lipids from animal tissues, *J. Biol. Chem.*, 226:497–509.

Gustafson, K.R. et. al., 1989, AIDS-antiviral sulfolipids from cyanobacteria (blue-green algae), *J. Natl. Cancer Inst.*, 81:1254–1258.

Haines, T.H., 1973, Sulfolipids and halosulfolipids, in *Lipids and Biomembranes of Eukaryotic Microorganisms*, Erwin, J.A. (ed.), Academic Press Inc., New York, pp. 197–232.

Harwood, J.L., 1980, Sulfolipids, in *The Biochemistry of Plants*, Academic Press Inc., New York, London, 4:301–320.

Kates M., 1986, Identification of individual lipids and lipid moieties, in *Techniques of Lipidology: Isolation, Analysis and Identification of Lipids*, Elsevier Science Publication, New York, pg. 381.

Rosenberg, A., and Pecker M., 1964, *Biochemistry*, Easton, 3:254–258.

PRACTICAL CONSIDERATIONS IN CYANOBACTERIAL MASS CULTIVATION

JoAnn C. Radway,[1] Joseph C. Weissman,[2] John R. Benemann,[3] and Edward W. Wilde[1]

[1]Westinghouse Savannah River Site
Aiken, South Carolina 29808
[2]Microbial Products, Inc.
705 SW 27th Avenue, No. 5, Vero Beach, Florida 32968
[3]Sea Ag, Inc.
343 Caravel Drive, Walnut Creek, California 94598

Key Words

Fischerella, nutrient removal, thermal mitigation, mass cultivation

1. SUMMARY

Bench-scale tests have been conducted on a process using cyanobacteria to remove nutrients and utilize waste heat in nuclear reactor cooling waters. Based on its tolerance of temperature extremes, *Fischerella* Strain 113 was selected for semicontinuous culture studies. Cultures maintained with 10 mg L^{-1} daily productivity, diurnally varying temperature (55 °C to 26–28 °C), 200 µE sec^{-1} m^{-2} illumination, and 50% biomass recycling into heated effluent at the beginning of each 12-h light period, removed >95% of NO_3^- + NO_2^--N, 71% of NH_4^+-N, and 70% of total P. Nutrient removal was not severely impaired under conditions simulating scaled-down reactor operation, increased insolation, or dissolved inorganic carbon (DIC) limitation. Recycling biomass at the end of the light period resulted in slower growth, unimpaired NO_3^- + NO_2^--N removal, 38–45% P removal, and no net NH_4^+ removal. Approximately 80% N removal and diurnally varying P removal (averaging 50–60%) are projected for the full-scale process.

2. INTRODUCTION

In large-scale cyanobacterial culture, one must consider physical/chemical factors (e.g., temperature, light, nutrient availability), biological factors (e.g., morphology, growth kinetics, growth requirements, and other physiological characteristics), and technological factors (e.g., requirements for harvesting, processing, and energy inputs needed to match physical/chemical factors to the organism's needs). An effective system exploits a successful match between the cyanobacterium and the available environmental conditions in such a way as to minimize the need for costly energy inputs. A unique opportunity in this regard is provided by cooling water effluents from fossil- or nuclear-fueled facilities, in which year-round thermal inputs are available as a waste product. Thermal loading of nutrient-rich receiving waters can result in eutrophication and other problems, requiring the construction of expensive cooling towers. Wilde et al. (1991) have described a conceptual process designed to mitigate combined thermal/nutrient impacts to L Lake, a nuclear reactor cooling reservoir at the Savannah River Site (Aiken, South Carolina). The process involves inoculation of cyanobacteria into the thermal plume, uptake of nutrients during plug flow and gradual cooling of the effluent, harvesting of biomass via microstrainers at a downstream location to remove nutrients, recycling of a portion of the biomass back to the inoculation site, and utilization of the remaining biomass for fertilizer or methane generation. To survive and remain dominant in such a system, the selected cyanobacterium must tolerate an extremely wide temperature range (55 °C – <30 °C, with occasional exposure to 60 °C during inoculation or 15 °C during winter reactor outages). It must be filamentous in order to allow selective harvesting and recycling (Weissman and Benemann, 1979), and should fix N in order to enhance P removal under otherwise N-limiting conditions. It must also be non-toxic and, most importantly, must remove most N and P during the anticipated 1-day residence time. The most likely candidates are thermophilic strains of the genus *Fischerella* (formerly *Mastigocladus laminosus*) (Castenholz, 1989). The present study examines the performance of *Fischerella* strains during simulations of the proposed process.

3. EXPERIMENTAL

3.1. Cultures and Culture Maintenance

Seven unialgal cultures, identified as *Fischerella* sp., were derived from cooling water reservoirs at the Savannah River Site (SRS). These were designated 2, 7, 8, 4L, 6L, 113, and 823. Three crude cultures (*Fischerella* sp. 1B, 6H, and T-7) were collected by Dr. L. Ralph Berger from a region of geothermal activity on the island of Hawaii. One unialgal strain (*Mastigocladus laminosus* UTEX B 1931) was obtained from the University of Texas culture collection. A final unialgal culture (14) appeared as a contaminant during transfer of the UTEX culture, but morphologically resembled the SRS-derived strains. Stock cultures were grown in ND medium, a combined-nitrogen-free modification of Medium D (Castenholz, 1982). They were maintained at 45 °C with continuous illumination (50 µE m^{-2} s^{-1}) and gentle shaking (150 rpm).

3.2. Thermal Tolerance Studies

Stock cultures were transferred to sterile L Lake water and allowed to grow 3–10 days at 45 °C to provide experimental inoculum. The material was then homogenized and used to inoculate replicate 75 mL aliquots of sterile L Lake water. Based on preliminary experiments, 3 mM NaHCO$_3$ was added as a carbon source for strains 1B, 6H, and 14, while other strains received 10 mM NaHCO$_3$. To test the effects of high-temperature exposure, cultures were incubated at 45 °C (stock culture conditions) for 24 h after inoculation, exposed to temperature treatments (45 °C control, 55 °C, 60 °C, 65 °C, 70 °C) for 30 min under ambient room illumination, then incubated at 45 °C for a further 47.5 h. Duplicate cultures were sacrificed for total dry weight measurements at each time point. Tolerance to suboptimal temperature was tested similarly; 24-h old cultures were exposed to 15 °C (50 µE sec^{-1} m^{-2}) for 48 h. Replicate cultures were wrapped in foil at 45 °C to test the effects of stopping growth without exposure to cold, while controls were kept at 45 °C under 50 µE sec^{-1} m^{-2} illumination. All cultures were then returned to stock culture conditions and their dry weight increases during the subsequent 48 h were measured.

3.3. Nutrient Removal Experiments

Fifty percent of culture biomass was harvested each day by filtration through 95 µm mesh screen cloth and resuspended in 4 L of filtered (3 µm), nonsterile L Lake water (usually at 55 °C), which was then allowed to cool to 25–30 °C over the ensuing 24 h. Cultures were illuminated diurnally (12 h light:12 h dark, typically 200 µE sec^{-1} m^{-2}), and recycling was typically performed at the beginning of the light period ("dawn"). Semicontinuous cultures were paired, one bubbled with air only and one bubbled with 0.1% CO_2/air. Experiments simulated baseline reactor operating conditions (daily 55 °C – <30 °C gradient), half-power operation conditions (daily 40 °C – <30 °C gradient), "dusk" recycling conditions (baseline conditions, but with recycling of 75% of biomass into hot water at the end of each light period), and high light conditions (baseline conditions with 400 µE sec^{-1} m^{-2} illumination). Nutrient analyses commenced after daily biomass production stabilized; replication was provided by sampling on 3–6 consecutive days.

3.4. Analytical Methods

Culture pH was measured electrometrically, alkalinity by the low alkalinity method of the U.S. Environmental Protection Agency (EPA, 1979), and temperature by calibrated thermometers. These data were used to calculate dissolved CO_2 levels. An LI-188B integrating quantum radiometer/photometer (LiCor, Inc.) was used to determine light intensity. Dry weight was measured according to EPA (1979). Nutrient analyses were performed on filtered (0.45 µm) samples. Ammonia was determined by the phenate method (Rand et al., 1979), nitrate + nitrite by the cadmium reduction/diazotization technique using Hach low-range nitrate kit reagents (Hach, Inc.), orthophosphate by the ascorbic acid method using Hach reactive phosphate kit reagents, and total dissolved P by the ascorbic acid method following persulfate digestion (Rand et al., 1979).

4. RESULTS AND DISCUSSION

4.1. Response to Temperature Extremes

Figure 1 shows dry weight increases during recovery from exposure to elevated temperatures. Exposure to 55 °C did not significantly affect subsequent growth compared to 45 °C controls. However, 60 °C exposure caused a significant (P = 0.0037) decline in growth, affecting all strains except 113 and T-7. Observations on pigment content and autofluorescence of Strain 4L suggest that this growth slowdown was associated with a temporary decline in Chl *a* content, but that viability was unimpaired (Radway et al., 1992). The observed differences between strains could indicate differences in the maximum tem-

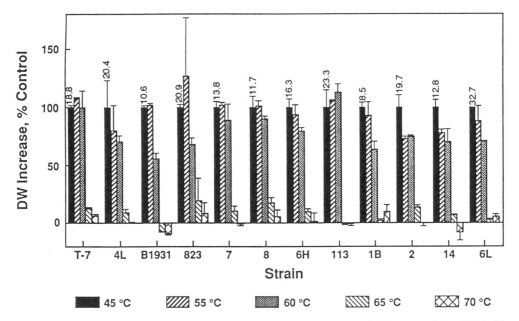

Figure 1. Recovery from heat shock treatments. Graph bars show mean dry weight increases during 47.5 h following a 30-min heat exposure period, expressed as a percentage of dry weight increase by 45 °C controls. Error bars indicate ranges of duplicate cultures. Numerals above bars indicate actual mean dry weight increases (mg L^{-1}) by control cultures.

perature tolerance, and show that strain selection for upper temperature tolerance is a practical approach in developing the proposed process.

All cultures visibly yellowed during a 30-min exposure to 65 °C, and subsequent growth declined by about 90% on average (significant at the P = 0.0036 level). These effects were even more pronounced after 70 °C exposure. This finding has clear implications for the choice of an inoculation point in the thermal plume.

The effects of a 48-h exposure to 15 °C or to darkness are shown in Figure 2. Both cold and dark treatments stopped growth during the exposure period (Radway et al., 1992). Subsequent dry weight increases by most strains did not differ significantly from those of 45 °C controls. Figure 2 suggests that strains T-7, 6H, and 14 recovered more slowly from cold treatment than did other strains. However, darkness had a similar effect, indicating that photooxidative or photodynamic damage was not involved. The cultures may simply have undergone a brief lag period before resuming growth. *Fischerella* sp. therefore demonstrates excellent potential for survival in the South Carolina climate during reactor shutdown periods.

4.2. Productivity and Nutrient Removal

Strain 113, which grew rapidly and showed good thermal tolerance, was chosen for semicontinuous culture studies. Productivity and nutrient removal by this strain are shown in Table 1. During a simulation of full power reactor operation by means of daily cooling from 55 °C to 26–28 °C, a DIC-enriched culture achieved an average daily biomass production of 10 mg L^{-1} d^{-1} and removed nearly all NO_3^- + NO_2^--N from L Lake water. Less complete removal was seen in DIC-limited cultures and in those subjected to simulated

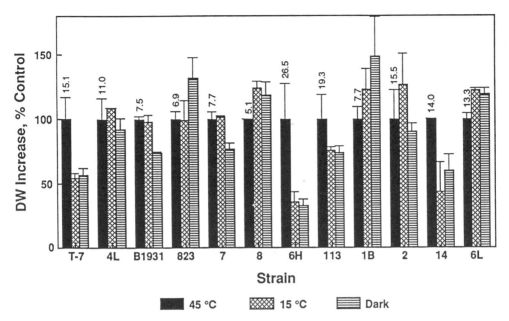

Figure 2. Recovery from cold or dark treatments. Graph bars show mean dry weight increases during 48 h following a 48-h exposure to 15 °C or darkness, expressed as a percentage of dry weight increase by 45 °C, illuminated controls. Error bars indicate ranges of duplicate cultures. Numerals above bars indicate actual mean dry weight increases (mg L^{-1}) by control cultures.

half-power operation (initial temperature 40 °C), but these differences were not significant. Ammonia and phosphate concentrations were also reduced to statistically similar levels under the two temperature regimes.

Since the proposed cultivation process involves continuous harvesting and recycling of *Fischerella*, we also investigated nutrient removal by cultures recycled into heated water at the end ("dusk") rather than the beginning ("dawn") of the daily light period (Table 1). It was necessary to recycle 75% of the biomass daily to avoid culture washout in this simulation. With or without DIC enrichment, NO_3^- + NO_2^--N removal efficiency averaged 96%. This significantly ($P<0.05$) exceeded removal by "dawn"-recycled cultures, probably as a result of increased biomass production. However, no ammonia was removed. Phosphorus removal was also significantly ($P<0.05$) less effective than during "dawn" recycling. Poor removal efficiency of these nutrients appeared to be related to their excretion by *Fischerella* under conditions of slower growth (Radway et al., 1994).

A final experiment measured the effect of increased illumination (400 µE s^{-1} m^{-2}) upon nutrient removal by "dawn"-recycled cultures (Table 1). Both a DIC-enriched and an unenriched culture reduced NO_3^- + NO_2^--N to below detection limits and were significantly ($P<0.05$) more effective than cultures illuminated at 200 µE s^{-1} m^{-1}. Ammonia, phosphate, and total phosphorus removal efficiencies statistically resembled those seen under lower light intensity and otherwise similar conditions.

Results indicate that the simulated process removed most dissolved inorganic N and P from L Lake water under "dawn" recycling conditions. The process appeared relatively insensitive to changes in temperature regimes, simulating changes in reactor operating parameters. It was also relatively unaffected by changes in illumination (within the range

Table 1. Biomass productivity and nutrient removal by semicontinuous cultures of Strain 113. Original nutrient levels in L Lake water used in the experiments were 308–390 μg L^{-1} NO$_3^-$ + NO$_2^-$-N, 18.7–79.1 μg L^{-1} NH$_3$-N, 23.9–63.0 μg L^{-1} PO$_4^{3-}$-P, and 54.9–66.3 μg L^{-1} total P

Initial temp. (°C)	CO$_2$ (μM)	Light (μEs^{-1}m^2)	Recycle time	Dry wt. prod. (mgL^{-1}·day^{-1})	NO$_3^-$ + NO$_2^-$-N (% removed)	NH$_3$-N (% removed)	PO$_4^{3-}$-P (% removed)	Total P (% removed)
55	39	200	Dawn	10.0	95	71	82	70
55	0.2	200	Dawn	7.0	73	76	80	63
40	48	200	Dawn	9.5	87	74	79	59
40	0.5	200	Dawn	8.5	79	79	70	61
55	33	200	Dusk	16.0	96	0	37	45
55	0.1	200	Dusk	10.3	96	0	41	38
55	105	400	Dawn	9.6	100	53	74	82
55	0.1	400	Dawn	10.0	100	89	76	83

tested), suggesting that it might be insensitive to changes in depth within the photic zone or diurnal and seasonal changes in light intensity. However, results indicate that it would be necessary to recycle varying percentages of biomass at different times of day, and that NH_3 and PO_4 removal will vary diurnally. Despite this constraint, N and P removal in a full-scale, continuously recycled system (with biomass recycling varying from 50% to 75%) could average 79–100% and 50–60% respectively, resulting in a substantial improvement in water quality.

ACKNOWLEDGMENTS

This paper was prepared in connection with work done under Contract No. DE-AC09-89SR18035 with the U.S. Department of Energy. We thank Dr. L. Ralph Berger for providing Hawaiian cultures and Mr. Christopher W. Jeter for technical assistance.

REFERENCES

Castenholz, R.W., 1982, Isolation and cultivation of thermophilic cyanobacteria, in *The Prokaryotes*, Springer-Verlag, New York, pp. 236–246.
Castenholz, R.W., 1989, Subsection V: Order stigonematales, in *Bergey's Manual of Systematic Bacteriology*, vol. 3, Williams & Wilkins, New York, pp. 1794–1799.
Radway, J.C., Weissman, J.C., Wilde, E.W., and Benemann, J.R., 1992, Exposure of *Fischerella* (*Mastigocladus*) to high and low temperature extremes: strain evaluation for a thermal mitigation process, *J. Appl. Phycol.*, 4:67–77.
Radway, J.C., Weissman, J.C., Wilde, E.W., and Benemann, J.R., 1994, Nutrient removal by thermophilic *Fischerella* (*Mastigocladus laminosus*) in a simulated algaculture process, *Biores. Technol.*, 50:227–233.
Standard Methods for the Examination of Water and Wastewater, Rand, M.C. et al. (eds.), 1979, American Public Health Association, Washington, D.C., p. 1193.
U.S. Environmental Protection Agency, 1979, *Methods for Chemical Analysis of Water and Wastes*, EPA-600/4-79-020, Cincinnati.
Weissman, J.C., and Benemann, J.R., 1979, Biomass recycling and species composition in continuous cultures, *Biotechnol. Bioeng.*, 21:627–648.
Wilde, E.W., Benemann, J.R., Weissman, J.C., and Tillett, D.M., 1991, Cultivation of algae and nutrient removal in a waste heat utilization process, *J. Appl. Phycol.*, 3:159–167.

58

SECRETED METABOLITE PRODUCTION IN PERFUSION PLANT CELL CULTURES

Wei Wen Su

Department of Biosystems Engineering
University of Hawaii at Manoa
3050 Maile Way
University of Hawaii
Honolulu, Hawaii 96822

Key Words

perfusion bioreactor, plant cells, secretion, high cell density

1. SUMMARY

Operation of a novel bioreactor that allows continuous perfusion cultivation of plant cell suspensions is described in this paper. This external-loop, air-lift bioreactor has an internal settling zone for cell separation. The settling zone is created by inserting a baffle plate into the upper portion of the downcomer. Using this bioreactor, an *Anchusa officinalis* suspension culture was cultivated to a cell density of 27.2 g DW L^{-1} in 14 days at a perfusion rate of 0.123 day^{-1}. The maximum total extracellular protein concentration attained was 1.11 g L^{-1}. Complete cell retention was achieved throughout the culture during which the maximum packed cell volume (PCV) exceeded 80%. In comparison, the maximum cell density and extracellular protein concentration in the batch culture were 12.6 g DW L^{-1} and 0.47 g L^{-1}, respectively. A very high acid phosphatase activity was observed in the spent medium of both batch and perfusion cultures. After 14 days of cultivation, the acid phosphatase activity in the batch and the perfusion cultures exceeded 2000 and 4000 units L^{-1}, respectively. The potential application of this perfusion bioreactor in culturing cyanobacteria and other photosynthetic microorganisms for hydrogen production is also discussed.

2. INTRODUCTION

Plants are important sources of many useful compounds. New technologies to obtain these compounds in large quantities through the use of in vitro cell culture techniques

BioHydrogen, edited by Zaborsky *et al.*
Plenum Press, New York, 1998

have shown enormous potential. Considering various forms of in vitro plant tissue cultures, cell suspension culture is most amenable to large-scale production of natural compounds, due primarily to its superior culture homogeneity. Technological feasibility of large-scale suspension plant cell cultures has already been demonstrated in Japan during the past decade in the commercial production of shikonin, ginseng, and berberine (Misawa, 1991).

For large-scale production, it is advantageous that the target metabolite is released into the culture medium, in which case downstream recovery efficiency will be greatly improved. To increase the productivity of secreted metabolic products, it is desirable to operate the culture at a high cell density in a continuous perfusion system (Su, 1995). By means of a cell-retention device, cells are retained in the perfusion bioreactor, while cell-free spent medium is constantly removed and the culture is simultaneously replenished with fresh medium. Cultivation of plant cell suspension to a high density with effective production of secreted proteins is reported in this paper. The potential application of the perfusion bioreactor in culturing immobilized cyanobacteria for hydrogen production is also discussed.

3. EXPERIMENTAL

3.1. Cell Culture

The stock culture of *A. officinalis* was maintained in a liquid Gamborg B5 medium supplemented with 1 mg L^{-1} 2,4-dichlorophenoxyacetic acid (2,4-D), 0.1 mg L^{-1} kinetin, and 30 g L^{-1} sucrose (De-Eknumkul and Ellis, 1984). The suspension was maintained on a gyratory shaker at 120 rpm and 25 °C and subcultured every 10 days using a 10% inoculum. The culture suspension is homogeneous with very few large cell aggregates.

3.2. Assays

Analytical procedures for cell dry weight and fresh weight, medium osmolarity and conductivity, and PCV have been described elsewhere (Su et al., 1993). For the measurement of total extracellular protein concentration, the culture sample was centrifuged, the supernatent filtered through a Whatman #5 filter, then dialyzed in a 3500 MWCO dialysis tube (Spectra/Por, Fisher Scientific) against 12.5 mM Tris buffer (pH 6.8) for 24 h at 4 °C prior to measurement using Bicinchoninic acid (BCA) protein assay (Smith et al., 1985). The dialysis step is necessary because certain component(s) of the B5 medium strongly interferes with the BCA assay. Cell viability was assessed by detecting the cell respiratory efficiency using 2,3,5-triphenyltetrazolium chloride (TTC) according to a protocol developed by Su et al. (1993). Dissolved oxygen (DO) was measured using a polarographic oxygen electrode (Ingold). Glucose 6-phosphate dehydrogenase activities in the cellular extract and the spent culture medium were analyzed based on the reduction of NAD(P) and the generation of NAD(P)H, which gives a maximum absorbance at 340 nm according to the method of Aitchison and Yeoman (1973). Acid phosphatase activity in the medium was measured by a colorimetric method using *p*-Nitrophenyl phosphate as the substrate (Nahas et al., 1982). Orthophosphate concentration was measured using the ascorbic acid method (Taras, 1971). To determine intracellular phosphate, a 1-g sample of cells in a test tube with 3 mL distilled water was frozen then thawed to rupture the cell membranes. The

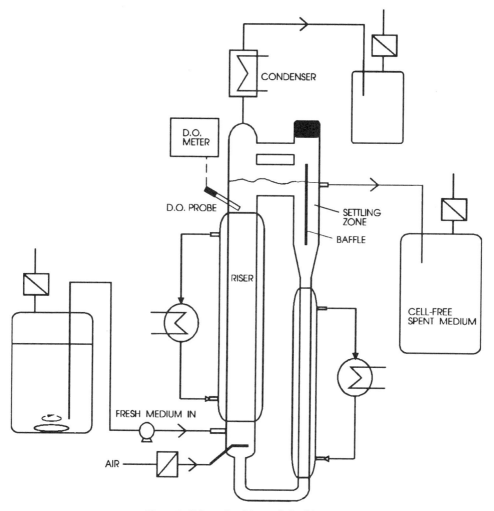

Figure 1. Schematic of the perfusion bioreactor.

sample was then centrifuged at 1500 rpm for 10 min. A 1-mL sample was taken from the supernatant and analyzed.

3.3. The Bioreactor

The perfusion air-lift bioreactor (PALB) system is shown schematically in Figure 1. The internal settling zone in the perfusion bioreactor is created by inserting a baffle plate into the upper portion of the downcomer. Clarified spent medium containing secreted metabolic products is continuously removed via the overflow from the settling zone, while fresh medium is fed into the well-mixed riser to replenish nutrients. This jacketed glass bioreactor has a working volume of 1.2 L. The volume of the settling zone is 108 mL. A sintered glass sparger tube with a mean pore opening of about 140 μm was used for gas dispersion. Diameter of the sparger tube is 0.5 cm. The delivery of perfusion medium was done using a ultra-low flow peristaltic pump (Model 77, Harvard Apparatus).

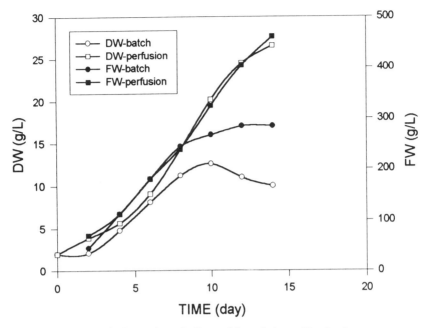

Figure 2. Comparison of cell growth in perfusion and batch cultures.

3.4. Cultivation of *A. officinalis* in the PALB

The bioreactor containing B5 medium supplemented with 1 mg L^{-1} 2,4-D, 0.1 mg L^{-1} kinetin, and 30 g L^{-1} sucrose was inoculated with a 10-day old culture of *A. officinalis*. For the perfusion run, the culture was allowed to grow in the batch mode until the medium osmolarity dropped below 100 mOSM L^{-1} (initially at 160 mOSM L^{-1}). The culture was then perfused using a B5 medium supplemented with 6% sucrose and reduced hormone concentrations (0.1 mg L^{-1} 2,4-D and 0.01 mg L^{-1} kinetin). The perfusion medium feed rate was set at 0.1 mL min^{-1}, which was equivalent to a perfusion rate of 0.123 day^{-1}. This protocol was found in a previous study to produce high cell density in the *A. officinalis* shake-flask culture (Su et al., 1993). The aeration rate was set at 400 mL min^{-1} (0.33 vvm; riser superficial gas velocity 20.4 cm min^{-1}). Foam control was achieved using a silicone-based antifoam solution (Antifoam C emulsion, Sigma Chemical). A batch cultivation was also conducted in the PALB for comparison to the perfusion cultivation. The batch run was operated under the same conditions (i.e., inoculum size, DO, aeration rate, medium) as those employed in the perfusion run, except no medium exchange was initiated. The DO level in both cultures was controlled at 30% of air saturation by adjusting the oxygen/air ratio in the gas feed stream using two mass flow controllers (Omega, Massachusetts) via a PID controller. The controller was implemented using a National Instruments Digital Signal Processor board driven by the LabVIEW software. Culture temperature in the bioreactor was kept at 25 °C.

4. RESULTS AND DISCUSSION

The time course of cell growth (expressed as dry weight and fresh weight) during the perfusion and the batch cultures of *A. officinalis* in the PALB is presented in Figure 2.

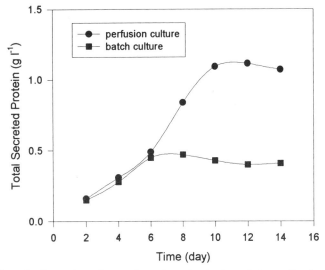

Figure 3. Comparison of secreted protein production in perfusion and batch cultures.

Culture perfusion was initiated on day 6 based on the medium osmolarity profile (Su et al., 1993). In the perfusion culture, cell dry weight and fresh weight attained 27.2 and 461 g L^{-1}, respectively. For comparison, the maximum dry weight attained was 12.6 g L^{-1} on day 10 in the batch culture. During the stationary phase in the batch culture, cell fresh weight continued to increase and reached 285 g L^{-1}. The increased fresh weight/dry weight ratio indicated elevated cellular water content and cell swelling, which was not observed in the perfused culture. Also noted from the stationary phase in the batch culture, both medium conductivity and osmolarity were very low, which were ca. 5 and 13% of the corresponding initial values.

In the perfusion culture, the total secreted protein concentration increased continuously and leveled off at ca. 1.1 g L^{-1} from day 10 on (Figure 3). This corresponded to a maximum volumetric productivity of 135 mg L^{-1} day^{-1}. The total secreted protein concentration in the batch culture peaked at 0.47 g L^{-1} on day 8, and gradually decreased to 0.4 g L^{-1} on day 14 (Figure 3). The maximum secreted protein productivity corresponding to protein concentration on day 6 was 71 mg L^{-1} day^{-1}. The proteins present in the medium resulting from secretion rather than cell leakage was confirmed by analyzing the specific activity of a cytosolic marker enzyme glucose 6-phosphate dehydrogenase as suggested by Sijmons et al. (1990). This intracellular enzyme catalyzes the irreversible oxidation of glucose 6-phosphate to gluconolactone-6-phosphate in the pentose phosphate pathway, and is present in almost all the higher organisms. The enzyme assay of the 14-day sample from the perfusion culture showed that the specific activity in the intracellular extract was 25 nmol NAD(P)H per min per mg protein, while no activity was detected in the cell-free spent medium.

One of the main design criteria for a perfusion bioreactor is the ability of the reactor to effectively retain cells so that a high cell concentration can be achieved and the reactor can operate functionally at that high cell loading. For plant cell cultures, the PALB with its simple cell separator design has been demonstrated in this study to fulfill that objective. The efficacy of the bioreactor is attributed in part to the unique location of the settling zone. By taking advantage of the well-defined flow pattern in the air-lift reactor, the bulk liquid flow in the downcomer drove the cells away from the settling region and thus improved the cell retention efficiency. Indeed, a totally clear settling zone was noted for the first eight days of the perfusion culture during which PCV was below 60%. It should be pointed out, however, that there were noticeable eddies which provoked some degrees of mixing at the entrance of the settling region. Since the settling zone for the most part was not stirred by the localized mixing at its entrance, the presence of eddies was found to have little effect on cell retention efficiency. Furthermore, since nearly all the air bubbles disengaged through the liquid/headspace interface, gas holdup in the downcomer was practically zero and the chance of bubbles getting into the settling zone to disturb cell sedimentation was nil. In this context, the external-loop, air-lift bioreactor design is better than the internal-loop design.

As the cell loading or the perfusion rate increased, it was evident that a cell-free settling zone could no longer be achieved as indicated from the later stage of the perfusion culture. In this case, counter-current sedimentation took place in the settler. Since the settling zone was essentially stagnant, lower DO than that in the riser was suspected. At the end of the perfusion culture, the DO concentration in the settling zone and the riser were 20 and 31% of air saturation, respectively. It should be noted, however, very similar cell viability (62% for the riser vs. 60% for the settler) and extracellular protein concentration (1.07 g L^{-1} for the riser vs. 1.02 g L^{-1} for the settler) were measured in the settling zone and the riser. To improve oxygen supply in the settler, membrane aeration tubing may be installed to provide bubble-free aeration (Su et al., 1992) without disturbing cell sedimentation. As far as the perfusion rate is concerned, due to the generally slow metabolic rates of plant cells (the maximum specific growth rate for *A. officinalis* is about 0.45 day^{-1}), it is believed that low perfusion rates (less than 1 day^{-1}) are sufficient to accomplish high-density cultivation, as demonstrated in the present study for the *Anchusa* system where a low perfusion rate of 0.123 day^{-1} was found satisfactory. Moreover, by operating at low perfusion rates, product in the spent medium is more concentrated, which is beneficial to downstream processing.

The application of a high sucrose concentration (6%) in the perfusion medium has enabled the maintenance of medium osmotic pressure under the low perfusion rate. The FW/DW ratio that served as an indicator of cellular water content was found to fluctuate between 17 and 20 throughout the culture period. Considering the accuracy in FW measurement, no significant increase in cellular water content was observed during the perfusion culture. On the contrary, cell water content increased substaintially during the stationary phase in the batch culture. As shown in a previous study, reduction of cellular water content in the *A. officinalis* culture led to a lower apparent culture viscosity (Su et al., 1994). According to Curtis and Emery (1993), both high viscosity and non-Newtonian behavior of dense plant cell suspension culture could be attributed to cell enlargement. In this connection, the ability to control cell enlargement or, more so, to reduce cell size becomes very crucial, especially in light of the impact of biotic phase volume on the cell retention efficiency in PALB. Further, mixing and oxygen transfer problems in the bioreactor are expected to be minimized as a result of the improved culture rheological characteristics.

The specific total extracellular protein productivity (q_p) of the perfused *Anchusa* culture was estimated to range from ca. 8–18 mg g^{-1} day^{-1} during the perfusion culture, while highest q_p was observed between days 6 and 8. For comparison, total protein secretion by the rice callus system reported by Simmons et al. (1991) was about 4 mg g^{-1} day^{-1}. Considering the spectrum and the specific productivity of extracellular proteins, *Anchusa* culture could be an effective host system for foreign protein production and secretion by using the promoter and signal sequence of the gene encoded for the most abundant secreted protein. Recently we have identified the predominant extracellular protein in the *Anchusa* culture to be an acid phosphatase (Su et al., 1998). The acid phosphatase activity in the 14-day batch and perfusion cultures was found to be in the range between 2000–3000 and 4000–5000 unit L^{-1}, respectively. It is well known that expression of phosphatases is stimulated by phosphate starvation. In the 14-day cultures, phosphate concentration in the medium was essentially zero, and the intracellular orthophosphate dropped below 0.1% of cell dry weight. Cloning of the promoter and signal sequence of the *Anchusa* acid phosphatase gene is currently underway in our lab.

With respect to biohydrogen production, application of immobilized cyanobacteria or other hydrogen-producing microorganisms has been explored by a number of researchers in recent years (Markove et al., 1993; Seon et al., 1993; Patel and Madamwar, 1995). The perfusion bioreactor described in this paper can be useful in culturing immobilized cells for biohydrogen production. With its advantageous heat and mass transfer characteristics, the perfusion bioreactor should be an attractive alternative to the packed-bed bioreactors (Su, 1995). Also, it is easier to operate, with superior hydrodynamic stability than the fluidized bioreactors. Further, the air-lift tubular reactor design has also been proven effective in photobioreactor cultivation of microalgae and other photosynthetic microorganisms (Lee, 1986).

5. CONCLUSION

This study has demonstrated the applicability of the proposed air-lift bioreactor for continuous perfusion cultivation of plant cells. Using this bioreactor operating at the perfusion mode with total cell retention, maximum cell density was increased more than twofold and total secreted protein productivity was about two times higher in comparison with the corresponding batch system. By incorporating internal or external illumination (Su and Kao, 1995), this perfusion bioreactor system may also be considered for cultivation of immobilized cyanobacteria for hydrogen production.

REFERENCES

Aitchison, P.A., and Yeoman, M.M., 1973, The use of 6-MP to investigate the control of glucose-6-phosphate dehydrogenase levels in cultured Artichoke tissue, *J. Expt. Bot.*, 24:1069–1083.

Curtis, W.R., and Emery, A.H., 1993, Plant cell suspension culture rheology, *Biotechnol. Bioeng.*, 42:520–526.

De-Eknamkul, W., and Ellis, B.E., 1984, Rosmarinic acid production and growth characteristics of *Anchusa officinalis* cell suspension cultures, *Planta Med.*, 51:346–350.

Lee, Y.K., 1986, Enclosed bioreactors for the mass cultivation of photosynthetic microorganisms: the future trend, *TIBTECH*, 4:186–189.

Markov, S.A., Lichtl, R., Rao, K.K., and Hall, D.O., 1993, A hollow fibre photobioreactor for continuous production of hydrogen by immobilized cyanobacteria under partial vacuum, *International Journal of Hydrogen Energy*, 18:901.

Misawa, M., 1991, Research activities in Japan, in *Plant Cell Culture in Japan*, Komamine, A. et al. (eds.), CMC, Tokyo, pp. 3–7.

Nahas, E., Terenzi, H.F., and Rossi, A., 1982, Effect of carbon source and pH on the production and secretion of acid phosphatase and alkaline phosphatase in *Neurospora crassa*, *J. Gen. Microbiol.*, 128:2017–2021.

Patel, S., and Madamwar, D., 1995, Continuous hydrogen evolution by an immobilized combined system of *Phormidium valderianum, Halobacterium halobium*, and *Escherichia coli* in a packed bed reactor, *International Journal of Hydrogen Energy*, 20:631.

Seon, Y.H., Lee, G.G., and Park, D.H., 1993, Hydrogen production by immobilized cells in the nozzle loop bioreactor, *Biotechnology Letters*, 15:1275.

Sijmons, P.C., Dekker, B.M.M., Schrammeijer, B., Verwoerd, T.C., van den Elzen, P.J.M., and Hoekema A., 1990, Production of correctly processed human serum albumin in transgenic plants, *Bio/technol.*, 8:217–221.

Simmons, C.R., Huang, N., Cao, Y., and Rodriguez, R.L., 1991, Synthesis and secretion of a-amylase by rice callus: evidence for differential gene expression, *Biotechnol. Bioeng.*, 38:545–551.

Smith, P.K., Krohn, P.I., Hermanson, G.T., Mallia, A.K., Gartner, F.H., Provenzano, M.D., Fujimoto, E.K., Goeke, N.M., Olson, B.J., and Klenk, D.C., 1985, Measurement of protein using bicinchoninic acid, *Anal. Biochem.*, 150:76–85.

Su, W.W., Caram, H.S., and Humphrey, A.E., 1992, Optimal design of the tubular microporous membrane aerator for shear sensitive cell cultures, *Biotechnol. Prog.*, 8:19–24.

Su, W.W., Lei, F., and Su, L.Y., 1993, Perfusion strategy for rosmarinic acid production by *Anchusa officinalis*, *Biotechnol. Bioeng.*, 42:884–890.

Su, W.W., Asali, E.C., and Humphrey, A.E., 1994, *Anchusa officinalis*: production of rosmarinic acid in perfusion cell cultures, in *Biotechnology in Agriculture and Forestry, vol. 26, Medicinal and Aromatic Plants VI*, Bajaj, Y.P.S. (ed.), Springer-Verlag, Berlin, pp. 1–20.

Su, W.W., and Kao, N.P., 1995, Circulation time distribution in an external-loop plant cell air-lift photobioreactor, Paper 208e presented at the AIChE Annual Meeting, Miami Beach, Florida, United States.

Su, W.W., 1995, Bioprocessing technology for plant cell suspension cultures, *Appl. Biochem. Biotechnol.*, 50:189–230.

Su, W.W., Liang, H., and Sun, S., 1998, Identification of the major secreted protein in *Anchusa officinalis* culture, unpublished.

Taras, M.J., 1971, *Standard Methods for the Examination of Water and Wastewater*, 13th edition, American Public Health Association, New York, pg. 874.

59

STRATEGIES FOR BIOPRODUCT OPTIMIZATION IN PLANT CELL TISSUE CULTURES

Susan C. Roberts and Michael L. Shuler

Chemical Engineering
Cornell University
120 Olin Hall
Ithaca, New York 14853

Key Words

plant cell tissue culture, secondary metabolite, Taxol, elicitation, medium optimization, immobilization, differentiation, metabolic engineering, hydrogen

1. SUMMARY

Cell cultures that can produce multiple products of commercial interest simultaneously have the potential to overcome economic barriers to commercialization. This paper reviews several strategies (elicitation, immobilization, medium optimization, differentiation, metabolic engineering, bioreactor considerations) for enhancing valuable product synthesis in plant cell cultures, with a focus on the production of Taxol from *Taxus* cell cultures (our laboratory research focus). While plants are capable of synthesizing valuable products (secondary metabolites), often undifferentiated tissue cultures cannot produce significant amounts of the target compound without the use of enhancement strategies. For example, cell immobilization cannot only protect cells from shear, but also promote secondary metabolism, in part due to increased cell interactions. The cell culture medium can be easily manipulated by changing sugar composition, type, and concentration of plant growth hormones, and the levels of phosphate and nitrogen. A two-stage process that promotes growth in the first stage and production in the second stage is often employed. Elicitation of enzyme systems can be used to achieve high levels of production. Methyl jasmonate is a signal transducer that has been shown to influence numerous plant species, including *Taxus*. Often, the synergistic application of these enhancement strategies is necessary to achieve acceptable product yields. There are also several issues to consider when designing a reactor for secondary metabolite production: shear stress, gas phase composition, and oxygen delivery. The generic strategies used to enhance secondary metabolite production in plant cell cultures can also be applied to increase hydrogen and valuable product yields in bacterial and algal systems.

2. INTRODUCTION

Plants are the source of many valuable products including pharmaceuticals, fragrances, food, and dyes. However, the supply of such products is often limited due to the slow-growing nature of the plant or variability in growing conditions. Cell culture can overcome these limitations by providing a manipulative environment to optimize both growth and production. Additionally, cell cultures that can produce multiple products of commercial interest simultaneously have the potential to overcome economic barriers. While plants are capable of producing these valuable chemicals, often undifferentiated tissue cultures cannot produce significant amounts of the target compound without the use of enhancement strategies. Such strategies include elicitation of enzyme systems, cell immobilization, medium optimization, obtaining more highly differentiated cultures, metabolic engineering, and bioreactor considerations. A recent review provides an overview of the strategies used to improve secondary metabolite accumulation in plant cell cultures (Dornenburg and Knorr, 1995). By synergistically applying enhancement strategies, yield improvements of 100–1000-fold can be obtained.

3. ELICITATION

Elicitation of enzyme systems using biotic or abiotic elicitors is often effective in enhancing secondary metabolite production. Some elicitors include bacterial/yeast extracts, heavy metals, or signal transduction compounds. Eliciting secondary metabolite accumulation in plant cell cultures (DiCosmo and Misawa, 1985; Nishi, 1993), induction of plant defense reactions through elicitation (Yoshikawa et al., 1993), and signal transduction pathways of plants (Jones, 1994) have been reviewed.

Methyl jasmonate (MJ) is a signal transducer that regulates the inducible defense genes in plants and is produced in response to fungal attack or wounding. Additionally, the exogenous application of MJ has been shown to increase the production of secondary metabolites (e.g., flavonoids, anthroquinone, and alkaloids) in 36 different plant cell culture systems (Gundlach et al., 1992). In the past few years, MJ has been shown to be an inexpensive, effective elicitor of secondary metabolism in many other systems including *Taxus*. The role of jasmonic acid in signal transduction pathways has recently been reviewed (Seo et al., 1997).

There have been several reports on the effectiveness of MJ at eliciting Taxol production in *Taxus* cell cultures (Mirjalili and Linden, 1996; Yukimune et al., 1997; Ketchum et al., 1998). Our laboratory has applied MJ to a variety of different *Taxus* cell cultures (Ketchum et al., 1998). The best results were observed with a MJ concentration of 100 µM, added on day 7 of the cell culture period. The highest Taxol concentrations were observed 7 days later (day 14 in the cell culture period). The addition of MJ decreased Taxol production time by 1–2 weeks, which would considerably reduce cost in an industrial large-scale process. Levels of 117 mg/L were achievable in 12 days.

Table 1. The effect of methyl jasmonate (MJ) on Taxol yield in several different *Taxus* cell lines. The Taxol concentrations presented are the maximum achieved in each experiment

Cell line	*Taxus* species	Total Taxol (mg/L) without MJ elicitation	Total Taxol (mg/L) with MJ elicitation	Fold increase
C93AD	canadensis	4.5	38	8.4
CO93P	canadensis	29	31	1.1
PC2	baccata	0.8	2.0	2.5

The effect of methyl jasmonate on several different *Taxus* cell lines in a single experiment is shown in Table 1. MJ-elicited cell cultures accumulate higher maximum Taxol concentrations in all cell lines. However, there was a variation in the effectiveness of MJ at enhancing Taxol accumulation between the cell lines (increases ranged from 1.1-fold to 8.4-fold). Similar variations in the elicitation of Taxol were observed by Yukimune et al. (1997) and Ketchum et al. (1998). Therefore, although MJ is an effective elicitor of Taxol accumulation, it is important to select a highly responsive cell line when designing an optimal cell culture process.

4. IMMOBILIZATION

Immobilization is the confinement of cells within particles under conditions favoring product synthesis, release, and cell reuse. Plant cell immobilization is a valuable strategy that can be used to increase volumetric productivity of undifferentiated cells. From an engineering perspective, immobilized cell systems provide many advantages, including protection from shear and the ease of cell and product recovery from a culture broth. Also, increased product release allows for the use of recovery methods in conjunction with immobilization to decrease product degradation or breakdown. Additional advantages of immobilization include: compatibility with non-growth associated product formation, high viability and production capability over an extended time, and facilitation of processing steps including rapid medium exchange. Figure 1 depicts different immobilization techniques including gel, foam, and natural aggregation. An overview of the methods for immobilization of plant cell cultures for optimizing secondary metabolite accumulation is provided by Deliu et al. (1994).

5. MEDIUM AND ENVIRONMENTAL OPTIMIZATION

Medium/environmental optimization studies have indicated the importance of certain compounds/conditions in the regulation of secondary metabolite biosynthesis. Hor-

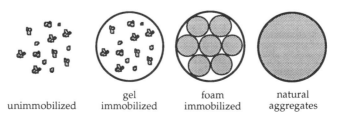

Figure 1. Common immobilization techniques utilized for plant cell culture systems.

mones, phosphate, nitrogen, carbohydrates, temperature, and precursor availability can regulate the transition from primary to secondary metabolism. Secondary metabolite production can be induced by transferring cells from a growth medium to a production medium, where optimal concentrations of key hormones and nutrients are present (see Table 2 for a summary). In fact, one of the few industrial plant cell culture processes, shikonin production from *Lithosperum erythrorhizon*, makes use of this strategy (Tabata and Fujita, 1985). When using bioreactors, media can be easily exchanged when cells are immobilized in beads or foam. This is a good example of the synergistic application of several enhancement strategies for optimizing secondary metabolite production.

6. DIFFERENTIATION

Often, undifferentiated plant cell cultures do not produce significant amounts of secondary metabolites. Differentiated cultures (such as roots or shoots) are usually capable of higher production. Differentiation is depicted in Figure 2. The importance of differentiation for improvements in secondary metabolite production can be illustrated by considering vindoline production by *Catharanthus roseus* cell cultures. The production of the alkaloids vincristine and vinblastine (both anti-cancer agents) has not been consistently achieved in undifferentiated plant cell tissue cultures due to the lack of vindoline biosynthesis. The production of vindoline requires chloroplast formation through differentiation (Constabel et al., 1982). However, vindoline production has been reported in cell cultures transformed with *Agrobacterium* (O'Keefe et al., 1997).

7. METABOLIC ENGINEERING

Metabolic engineering is loosely defined as "the manipulation of the metabolic pathways of an organism to achieve a desired goal," (Farmer and Liao, 1996), that is, product

Table 2. Factors that need to be considered when optimizing medium/environmental conditions for growth and secondary metabolite production in plant cell tissue cultures

Medium/environmental variables	Effect on 2° metabolism
Carbohydrates	Alkaloid production, as well as other 2° metabolites, is increased by high sugar levels. Type of sugar utilized can also affect production.
Hormones (auxins, cytokines)	Choice of hormones can affect the degrees of differentiation and production enhancement—for example, low levels of 2,4-D inhibit the production of many 2° products.
Phosphate	High phosphate has been shown to inhibit enzymes involved in 2° metabolite biosynthesis; therefore, low levels are usually preferable.
Nitrogen	The total nitrogen concentration, as well as the nitrate:ammonia ratio, are important factors in optimizing production. Mixed effects are observed in different systems, but typically low nitrogen levels are preferable.
Temperature	Optimal growth remperature has been shown to be higher than optimal production temperature for alkaloid biosynthesis.
Precursor feeding	The precursor pool is sometimes the rate-limiting step in the production of 2° metabolites. By adding precursors to the medium, this limitation can be overcome.

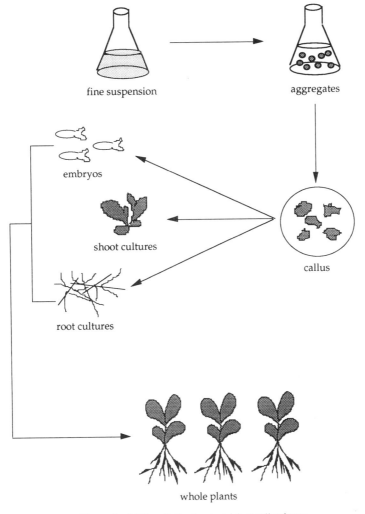

Figure 2. Differentiation in plants/plant cell cultures.

synthesis. A major problem in the advancement of industrial plant cell culture for the production of secondary metabolites is the lack of knowledge about the biosynthetic pathways and regulation of cell metabolism. In plant cells, compartmentation plays a role in regulating synthesis due to the availability of appropriate precursors and transport of necessary compounds. The most well-studied system has been the production of alkaloids (Roberts and Shuler, 1997). Some pathways have been characterized and efforts have been directed toward optimizing these pathways using appropriate genetic engineering techniques. There have been many excellent reviews on general metabolic engineering as well as the application of metabolic engineering to plant systems (Bailey, 1991; Stephanopoulos and Sinskey, 1993; Stephanopoulos, 1994; Kutchan, 1995; Bruce et al., 1995; Farmer and Liao, 1996).

Traditional metabolic engineering involves genetic manipulations; however, there are non-genetic approaches for studying metabolism. The use of inhibitors, precursor feeding, and monitoring kinetic profiles of pathway products can all be used to determine metabolic information without the use of genetic engineering. For example, our laboratory has been studying the metabolism of Taxol in cell culture (biosynthesis, transport, storage, and degradation) to determine which steps are limiting in the production of Taxol.

8. BIOREACTOR CONSIDERATIONS

In our laboratory, we have studied the cultivation of plant cell cultures in bioreactors for the production of secondary metabolites. Reactor configuration, gas concentrations (oxygen, carbon dioxide, and ethylene), and shear can all affect secondary metabolite production, particularly in large-scale systems. The characterization of these parameters on a small scale can facilitate scale-up. There are several excellent reviews addressing these bioreactor considerations in the scale-up of plant cell culture processes (Hua et al., 1993; ten Hoopen et al., 1994; Tyler et al., 1995; Namdev and Dunlop, 1995; Zhong et al., 1995).

8.1. Gas Phase Composition

Gas phase composition was shown to influence the timing and levels of Taxol production (Mirjalili and Linden, 1995), with low oxygen concentrations promoting early production and high carbon dioxide inhibiting Taxol production. The optimal gas mixture was 10% v/v oxygen, 0.5% v/v carbon dioxide, and 5 ppm ethylene. Volatile compounds are important in signal transduction. Recycle reactors, where a fraction of the exhaust gas is recirculated to mimic the conditions in shake flask cultures, are often effective in enhancing secondary metabolite production (Schlatmann et al., 1993).

8.2. Bioreactor Configuration

Unlike microbial suspension culture, plant cell suspensions are highly aggregated, and as a consequence mixing must be continuous to prevent settling out of cells. Plant cells have been grown in a variety of bioreactors, including stirred tank, air-lift, bubble column, rotating drum, photo, and trickling film/mist for root culture. Some of these are depicted in Figure 3. Plant cells are also sensitive to hydrodynamic stress. Impeller design is crucial to provide adequate mixing while minimizing shear. Shear stress can be reduced by using a helical-ribbon impeller with gas sparging (Sun, 1997).

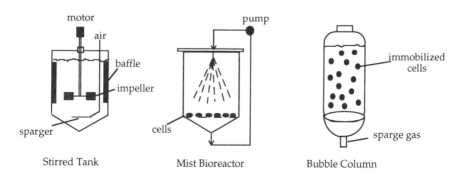

Figure 3. Various reactor types used for the cultivation of plant cells.

9. HYDROGEN PRODUCTION FROM BACTERIAL OR ALGAL SYSTEMS

The generic strategies outlined in this paper for secondary metabolite yield enhancement in plant culture systems could be applied to enhancing hydrogen and/or co-product yields from hydrogen-producing microalgae or cyanobacteria. For instance, immobilization of bacterial or algal cell cultures can result in high cell densities and increases in process time, which could lead to an enhancement of hydrogen production. Additionally, because methyl jasmonate is an effective elicitor of secondary metabolite production in plant cell systems, it may be effective at eliciting valuable product synthesis in algal cell systems. The coupling of valuable product synthesis with hydrogen production in algal cell systems can make the process more economically feasible.

10. HYDROGEN PRODUCTION FROM PLANTS?

10.1. Background

Plants lack the hydrogenase enzyme needed for hydrogen production, so endogenous synthesis is not feasible. Subcellular extracts prepared from higher plants evolve hydrogen if an active preparation of hydrogenase is added (Plaskett, 1979). The enzyme nitrogenase is also capable of synthesizing hydrogen under some circumstances (i.e., the absence of nitrogen). Eukaryotic plants cannot fix nitrogen alone, but some plants form symbiotic relationships with various heterotrophic organisms (usually bacteria). Both the hydrogenase and nitrogenase enzymes are sensitive to oxygen, and the symbiotic relationship between the plant and bacteria is effective in reducing this inhibition. The symbiotic relationship results in the production of leghemoglobin which can bind oxygen in the vicinity of the nitrogenase and reduce oxygen inhibition. The plant host synthesizes the globin portion while most of the heme is produced by the bacteria.

10.2. Prospects

10.2.1. Genetic Engineering. Transformation of the hydrogenase gene into plants could render them hydrogen producers in cell culture. This option would be attractive if it could be coupled with the production of another valuable product. Since plant cell cultures are capable of producing valuable secondary metabolites, this idea is theoretically feasible.

10.2.2. Fusion of Algal and Plant Cells. The fusion of plant and algal protoplast has been achieved (Fowke et al., 1979; Fowke et al., 1981). Algae have hydrogen production capability and plants produce many valuable compounds. These systems could be coupled for co-product synthesis.

10.2.3. Cell-Free Extracts. Plant cell-free extracts have proven to synthesize hydrogen when an active hydrogenase preparation is added. This capability could be exploited to produce greater quantities of hydrogen.

Although the production of hydrogen from plants is theoretically feasible, we do not believe plant systems can compete with bacterial or algal systems for the supply of hydrogen. Plants do not synthesize the hydrogenase enzyme needed for hydrogen production, and although the hydrogenase gene could theoretically be transformed into plants/cell cultures, the complexity and expense would limit this possibility. Additionally, plant systems grow at significantly slower rates that bacterial or algal systems, again increasing process costs.

11. CONCLUSIONS

A variety of strategies can be used to enhance the production of valuable products from plant cell tissue cultures. The synergistic combination of these strategies can result in yield increases of up to 1000-fold. Although plant cell culture processes are not commonly used industrially, recent and significant contributions offer promise to the viability of large-scale plant cell tissue culture systems. The production of hydrogen from plants is theoretically possible, but currently not economically feasible. However, the strategies developed for enhancement of secondary metabolite yields in plant cell culture systems can be applied to enhance hydrogen and/or valuable product yields in bacterial and algal systems.

REFERENCES

Bailey, J., 1991, Toward a science of metabolic engineering, *Science*, 252:1668–1675.
Bruce, N., French, C., Hailes, A., Long, M., and Rathbone, D., 1995, Engineering pathways for transformations of morphine alkaloids, *Trends in Biotechnology*, 13:200–205.
Constabel, F., Gaudet-LaPrairie, P., Kurz, W., and Kutney, J., 1982, Alkaloid production in *Catharanthus roseus* cell cultures, *Plant Cell Reports*, 1:139–142.
Di Cosmo, F., and Misawa, F., 1985, Eliciting secondary metabolism in plant cell cultures, *Trends in Biotechnology*, 3(12):318–322.
Deliu, C., Ispas, G., and Munteanu-Deliu, C., 1994, Methods for immobilization of plant cell cultures producing secondary metabolites, *Studia Universitatis Babes-Bolyai Biologia*, 39(1):59–68.
Dornenburg, H., and Knorr, D., 1995, Strategies for the improvement of secondary metabolite production in plant cell cultures, *Enzyme and Microbial Technology*, 17:674–684.
Farmer, W., and Liao, J., 1996, Progress in metabolic engineering, *Current Opinion in Biotechnology*, 7:198–204.
Fowke, L., Gresshoff, P., and Marchant, H., 1979, Transfer of organelles of the alga *Chlamydomonas reinhardii* into carrot cells by protoplast fusion, *Planta*, 144:341–347.
Fowke, L., Marchant, H., and Gresshoff, P., 1981, Fusion of protoplasts from carrot cell cultures and the green alga *Stigeoclonium*, *Canadian Journal of Botany*, 59:1021–1025.
Gundlach, H., Muller, M., Kutchan, T., and Zenk, M., 1992, Jasmonic acid is a signal transducer in elicitor-induced plant cell cultures, *Proceedings of the National Academy of Sciences*, 89:2389–2393.
Hua, J., Erickson, L., Yiin, T., and Glasgow, L., 1993, A review of the effects of shear and interfacial phenomena on cell viability, *Critical Reviews in Biotechnology*, 13(4):305–328.
Jones, A., 1994, Surprising signals in plant cells, *Science*, 263:183–184.
Ketchum, R., Gibson, D., Croteau, R., and Shuler, M., 1998, The response of suspension cell cultures of *Taxus* following elicitation with methyl jasmonate, submitted.
Kutchan, T., 1995, Alkaloid biosynthesis—the basis for metabolic engineering of medicinal plants, *The Plant Cell*, 7:1059–1070.
Mirjalili, N., and Linden, J., 1995, Gas phase composition effects on suspension cultures of *Taxus cuspidata*, *Biotechnology and Bioengineering*, 48:123–132.
Mirjalili, N., and Linden, J., 1996, Methyl jasmonate induced production of Taxol in suspension cultures of *Taxus cuspidata*: ethylene interaction and induction models, *Biotechnology Progress*, 12(1):110–118.
Namdev, P., and Dunlop, E., 1995, Shear sensitivity of plant cells in suspensions, *Applied Biochemistry and Biotechnology*, 54:109–131.
Nishi, A., 1993, Elicitor-induced production of secondary metabolites in higher plants, *Yakugaku Zasshi*, 113(12):847–860.
O'Keefe, B., Mahady, G., Gills, J., and Beecher, C., 1997, Stable vindoline production in transformed cell cultures of *Catharanthus roseus*, *Journal of Natural Products*, 60:261–254.

Plaskett, L., 1979, The generation of hydrogen gas using the photosynthetic mechanism of green plants and solar energy (biophotolysis), *ETSU Publication CR/20*, March.

Roberts, S., and Shuler, M., 1997, Large-scale plant cell culture, *Current Opinion in Biotechnology*, 8:154–159.

Schlatmann, J., Nuutila, A., Van Gulik, W., ten Hoopen, H., Verpoorte, R, and Heijnen, J., 1993, Scale-up of ajmalicine production by plant cell cultures of *Catharanthus roseus*, *Biotechnology and Bioengineering*, 41:253–262.

Seo, S., Sano, H., and Ohashi, Y., 1997, Jasmonic acid in wound signal transduction pathways, *Physiologia Plantarum*, 101(4):740–745.

Stephanopoulos, G., and Sinskey, A., 1993, Metabolic engineering—methodologies and future prospects, *Trends in Biotechnology*, 11:392–396.

Stephanopoulos, G., 1994, Metabolic engineering, *Current Opinion in Biotechnology*, 5:196–200.

Sun, Y., 1997, M.S. thesis, Cornell University.

Tabata, M, and Fujita, Y., 1985, Production of shikonin by plant cell cultures, in *Biotechnology in Plant Science* (Zaitlin, M., ed.).

ten Hoopen, H., Gulik, W., Schlatmann, J., Moreno, P., Vinke, J., Heijnen, J., and Verpoorte, R., 1994, Ajmalicine production by cell cultures of *Catharanthus roseus*: from shake flask to bioreactor, *Plant Cell, Tissue and Organ Culture*, 38:85–91.

Tyler, R., Kurz, W., Paiva, N., and Chavadej. S., 1995, Bioreactors for surface-immobilized cells, *Plant Cell, Tissue and Organ Culture*, 42:81–90.

Yoshikawa, M., Yamaoka, N., and Takeuchi, Y., 1993, Elicitors: their significance and primary modes of action in the induction of plant defense reactions, *Plant Cell Physiology*, 34(8):1193–1173.

Yukimune, Y., Tabata, H., Higashi, Y., and Hara, Y., 1997, Methyl jasmonate-induced overproduction of paclitaxel and baccatin III in *Taxus* cell suspension cultures, *Nature Biotechnology*, 14:1129–1132.

Zhong, J., Yu, J., and Yoshida, T., 1995, Recent advances in plant cell cultures in bioreactors, *World Journal of Microbiology and Biotechnology*, 11:461–467.

60

THE *Renilla* LUCIFERASE-MODIFIED GFP FUSION PROTEIN IS FUNCTIONAL IN TRANSFORMED CELLS

Yubao Wang,[1] Gefu Wang,[1] Dennis J. O'Kane,[2] and Aladar A. Szalay[1]

[1]Center for Molecular Biology and Gene Therapy
School of Medicine
Loma Linda University
Loma Linda, California 92350
[2]Department of Pathology and Molecular Medicine
Mayo Clinic
Rochester, Minnesota 55905

Key Words

Renilla luciferase, GFP, fusion protein, gene expression

1. SUMMARY

The cDNA of *Renilla reniformis* luciferase (ruc) has been cloned and used successfully as a marker gene in a variety of transgenic species. Similarly, the transfer and expression of green fluorescent protein (GFP) cDNA (gfp) and its mutants from *Aequorea victoria* resulted in high levels of GFP in transformed cells, allowing convenient visualization of gene expression under the microscope.

Here we present the construction of four fusion genes from the cDNAs of *Renilla* luciferase and *Aequorea* GFP mutants (gfp2 and gfph, which have been engineered specifically for expression in prokaryotic organisms and mammalian cells, respectively). The fusion gene I (rg2) contains the *Renilla* luciferase cDNA linked through a 15-nucleotide (5 amino acid) spacer added to its 3' end to the 5' end of the intact gfp2. When the gfph fragment replaces gfp2, fusion gene II (rg) was formed. In fusion gene III (g2r), the positions of the ruc and gfp2 are reversed with a linker composed of seven amino acids in length. In fusion gene IV (gr), gfph replaced gfp2 with a linker of nine amino acids. The fusion gene cassettes I and III were placed into pBluescript KS II (+) and the fusion gene II and IV into mammalian expression vector pCEP4. The above plasmids were transformed into *E.*

coli, different mammalian cell lines, and mouse embryos (microinjection). Fusion proteins with an apparent molecular weight around 65 kDa were detected by Western blotting using either anti-*Renilla* luciferase antibody or anti-*Aequorea* GFP antibody. Proteins RG2 and G2R extracted from the transformed *E. coli* have both green fluorescence activity and luciferase activity when expressed in *E. coli*. RG and GR are active in mammalian cells, ES cells, and mouse embryos. Fluorescence resonance energy transfer (FRET) between *Renilla* luciferase (emission at 478 nm) and *Aequorea* GFP (emission 510 nm) was detected by spectrofluorimetry, only if the two proteins were linked.

The *Renilla* luciferase-GFP fusion proteins offer a novel marker system for photosynthetic microorganisms and plants. This fusion protein helps to overcome the problems in quantifying GFP fluorescence. The determination of fusion protein in cells can be quantified based on luciferase activity. Furthermore, the system may be useful in the study of protein-protein interactions in vivo.

2. INTRODUCTION

The cDNA from *Renilla* luciferase (*ruc*) has been isolated and sequenced (Lorenz et al., 1991). By providing appropriate promoters, the cDNA gene cassettes were expressed in bacteria, transformed plant cells, and mammalian cells (Mayerhofer et al., 1995; Lorenz et al., 1995). The documented high efficiency of the *Renilla* luciferase is a useful and novel trait for a marker enzyme for gene expression studies. The usefulness of the green fluorescent protein from the jellyfish *Aequorea victoria* was documented as a reporter in prokaryotes and animal cell systems (Chalfie et al., 1994). The UV light stimulated GFP fluorescence does not require cofactors and the gene product alone is sufficient to allow detection of living cells under the light microscope. Bioluminescence in *Renilla reniformis* is produced by the reaction between coelenterazine (luciferin) and oxygen, a reaction that is catalyzed by *Renilla* luciferase. The reaction yields blue light with an emission wavelength maximum of 478 nm using the purified enzyme. In *Renilla reniformis* cells, this reaction is shifted toward the green with a maximum of 510 nm. This wavelength transition is due to an energy transfer to a green fluorescent protein. In this paper, we describe the engineering of a novel protein with dual functions combining characteristics for *Renilla* luciferase and GFP molecules. This task is accomplished by construction of a fusion gene between the cDNA of *Renilla* and the cDNA of the modified *Aequorea* GFP molecules (Zolotukhin et al., 1996; Cormack et al., 1996).

3. EXPERIMENTAL

3.1. Vectors and Cells

The vectors used for cloning and expression of the gene constructs in *E. coli* and mammalian systems were pBluescriptKS II (Strategen) and pCEP4, respectively. The cDNA of *Renilla* luciferase and gfp_2 were in plasmids designed pCEP4-RUC provided by Dr. Cormier and pBcGFP2 by Dr. Cormack, and GFP_h was in pTR-βcatin-GFP. *E. coli* strains used included DLT101 and DH5α. Mammalian cell line LM-TK⁻ was used as a recipient for gene expression.

3.2. Primers

The following six primers were designed for cloning of RG and GR gene constructs.

RUC5:   CTGCAGAGGAGGAATTCAGCTTAAAGATG   for 5' luciferase cDNA;
RUC3:   GCGGCCGCTTGTTCATTTTTGAGAAC   for 3' luciferase cDNA;
GFP$_2$–5: GCGGCCGCACCGCATATGAGTAAAGGAGAA   for 5' GFP$_2$ cDNA;
GFP$_2$–3: GAATTCTTGCGGTTTGTATAGTTCATCCA   for 3' GFP$_2$ cDNA;
GFP$_h$-5: GGGGTACCCCATGAGCAAGGGCG   for 5' GFP$_h$ cDNA; and
GFP$_h$-3: GGGGTACCCCTTGTACAGCTCGTCCATGCCA   for 3' GFP$_h$ cDNA.

3.3. Construction of *Renilla* GFP Fusion (RG or RG$_2$ Cassette) and GFP *Renilla* Fusion (GR or G$_2$R Cassette)

The fusion gene I (rg2) contains the *Renilla* luciferase cDNA linked through a 15-nucleotide spacer added to its 3' end to the 5' end of the intact gfp2 (Figure 1a). When the gfph fragment replaces gfp2, fusion gene II (rg) was formed (Figure 1b). In fusion gene III (g2r), the positions of the ruc and gfp2 are reversed with a linker composed of seven amino acids in length (Figure 1a). In fusion gene IV (gr), gfph replaced gfp2 with a linker of nine amino acids (Figure 1b). The fusion gene cassettes I and III were placed into prokaryotic expression vector pBluescript KS II (+) and the fusion gene II and IV into mammalian expression vector pCEP4.

3.4. Luciferase Assay

Before and after IPTG induction, an aliquot of transformed *E. coli* DLT101 was used for luciferase assay in a Turner TD 20e luminometer. The results were recorded as relative light units. The mammalian cells were harvested 36 h after transfection and measured for luciferase activity.

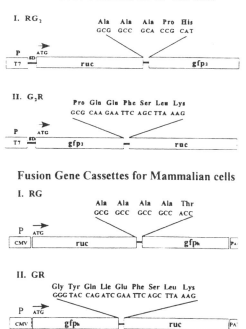

Figure 1. Construction of *Renilla* luciferase and gfp fusion gene cassettes for gene expression in *E. coli* (a) and in mammalian cells (b).

3.5. Assay of GFP Activity in Vivo

Transformed *E. coli* positive colonies were identified and cultured in 37 °C LB medium with 100 µg/mL of ampicillin selection. After 12 h incubation, one drop of *E. coli* culture was put on a slide and visualized by fluorescent microscopy at 1000 × magnification. LM-TK⁻ cells were transfected with plasmids pCEP4-RG and pCEP4-GR using calcium phosphate methods. The culture dishes were monitored using an inverted fluorescent microscope 12 h after the transfection. GFP activity was analyzed and detected in cells and embryos using an FITC filter system combined with a Xenon light as a UV excitation source. The resulting fluorescence image was directly photographed or sent through a video camera to an image analysis system.

3.6. Spectrofluorimetry

An SPEX fluorolog spectrofluorimeter operated in the ratio mode was used to detect the corrected emission spectra. Fluorescence emission was measured by UV excitation at 390 nm. Bioluminescence emission was recorded with the excitation beam blocked following the addition of 0.1 µg of coelenterazine in acidified methanol. Five spectra were averaged for each sample over the wavelength range from 400–600 nm.

3.7. Protein Isolation and Immuno Detection

E. coli 1 mL (OD_{600} = 1.0) was harvested and 400 µL of cell suspension buffer (0.1M NaCl; 0.01 M Tris-HCl pH 7.6; 0.001M EDTA; 100 µg/mL PMSF) and 100 µL of loading buffer (50mM Tris-HCl pH 6.8; 2% SDS; 10% glycerol; 5% 2-mercaptoethanol) were added. The samples were boiled for 4 min and loaded on a 7.5–20% gradient, SDS-polyacrylamide gel. Monoclonal anti-*Renilla* luciferase and polyclonal anti-*Aequorea* GFP antibody were used as the primary antibody, and goat peroxidase conjugated anti-IgG (anti-mouse for monoclonal and anti-rabbit for polyclonal) were employed as the secondary antibody.

3.8. Generation of Transformed ES Cells

The embryonic stem cells (ES) used in this study were from Dr. J. Mann. The STO feeder cells for supporting the growth of ES cells were from L. Robertson. Hygromycin B resistant feeder cell line was generated by transfecting STO with plasmid pCEP4, which contains the hygromycin B resistance gene. ES cells were cultured by standard procedures. After transformation, ES cells were cultured for two days in medium without selection, then placed into hygromycin B (200 µg/mL) containing selection medium.

3.9. Microinjection of Mammalian Cells and Mouse Embryos

DNA for microinjection was prepared as follows: Plasmid DNA was first purified to remove endotoxic compounds (EndoFree Plasmid Maxi kit, Qiagen), then linearized with a restriction endonuclease. The linearized plasmid DNA was further diluted to 1 ng/µL with injection buffer, filtered, and used directly for microinjection. Microinjections of mammalian cells and mouse embryos were performed using an automatic micromanipulator (Model 5171, Eppendorf) and a transjector (Model 5246, Eppendorf) system. Injection needles used were either purchased from Eppendorf Company (Femtotips & Femtotips II) or self-made (Borosilicate glass with filament, ID 0.78 mm, Sutter Instrument Company)

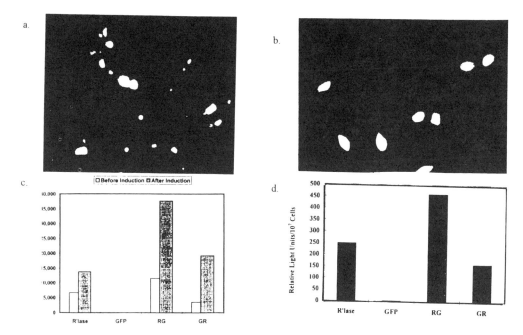

Figure 2. Analysis of GFP activity in transformed cells by fluorescence microscopy and fluorescence image system: (a) in *E. coli* cells; (b) in LM-TK⁻ mouse fibroblast cells; (c) luciferase activity with and without promoter induction in RG and GR transformed in *E. coli* cells; and (d) in RG and GR transformed LM-TK⁻ mouse fibroblast cells.

using a micropipette puller (Sutter Instrument Company, MP-97). The injection volume ranged from 2 to 10 picoliter upon each injection.

3.10. Image Analysis System

The image analysis system for detection of GFP in living organisms consisted of an inverted fluorescent microscope (Zeiss, Axiovert 100TV), CCD video camera (Hamamatsu, C2400), monitor (Sony), image intensifier (Hamamatsu, M4314), computer connected to a digital color printer (Sony, UP-D8800), and Polaroid Digital Palette. Both video images and fluorescence images were collected and analyzed using MetaMorph software and finally montaged using Photoshop software programs.

4. RESULTS AND CONCLUSIONS

The construction of two fusion genes is described in detail in Figure 1. Both gene cassettes 1 and 2 were placed into a prokaryotic expression vector pBluescriptKS II (Figure 1a) and into the pCEP4 eukaryotic expression vector (Figure 1b) and transformed into *E. coli* and into different mammalian cell lines. Cells containing the fusion gene cassettes gave strong fluorescence under the light microscope. Figure 2a shows individual *E. coli* cells transformed with the RG_2 construct exhibiting strong green fluorescence under oil immersion. Figure 2b shows the strong green fluorescence emitted by mammalian cells transformed with the RG construct and GR construct. In contrast, luciferase measurements

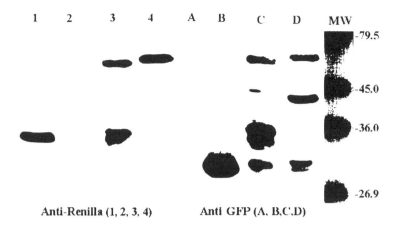

Figure 3. Detection of fusion protein in cell extracts of *E. coli* by Western blot analysis using anti- sera against *Renilla* luciferase and *Aequorea* GFP. Lane 1&A: *E. coli* cells containing *Renilla* luciferase gene cassette; Lane 2&B: *E.coli* cells containing GFP gene cassette; Lane 3&C: *E. coli* cells containing G_2R fusion gene cassette; Lane 4&D: *E. coli* cells containing GR_2 fusion gene cassette.

(Figures 2c, d) show that cells transformed with the RG construct have significant luciferase activity, which is reduced by threefold in the GR construct containing cells. Analysis of the fusion proteins in SDS polyacrylamide gels, followed by immunodetection of luciferase and GFP, confirms the presence of a 65-KDa band indicative of the full length fusion protein (Figure 3).

Data obtained from spectrofluorimetric measurements indicate that there is energy transfer (from 470–510 nm) between *Renilla* luciferase and GFP in the RG fusion contain-

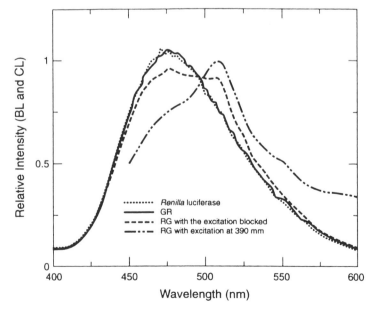

Figure 4. Light emission spectra from transformed *E. coli* cells.

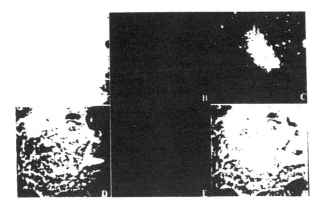

Figure 5. Expression of RG fusion gene in mouse embryonic stem cells based on fluorescence image analysis.

ing bacterial as well as mammalians cells (Figure 4). Cells containing the GR fusion show one emission peak at 470 nm, which indicates *Renilla* luciferase activity. In addition, RG fusion-containing cells show a 510-nm emission peak upon excitation at 390 nm. This activity is not found in GR construct-transformed cells when "humanized" GFP was used. Figure 5 shows ES cells transformed with the RG construct. High level of GFP activity was detected in vivo under the fluorescent microscope.

Based on the above data, we conclude that the 65-KDa, RG, and GR polypeptides exhibit both *Renilla* luciferase and GFP activity in living cells. These bifunctional polypeptides may become useful tools for identification of transformed cells at the single-cell level based on fluorescence. Simultaneously, it may allow quantifying of promoter activation in transformed tissues and transgenic organisms by measurement of luciferase activity.

REFERENCES

Chalfie, M., Tu, Y., Euskirchen, G., Ward, W.W., and Prasher, D.C., 1994, Green fluorescent protein as a marker for gene expression, *Science*, 263:802–805.

Cormack, B.P., Valdivia, R.H., and Falkow, S., 1996, FACS-optimized mutants of the green fluorescent protein (GFP), *Gene*, 173:33–38.

Lorenz, W.W., McCann, R.O., Longiaru, M., and Cormier, M.J., 1991, Isolation and expression of a cDNA encoding *Renilla reniformis* luciferase, *Proceedings of the National Academy of Sciences (USA)*, 88:4438–4442.

Lorenz, W.W., Cormier, M.J., O'Kane, D.J., Hua, D., Escher, A., and Szalay, A.A., 1995, Expression of the *Renilla reniformis* luciferase gene in mammalian cells, *J. Biolumin. Chemilumin.*, 11:31–37.

Mayerhofer, R., Langridge, W.H.R., Cormier, M.J., and Szalay, A.A., 1995, Expression of recombinant *Renilla* luciferase in transgenic plants results in high levels of light emission, *The Plant Journal*, 7:1031–1038.

Wang, Y., Wang, G., O'Kane, D.J., and Szalay, A.A., 1996, The *Renilla* luciferase-modified fusion protein is functional in transformed cells, in *Bioluminescence and Chemiluminescence: Molecular Reportting with Photons*, Hastings, J.W. et al. (eds.), pp. 419–422.

Wang, G., Csch, S., Wang, Y., and Szalay, A.A., 1996, Imaging of luciferase fusion gene expression in transformed cells and embryos, in *Bioluminescence and Chemiluminescence: Molecular Reporting with Photons*, Hastings, J.W. et al. (eds.), pp. 415–418.

Zolotukhin, S., Potter, M., Hauswirth, W.W., Guy, J., and Muzyczka, N., 1996, A "humanized" green fluorescent protein cDNA adapted for high-level expression in mammalian cells, *J. Virology*, 70:4646–4654.

RITE BIOLOGICAL HYDROGEN PROGRAM

Tadaaki Kawasugi, Paola Pedroni, Masato Miyake, Akio Toriyama,
Sakae Fukunaga, Koichi Takasaki, and Teruaki Kawamoto

Project Center for CO_2 Fixation and Utilization
Research Institute of Innovative Technology for the Earth
Toyokaiji Building, No. 7, 8F
2-8-11 Nishishinbashi
Minato, Tokyo 105, Japan

1. INTRODUCTION

The Research Institute of Innovative Technology for the Earth (RITE) is developing hydrogen production technology based on biological processes. The objective of the project is to develop technologies for the efficient production of hydrogen, utilizing solar energy and photosynthetic microorganisms. This project, named "Development of Environmentally Friendly Technology for the Production of Hydrogen," began in 1991 and will continue until March 1999 (total of eight years).

The research project is delegated to RITE by the New Energy and Industrial Technology Development Organization (NEDO), under the guidance of the Ministry of International Trade and Industry (MITI).

RITE is conducting research and development activities in cooperation with six private companies, including an Italian one. RITE also is conducting joint research with the National Institute of Bioscience and Human Technology (NIBH) in Tsukuba. Moreover, supporting research is being conducted at two domestic universities, Tokyo University of Agriculture and Technology and Ibaraki University, and two overseas academic institutions, the University of Hawaii and King's College London.

The project consists primarily of the following five topics:

- screening and breeding improvement of photosynthetic microorganisms,
- large-scale cultivation techniques,
- survey on hydrogen separation techniques,
- survey on by-product recovery techniques, and
- system integration.

These subjects have been proceeding concurrently and utilizing the outcome of other research.

BioHydrogen, edited by Zaborsky *et al.*
Plenum Press, New York, 1998

2. SCREENING AND BREEDING IMPROVEMENT OF PHOTOSYNTHETIC MICROORGANISMS

Research and development on screening and breeding improvement is being conducted in Italy by Eniricerche and in Japan by NIBH (in cooperation with private companies).

2.1. Research and Development Activities in Italy

Research and development activities at Eniricerche are in two main areas: hydrogenases and photosynthetic bacteria. Work also involves three areas of applied research: the construction of uptake hydrogenase and polyhydroxyalkanoate (PHA) mutants of *Rhodobacter sphaeroides* RV; the overexpression of negative and positive regulators of the two enzymes involved in its hydrogen metabolism; and the development of an in vitro system using hydrogenase as a catalyst.

Hydrogenases we have characterized were selected on the basis of two main features: high H_2 evolution activity and solubility. Three evolution hydrogenases, *Pyrococcus furiosus*, *Acetomicrobium flavidum*, and *Clostridium pasteurianum*, were purified and the corresponding coding genes were cloned. Expression in *E. coli*, using the pBtacl vector, gave for *P. furiosus* a recombinant enzyme both in the soluble and insoluble fractions, inactive probably because it was not processed. Recombinant *A. flavidum* hydrogenase was soluble and inactive, and its non-processing was experimentally demonstrated. Expression of *C. pasteurianum* hydrogenate gave an insoluble protein, inactive but nothing is known about the processing of [Fe] hydrogenases. The next step was the expression in *R. sphaeroides* RV. For *P. furiosus* and *A. flavidum* enzymes, no expression was detected by Western blot using the vector pKT230, while for *C. pasteurianum*, the expression was detected by the same method using a vector suitable for *R. sphaeroides* RV.

The study of the photosynthetic bacterium *R. sphaeroides* RV is the second area of basic research. The objectives are to establish the growth conditions for H_2 evolution, optimize methods for heterologous DNA transfer, and develop pKT230-derived vectors suitable for the expression of foreign genes in this photosynthetic bacterium. Starting from a size-reduced derivative, named pSM802, three different plasmids were developed.

Mutagenesis experiments were performed to obtain engineered *R. sphaeroides* RV strains impaired in enzymatic activities related to or competing with the H_2 evolution pathway. The isolation of uptake hydrogenase mutants by interposon mutagenesis of the hupL gene is in progress, while we have already obtained PHA mutants by interposon mutagenesis of the phaC synthase gene. Three clones were genetically characterized and they were the result of different crossover events.

The most recent activity involved obtaining engineered *R. sphaeroides* RV strains overexpressing regulatory genes of the two enzymes involved in their hydrogen metabolism. In particular, negative regulators of the uptake hydrogenase, such as hypA and hupTUV, were indicated as the repressor system. On the other hand, positive regulators of the major nitrogenase, such as nifR2/R1, were indicated as the nitrogen-sensing system, and nifA1/A2 were indicated as the central regulators.

The in vitro hydrogen production system was developed by coupling semi-conductors with hydrogenases used as catalysts. Some results were obtained with the three hydrogenases tested, using the semiconductor titanium dioxide in the presence of methylviologen as an electron carrier and Tris-HCl as an electron donor.

2.2. Research and Development Activities in Japan

Four research projects are being carried out at NIBH: modification of photosynthetic pigment composition, expression of bacterial hydrogenase in cyanobacteria, simulation of outdoor cultivation for evaluation of hydrogen productivity, and the metabolic control of the production of hydrogen and by-products.

The improvement of light penetration into a bioreactor would be a breakthrough in the enhancement of hydrogen production. For this purpose, we are studying the genetic modification of pigment composition by random mutagenesis, genetic control by antisense RNA, and promoter competition. The imbalanced light-harvesting complex mutant P3 was obtained by UV mutagenesis. The mutant showed improved hydrogen production rate and efficiency. The promoter competition method was performed for reducing light-harvesting complexes. We succeeded in reducing LH1 complex selectively.

The advantage of cyanobacteria is that they can produce hydrogen from water by using sunlight. To enhance the hydrogen production rate or quantum requirement, the expression of clostridial hydrogenase in cyanobacteria was investigated. Recently, we succeeded in expressing the clostridial hydrogenase peptide in cyanobacteria. The expression required a strong promoter and the modification of the ribosomal binding site.

The evaluation of the genetically engineered strain is very important. For a reasonable evaluation, we are studying the simulation of outdoor culture and have focused on the effects of sunlight wavelength and the light irradiation cycle on hydrogen production.

Photosynthetic bacteria produce not only hydrogen but also some by-products, such as polyhydroxybutyrate (PHB). Control of the production of hydrogen and PHB is under investigation.

These research activities will be combined to develop efficient fusion technology for maximum hydrogen production.

3. LARGE-SCALE CULTIVATION

3.1. Evaluation of the Substrate Characteristics and Engineering Elemental Techniques

Current research and development on large-scale cultivation techniques include the evaluation of the substrate characteristics, engineering elemental techniques for hydrogen production, and photobioreactor development.

One of the current challenges is to search for a wide range of raw materials and develop methods that will facilitate the conversion of those raw materials into substrates suitable for photosynthetic bacteria. In selecting raw materials, two points should be considered: First, the raw materials should be easily convertible into organic acids suitable for hydrogen production. Second, the raw materials should be obtainable in large quantities.

In preparation for using these undetermined substrates, we have proposed further evaluation and experimentation on substrate characteristics to explore their hydrogen-generating potential. However, at this stage, the experimentation is not intended to determine whether a substrate is good or bad. Instead, the objective is to successfully carry out the experiment.

In the experimental set-up, a Roux flask having a capacity of 1.5 L and depth of 6 cm is connected by a plastic tube to a glass syringe for collecting and measuring generated hydrogen. It is submerged in a water bath kept at a temperature of 30 °C. Precultivated photosynthetic bacteria is inoculated in 1.4 L of substrate so that a predetermined OD concentration can be achieved. Light is then radiated from a halogen lamp to achieve a constant quantity of light (0.58 kW/m^2) on the surface of the flask.

Research and development on large-scale cultivation techniques is now being carried out by four participating companies, where four different kinds of substrates are being used. The first substrate is a medium-temperature anaerobic fermentation liquid of liquid sugar waste; the second is a high-temperature anaerobic fermentation liquid of waste liquid from a sugar refinery; the third is heat treatment liquor-extract of sewage sludge; and the fourth is lactic acid fermentation liquid from organic waste.

Based on these substrates, the evaluation experiments on substrate characteristics are being carried out. An one example, the use of heat treatment liquor extract of sewage sludge generated about 1.1 L hydrogen/L substrate and 1.0 L hydrogen/g TOC.

The subjects to be studied in the future are accumulation of experimental data using the actual substrate and a review of the evaluating method on the hydrogen-generating potential, taking into consideration the carbon mass balance.

Another challenge is obtaining knowledge about both cultivation and control techniques of the photosynthetic process under various practical conditions, for example, when a reactor is installed in a place exposed to sunlight. The technical factors for enabling photosynthetic bacteria to produce hydrogen under practical conditions are light intensity, C/N ratio, organic acid metabolism and hydrogen generation, dilution rate, and so forth. They are roughly divided into four groups: factors related to sunlight, constituents of substrates, the photosynthetic bacteria itself, and operating conditions. These engineering elemental techniques are now being studied by the participating companies. We will be able to develop effective photobioreactors from the collective results.

The hydrogen-generating potential of substrates and engineering elemental techniques will continue to be topics of many future meetings.

3.2. Manufacturing and Operation of Photobioreactors

It is important to manufacture and operate reactors that contain photosynthetic bacteria. The objective is the development of photobioreactors that can produce environmentally friendly hydrogen without a significant decrease in efficiency, compared to laboratory results. Creation of total systems, including photobioreactors, has been attempted. The first matter we studied was the development of reactors enabling increased efficiency of light utilization and space reduction. The second was the examination of the prevention of excess warming caused by sunlight. The third topic was easier maintenance of the reactor.

Five different reactors are under development. Each reactor has one or more advantages over standard flat-type reactors. The floating plane-type reactor is similar to the flat-type in structure, but it is settled below the water surface; therefore, it is naturally cooled by the surrounding water without consuming energy for cooling. The tubular-type reactor consists of parallel tubes. One of its advantages is that the inside of the reactor can be easily cleaned using a special brush. Also, the arrangement of the tubes allows light to be utilized efficiently. The thin-layer plane reactor can be positioned along the wall of a building, thereby saving space. It consists of many planes at angles that can be adjusted according to the direction of sunlight. The induced and diffused reactor uses ID plates made of polyacrylate and polyester, which are positioned vertically and serve to induce and diffuse uniform light into the reactor. This enables the efficient production of hydrogen. In the internal irradiation reactor, light is collected by sunlight energy harvesting equipment and introduced into the reactor through optical fibers. Therefore, vertical arrangement of the reactors is possible. This reactor also filters out infrared light.

In studies of reactor development, the reactors ranged in size from 5 to 10 L. Halogen lamps were used except in the test of the internal irradiation type. The temperature was between 30 and 35 °C. Various organic acids were used as substrates. The feed pattern was batch or continuous.

The floating plane reactor was operated continuously for approximately 70 days. Steady hydrogen production was maintained throughout the period. The average hydrogen production rate by reactor volume was 200–400 mL/L/day.

In the induced and diffused reactor, the ID plates are cylindrically arranged, although other arrangements are also possible. Operation of a conventional flat-type reactor with a depth of 20 cm was not successful. However, very efficient hydrogen production was observed after the ID plates were placed in the reactor. Seven thousand milliliters of gas were produced in 15 days of operation of a batch culture.

In the internal irradiation photobioreactor, sunlight was introduced through the use of optical fibers. The data was obtained with real sunlight. The time-course of light energy and cumulative gas production were obtained. The lag-time during the start-up period was the result of a high ammonium concentration. It is clear that gas production was low on the day with little sunshine.

In a system study, an acidogenic reactor used for pre-treatment was combined with photobioreactors. The content of the two photobioreactors was recycled. Vegetable waste was supplied into the acidogenic reactor and the system was operated continuously. The volumetric hydrogen production rate by reactor volume was 920 mL/L/day.

4. SUMMARY

- Five different types of photobioreactors were developed. Each had advantages in comparison to conventional plane reactors.
- Bench-scale experiments on three of the above reactors, as well as an experiment on a total system, were conducted.
- The produced hydrogen energy per supplied light energy in the three photobioreactors ranged from 0.3 to 2.2%.
- We are proceeding with the study of photobioreactors and total systems to evaluate the technology of biological hydrogen production.

62

STANDARDS WORKSHOP

Chair: David O. Hall

Panelists: John Benemann, Elias Greenbaum, Peter Lindblad,
Tadashi Matsunaga, Jun Miyake, Mario Tredici, and Oskar Zaborsky

INTRODUCTION

The expanding international body of knowledge on biohydrogen, and the work being conducted by researchers throughout the world, led to a workshop on the development of standards for the meaningful exchange and comparison of information and results across borders, cultures, languages, and other potential obstacles.

David Hall launched the discussion by posing the question: How should H_2 production rates and light conversion efficiencies be reported for meaningful comparisons of data from different labs?

For example:

- Radiation: total solar spectrum, visible spectrum or PAR; incident radiation or absorbed radiation; standards for lamps used for irradiation?
- Surface of bioreactor: ground area, light-intercepting area? (photostage)
- Volume: total volume of reactor, total volume of liquid in the reactor?
- Biomass: dry weight, wet weight, protein, cell content?
- Chlorophyll: total chlorophyll, Chl.a, assay method?
- H_2 production: μmoles or μL/time; μmoles or μg/Chl.a; mg/mg protein, μg/mg biomass; moles (liters)/volume of reactor?
- Time: sec, min, h, day, year; illumination period only?
- Efficiency: based on incident or absorbed radiation? Measurement of radiation: solar, simulated sun, lamps?

David Hall introduced the subjects above, explained the need for standards in reporting photobiological H_2 production, and invited comments and suggestions first from the panel and then from the attendees. As a basis of discussion of conversion efficiencies he put forward Jim Bolton's equation (Solar Energy 57 [1996] 37–50):

$$\eta_c = \frac{\Delta G_p^o R_p}{E_s A}$$

ΔG_p^o = Standard Gibb's free energy (237.2 kj mol^{-1} at 298 k)
R_p = rate (mol sec^{-1}) of generation of product p (H$_2$) in its standard state
E_s = total incident solar radiation Wm^{-2}
A = irradiated area in m^2

These suggestions were applied in an example.

A suspension of algae is irradiated with sunlight (irradiance 30 Wm^{-2}), where the area exposed to the sunlight is 10 cm^2. Highly purified helium gas (saturated with water vapor) is passed through the suspension at 25 °C at a rate of 100 mL min^{-1} and analyzed for hydrogen and oxygen, which are found to be evolved in a 2:1 ratio with the hydrogen concentration at 100 ppmv. What is the solar photoproduction efficiency?

100 mL min^{-1} corresponds to 6.81 × 10^{-5} mol s^{-1} of carrier gas; hence, the rate of production of hydrogen is

$$R_{H_2} = (1.0 \times 10^{-4})(6.81 \times 10^{-5}) = 6.81 \times 10^{-9} \text{ mol s}^{-1}$$

The hydrogen gas is produced at 10^{-4} atm; therefore, the ΔG^O must be replaced by

$$\Delta G = \Delta G^0 - RT \ln (P^o/P)$$
$$= 237,200 - 8.3145 \times 298.15 \ln (1/10^{-4})$$
$$= 237,200 - 22,832 = 214,368 \text{ J}$$

The incident solar power (E_sA) is 30 × 0.0010 = 0.030 W. Thus, from Equation 3

$$\eta_c = \frac{214,367 \times 6.81 \times 10^{-9}}{0.030} = 0.0487$$

or 4.87%.

In literature, there have been some confusion and inconsistencies in the reporting of solar hydrogen efficiencies. In some cases, E_s has included only that part of the solar spectrum absorbed by the system rather than the E_s for the total solar spectrum. This is particularly true in the case of photosynthetic systems, where often efficiencies are reported on the basis of "photosynthetically active radiation" (defined as solar radiation in the 400–700 nm region–[PAR]). In other cases, efficiencies have been calculated only on the basis of the solar light absorbed by the system, rather than the incident broadband total irradiance.

PANELISTS' COMMENTS

Comments and observations from the panelists are summarized below.

John Benemann: Researchers should avoid making exorbitant claims on rates and efficiencies of H$_2$ production. It is not practical in the lab to use a standard lamp for all H$_2$ production experiments. Ordinary lamps (680–700 nm) should be standardized to convert their irradiance to solar radiation 1.5 AM.

Elias Greenbaum: One should stick to Jim Bolton's units of energy. Create a Web page for the hydrogen community and write the energy conversion units. David Hall agreed to put the units in the KCL Life Sciences Web page (once agreed upon).

Peter Lindblad: It is difficult to compare the biological H_2 production data in the literature. H_2 production should be reported on the basis of biomass units. Jim Bolton's SI units are sufficient.

Tadashi Matsunaga: Suggests that the recommendations be published in the Biohydrogen proceedings. The H_2 production atmosphere (N_2, Ar, air, CO_2, etc.) should be mentioned.

Jun Miyake: In reporting conversion efficiencies, the source of irradiance (type of lamp) should be mentioned, for example, W, Xe, solar similator, optical fibres, etc. Xenon lamp ERDA = 1.5; W lamp = 2950 K.

Mode of irradiation:
    Stage 1: Halogen solar simulator. 1 kw/m^2, 12–24 h operation.
    Stage 2: Halogen solar simulator. on-off cycle = 7.0 kw h/m^2/day, several days.
    Stage 3: Solar simulator in continuous operation.

Efficiency:
    (a) Input should be light energy + substrate energy.
    (b) Output-combustion enthalpy of H_2 or ΔG^0 H_2.

Mario Tredici: There are problems in reporting production rates and efficiency of photobioreactors (PhBR). In a closed PhBR, volumetric productivity (gl^{-1}) is related to surface-to-volume ratio of the reactor. Therefore, volume productivity data alone is not sufficient to gauge the performance of the PhBR.

Other questions include how to measure the area of a PhBR, determine the productivity per unit of the illuminated surface area, and the illuminated area. Generally, the reports underestimate the efficiency of the system.

$$Efficiency = \frac{Combustible\ energy\ of\ biomass\ produced}{Light\ incident\ on\ the\ surface}$$

The shaded area (horizontal section of the PhBR) should be taken as the light-incidence area.

Oskar Zaborsky: High priority areas of uncertainty should be identified. A preliminary draft should be prepared by the chairman and distributed to the H_2 community through a Web site or e-mail.

T. Hattori (RITE): RITE is sponsoring standardization work for biohydrogen research. A leaflet describing the objectives, techniques, equipment, and data analyses was distributed to the participants.

PARTICIPANTS' COMMENTS

Michael Seibert: The U.S. Department of Energy has a Web site. A wide spectrum of fluorescent VIJA lamps are commercially available and can be used as a standard source of irradiance.

James Ogbonna: Productivity depends on the aim of the research undertaken and the product to be obtained. Efficiency is dependent on the season for outdoor bioreactors. In the expression of H_2 production, volume mass of H_2 per dry wt of cells is better than H_2 per wt of Chl.

Yao-Hua Luo: Rates of H_2 evolution and efficiency of H_2 production are different concepts that are not always compatible.

S. Boussiba: In evaluating the efficiency of PhBRs, the views of other groups operating different types of reactors should be considered. Why not consider radiation absorbed by the reactor and contents rather than measuring the total incident raidaiton?

Yasuo Asada: Conversion efficiencies based on a Chlorophyll basis are tricky but this is the only way to compare the activities of different microorganisms, particularly cyanobacteria.

Elias Greenbaum: Chlorophyll-based activity is better for 'greens,' too.

Masayoshi Morimoto: The precise experimental conditions should be described (light/dark: aerobic/anaerobic/Ar; batch/continuous; small scale in the lab/outdoors) in expressing productivity.

PROVISIONAL CONCLUSIONS

Workshop rapporteur K.K. Rao offered the following provisional conclusions, based on the comments and discussion.

- For comparison with H_2 production activities from PV, PC, and PEC systems, photobiological H_2 production activity should be expressed in units recommended by Jim Bolton.
- For bacterial cells (including green alga), activity should be expressed in terms of total chlorophyll or ash-free dry weight.
- Do not express activity in terms of OD (optical density).
- Productivity of a PhBR can be expressed as g per surface area per day or g per bioreactor volume per day, etc.

SUMMARY

- Total incident radiation (E_s) → Wm^{-2}
- Irradiated area (A) → m^2
- Rate of H_2 production (R_p) → $mol\ s^{-1}$
- Standard Gibbs free energy (ΔGp^o) E_s 237.2 kJ mol^{-1} at 298 K
- Efficiency (η_c) → moles product per total incident radiation per area per second
- Biomass → oven dry weight (ideally ash free)

ROUNDTABLE

The Roundtable featured a number of speakers offering their perspectives on future directions for biohydrogen research and development.

INTERNATIONAL BIOHYDROGEN PROGRAM—A PROPOSAL

Morton Satin
Food and Agriculture Organization, United Nations, Rome, Italy

We have heard many speakers, including Dr. Edward David, who delivered the keynote address, talk of the necessity for international cooperation in the field of biohydrogen production. Indeed, he specifically talked of the need to develop functional partnerships incorporating elements of government, industry, and academia. Prior to the end of the Cold War, it was just these types of alliances that were so productive in turning theoretical technologies into practical applications. Dr. David went on to say that the future of biohydrogen would be dependent upon similar collaborations, but on an international basis. Many others here have expressed similar sentiments.

Although we no longer face the Cold War, collectively, we face problems of similar gravity and magnitude. The last century has seen us become willing captives to an unsustainable future. The economically developed world is addicted to high energy consumption and global economic development will be reflected by the ever-expanding use of fossil fuels. The predicted growth in world population, supposedly peaking somewhere 8 to 10 billion people, is not the critical issue because the majority of these people are currently not major energy consumers. However, as the less developed countries of the world develop their economies and their citizens strive to enjoy their full and fair measure of the biosphere's renewable and non-renewable resources, the insolvency of our way of life will become only too apparent.

The struggle we will face in the future is not at all like the Cold War; our enemy is neither from the North, South, East, nor West. We do not face a fight characterized by a single battle or a focused apocalyptic event. If we continue with our current lifestyle, we will experience a slow, protracted, diminution of quality of life. In how many capital cities of the world today do we see traffic police wearing masks to protect them from pollution. In my own city of Rome, they are cleaning up all the monuments in preparation of the Papal Year 2000. It is difficult to describe to you how beautiful the city is underneath the malignant coat of the pollution. The problem is that no one is sure if the monuments will remain white for the next 30 months. I will spare you all the other scenarios predicated upon our continued use of fossil fuels. If I may borrow from Dr. Miyake's talk, ultimately, our enemy is entropy mismanagement.

All this to say that the work you are all engaged in is of extreme importance, now and for future generations of the Family of Man.

Prior to attending this meeting, I was not certain whether an international body such as FAO or any other United Nations entity could serve a useful purpose in establishing an international endeavor to promote the biological production of hydrogen. I am still not certain whether that could be the case. As with any form of advanced technology, politics and national and commercial interests all seem to discourage many forms of international collaboration. Be that as it may, I would like to place before you a modality for international cooperation that you may be interested in considering. Even if such a proposal is not of interest, it may stimulate you to form alternative ideas for international cooperation in this field.

I would like to propose the establishment, under an international aegis, of a program which, for the sake of simplicity, would be called the International Biohydrogen Program. The purpose of the program would be the following:

- Promote an international understanding of the importance of biohydrogen for the future sustainability and well-being of the earth.
- Develop international cooperative programs on basic hydrogen research to be contracted out to established centers of expertise around the world.
- Create and carry out training programs on all aspects of biohydrogen production.
- International harmonization of standards for terminology, analytical measurement, equipment, etc., that are used in biohydrogen research.
- Establishment of an international biohydrogen database incorporating all the published literature and the equally important gray, unpublished literature that sits neglected in file cabinets around the world doing nothing but contributing to our intellectual entropy.
- Generate new sources of funding in support of research, training, and information dissemination.
- Setting of international goals or targets to achieve the realization of biohydrogen as an economically practical source of energy. This would not only incorporate technical milestones, but economic and political ones as well. Once we truly believe that the world can feasibly live with biohydrogen, it will be that much easier to establish an international disincentive for fossil fuel use.

Normally, such endeavors are sponsored exclusively by member governments who agree to support an international secretariat mandated to carry out the program. In this particular instance, I would also suggest that industrial membership be permitted. It is clearly in the interests of industry to be fully aware of all the progress going on in the development of their future source of energy. Annual meetings are held to set out the year's program and focused technical meetings to deal with specific subjects are held throughout the course of the year. In similar programs within the UN system, each member government establishes a core group of scientists to represent them in the program. Although the scientists represent their member governments, they do so in their personal capacities as experts in the field. This allows for a more open and frank exchange of views.

To recapitulate the idea, I proposed the consideration of an International Biohydrogen Program under the aegis of the United Nations flag. Key member governments would be asked to support the establishment of an international secretariat mandated to carry out an agreed program of research, training, harmonization of standards, information dissemination, and biohydrogen promotion. Industries would be invited to become full supporting members. Governments would establish a core group of experts working in this area to represent them at annual meetings in order to develop an agreed program.

Such a program would meet the key criteria set out by Dr. David in his keynote address. With reference to a champion of the biohydrogen cause, the program would have its international champion in the guise of the United Nations; with reference to the coalition of research laboratories to carry out international research, these would be the laboratories to which the above-mentioned research would be contracted to; and finally to his stated need for an international recognition that biohydrogen is a sustainable option, this would be the goal of the program's important promotional work.

Details such as the length of the initial mandate for such a program, size and location of the secretariat, annual supporting fees for various governments, industry, etc., require further consideration.

As I originally stated, I really do not know if we can get over all the bureaucratic and political barriers that so often accompany endeavors such as this. What I do know is that the work you are doing warrants every effort to establish a long-term international cooperative effort in order to ensure its fulfillment. Thank you.

INTERNATIONAL BIOHYDROGEN CENTER (IBC): A CONCEPT FOR AN INTERNATIONAL COOPERATIVE PROGRAM

Oskar R. Zaborsky
School of Ocean and Earth Science and Technology, University of Hawaii

Background

Hydrogen is produced in nature by several biological processes. To date, however, hydrogen is not produced commercially by any biological process. Yet, at the same time, unparalleled capabilities to tackle this opportunity exist through almost daily advances in biotechnology.

Problem

There are several major scientific and technical problems associated with the realization of producing hydrogen commercially by biological means. There are also institutional problems that prevent effective actions, among which are the lack of coupling biology with engineering and the lack of a facility dedicated to engineering research, demonstration, and technology validation. Engineering research is critically needed to provide a framework for placing the biological research, often quite fundamental, into a perspective of reality that is familiar to industry. Currently, there is no center or facility in the world that is geared to the integration of biology and engineering.

Approach

The approach advocated is to establish a facility dedicated to engineering research, demonstration, and validation, most appropriately named the International Biohydrogen

Center (IBC). Initially, this center should be a U.S.-Japan initiative but permitting any other country to join if funding is provided either by that country or another international organization on behalf of that country (e.g., the United Nations). The location of this facility must be in a suitable part of the globe having the right environmental conditions—abundant sunlight, warm temperatures, and water—and a technology infrastructure that is already in existence. Such a place, of course, is Hawaii, in particular the island of Hawaii or the Big Island.

Organizational Structure

There are several organizational structures that can be considered, with perhaps the most appropriate one being an independent non-profit center associated with the State of Hawaii and the University of Hawaii. An alternative structure is to place it within an existing organization such as the School of Ocean and Earth Science and Technology. IBC would be administered by a staff, with an executive director, and governed by a board of trustees selected from member countries and key organizations. IBC would be dedicated to research that has relevance to commercialization and would include engineering research, scale-up, demonstration projects, systems and economic analyses, and technology validation. A key goal would be to actually generate and use hydrogen at the site—proof that a particular approach or technology is actually able to produce valuable goods and services.

Initial Steps

Initial steps in establishing IBC are to enlist key people from the U.S. and Japan in a planning committee that would prepare a strategic plan of action. This committee should be composed of 3–5 individuals from each country—engineers, scientists, and administrators from academia, industry, and government.

Funding

The initial planning committee should have a budget of $100,000–150,000 for convening several meetings (at least three) and to produce the strategic plan document. Given the nature of centers dedicated to engineering research, demonstration projects, and technology validation, the estimated yearly operating budget for such a center with 1–2 major projects would be in the range of $3–8 million. Of course, initially the funding for such a center would be less, ranging from $1–2 million. Funds would be derived from the federal and state governments and industry.

CLIMATE CHANGE AND BIOHYDROGEN

Perry Bergman
Federal Energy Technology Center, U.S. Department of Energy

The development of BioHydrogen technologies and the Office of Fossil Energy of the U.S. Department of Energy share the common goal of providing environmentally clean sources of power for industry and individual users. An area of growing emphasis and mutual interest is the avoidance of changes in global climate, which will require a coordinated energy strategy involving both renewables and fossil fuels. Shortly after the adoption of the Framework Convention on Climate Change, President Clinton announced the U.S. commitment to stabilize greenhouse gas emissions. The Office of Fossil Energy was charged with the difficult responsibility of reducing the amount of heat-trapping gas species, notably CO_2, released into the atmosphere as a result of coal, oil, and gas usage, while ensuring an abundant supply of inexpensive fossil energy.

Although renewables and nuclear sources may be preferable from the perspective of emissions, economics demand that, over the near- and mid-terms, fossil fuels play a central role in meeting the ever-expanding demand for energy worldwide. Utilization of these fuels currently releases in excess of 6 gigatonnes of carbon as CO_2 annually, with emissions increasing in concert with economic growth. Natural sinks have been incapable of absorbing the anthropogenic increment to the carbon budget and CO_2 in the atmosphere has consequently risen by about 30% from pre-industrial levels. There is concern that continued increases could alter the radiative forcing of the earth to an extent that would incur undesirable changes in global climate.

In response to this potential problem, several programs have been initiated under the U.S. Climate Change Action Plan (CCAP) to stabilize CO_2 and other greenhouse gas emissions generated by the production or use of fossil fuels. As an example, the Climate Challenge Program, which is a cooperative effort between the U.S. Department of Energy and the electric utility industry, has achieved substantial success during the less than four years it has existed. The program pursues voluntary commitments from the utilities to reduce greenhouse gas emissions. About 600 utilities, representing approximately 60% of the 1990 U.S. utility carbon emissions, have signed 114 participation accords with the DOE as of the fall of last year. Pledged emissions reductions are being pursued through a broad range of activities ranging from improvements in power cycle heat rates to reforestation and increased utilization of biofuels and other renewable energy sources.

The Office of Fossil Energy also is actively involved in R&D on innovative technologies to recover, reuse, and dispose of CO_2. Some topics that currently are being supported include storage of CO_2 captured from combustors in the deep ocean and subterranean structures and biological and chemical reuse options. The latter area may be relevant to the BioHydrogen initiative. Some of this work is, or will be, pursued under several international programs such as an agreement between the DOE and the Ministry of International Trade and Industry (MITI) of the Japanese Government, the IEA Greenhouse Gas R&D Programme, and Task Force 7 of the Climate Technology Initiative (CTI). International cooperation is particularly appropriate, given the global nature of the problem being confronted.

It should be noted that both fossil fuel greenhouse gas emissions control and BioHydrogen are incorporated as parts of Task Force 7 of the CTI, which currently is referred to as the "Task Force on Greenhouse Gas Capture and Disposal and Hydrogen Production." To date, the Office of Fossil Energy has assumed a leading role in the negotiations on the CTI and has spearheaded the effort to enlist U.S. commitment to this program.

To summarize, a practical and effective strategy to address the problem of climate change acknowledges the important roles played by both fossil and renewable energy sources. The growing initiative on BioHydrogen is expected to make substantial future contributions to the common goal of stabilizing greenhouse gas levels. The Fossil Energy Program of the U.S. Department of Energy commends this effort and anticipates the possibility of expanding cooperation through its CO_2 reuse technology development projects and under the various international programs, such as CTI, in which it is participating. Moreover, it should be emphasized that the model of intense industrial participation which BioHydrogen is pursuing epitomizes the spirit of the CCAP as enunciated by President

Clinton in his April 1993 address. During that address he stated that the plan "must be a clarion call, not for more bureaucracy or regulation or unnecessary costs, but instead for ...ingenuity and creativity to produce the best and most cost-effective technology."

BIOHYDROGEN: ONE APPROACH TO SUSTAINABLE DEVELOPMENT

Dr. Rudolf Straub
Biotechnology Division, Project Management Organization, Biology, Energy, Ecology (BEO), Federal Ministry for Education, Science, Research and Technology, Research Centre Juelich GmbH, Germany

Since the oil crisis in 1973, hydrogen has been touted as the "fuel of the future." As oil prices fell and the economics of implementing hydrogen as a fuel became non- competitive, one reassessed the statement to "fuel of the distant future." But it did not remain this way. At the end of the 1980s, it was recognized that fossil fuels were contributing to the greenhouse effect and, as a result, hydrogen regained some of its immediacy.

The advantage of hydrogen is that it is a convenient method of transporting energy from areas with an abundance of solar energy to those regions requiring energy. The properties that make hydrogen such an interesting material is that—based on solar photoproduction and using photobiological systems—it is simultaneously a raw material, an energy carrier, and ecologically friendly. Its contribution to an economy based on sustainable development would be considerable. As a product of a natural process, hydrogen can be passively generated under ambient conditions. We have known for about 50 years that under certain conditions algae and bacteria can produce hydrogen. In order to ascertain if this process can be harnessed, if bottlenecks can be widened, and if it eventually can be converted into a technologically and economically viable procedure, certain basic information about the molecular and physiological characteristics has to be acquired.

For this reason, the German Federal Ministry for Education, Science, Research, and Technology (BMBF) launched a collaborative research program involving researchers from many disciplines to examine the problem. From 1989 until 1994, these researchers formed a pool of knowledge and resources that was one-of-a-kind in the world. Some 45 groups representing such fundamental disciplines as molecular biology, microbiology, physiology, biochemistry, biophysics, bionics, and bioprocessing were received funding of about 22 million DM ($13 million US).

The major highlights of this research effort are mentioned below: The first topic was biophotolytic hydrogen production. In this area, considerable progress has been made in the characterization of the processes involved. For example, the partial sequencing of hydrogenase that had been isolated from green algae has been achieved. This information will be employed in the cloning of hydrogenase genes. For cyanobacteria, genes governing the production of reversible hydrogenase were successfully cloned for the first time. The important genetic elements involving the related nitrogenase were also identified.

The second topic encompassed the photoproduction of hydrogen from biomass using purple bacteria. This subject was studied, possibly for the first time, from a combined molecular genetics approach and a physiological approach. Based on the results of this study concerning the metabolism of hydrogen, it was possible to develop a strategy to increase the hydrogen production by blocking the hydrogen-consuming hydrogenase. This was accomplished using both molecular genetics and a biochemical/physiological method. Bacteria cultures that were capable of producing three times the original amount of hydrogen were cultivated. These results have shown that it is possible to specifically optimize the production of hydrogen.

The third topic concerned the structure and function of hydrogenase. It was shown that it was possible to isolate the hydrogenase gene from numerous organisms. This made it possible to collect an exhaustive amount of information regarding the molecular genetics of hydrogenase. The final highlight related to biomimetic hydrogenase and water-splitting systems. It was possible to prepare a number of models relating to hydrogenase and water-splitting, which permitted the incorporation of factors such as type and quantity of ligands or cluster geometries. Of particular interest are models describing the exchange of deuterium and proton reduction.

The BMBF-Division for Biological Research and Technology sees the following points as essential for the further development of this subject:

1. The results from these fundamental examinations should be transferred to an application. Although public opinion is favorable to biophotolysis, the economics, in short, the cost-effectiveness, is not. This has not changed in the last three years. The solar production of hydrogen from a biomass is well-advanced from an R&D point-of-view, yet it still requires considerable engineering to make the process commercially feasible.
2. In the middle- to long-term view, hydrogen must replace to some extent fossil fuels as a source of energy. This is important for both ecological and political reasons. Research has shown that biophotolysis has the potential to bring this about. By strengthening the methods and tools of molecular biology and genetics, the development of hydrogen-producing organisms, which can close the commercial gap, can be realized. In this respect, research into the enzymes hydrogenase and nitrogenase must be intensified to further examine their function and selectivity in the production of hydrogen.
3. In biophotolysis, certain enzymes are responsible for catalyzing the reversible reduction of protons to molecular hydrogen and the irreversible oxidation of hydroxide ions to oxygen. A detailed examination of their structure and the mechanism involved in the catalytic processes would be invaluable in the development of related biomimetic catalysts. The catalysts could be incorporated in other hydrogen-related technologies, such as fuel cells. These topics raise fundamental and multi-disciplinary questions that could be answered by continued basic and interdisciplinary joint research.

On the other hand, we have to force energy and ecological issues. The Climate Technology Initiative (CTI) could give an umbrella for these purposes. This is why the BMBF would appreciate support for a continued and modified joint basic research program, aimed at technical applications, which is based on the participation of internationally renowned working groups all over the world. This comprehensive initiative could be performed by a network of concrete and time-limited research projects.

For the realization of such a program, we propose the following procedure:

1. Each country that is interested in such a program should name one (maximum of two) senior scientist who elaborates, in cooperation with the science community and the national ministry engaged, a national research proposal.
2. After this would be elaborated a draft program that considers more or less the national aims.

3. These steps should be backed by a discussion forum on a Website.
4. The program should be "streamlined" top-down by definition of applied aims, definition of research topics, derived from the general aims, and definition of basic needs and questions. In short, the program must be focused on specific applied aims and be limited on a selected number and species of organisms.
5. The harmonized draft program shall be presented to the national governments and the CTI. We are prepared to support the elaboration of such an international cost-shared program. After elaboration of a program, the German government is ready to consider the possibility of a suitable national contribution. We expect recommendations from the scientific community for the programmatic or organizational realization of such a cooperation and are open to them.

BIOHYDROGEN 97 PROGRAM

Pre-Conference Day - Sunday, June 22, 1997
4:00 - 9:00 pm - Registration (King's Ballroom foyer)
6:00 - 7:30 pm - Reception (Sunset Terrace) - Hosted by Cyanotech Corporation
7:30 pm - Dinner (participants on own)

Day 1 - Monday, June 23, 1997
7:00 am - Registration (King's Ballroom foyer)
7:30 - 8:30 am - Breakfast and Registration (King's Ballroom foyer)

Plenary Session (Ekahi Ballroom)
8:30 - 9:00 am - Opening Remarks
    Oskar R. Zaborsky, BioHydrogen '97 General Chairman, University of Hawaii (UH)
    C. Barry Raleigh, Dean, School of Ocean and Earth Science and Technology, UH
    Neil Rossmeissl, U.S. Department of Energy
    David Boron, National Science Foundation
    Yuko Kurashige, New Energy and Industrial Technology Development Organization, Japan
9:00 - 9:30 am - Keynote Address
    Edward E. David, Jr., EED, Inc., former Science Advisor to the President of the United States
9:30 - 10:00 am - **Plenary Lecture - The Science of Biohydrogen**
    Jun Miyake, National Institute for Advanced Interdisciplinary Research, National Institute of Bioscience and Human-Technology, AIST/MITI, Japan
10:00 - 10:30 am - **Plenary Lecture - The Technology of Biohydrogen**
    John Benemann, Consultant
10:30 - 11:00 am - Break

11:00 am -12:30 pm - **Session 1: Fundamentals** (Ekolu Ballroom)
    Co-chairs: Shigetoh Miyachi, Marine Biotechnology Institute Co., Ltd., Japan
                     Anthony San Pietro, Indiana University
    John Neidhardt, John R. Benemann and <u>Anastasios Melis</u>, University of California, Berkeley
       "Maximizing Photosynthetic Productivity and Light Utilization in Microbial Cultures by Minimizing the Light-Harvesting Chlorophyll Antenna Size of the Photosystems"
    <u>Peter Lindblad</u>, Alfred Hansel, Fredrik Oxelfelt, Paula Tamagnini, and Olga Troshina, Uppsala University, Sweden
       "*Nostoc* PCC 73102 and H_2: Knowledge, Research, and Biotechnological Challenges"
    C. Tosi, E. Franchi, F. Rodriguez, A. Selvaggi, and <u>P. Pedroni</u>, Eniricerche S.p.A., Italy
       "Molecular Biology of Hydrogenases"
    <u>Koji Sode</u>[1], Mika Watanabe[1], Hiroshi Makimoto[1], and Masamitsu Tomiyama[2] ([1]Tokyo University of Agriculture & Technology, [2]National Institute of Agrobiological Resources, Japan)
       "Metabolic Engineering of *Escherichia coli* Anaerobic Pathway for the Enhancement of Bacterial Hydrogen Production"

12:30 - 1:30 pm - Lunch (Garden Court)
    Maurice H. Kaya, Department of Business, Economic Development, and Tourism, State of Hawaii

(underscore indicates presenter)

Day 1 - Monday, June 23, 1997 (continued)

1:30 - 3:00 pm - **Session 2: Photosynthetic Bacteria** (Ekolu Ballroom)
    Co-chairs: Hiroshi Kitamura, Tokyo Metropolitan University, Japan
                   J. Grant Burgess, Heriot-Watt University, United Kingdom

Masato Miyake[1], Makoto Sekine[1], Lyudmila G. Vasilieva[2], Eiju Nakada[2], Tatsuki Wakayama[2], Yasuo Asada[1], and Jun Miyake[3] ([1]National Institute of Bioscience & Human Technology, [2]Research Institute of Innovative Technology for the Earth, NIBH, [3]National Institute for Advanced Interdisciplinary Research)
    "Effect of Light on Hydrogen Production by *Rhodobacter sphaeroides*: Contribution of the Pigment Composition"

Masayoshi Minami, Kubota Corporation, Japan
    "Biohydrogen Production Using Sewage Sludge by Photosynthetic Bacteria"

Reda M.A. El-Shishtawy[1], Yoji Kitajima, Seiji Otsuka, Shozo Kawasaki, and Masayoshi Morimoto, Kajima Laboratory, RITE, Japan ([1]National Research Centre, Egypt)
    "Study on the Behavior of Production and Uptake of Photobiohydrogen: Fermentation and CO_2 Fixation by Photosynthetic Bacterium *Rhodobacter sphaeroides* RV"

J. Grant Burgess, Kenneth G. Boyd, and Andrew Mearns-Spragg, Heriot-Watt University, United Kingdom
    "Microbial Ecology of Living Surfaces in the Sea: A Novel Source of Photosynthetic Bacteria for Hydrogen Production"

3:00 - 3:30 pm - Break

3:30 - 5:00 pm - **Session 3: Cyanobacteria** (Ekolu Ballroom)
    Co-chairs: D.O. Hall, University of London, United Kingdom
                   Takahira Ogawa, Kumamoto Institute of Technology, Japan

Yasuo Asada,[1] Masato Miyake,[1] Youji Koike,[2] Katsuhiro Aoyama,[3] Ieaki Uemura,[3] and Jun Miyake,[4] ([1]National Institute of Bioscience and Human-Technology, AIST/MITI, [2]Research Institute of Innovative Technology for the Earth (RITE), [3]Tokyo Gas Co., Ltd, Frontier Technology Research Institute, [4]National Institute for Advanced Interdisciplinary Research)
    "Hydrogenase-mediated Hydrogen Metabolism and Construction of New System in Cyanobacteria"

Jyothirmai Gubili and Dulal Borthakur, University of Hawaii
    "Identification of an Uptake Hydrogenase Gene Cluster from *Anabaena* sp. Strain PCC7120"

Jens Appel, Saranya Phunpruch, and Rüdiger Schulz, Philipps-University, Germany
    "Hydrogenase(s) in *Synechocystis*: Tools for Photohydrogen Production?"

Elisha Tel-Or, Sigal Lechno and Noa Barnea, The Hebrew University of Jerusalem; and Sammy Boussiba, Ben Gurion University of the Negev, Israel
    "Improvement of H_2 Photoproduction by *Nostoc muscorum*, *Azolla filiculoides* and *Anabaena siamensis*"

Aladar A. Szalay, Loma Linda University
    "Transfer of Large Segments of DNA into Cyanobacteria: Construction of a Cyanobacterial Artificial Chromosome CBACs"

Sergei A. Markov, National Renewable Energy Laboratory and M.V. Lomonosov Moscow State University
    "Bioreactors for Hydrogen Production"

6:00 - 7:30 pm - Dinner (Garden Court)

7:30 - 9:30 pm - **Poster Session A** (King's Suites)

Day 2 - Tuesday, June 24, 1997

7:30 am - Registration (King's Ballroom foyer)
7:30 - 8:30 am - Breakfast (King's Ballroom foyer)

8:30 - 10:00 am - **Session 4: Green Algae** (Ekolu Ballroom)
    Co-chairs: Elias Greenbaum, Oak Ridge National Laboratory
                    Kazuhisa Miyamoto, Osaka University, Japan
    Michael Seibert[1], Timothy Flynn[1], Anastasios Melis[2], and Maria L. Ghirardi[1,2] ([1]National Renewable Energy Laboratory, [2]University of Califorina, Berkeley)
        "Prospects for Developing Algae that Produce Hydrogen Under Aerobic Conditions"
    Elias Greenbaum and James Weifu Lee, Oak Ridge National Laboratory
        "Photosynthetic Hydrogen Production by Green Algae"
    Rüdiger Schulz, Jörg Schnackenberg, Kerstin Stangier, Röbbe Wünschiers, Thomas Zinn and Horst Senger, Philipps-University, Germany
        "Light-Dependent Hydrogen Production of the Green Alga *Scenedesmus obliquus*"
    Kazuhisa Miyamoto, Akiko Ike, Ken-ichi Yoshihara, Hiroyasu Nagase, and Kazumasa Hirata, Osaka University, Japan
        "Algal CO_2 Fixation and H_2 Photoproduction"

10:00 - 10:30 am - Break

10:30 am -12:00 pm - **Session 5: Bioreactor Engineering and Scale-Up** (Ekolu Ballroom)
    Co-chairs: Masayoshi Morimoto, Kajima Corporation, Japan
                    Mario Tredici, University of Florence
    Y. Ikuta[1], T. Akano[2,3], N. Shioji[4], I. Maeda[4] ([1]Mitsubishi Heavy Industries, Ltd., Engineering and Construction Center, [2]Kansai Electric Power Co., [3]RITE Amasgasaki 2nd and Nankoh Laboratory, [4]Mitsubishi Heavy Industries, Ltd. Hyogo, [4]Osaka University, Japan)
        "Hydrogen Production by Photosynthetic Microorganism"
    Hideo Tanaka and James C. Ogbonna, University of Tsukuba, Japan
        "Development of Efficient Large-scale Photobioreactors - A Key Factor for Commercial Production of Biohydrogen"
    Sabine Tramm-Werner, Umweltbiotechnologie, Germany
        "Photobioreactor for Large-Scale Hydrogen-Production"
    Mario R. Tredici, Graziella Chini Zittelli, University of Florence, Italy ; and John R. Benemann, University of California, Berkeley
        "A Tubular Internal Gas Exchange Photobioreactor for Biological Hydrogen Production"

12:00 - 1:00 pm - Lunch (Garden Court)
    Gerald Cysewski, Cyanotech Corporation

1:00 - 5:00 pm

    Technical Field Tour 1 - Natural Energy Laboratory of Hawaii Authority (NELHA)
        (meet at Porte Cochère at 12:45 pm)
    Discussion and Discovery Time - Ad hoc discussions among participants on topical issues.

6:00 - 7:00 pm - Dinner (Garden Court)

Day 2 - Tuesday, June 24, 1997 (continued)

7:30 - 9:00 pm - **Session 6: Young Investigator Session** (Ekolu Ballroom)
    Co-Chairs: Oskar Zaborsky, University of Hawaii
                     James Frank, Argonne National Laboratory

Eiju Nakada[1], Satoshi Nishikata[1], Yasuo Asada[2], and Jun Miyake[3] ([1]Fuji Electric Corporate R&D, Ltd., [2]National Institute of Bioscience and Human-Technology, AIST/MITI, [3]National Institute of Advanced Interdisciplinary Research, AIST/MITI)
    "Light Penetration and Wavelength Effect on Photosynthetic Bacteria Culture for Hydrogen Production"

Haruko Takeyama, Tokyo University of Agriculture and Technology
    "Rapid Detection of Marine Nitrogen-Fixing Cyanobacteria Using Direct Nested PCR on Single Cells"

J. Schnackenberg[1,2], H. Ikemoto[2], and S. Miyachi[2,3] ([1]Max-Planck-Institut fuer Biochemie, Germany; [2]Marine Biotechnology Institute, Kamaishi Laboratories; [3]Marine Biotechnology Institute, Head Office, Japan)
    "Photohydrogen Evolution in the Marine Green Alga *Chlorococcum littorale*"

Lyudmila Vasilyeva[1,2,3], Masato Miyake[1], Eiji Nakada[2,3], Yasuo Asada[1], and Jun Miyake[1] ([1]National Institute of Bioscience and Human Technology, AIST/MITI; [2]Research Institute of Innovative Technology for the Earth (RITE); [3]Fuji Electric Co. R&D, Ltd., Japan)
    "Regulation of Bchl Level in *Rb. sphaeroides* for Optimization of Hydrogen Production, Genetic Approach"

Gregory L. Rorrer, Oregon State University
    "Photobioreactor Design Considerations for Macroalgal Suspension Cultures"

9:00 - 10:00 pm - **Poster Session B** (King's Suites)

Day 3 - Wednesday, June 25, 1997

7:30 am - Registration (King's Ballroom foyer)

7:30 - 8:30 am - Breakfast (King's Ballroom foyer)

8:30 am -12:00 pm
    Discussion and Discovery Time
    Organizational Meetings

8:30 - 9:30 am - **Workshop A: Research Institute of Innovative Technology for the Earth (RITE) Program** (Ekolu Ballroom)
    Co-chairs: Teruaki Kawamoto, RITE Ishikawajima-Harima Heavy Industries Co., Ltd., Japan
                  Koichi Takasaki, RITE, Japan

9:30 - 10:00 am - Break

10:00 - 11:00 am - **Workshop B: Standardization** (Ekolu Ballroom)
    Chair: D.O. Hall, University of London, United Kingdom
    Panelists: John Benemann, Consultant
                   Elias Greenbaum, Oak Ridge National Laboratory
                   Peter Lindblad, Uppsala University
                   Tadashi Matsunaga, Tokyo University of Agriculture & Technology
                   Jun Miyake, National Institute for Advanced Interdisciplinary Research
                   Mario Tredici, University of Florence
                   Oskar Zaborsky, University of Hawaii
                   K. Krishna Rao, University of London (Rapporteur)

12:00 - 1:00 pm - Lunch (Garden Court)

Day 3 - Wednesday, June 25, 1997 (continued)

1:00 - 2:30 pm - **Session 7: Fermentations and Other Production Systems** (Ekolu Ballroom)
    Co-chairs: Paul Weaver, National Renewable Energy Laboratory
                   Roger Prince, Exxon Research & Engineering Co.

Shigeharu Tanisho, Yokohama National University, Japan
    "Hydrogen Production by Facultative Anaerobe *Enterobacter aerogenes*"

P. F. Weaver, P.-C. Maness, and S. A. Markov, National Renewable Energy Laboratory
    "Anaerobic, Dark Conversion of CO into H_2 by Photosynthetic Bacteria"

Sakae Fukunaga, Kichi Fujiki, Kiyoshi Nagai, Shigeru Uchiyama, and Toshi Otsuki, Ishikawajima-Harima Heavy Industries Co., Ltd., RITE, Japan
    "Hydrogen Production by Anaerobic Bacteria"

Jay Keasling, Jaya Paramanik, and John Benemann, University of California at Berkeley
    "Metabolic Engineering for Hydrogen Fermentations"

2:30 - 3:00 pm - Break

3:00 - 4:30 pm - **Session 8: Institutional Issues** (Ekolu Ballroom)
    Co-chairs: Catherine Gregoire Padro, National Renewable Energy Laboratory
                   Tatsuya Mizukoshi, Showa Denko, Japan

Neil Rossmeissl, U.S. Department of Energy
    "International Collaboration in BioHydrogen - An Opportunity"

Margaret Mann, National Renewable Energy Laboratory
    "Process Analysis of Algal and Bacterial Hydrogen Production Systems"

Lois D. Blaine, American Type Culture Collection, and Oskar R. Zaborsky, University of Hawaii
    "Principles of Bioinformatics as Applied to Hydrogen-Producing Microorganisms"

4:30 - 5:30 pm   Organizational Meetings - IEA

5:30 - Luau - Banquet (Luau Grounds)

Day 4 - Thursday, June 26, 1997

7:00 am - Registration (King's Ballroom foyer)

7:00 - 8:00 am - Breakfast (King's Suites foyer)

8:00 - 9:30 am - **Session 9: Hybrid Systems and Coproducts** (King's Suites)
    Co-chairs: Tadashi Matsunaga, Tokyo University of Agriculture & Technology
                   Michael Seibert, National Renewable Energy Laboratory

Ingo Rechenberg, Technical University of Berlin, Germany
    "Artificial Bacterial Algal Systems (ArBAS): Sahara Experiments"

Jonathan Woodward and Kimberley A. Cordray, Oak Ridge National Laboratory
    "Utilization of Extremozymes for the Oxidation of Renewable Carbohydrates to Hydrogen and Sugar Acids"

Ken Sasaki, Hiroshima-Denki Institute of Technology, Japan
    "Hydrogen and 5-Aminolevulinic Acid Production by Photosynthetic Bacteria"

V. Sediroglu, M. Yücel, U. Gündüz, L. Türker, and I. Eroglu, Middle East Technical University, Turkey
    "The Effect of *Halobacterium halobium* on the Photoelectrochemical Hydrogen Production"

Katsuhiro Aoyama[1], Ieaki Uemura[1], Jun Miyake[2], and Yasuo Asada[2] ([1]Frontier Technology Research Institute, Tokyo Gas Co., Ltd.; [2]National Institute of Bioscience and Human-Technology, AIST/MITI)
    "Photosynthetic Bacterial Hydrogen Production by Fermentation Products of Cyanobacterium, *Spirulina platensis*"

Thursday, June 26, 1997 (continued)

9:30 - 10:15 am - **Plenary Lecture - Marine Genome** (King's Suites)
  Tadashi Matsunaga, Tokyo University of Agriculture and Technology, Japan

10:15 -10:45 am - Break

10:45 am -12:00 noon - **Session 10: Needs and Future Directions - A Panel Discussion** (King's Suites)
  Co-chairs: Jun Miyake, Natl. Inst. for Advanced Interdisciplinary Research, Japan
               Oskar Zaborsky, University of Hawaii
  Panelists: Roger Prince, Exxon Research and Engineering Co.
               Perry Bergman, Department of Energy
               John Benemann, Consultant
               Bjorn Gaudernack, Institute for Energy Technology, Norway
               Morton Satin, Food and Agriculture Organization of the United Nations, Italy
               Ichiro Shimizu, Research Institute of Innovative Technology for the Earth, Japan
               Rudolf Straub, Forschungszentrum Juelich GmbH, Germany (invited)

12:00 - 12:20 pm - **Closing Ceremony** (King's Suites)
  Neil Rossmeissl, U.S. Department of Energy
  Ichiro Shimizu, Research Institute of Innovative Technology for the Earth, Japan
  Patrick Takahashi, University of Hawaii

12:20 - 1:15 pm - Lunch (Garden Court)

1:15 pm - Departure

Post Conference Day - Friday, June 27, 1997 (on Oahu)

Visitations to University of Hawaii at Manoa
  - Pacific Ocean Science and Technology (POST) Building
  - Hawaii Natural Energy Institute - Marine Biotechnology Center
  - Hawaii Institute of Marine Biology - Coconut Island

POSTER SESSION A (Monday, June 23, 1997)

7:30 - 9:30 pm (King's Suites)

1. <u>Takaaki Arai</u>, Tatsuki Wakayama, Satoshi Okano and Hiroshi Kitamura, Nihon University, Japan
  "Open Air Hydrogen Production by Photosynthetic Bacteria used Solar Energy during Winter Season in Central Japan"

2. Roger Babcock, Jr., University of Hawaii
  "Environmental Aspects of Large-Scale Microalgae Cultivation: Implications for Biological Hydrogen Production"

3. John Braman, Tucson, Arizona
  "Sturdy Biological Proton Exchange Membranes"

4. <u>Seher Dagdeviren</u>, Shivaun Archer, Karen McDonald and Alan Jackman, University of California, Davis
  "Production of Sulfolipids from Cyanobacteria in Photobioreactors"

5. <u>Reda M.A. El-Shishtawy</u>[1], Shozo Kawasaki and Masayoshi Morimoto, Kajima Laboratory, RITE ([1]present address: National Research Centre, Cairo, Egypt)
  "Cylindrical Type Induced and Diffused Photobioreactor: A Novel Phtoreactor for Large-Scale H_2 Production"

POSTER SESSION A (continued)

6. Maria L. Ghirardi, Timothy Flynn, Edwin Tracy, Dave Benson, and Michael Seibert, National Renewable Energy Laboratory
   "Selective Procedures for Isolating Oxygen-Tolerant, Hydrogen-Producing Mutants of *C. reinhardtii*"

7. D.O. Hall, A.A. Tsygankov, J.G. Liu and K.K. Rao, King's College London
   "An Automated Helical Photobioreactor Incorporating Cyanobacteria for Continuous Hydrogen Production"

8. Yoji Kitajima[1], Reda M.A. El-Shishtawy[1,4], Yoshiyuki Ueno[2], Seiji Otsuka[1], Jun Miyake[3], and Masayoshi Morimoto[1] ([1]Kajima Laboratory, RITE; [2]Marine Biotechnology Institute; [3]National Institute of Bioscience and Human-Technology, AIST/MITI, Japan; [4]present address: National Research Centre, Cairo, Egypt)
   "Analysis of Compensation Point of Light Using Plane-type Photosynthetic Bioreactor"

9. H. Kuriaki, T. Nishishiro, M. Kimura, K. Sano, M. Minami, and A. Toriyama, Kubota Corp., Japan
   "Continuous Culture Techniques for Hydrogen Production by *Rhodobacter sphaeroides* Strain RV"

10. Steven J. Meier, T. Alan Hatton, and Daniel I.C. Wang, Massachusetts Institute of Technology
    "Mechanism and Implications of Biological Cell Attachment to Gas Bubbles in Bioreactors"

11. Toshi Otsuki, Shigeru Uchiyama, Kiichi Fujiki, Sakae Fukunaga, RITE IHI Laboratory Research Institute, Ishikawajima-Harima Heavy Industries Co., Ltd.
    "Hydrogen Production by Floating-type Photobioreactor"

12. I. Eroglu, K. Aslan, U. Gündüz, M. Yücel, and L. Türker, Middle East Technical University, Turkey
    "Continuous Hydrogen Production by *Rhodobacter sphaeroides* O.U.001"

13. JoAnn C. Radway[1], Joseph C. Weissman[2], John R. Benemann[3], and Edward W. Wilde[1] ([1]Westinghouse Savannah River Company, [2]Microbial Products, Inc., [3]Sea Ag, Inc.)
    "Practical Considerations in Cyanobacterial Mass Cultivation"

14. Wei Wen Su, University of Hawaii
    "Secreted Metabolite Production in Perfusion Plant Cell Cultures"

15. James P. Szyper[1], Brandon A. Yoza[1], John R. Benemann[2], Mario R. Tredici[3], and Oskar R. Zaborsky[1] ([1]University of Hawaii, [2]University of California, Berkeley, [3]University of Florence)
    "Tubular Photobioreactor Technology: Development and Testing in Hawaii"

16. Serdar Türkarslan, Deniz Ö. Yigit, Kadir Aslan, Inci Eroglu, and Ufuk Gündüz, Middle East Technical University, Turkey
    "Photobiological Hydrogen Production by *Rhodobacter sphaeroides* O.U.001 by Utilization of Waste Water from Milk Industry"

17. T. Wakayama[1], A. Toriyama[2], T. Kawasugi[2], Y. Asada[3], and J. Miyake[4] ([1]RITE, [2]Kubota Co., [3]NIBH, [4]NAIR, AIST/MITI)
    "Simulation of Light Cycle of Sunlight for Photo-Hydrogen Production by a Photosynthetic Bacterium *Rhodobacter sphaeroides* RV"

POSTER SESSION B (Tuesday, June 24, 1997)
9:00 - 10:00 pm (King's Suites)

1. Takaaki Arai, Tatsuki Wakayama, Makoto Sekine, and Hiroshi Kitamura, Nihon University, Japan
   "Hydrogen Production from Mixed Organic Acids by Photosynthetic Bacteria"

2. Claudio D. Carrasco, Joleen S. Garcia, and James W. Golden, Texas A&M University
   "Programmed DNA Rearrangement of a Hydrogenase Gene During *Anabaena* Heterocyst Development"

3. Alfred Hansel and Peter Lindblad, Uppsala University, Sweden
   "Uptake Hydrogenase in *Nostoc* PCC 73102: Genetic Structure and Insertional Inactivation"

4. I.N. Gogotov, L.T. Serebryakova, and N.A. Zorin, Institute of Soil Science and Photosynthesis, Russia
   "Reversible Hydrogenases from Phototrophic Bacteria - Characteristics and Application for Hydrogen Production"

5. Patrick C. Hallenbeck, Alexander F. Yakunin, and Giuseppa Gennaro, University of Montreal
   "Electron Transport as a Limiting Factor in Biological Hydrogen Production"

6. Hugh McTavish, University of Minnesota
   "Hydrogen Evolution by Direct Electron Transfer from Photosystem I to Hydrogenases"

7. Hugh McTavish, University of Minnesota
   "Reconstitution of an Iron-Only Hydrogenase"

8. Akiko Ike, Tomoko Murakawa, Kazumasa Hirata, and Kazuhisa Miyamoto, Osaka University
   "H_2 Production from Starch by a Marine Bacterial Community from Night Soil Treatment Sludge"

9. Emir Khatipov[1,2], Masato Miyake[1], Jun Miyake[1], and Yasuo Asada[1] ([1]National Institute of Bioscience and Human Technology, AIST/MITI; [2]Research Institute of Innovative Technology for the Earth (RITE))
   "The Effect of Nitrogen Availability on Polyhydroxybutyrate Accumulation in *Rhodobacter sphaeroides* RV: A Relation to Hydrogen Evolution"

10. Y. Koike[1], K. Aoyama[2], I. Uemura[2], M. Miyake[3], J. Miyake[4], and Y. Asada[3] ([1]Research Institute for Innovative Technology for the Earth, [2]Tokyo-GAS, [3]NIBH, [4]NAIR)
    "Heterologous Expression of Clostridial Hydrogenase in Cyanobacteria"

11. Yao-Hua Luo, University of Miami
    "Preferential Degradation of Endogenous Substance for H_2 Production by Cyanobacteria during Biosolar Conversion"

12. Mitsufumi Matsumoto, Brandon Yoza, and Oskar R. Zaborsky, University of Hawaii
    "Photosynthetic Bacteria of Hawaii: Potential for Hydrogen Production"

13. Tadashi Matsunaga, Akiyo Yamada and Tsuyoshi Tanaka, Tokyo University of Agriculture and Technology
    "Conversion Efficiencies of Light Energy to Hydrogen by a Novel *Rhodovulum* sp. and its Uptake-Hydrogenase Mutant"

14. Susan C. Roberts and Michael L. Shuler, Cornell University
    "Strategies for Bioproduct Optimization in Plant Cell Tissue Cultures"

15. Y.B. Wang, G. Wang, and A.A. Szalay, Loma Linda University
    "The *Renilla* Luciferase-Modified GFP Fusion Proteins are Functional in Transformed Cells"

PARTICIPANT ROSTER

Dr. Michael J. Antal, Jr.
University of Hawaii
Hawaii Natural Energy Institute
2540 Dole Street, Holmes Hall 246
Honolulu, HI 96822
Phone: (808) 956-7267
FAX: (808) 956-2336
E-mail: antal@wiliki.eng.hawaii.edu

Mr. Katsuhiro Aoyama
Frontier Technology Research Institute
Tokyo Gas Co., Ltd.
1-7-7 Suehiro-cho, Tsurumi-ku, Yokohama
Kanagawa 230, Japan
Phone: +81 45-505-8817
FAX: +81 45-505-8821
E-mail: k_aoyama@tokyo-gas.co.jp

Mr. Takaaki Arai
Nihon University
College of Industrial Technology
1-2-1, Izumi-cho, Narashino
Chiba 275, Japan
Phone: +81-474-2561
FAX: +81-474-2579
E-mail: t5arai@ccu.cit.nihon-u.ac.jp

Dr. Yasuo Asada
Natl. Inst. of Bioscience & Human Technology
Molecular Bioenergetics Laboratory
1-1-3 Higashi, Tsukuba
Ibaraki 305, Japan
Phone: +81-298-54-6052
FAX: +81-298-56-4740
E-mail: yasada@ccmail.nibh.go.jp

Dr. Roger Babcock
University of Hawaii
Department of Civil Engineering/WRRC
2540 Dole Street, Holmes 383
Honolulu, HI 96822
Phone: (808) 956-7298
FAX: (808) 956-5014
E-mail: babcock@wiliki.eng.hawaii.edu

Dr. John Benemann
Consultant
343 Caravelle Drive
Walnut Creek, CA 94598
Phone: (510) 939-5864
FAX: (510) 939-5864
E-mail: JBENEMANN@aol.com

Dr. Perry Bergman
U.S. Department of Energy
Federal Energy Technology Center
P.O. Box 10940
Pittsburgh, PA 15236-0940
Phone: (412) 892-4890
FAX: (412) 892-5917
E-mail:

Ms. Lois D. Blaine
American Type Culture Collection
Information systems/Bioinformatics
12301 Parklawn Drive
Rockville, MD 20852
Phone: (301) 816-4370
FAX: (301) 816-4379
E-mail: lblaine@atcc.org

Dr. David Boron
National Science Foundation
4201 Wilson Blvd.
Arlington, VA 22230
Phone: (703) 306-1319
FAX: (703) 306-0312
E-mail: DBORON@NSF.GOV

Dr. Dulal Borthakur
University of Hawaii
Plant Molecular Physiology
3190 Maile Way, St. John 503
Honolulu, HI 96822
Phone: (808) 956-6600
FAX: (808) 956-3542
E-mail: dulal@hawaii.edu

Prof. Sammy Boussiba
Ben-Gurion University of the Negev
Microalgal Biotechnololgy
Sede-Boker Campus
84990 Israel
Phone: 972-7-6596795
FAX: 972-7-6596802
E-mail: Sammy@bgumail.bgu.ac.il

Mr. John Braman
4751 W. Waterbuck Drive
Tucson, AZ 85742
Phone: (520) 579-5578
FAX:
E-mail: johnbraman@mail.davis.uri.edu

Dr. J. Grant Burgess
Heriot-Watt University
Department of Biological Sciences
Edinburgh EH14 4AS
United Kingdom
Phone: +44-131-451-3463
FAX: +44-131-451-3009
E-mail: j.g.burgess@hw.ac.uk

Dr. Paolo Carrera
Eniricerche SpA
Via Maritano, 26
20097 S. Donato Milanese, Italy
Phone: 39 2 520 35536
FAX: 39 2 520 4957
E-mail:

Dr. Gerald R. Cysewski
Cyanotech Corporation
Hawaiian Ocean Science & Technology Park
73-4460 Queen Kaahumanu Hwy, #102
Kailua-Kona, HI 96740
Phone: (808) 326-1353
FAX: (808) 329-3597
E-mail:

Ms. Seher Dagdeviren
University of California at Davis
Dept. of Chemical Engineering & Materials Science
Davis, CA 95616
Phone: (916) 754-9452
FAX: (916) 752-1031
E-mail: sdeviren@ucdavis.edu

Dr. Tom Daniel
Natural Energy Laboratory of Hawaii Authority
73-4460 Queen Kaahumanu Hwy, #101
Kailua-Kona, HI 96740
Phone: (808) 329-7341
FAX: (808) 326-3262
E-mail:

Dr. Edward David, Jr.
EDD, Inc.
P.O. Box 435
Bedminster, NJ 07921
Phone: (908) 234-9319
FAX: (980) 234-2956
E-mail: eddavid@media.mit.edu

Dr. Reda M. A. El-Shishtawy
National Research Centre
Textile Division
Dokki, Cairo, Egypt
Phone: 202 38 28 779
FAX: 202 33 70 931
E-mail: shishtawy@FRCU.EUN.EG

Carolyn Elam
International Energy Agency
Executive Secretariat
1617 Cole Boulevard
Golden, CO 80401
Phone: (303) 275-3780
FAX: (303) 275-3782
E-mail: elamc@tcplink.nrel.gov

Participant Roster

Dr. Inci Eroglu
Middle East Technical University
Department of Chemical Engineering
Ankara 06531, Turkey
Phone: 90 312 210 2609
FAX: 90 312 210 1264
E-mail: ieroglu@rorqual.cc.metu.edu.tr

Dr. James Frank
Argonne National Laboratory
Waste Management and Bioengineering
9700 S. Cass Ave, ESD362/C309
Argonne, IL 60439
Phone: (630) 252-7693
FAX: (630) 252-9281
E-mail: james_frank@qmgate.anl.gov

Mr. Sakae Fukunaga
Ishikawajima-Harima Heavy Industries Co., Ltd.
1, Shin-nakahara-cho
Isogo-ku, Yokohama
Kanagawa 235, Japan
Phone: +81 45-759-2165
FAX: +81 45-759-2149
E-mail: sakae_fukunaga@ihi.co.jp

Mr. Bjorn Gauderneck
Institute for Energy Technology
P.O. Box 40
N-2007 Kjeller, Norway
Phone: +47 63806179
FAX: +47 63812905
E-mail: bjorng@ife.no

Ms. Maria Ghirardi
National Renewable Energy Laboratory
1617 Cole Boulevard
Golden, CO 80401
Phone: (303) 384-6312
FAX: (303) 384-6150
E-mail: ghirardm@tcplink.nrel.gov

Dr. James Golden
Texas A&M University
Department of Biology
College Station, TX 77843-3258
Phone: (409) 845-9823
FAX: (409) 862-7659
E-mail: jgolden@tamu.edu

Dr. Eli Greenbaum
Oak Ridge National Laboratory
P.O. Box 2008, Bldg. 4500N
Oak Ridge, TN 37831-6194
Phone: (423) 574-6835
FAX: (423) 574-1275
E-mail: exg@ornl.gov

Ms. Adrianne N. Greenlees
Natural Energy Laboratory of Hawaii Authority
73-4460 Queen Kaahumanu Hwy., #101
Kailua-Kona, HI 96740
Phone: (808) 329-7341
FAX: (808) 326-3262
E-mail: nelha@ilhawaii.net

Prof. Ufuk Gunduz
Middle East Technical University
Biology Department
Ankara 06531, Turkey
Phone: 90-312-2105183
FAX: 90-312-2101113
E-mail: ufukg@rorqual.cc.metu.edu.tr

Prof. D.O. Hall
King's College London, Univ. of London
Div. of Life Sciences
Campden Hill Road
London W8 7AH, United Kingdom
Phone: +44 171 333 4317
FAX: +44 171 333 4500
E-mail: d.hall@kcl.ac.uk

Dr. Patrick C. Hallenbeck
University of Montreal
Dept. of Microbiology & Immunology
CP 6128, Centre-ville
Montreal, Quebec H3C 3J7, Canada
Phone: 514-343-6278
FAX: 514-343-5701
E-mail: hallenbe@ere.umontreal.ca

Dr. Richard Hamilton
Oxford Bioscience Partners
650 Town Center Drive, Suite 810
Costa Mesa, CA 92626
Phone: (714) 754-5719
FAX: (714) 754-6802
E-mail: rwh@deltanet.com

Dr. Alfred Hansel
Uppsala University
Department of Physiological Botany
Villavagen 6
S-75236 Uppsala, Sweden
Phone: +46 - 18 471 28 14
FAX: +46 - 18 471 28 26
E-mail: alfred.hansel@fysbot.uu.se

Mr. Randy Harr
Hawaii Institute of Marine Biology
P.O. Box 1346
Kaneohe, HI 96744
Phone: (808) 236-7456
FAX: (808) 236-7443

Mr. Tatsuo Hattori
Tokyo Gas Co., Ltd.
1-5-20 Kaigan, Minato-ku
Tokyo 105, Japan
Phone: +81-3-5400-7597
FAX: +81-3-3432-7629
E-mail: HBD00741@niftyserve.or.jp

Mr. Douglas Hooker
U.S. Department of Energy
1617 Cole Boulevard
Golden, CO 80401
Phone: (303) 275-4780
FAX: (303) 275-4753
E-mail:

Dr. Mark E. Huntley
Aquasearch, Inc.
3222 Diamond Head Road
Honolulu, HI 96815
Phone: (808) 923-4627
FAX: (808) 923-4719
E-mail: aqse@aol.com

Ms. Akiko Ike
Osaka University Faculty of Pharmaceutical Sciences
1-6 Yamada-oka, Suita
Osaka 565, Japan
Phone: 81 6-879-8237
FAX: 81 6-879-8239
E-mail: ike@em.phs.osaka-u.ac.jp

Mr. Yoshiaki Ikuta
Mitsubishi Heavy Industries, Ltd.
Engineering & Construction Center
3-3-1 Minato-Mirai
Nishi-ku, Yokohama 220, Japan
Phone: +81 045-224-9399
FAX: +81 045-223-1705
E-mail: RXE20634@niftyserve.or.jp

Ms. Eileen Kalim
National Renewable Energy Laboratory
1617 Cole Blvd.
Golden, CO 80401
Phone: (303) 275-3864
FAX: (303) 275-3885
E-mail: kalime@tcplink.nrel.gov

Mr. Teruaki Kawamoto
Ishikawajima-Harima Heavy Industries Co., Ltd
3-1-15 Toyosu, Kohto
Tokyo 135, Japan
Phone: +81-3-3534-3500
FAX: +81-3-3534-3631
E-mail: teruaki_kawamoto@ihi.co.jp

Dr. Tadaaki Kawasugi
Research Inst. of Innovative Technology for the Earth
7 Toyokaiji-Bldg., 2-8-11 Nishishinbashi
Minato-ku
Tokyo 105, Japan
Phone: +81 3-3503-5666
FAX: +81 3-3503-4533
E-mail: ta-kawa@kubota.co.jp

Mr. Maurice Kaya
Dept. of Business, Economic Dev. and Tourism
Energy, Resources, and Technology Division
P.O. Box 2359
Honolulu, HI 96804
Phone: (808) 587-3812
FAX: (808) 586-2536
E-mail: mkaya@pixi.com

Dr. Jay Keasling
University of California
Department of Chemical Engineering
201 Gilman Hall
Berkeley, CA 94720-1462
Phone: (510) 642-4862
FAX: (510) 642-4778
E-mail: keasling@socrates.berkeley.edu

Dr. Emir-Asan A. Khatipov
RITE NIBH Lab.
1-1-3 Higashi, Tsukuba
Ibaraki 305, Japan
Phone: +81-298-56-5103
FAX: +81-298-56-4740
E-mail: ekhatipov@ccmail.nibh.go.jp

Mr. William Y. Kikuchi
WYK Visual Communications
72 Hale Manu Drive
Hilo, HI 96720
Phone: (808) 959-6021
FAX: (808) 959-3681
E-mail:

Dr. Mi-Sun Kim
Korea Institute of Energy Research
Biomass Research Team
71-2 Jang-Dong, Yusung-ku
Taejon 305-343, Korea
Phone: 82-42-860-3554
FAX: 82-42-860-3132
E-mail: BMMSKIM@SUN330.KIER.RE.KR

Mr. Yoji Kitajima
Kajima Corporation
Kajima Technical Research Institute
19-1, Tobitakyu 2-chome, Chofu
Tokyo 182, Japan
Phone: +81-424-89-7147
FAX: +81-424-89-2896
E-mail: LEG06702@niftyserve.or.jp

Dr. Hiroshi Kitamura
Tokyo Metropolitan University
5-3-4 Nishiohi, Shinagawa
Tokyo 140, Japan
Phone: 81 3-3772-1914
FAX: 81 3-3776-7125
E-mail:

Mr. Yoji Koike
RITE NIBH Lab.
Natl. Inst. of Bioscience and Human Technology
1-1-3 Higashi, Tsukuba
Ibaraki 305, Japan
Phone: +81 298-56-5103
FAX: +81 298-56-4740
E-mail: yokoike@ccmail.nibh.go.jp

Mr. Yuko Kurashige
New Energy & Industrial Technology Dev. Org.
Sunshine 60, 29F
1-1, 3-Chome, Higashi, Toshima-ku
Tokyo 170, Japan
Phone: +81 3-3987-9368
FAX: +81 3-5391-1744
E-mail: kurashigeyuk@nedo.go.jp

Mr. Hajime Kuriaki
Kubota Corporation
Advanced Technology Laboratory
5-6, Koyodai, Ryugasaki
Ibaraki 301, Japan
Phone: +81-297-64-7475
FAX: +81-297-64-7266
E-mail: hajime-k@kubota.co.jp

Dr. Peter Lindblad
Uppsala University
Dept. Physiological Botany
Villavagen 6
Uppsala S-75236, Sweden
Phone: +46 - 18 471 28 26
FAX: +46 - 18 471 28 26
E-mail: Peter.Lindblad@Fysbot.uu.se

Dr. Yao-Hua Luo
University of Miami
MBF/RSMAS
884 NE 205 Terr. Miami, FL 33179
Phone: (305) 652-1381
FAX: (305) 652-1381
E-mail: yaluo@juno.com

Ms. Margaret K. Mann
National Renewable Energy Laboratory
1617 Cole Blvd.
Golden, CO 80401
Phone: (303) 275-2921
FAX: (303) 275-2905
E-mail: mannm@tcplink.nrel.gov

Dr. Sergei A. Markov
National Renewable Energy Laboratory
1617 Cole Boulevard
Golden, CO 80401
Phone: (303) 384-6277
FAX: (303) 384-6150
E-mail: markovs@tcplink.nrel.gov

Dr. Stephen Masutani
Hawaii Natural Energy Institute
University of Hawaii
2540 Dole Street, Holmes Hall 246
Honolulu, HI 96822
Phone: (808) 956-7388
FAX: (808) 956-2335
E-mail: masutan@wiliki.eng.hawaii.edu

Mr. Mitsufumi Matsumoto
Tokyo University of Agriculture & Technology
Department of Biotechnology
2-24-16 Nakacho, Koganei
Tokyo 184, Japan
Phone: (808) 587-3490
FAX: (808) 587-3494
E-mail:

Dr. Tadashi Matsunaga
Tokyo University of Agriculture & Technology
Department of Biotechnology
2-24-16 Nakacho, Koganei
Tokyo 184, Japan
Phone: +81 423 88 7021
FAX: +81 423 85 7713
E-mail: tmatsuna@cc.tuat.ac.jp

Mike May
Hawaiian Electric Co.
P.O. Box 2750
Honolulu, HI 96840-0001
Phone: (808) 543-4433
FAX: (808) 543-4445
E-mail:

Mr. James McElvaney
Sustainable Technologies, Inc.
690 K Naele Road
Kula, HI 96790
Phone: (808) 876-0602
FAX: (808) 876-0134
E-mail: jmce@sustainable.com

Dr. Hugh McTavish
University of Minnesota
Department of Biochemistry
1479 Gortner Avenue
St. Paul, MN 55108
Phone: (612) 624-4278
FAX: (612) 625-5780
E-mail: mctavish@biosci.cbs.umn.edu

Mr. Steven J. Meier
Massachusetts Institute of Technology
77 Massachusetts Ave., Room 20A-207
Cambridge, MA 02139-4307
Phone: (617) 641-2583
FAX: (617) 253-2400
E-mail: sjmeier@mit.edu

Prof. Tasios Melis
University of California, Berkeley
Plant and Microbial Biology
411 Koshland Hall
Berkeley, CA 94720-3102
Phone: (510) 642-8166
FAX: (510) 642-4995
E-mail: melis@nature.berkeley.edu

Dr. H. Harvey Michels
United Technologies Research Center
411 Silver Lane
East Hartford, CT 06108
Phone: (860) 610-7489
FAX: (860) 610-7909
E-mail: MichelHH@utrc.utc.com

Mr. Masayoshi Minami
Kubota Corporation
Advanced Technology Laboratory
5-6, Koyodai, Ryugasaki
Ibaraki 301, Japan
Phone: +81 297-64-7475
FAX: +81 297-64-7266
E-mail: m-minami@kubota.co.jp

Ms. Kimii Mitsui
Phone: (516) 626-0763
FAX:
E-mail:

Dr. Shigetoh Miyachi
Marine Biotechnology Institute Co., Ltd.
1-28-10 Hongo, Bunkyo-ku
Tokyo 113, Japan
Phone: 81-3-5684-6211
FAX: 81-3-5684-6200
E-mail: miyachi@super.win.or.jp

Dr. Jun Miyake
Natl. Inst. for Advanced Interdisplinary Research
Higashi 1-1-4, Tsukuba
Ibaraki 305, Japan
Phone: 81-298-54-2558
FAX: 81-298-54-2565
E-mail: miyake@nair.go.jp

Mr. Masato Miyake
Natl. Inst. of Bioscience & Human Technology
Higashi 1-1-4, Tsukuba
Ibaraki 305, Japan
Phone: +81 298-54-6052
FAX: +81 298-56-5138
E-mail: mmiyake@ccmail.nibh.go.jp

Dr. Kazuhisa Miyamoto
Osaka University Faculty of Pharmaceutical Sciences
1-6 Yamada-oka, Suita
Osaka 565, Japan
Phone: +81 6 879 8235
FAX: +81 6 879 8239
E-mail: miyamoto@phs.osaka-u.ac.jp

Mr. Tatsuya Mizukoshi
SHOWA DENKO K.K.
13-9, Shiba Daimon 1-chome
Minato-ku
Tokyo 105, Japan
Phone: +81-3-5470-3498
FAX: +81-3-3431-2944
E-mail: BZR04146@niftyserve.or.jp

Dr. Masayoshi Morimoto
Michigan Corporation
4-29-11 Shimo-Takaido
Suginami-ku
Tokyo, 168-0073, Japan
Phone: +81-3-3306-5662
FAX: +81-3-3306-5662
E-mail: masamorimoto@classic.msn.com

Mr. Hisashi Nagadomi
Nagao Co. Ltd.
Research & Development Center
221, Miyaura
Okayama 702, Japan
Phone: +81-86-267-1366
FAX: +81-86-267-3487
E-mail:

Mr. Eiju Nakada
Fuji Electric Corporate R&D, Ltd.
2-2-1 Nagasaka, Yokosuka
Kanagawa 240-1, Japan
Phone: +81 468-57-6732
FAX: +81 468-57-6946
E-mail: nakada-eiju@fujielectric.co.jp

Mr. Satoshi Nishikata
Fuji Electric Corp. Research and Development, Ltd.
2-2-1 Nagasaka, Yokosuka
Kanagawa 240-1, Japan
Phone: +81 468-57-6732
FAX: +81 468-57-6946
E-mail: nishikata-satoshi@fujielectric.co.jp

Dr. David O'Keefe
Full Circle Solutions, Inc.
3931 SE 37th Street
Gainesville, FL 32641
Phone: (352) 373-9313
FAX: (352) 371-6027
E-mail: fullcircle@delphi.com

Dr. Takahira Ogawa
Kumamoto Institute of Technology
4-22-1 Ikeda, Kumamoto
Kumamoto 860, Japan
Phone: +81 96-326-3111,x5136
FAX: +81 96-326-3000
E-mail: ogawa@bio.kumamoto-it.ac.jp

Dr. James C. Ogbonna
University of Tsukuba
Institute of Applied Biochemistry
1-1-1 Tennodai, Tsukuba
Ibaraki 305, Japan
Phone: +81-298-53-6646
FAX: +81-298-53-4605
E-mail: jogbonna@sakura.cc.tsukuba.ac.jp

Mr. Seiji Otsuka
Kajima Corporation
19-1, Tobitakyu 2-chome
Chofu
Tokyo 182, Japan
Phone: +81-424-89-7147
FAX: +81-424-89-2896
E-mail: LEG06701@niftyserve.or.jp

Mr. Toshi Otsuki
Ishikawajima-Harima Heavy Industries Co., Ltd.
1, Shin-nakahara-cho
Isogo-kuYokohama 235, Japan
Phone: +81 45-759-2162
FAX: +81 45-759-2149
E-mail: toshi_ootsuki@ihi.co.jp

Participant Roster

Ms. Catherine Gregoire Padro
National Renewable Energy Laboratory
1617 Cole Boulevard
Golden, CO 80401
Phone: (303) 275-2919
FAX: (303) 275-2905
E-mail: gregoirc@tcplink.nrel.gov

Dr. Paola Pedroni
ENIRICERCHE S.p.A.
Via F. Maritano
26-S. Donato Milanese
Milan 20097, Italy
Phone: +39-2-52046615
FAX: +39-2-52022974
Email: ppedroni@eniricerche.eni.it

Dr. Roger Prince
Exxon Research & Engineering Co.
Route 22E
Annandale, NJ 08801
Phone: (908) 730-2134
FAX: (908) 730-3279
E-mail: rcprinc@erenj.com

Dr. JoAnn C. Radway
Westinghouse Savannah River Site
Aiken, SC 29808
Phone: (803) 557-7095
Fax: (803) 557-7223
E-mail: jradway@csra.net

Dr. C. Barry Raleigh
University of Hawaii
School of Ocean and Earth Science and Technology
1000 Pope Road, MSB 205
Honolulu, HI 96822
Phone: (808) 956-6172
FAX: (808) 956-9152
E-mail:

Dr. K. Krishna Rao
King's College London
Division of Life Sciences
Campden Hill Road
London W8 7AH, United Kingdom
Phone: 44-171-333-4323
FAX: 44-171-333-4500
E-mail: krishna.rao@kcl.ac.uk

Prof. Dr. Ingo Rechenberg
Technical University of Berlin
TU Berlin, Bionik Ackerstr. 71-76
D-13355 Berlin, Germany
Phone: 49 30 314 72655
FAX: 49 30 314 72658
E-mail: rechenberg@fb10.tu-berlin.de

Ms. Susan Roberts
Cornell University
Dept. of Chemical Engineering
120 Olin Hall
Ithaca, NY 14853
Phone: (607) 255-5204
FAX: (607) 255-9166
E-mail: smoser@cheme.cornell.edu

Dr. Rick Rocheleau
Hawaii Natural Energy Institute
2540 Dole Street, Holmes Hall 246
Honolulu, HI 96822
Phone: (808) 956-2337
FAX: (808) 956-2336
E-mail: rochele@wiliki.eng.hawaii.edu

Dr. Gregory L. Rorrer
Oregon State University
Department of Chemical Engineering
Corvallis, OR 97331
Phone: (541) 737-3370
FAX: (541) 737-4600
E-mail: rorrerg@ccmail.orst.edu

Dr. Neil Rossmeissl
U.S. Dept. of Energy
Hydrogen Program
1000 Independence Ave.
Washington, DC 20585
Phone: (202) 586-8668
FAX: (202) 586-0784
E-mail: neil.rossmeissl@hq.doe.gov

Mr. Kiyoshi Saito
NEDO
Sunshine 60, 29 Fl, 1-1, 3-Chome
Higashi Ikebukuro, Toshima-ku
Tokyo 170, Japan
Phone: 81-3-3987-9368
FAX: 81-3-5391-1744
E-mail: saitokys@nedo.go.jp

Dr. Anthony San Pietro
Indiana University
Department of Biology
Jordan Hall 142
Bloomington, IN 47405-6801
Phone: (812) 855-4115
FAX: (812) 855-6705
E-mail: sanpietro@indiana.edu

Dr. Ken Sasaki
Hiroshima-Denki University
6-20-1, Nakano, Akiku
Hiroshima 739-03, Japan
Phone: +81-82-893-0381
FAX: +81-82-893-5012
E-mail: sasaki@g.hiroshima-dit.ac.jp

Dr. Morton Satin
Food and Agriculture Organization of the United
Nations
Agro-Industries and Post-Harvest Management
Via Terme di Caracalla
Rome 00100, Italy
Phone: (396) 5225-3334
FAX: (396) 5225-4960
E-mail: morton.satin@fao.org

Dr. Joerg Schnackenberg
Max-Planck-Institut fuer Biochemie
Abt. Strukturforschung
Am Klopferspitz 18a
D-82152 Muenchen-Martinsried, Germany
Phone: +49 89-85782704
FAX: +49 89-85783516
E-mail: schnacke@biochem.mpg.de

Dr. Ruediger Schulz
Philipps-University
FB Biology/Botany
Karl-von-Frisch-Str.
D-35032 Marburg, Germany
Phone: +49-6421-282484
FAX: +49-6421-282057
E-mail: schulzr@mailer.uni-marburg.de

Dr. Michael Seibert
National Renewable Energy Laboratory
1617 Cole Boulevard
Golden, CO 80401
Phone: (303) 384-6279
FAX: (303) 384-6150
E-mail: seibertm@tcplink.nrel.gov

Mr. Ichiro Shimizu
Research Inst. of Innovative Technology for the Earth
7 Toyokaiji-Bldg
2-8-11 Nishishinbashi
Minato-ku, Tokyo 105, Japan
Phone: +81 3-3503-5666
FAX: +81 3-3503-4533
E-mail: XLL06167@niftyserve.or.jp

Dr. Robert Shleser
The `Aina Institute
P.O. Box 560
Waimanalo, HI 96795
Phone: (808) 259-5042
FAX: (808) 259-8049
E-mail: shleser@aloha.net

Participant Roster

Dr. Koji Sode
Tokyo University of Agriculture & Technology
Department of Biotechnology
Koganei Tokyo 184, Japan
Phone: +81 423-88 7027
FAX: +81 423-88 7027
E-mail: sode@cc.tuat.ac.jp

Dr. Rudolf Straub
Research Centre Juelich
Project Management BEO 22
P.O. Box 1913
D-52425 Juelich, Germany
Phone: 49 2461-614460
FAX: 49 2461-612730
E-mail: r.straub@fz-juelich.de

Dr. Wei Wen Su
University of Hawaii
Department of Biosystems Engineering
3050 Maile Way
Honolulu, HI 96822
Phone: (808) 956-3581
FAX: (808) 956-9269
E-mail: wsu@hawaii.edu

Dr. A.A. Szalay
Loma Linda University
Center for Molecular Biology & Gene Therapy
School of Medicine
Loma Linda, CA 92350
Phone: (909) 478-8777
FAX: (909) 478-4177
E-mail:

Dr. James Szyper
University of Hawaii
c/o Hawaii Natural Energy Institute
2540 Dole Street, Holmes 246
Honolulu, HI 96822
Phone: (808) 587-3490
FAX: (808) 587-3494
E-mail: jszyper@hawaii.edu

Dr. Patrick K. Takahashi
Hawaii Natural Energy Institute
University of Hawaii
2540 Dole Street, Holmes Hall 246
Honolulu, HI 96822
Phone: (808) 956-8346
FAX: (808) 956-2336
E-mail: patrick@ wiliki.eng.hawaii.edu

Koichi Takasaki
Research Inst. of Innovative Technology for the Earth
7 Toyokaiji-Bldg.
2-8-11 Nishishinbashi, Minato
Tokyo 105, Japan
Phone: +81-3-3503-5666
FAX: +81-3-3503-4533
E-maiL LEG05253@niftyserve.or.jp

Dr. Haruko Takeyama
Tokyo University of Agriculture & Technology
Department of Biotechnology
2-24-16, Naka-cho, Koganei
Tokyo 184, Japan
Phone: +81 423-88-7021
FAX: +81 423-85-7713
E-mail: haruko@cc.tuat.ac.jp

Dr. Shigaharu Tanisho
Yokohama National University
79-5 Hodogaya-ku
Yokohama 240, Japan
Phone:
FAX: +81 45-339 3996
E-mail: tanisho@chemeng.bsk.ynu.ac.jp

Mr. David A. Tarnas
State of Hawaii House of Representatives
State Capitol, Room 326
Honolulu, HI 96813
Phone: (808) 586-8510
FAX: (808) 586-8514
E-mail: reptarnas@capitol.hawaii.gov

Dr. Elisha Tel-Or
Hebrew University of Jerusalem
Faculty of Agriculture
P.O. Box 12
Rehovot 76100, Israel
Phone: 972 8 9481262
FAX: 978 8 9467763
E-mail: telor@agri.huji.ac.il

Dr. Akio Toriyama
Kubota Corporation
Advanced Technology Laboratory
5-6, Koyodai, Ryugasaki
Ibaraki 301, Japan
Phone: +81-297-64-7472
FAX: +81-297-64-7266
E-mail: to-akio@kubota.co.jp

Dr. Sabine Tramm-Werner
Umweltbiotechnologie
Ritterstrabe 12a
D-52072 Aachen, Germany
Phone: 49 241-870093
FAX: 49 241-870091
E-mail:

Prof. Mario Tredici
University of Florence
P.le delle Cascine, 27
Firenze 50144, Italy
Phone: 39-55-3288306
FAX: 39-55-330431
E-mail: tredici@csma.fi.cnr.it

Dr. Anatoly Tsygankov
Institute of Soil Science and Photosynthesis
Pushchino
Moscow Region 142292
Russia
Phone:
FAX: 7-967-79-0532
E-mail: TTT@issp.serpukhov.su

Dr. Lemi Turker
Middle East Technical University
Department of Chemistry
Ankara 06531, Turkey
Phone: 90 312 210 3244
FAX: 90 312 210 3280
E-mail: lturker@rorqual.cc.metu.edu.tr

Mr. Shigeru Uchiyama
Ishikawajima-Harima Heavy Industries Co., Ltd.
1, Shin-nakahara-cho, Isogo-ku
Yokohama
Kanagawa 235, Japan
Phone: +81-45-759-2165
FAX: +81-45-759-2149
E-mail: shigeru_uchiyama@ihi.co.jp

Dr. Lyudmila Vasilyeva
RITE NIBH Lab.
National Inst. of Bioscience and Human Technology
1-1-3 Higashi, Tsukuba
Ibaraki 305, Japan
Phone: +81-298-56-5103
FAX: +81-298-56-4740
E-mail: lvasilyeva@ccmail.nibh.go.jp

Dr. Youji Wachi
Tokyo University of Agriculture & Technology
Department of Biotechnology
2-24-16 Nakacho, Koganei
Tokyo 184, Japan
Phone:
FAX: +81 423 85 7713
E-mail:

Mr. Tatsuki Wakayama
RITE NIBH Lab.
National Inst. of Bioscience and Human Technology
1-1-3 Higashi, Tsukuba
Ibaraki 305, Japan
Phone: +81-298-56-5103
FAX: +81-298-56-4740
E-mail: twakayama@ccmail.nibh.go.jp

Dr. Gefu Wang
Loma Linda University
Center for Molecular Biology & Gene Therapy
11085 Campus Street
Loma Linda, CA 92350
Phone: (909) 478-8777
FAX: (909) 478-4177
E-mail: wanggef@sc.llu.edu

Dr. Ekkehard Warmuth
Federal Ministry for Research & Technology Division
for Biological Research and Technology
P.O. Box 200240
D-53170 Bonn, Germany
Phone: 49 228 59-3669
FAX: 49 228 573-605
E-mail:

Dr. Paul Weaver
National Renewable Energy Laboratory
1617 Cole Blvd.
Golden, CO 80401
Phone: (303) 384-6288
FAX: (303) 384-6150
E-mail: weaverp@tcplink.nrel.gov

Dr. Jonathan Woodward
Oak Ridge National Lab
P.O. Box 2008
Oak Ridge, TN 37831-6194
Phone: (423) 574-6826
FAX: (423) 574-6843, 6442
E-mail: oop@ornl.gov

Mr. Brandon Yoza
University of Hawaii
Hawaii Natural Energy Institute
2540 Dole Street, Holmes Hall 246
Honolulu, HI 96822
Phone: (808) 587-3490
FAX: (808) 587-3994
E-mail:

Dr. Oskar R. Zaborsky
University of Hawaii
Hawaii Natural Energy Institute
2540 Dole Street, Holmes Hall 246
Honolulu, HI 96822
Phone: (808) 956-8146
FAX: (808) 956-2335
E-mail: ozabo@hawaii.edu

INDEX

Acetate
    in bioreactors and photobioreactors, 363–365, 369, 370, 371, 373
    polyhydroxybutyrate accumulation and, 159, 160
Acetic acid, 13, 23, 133, 308, 323, 326
Acetomicrobium flavidum, 65, 68–69, 502
Acid phosphatase, 475, 476, 481
Adenosine triphosphate (ATP), 21, 112, 121, 124
    Anabaena 7120 and, 182
    Chlamydomonas reinhardtii and, 228
    light energy conversion and, 14, 15–16
    metabolic engineering and, 88, 92
Advanced Technology Program (ATP), 2
Aequorea, 493, 494
Alcaligenes eutrophus, 190, 191, 192
Algae, 489; *see also specific types*
    in Erg Chebbi, 290
    fusion of plant cells with, 489
    green: *see* Green algae
    hydrogenase-based systems, 420–421
    macro: *see* Macroalgae
    micro: *see* Microalgae
    pond-based systems, 415–416, 421, 423–424
    technoeconomic analysis of production systems, 415–416, 420–421, 423–424
Algal biomass
    biophotolysis and, 26
    lactic acid fermentation and, 267, 269–271
Alkaline phosphatase (AP), 92, 93
Alkaloids, 487
5-Aminolevulinic acid (ALA), 133–141
Ammonia, 181, 182
Anabaena azollae, 431–432, 437
Anabaena cylindrica, 20, 112, 176
Anabaena MiamiBG 060002, 198, 202
Anabaena MiamiBG 094001, 198, 202
Anabaena PCC 7120, 53, 54, 55, 56, 58–60, 190, 209, 210
    direct nested PCR and, 197–202
    DNA gene rearrangement during heterocyst development in, 203–206

Anabaena PCC 7120 (*cont.*)
    sulfolipid production from, 464, 465
    uptake hydrogenase gene cluster from, 181–187
Anabaena variabilis, 55, 58–60, 190, 192, 206, 210, 386
Anabaena variabilis ATCC 29413, 213, 432, 435, 438–439
Anacystis nidulans, 55, 190
Anaerobic adaptation, 246, 255
Anaerobic bacteria, 11, 12, 13
Anaerobic fermentation: *see* Dark fermentation
Anaerobic fermentation bioreactors, 388
Anchusa officinalis, 475–481
Anoxygenic photosynthesis, 244
Antenna size, 240
    electron transport and, 237–239
    minimizing, 41–51
Anthoceros punctatus, 210
*ara*C gene, 90
Arthrospira platensis, 392, 394, 395, 443–444; *see also Spirulina platensis*
Artificial bacterial algal symbiosis (ArBAS) project, 281–293
Ascorbic acid method, 476
A-type photobioreactors, 346–347, 348–351
Average light intensities, 331, 336
Azotobacter chroococcum, 182, 183, 187
Azotobacter vinelandii, 55, 102, 182, 183, 187

Bacteria, 489; *see also specific types*
    anaerobic, 11, 12, 13
    cyano: *see* Cyanobacteria
    in energy conversion, 10–11
    in Erg Chebbi, 290
    halotolerant, 311–317
    photosynthetic: *see* Photosynthetic bacteria
    purple: *see* Purple bacteria
    technoeconomic analysis of production systems, 417–420, 423
    toolkit for metabolic engineering of, 87–96
Bacterial water-gas shift process, 417–420, 423

Bacteriochlorophyll (Bchl), 124, 125
Bacteriorhodopsin (BR), 295, 296, 298–299, 300–302; *see also* Purple membrane fragments
Batch cultures
  of *Anchusa officinalis*, 478–479, 480, 481
  of cyanobacteria, 436–437, 461
Batch operation
  marine macroalgal suspension cultures in, 404, 406, 412
  Rhodobacter sphaeroides O.U.001 and, 143, 145–147
  waste water treatment and, 154
Bench-scale units, 326, 426
Bench-top photobioreactors (PhBR), 431–439
Bicarbonate, 235, 237–239; *see also* Sodium bicarbonate
Bidirectional hydrogenase, 55, 58–60
Biocoil reactors, 442
Biofilm formation, 434
BioHydrogen 97 (keynote address), 1–6
BioHydrogen Program, University of Hawaii, 442
Bioinformatics, 451–457
Biological energy conversion, 10–12
Biomass, 516–517
  algal: *see* Algal biomass
  microalgal: *see* Microalgal biomass
Bio-photoelectrochemical (bio-PEC) reactor, 295–303
Biophotolysis, 24–27, 392, 442, 517
  direct, 19, 25–26, 392, 396, 449
  indirect, 19–20, 21, 392, 396, 449
  prior cost analysis of processes, 393
Bioproduct optimization, 483–490
Biopterin glucoside (BG), 34–35
Bioreactors, 248, 383–389
  anaerobic fermentation, 388
  in ArBAS project, 287, 290–291
  bio-photoelectrochemical, 295–303
  dark, 387–388, 389
  laser dye, 290–291
  membrane, 387–388
  open pond systems vs., 449
  perfusion air-lift, 475–481
  photo: *see* Photobioreactors
  plane-type photosynthetic, 359–367
  plant cell cultures and, 483, 488–489
  shift reaction, 387–388, 389
  spiral PVC tubular bubble-train, 388
Bradyrhizobium japonicum, 55, 182, 183, 187, 206, 246–247
B-type photobioreactors, 346, 347, 348, 349, 350, 351
Butylated hydroxy toluene (BHT), 462
Butyrate, 363–365, 369, 370, 371, 373

Cancer/anti-tumor agents, 140, 459, 460; *see also* specific agents
Carbohydrates, 288–290
Carbon dioxide, 235, 268
  bacterial water-gas shift process and, 417, 419, 423
  Dunaliella salina grown with supplemental, 47–48
  oxygen exchange with in photobioreactors, 441–446

Carbon dioxide (*cont.*)
  proposed reduction in emissions, 450
  simultaneous recovery of nitric oxide and, 265, 268–269
  splitting of lactate into hydrogen and, 281–282
Carbon dioxide fixation, 19–20
  Anabaena 7120 and, 182
  biophotolysis and: *see* Direct biophotolysis; Indirect biophotolysis
  Chlamydomonas reinhardtii and, 239–240, 265–271
  Chlorella pyrenoidosa and, 340–341
  microalgae and, 265–271
  microalgae biomass and, 311–317
  Rhodobacter sphaeroides RV and, 117, 118, 121–122, 311, 316–317
Carbon monoxide
  bacterial water-gas shift process and, 417, 420, 423
  in bioreactors, 387–388
  conversion to hydrogen, 22–23
Carbon/nitrogen ratio, 288–290
b-Carotene, 34
cat gene, 111, 113–114
Catharanthus rosens, 486
Cell-free extracts, 175–176, 490
Cellobiose, 314
Cellulose, 314
Cellulose acetate membranes (CAM), 296, 298–299, 301–303
Chlamydomonas, 306, 319, 321–323
Chlamydomonas oblonga, 281, 288, 289, 292
Chlamydomonas reinhardtii, 42, 248, 254
  CO_2 fixation in, 239–240, 265–271
  oxygen-tolerant mutants of, 227–233
  starch from, 311, 312–313, 314, 315, 317
Chlorella, 460
  photosynthetic hydrogen and oxygen production by, 235–240
Chlorella NKG 042401, 32
Chlorella pyrenoidosa, 312–313, 315, 334, 335, 337, 338–339
  CO_2 fixation by, 340–341
Chlorella vulgaris, 42, 138
Chlorophyceae, 245
Chlorophyll, 140, 437, 462, 465
  bacterio: *see* Bacteriochlorophyll
Chlorophyll antenna size: *see* Antenna size
Chloroplasts, 460, 486
Cis-palmitoleic acid, 32
Clamydomonas moewsuii, 21
Climate Challenge Program, 515
Climate change, 514–516
Climate Change Action Plan, U.S. (CCAP), 515
Climate Technology Initiative (CTI), 447, 450, 515, 517, 518
Cloning: *see* Polymerase chain reaction
Closed photobioreactors, 393–394
Clostridium, 13
Clostridium butyricum, 274
Clostridium kluyveri, 20

Index   545

Clostridium pasteurianum, 502
  heterologous expression of hydrogenase in, 111–114
  reconstitution of iron-only hydrogenase in, 105–109
Commencement challenge (address), 39–40
Complement, 83
Complex I, 191–193
Continuous operation, 143, 145, 148–149
Continuous-recycle operation, 404, 406, 412
crtI gene, 34
crtY gene, 34
C-type photobioreactors, 346, 347, 348, 351
Cyanobacteria, 12, 21, 503, 516; *see also specific types*
  in energy conversion, 11
  fermentation products of, 305–309
  helical tubular photobioreactor incorporating, 431–439
  heterologous expression of clostridial hydrogenase in, 111–114
  in internal gas exchange photobioreactors, 442
  light energy conversion and, 14, 16
  marine: *see* Marine cyanobacteria
  mass cultivation of, 467–473
  nitrogen-fixing: *see* Nitrogen-fixing cyanobacteria
  non-nitrogen-fixing: *see* Non-nitrogen-fixing cyanobacteria
  in photobioreactors, 384–387, 459–466
  sulfolipid production from, 459–466
Cytochrome b-c, 273, 275
Cytochrome c_3, 260
Cytochrome c_6, 253, 256, 258, 259, 260–262
Cytochrome c_{550}, 177

D5, 228–233
Dark bioreactors, 387–388, 389
Dark fermentation, 19, 21, 23, 26, 449
  Escherichia coli and, 87–96
  microalgae and, 319, 320, 321–323, 324, 326, 327
  Microcystis aeruginosa and, 174–175
  Rhodobacter sphaeroides RV and, 117, 118, 121–122
DBMIB, 173, 176, 177
DCMU, 173, 176, 177, 228, 231
Department of Defense, U.S., 2, 3
Department of Energy, U.S., 4, 22, 416, 442, 514, 515
Desulfovibrio fructosovorans, 204
Desulfovibrio gigas, 205
Desulfovibrio vulgaris, 108, 248, 260
2,4-Dichlorophenyl-4-nitrophenyl ether (NIP), 140
Differentiation, 483, 486
Diffused light intensities, 356–357
Direct biophotolysis, 19, 25–26, 392, 396, 449
Directed evolution, 282–286
Direct nested polymerase chain reaction (PCR), 197–202
Dissolved oxygen concentration, 276, 277, 278, 339
DNA gene rearrangement, programmed, 203–206
Docosahexaenoic acid (DHA), 32–34
Double-crossover experiments, 216

Dunaliella, 239
Dunaliella salina, 41–51
Dunaliella tertiolecta, 42, 265–271, 313, 317

Eicosapentaenoic acid (EPA), 32–34
Electric Power Research Institute (EPRI), 4
Electron transport
  antenna size and, 237–239
  as a limiting factor in hydrogen production, 99–103
  Scenedesmus obliquus hydrogenase and, 248, 253–262
Elicitation, 483, 484–485
Energy conversion
  biological, 10–12
  light: *see* Light energy conversion
Energy system, 7–8
Enterobacter aerogenes, 388
Enterobacter aerogenes E.82005, 273–279
Entropy, 10–17
Entropy engineering, 7–10
Entropy problem, 7–8
Environmental optimization, 485–486
Erg Chebbi, 286–287, 290
Erythrobacter longus OCh101, 34
Escherichia coli, 108, 144, 183, 185, 187, 191, 247, 502
  clostridial hydrogenase expression in, 111, 112–113
  genes encoding hydrogenase of, 67, 69
  Halobacterium halobium coupled with, 296
  hydrogenase 3 overexpression in, 73, 74, 75–77
  metabolic engineering of, 73–78, 87–96
  nitrate reductase deficiency in, 73–74, 77–78
  Nostoc 73102 conjugation with, 210, 211, 212, 216
  Renilla GFP fusion protein and, 493–494, 496, 497
Ethanol, 10, 23–24, 308, 323, 326, 369, 373
Euglena, 460
Euglena gracilis, 334, 335, 341
Euglenophyceae, 245
Exogenous substrates, 219–224

FAD, 279
$FADH_2$, 275, 278–279
Fatty acids, 32–34
FdI protein, 100, 101–103
fdn gene, 77
fdxN gene, 203
Fed-batch photobioreactors, 461
Fermentation, 244
  dark: *see* Dark fermentation
  lactic acid: *see* Lactic acid fermentation
  metabolic engineering and, 73–78, 87–96
  photo, 21, 392
  Spirulina platensis NIES-46 and, 305–309
Fermentative hydrogen evolution, 273–274
Ferredoxin, 20, 100, 228
  light energy conversion and, 15
  Scenedesmus obliquus and, 253, 254, 258, 259–260, 261
Ferricyanide, 259

Fischerella strain 113, 467–473
Fish, genome analysis in, 36–37
Flavodoxin, 100
Floating-type photobioreactors, 369–373, 504, 505
Fluorescence resonance energy transfer, 494
Formate dehydrogenase-N (FDH-N), 74, 77, 78
Formate hydrogenlyase system (FHL), 74, 76–77
Formic acid, 308
Fossil fuels, 8, 9, 11, 296, 450, 511, 514–516
Framework Convention on Climate Change, 515

Gene clusters, 66–67, 69, 71
   Anabaena 7120, 181–187
   Synechocystis 6803, 190–191
Genes, *see also* Specific genes
   hydrogenase, 66–67, 69, 71
   improved control of expression, 92–93
   programmed DNA rearrangement of, 203–206
   salinity stress responsive, 35–36
   Synechocystis, 454–455
   UV-A stress responsive, 34–35
Genetic engineering, 489
Genomes, marine, 31–37
German Federal Ministry for Education, Science, Research, and Technology (BMBF), 516, 517
Germany, 5, 21–22, 449, 450
Glasstube photobioreactors, 384–385
Global Positioning Satellites (GPS), 2
Gluconic acid, 23
Glucose, 10, 23
   CO_2-fixing microalgal biomass and, 314, 316–317
   Enterobacter aerogenes E.82005 and, 273, 274
   in floating-type photobioreactors, 370
   metabolic engineering and, 88, 89
   Synechococcus 043511 and, 219–222, 223
   waste water treatment and, 152
Glucose 6-phosphate dehydrogenase, 476, 479
Glutamate, 158, 159; *see also* Sodium glutamate
Glycerol, 323, 326, 496
Glycogen, 175, 176, 307–308
Greece, 5
Green algae, 12, 20–21, 27, 516
   oxygen tolerance in mutants of, 227–233
   photosynthetic hydrogen and oxygen production by, 235–240
Green fluorescent protein (GFP), 493–499
Greenhouse gases, 311–312, 450, 515
groEL gene, 35
GR protein, 494–495, 497–499
Gunnera manicata, 210

H-1, 167–171
Hairpins, 88, 92
Halobacterium halobium, 144
   bio-photoelectrochemical reactor and, 295–303
Halotolerant bacterial communities, 311–317
Hawaii, 163–166, 514
Hawaii Natural Energy Institute (HNEI), 514
II cluster, 66

Heat-hydrochloric acid treatment, 311, 313–314, 315, 316–317
Heavy-metal poisoning, 140
Helical tubular photobioreactors, 431–439
Herbicides, 139–140
Heterocysts, 182, 203–206, 210, 281
hox genes, 55, 58–60, 189, 190–191, 194, 213
HTML codes, 453
Human immunodeficiency virus (HIV), 459, 460
hup genes, 56–58, 181–187
hupB gene, 181, 183, 185
hupE gene, 71
hupF gene, 185
hupL gene, 53, 54, 56, 62, 71, 210, 212, 216
   programmed DNA rearrangement of, 203–206
hupS gene, 71, 185, 203, 205, 206, 210, 213, 216
hupSL gene, 53, 56, 203, 206, 209, 210, 213
hupT gene, 56
hupTUV gene, 206, 502
hupU gene, 203, 206
hupUV gene, 56, 206
hupV gene, 206
hya genes, 75
hyb genes, 75
hyc genes, 73, 76
hydD gene, 69
Hydrogenase, 16, 19, 20, 22, 27, 55, 73, 74, 75–77, 105–109, 516
   bidirectional, 55, 58–60
   Chlamydomonas reinhardtii, 227, 228, 231, 254
   DNA gene rearrangement in *Anabaena* 7120, 203–206
   green algae, 245
   heterologous expression of clostridial, 111–114
   hydrogen metabolism mediation in *Microcystis aeruginosa*, 173–178
   inducible, 432
   iron-only: *see* Iron-only hydrogenase
   light energy conversion and, 14
   metal-free, 244
   molecular biology of, 65–71
   nickel-iron: *see* Nickel-iron hydrogenase
   nickel-iron-selenium, 244
   nitrogenase vs., 448–449
   occurrence and characteristics of, 244
   oxygen sensitivity of, 227, 228, 236–237, 449
   reversible, 25, 55, 192–193
   in RITE program, 502, 503
   Scenedesmus obliquus, 243–249, 253–262
   structure and function of, 517
   sulf, 67–68
   Synechocystis, 454–455
   Synechocystis 6803, 189–194
   uptake: *see* Uptake hydrogenase
Hydrogenase-based algae systems, 420–421
Hydrogen evolution
   by cell-free extracts, 175–176
   under dark and anaerobic conditions, 174–175

Hydrogen evolution (*cont.*)
  Enterobacter aerogenes E.82005 and, 273–279
  fermentative, 273–274
  as a function of nitrogen availability, 157–160
  photo, 245, 273–274
Hydrogen metabolism, 173–178
Hydrogen production
  algae in: *see* Algae
  5-aminolevulinic acid production and, 133–141
  bacteria in: *see* Bacteria
  bioinformatics and, 451–457
  bioreactors for: *see* Bioreactors; Photobioreactors
  electron transport as a limiting factor in, 99–103
  hydrogenase 3 overexpression and, 73, 74, 75–77
  mechanisms of, 11–13
  metabolic engineering and: *see* Metabolic engineering
  nitrate reductase deficiency and, 73–74, 77–78
  plants in, 489–490
  selective pressure on, 230–231
  technoeconomic analysis of systems for, 415–424
  uptake-hydrogenase mutant in, 167–171
Hydrogen Program, U.S., 449, 450
Hydrogen uptake
  by light irradiation, 176
  photo: *see* Photohydrogen uptake
  in plane-type photosynthetic bioreactors, 359–360, 361–362, 363–365
  selective pressure on, 231–232
hydSL gene, 69
hyf genes, 76
hyp genes, 61
hypA gene, 502
hypB gene, 183
hypF gene, 71, 185

Ibaraki University, 501
Illuminated volume fraction, 332
Immobilization, 483, 485, 489
Incident light intensities, 331, 333, 336
India, 152
Indirect biophotolysis, 19–20, 21, 26, 392, 396, 449
  single-stage, 392
  two-stage, 392
Induced and diffused photobioreactors (IDPBR), 353–358, 504–505
Inducible hydrogenase, 432
Institute of Gas Technology, 419
Internal gas exchange photobioreactors, 441–446
Internal irradiation photobioreactors, 505
International Biohydrogen Center (IBC), 513–514
International Biohydrogen Program, 511–513
International collaboration, 4–5, 447–450
International Energy Agency (IEA), 450, 456, 515
Iron-only hydrogenase, 66, 190, 243, 248, 502
  characteristics of, 244
  reconstitution of, 105–109
Iron-protein, 101–103
Iron-sulfur clusters, 66, 189, 190, 191, 244
Italy, 502

Japan, 2–3, 5, 21, 67, 450, 476, 514, 515
  research and development in, 503
  RITE program in: *see* Research Institute of Innovative Technology for the Earth
  waste water treatment in, 152

King's College London, 501
Klebsiella pneumoniae, 102, 103
Korea, 5

Lactate, 12, 159, 160, 281–282; *see also* Sodium D,L-lactate
Lactic acid, 282, 287–288, 308
Lactic acid fermentation
  algal biomass and, 267, 269–271
  microalgae and, 265, 266, 267
  microalgal biomass and, 311, 312, 313–314, 315, 317
Lactobacillus amylovorus, 265, 267, 313, 315, 317
lacZ gene, 90
Lambda genomic library, 183–185
Laminaria saccharina, 403, 405, 406–410
Laser dye bioreactors, 290–291
Light compensation point, 359–367
Light-distribution coefficient, 332, 337
Light energy conversion, 13–16
  in photobioreactors, 350–352, 356
  by uptake-hydrogenase mutant, 167–171
Light energy per unit volume, 331–332, 336–337
Light-harvesting complexes, 82, 123, 124, 126, 128–130
Light-harvesting mutants, 123–130
Light intensities, 136
  average, 331, 336
  diffused, 356–357
  incident, 331, 333, 336
Light irradiation, 176
Light penetration, 345–352, 503
Light-saturation curve of photosynthesis, 44, 45, 47, 237–239
Light shift experiments, 48, 49–50
Light-supply coefficient, 335, 338, 339–340, 341
Light supply index, 331–333
Linoleic acid, 32
Lithospermum erythrorhizon, 486
L Lake, 468–469, 470, 471
Low-copy plasmids, 87–88, 90–91

Macroalgae, 403–413
Macrozamia, 210
Malic acid, 133, 143, 144–145, 147, 148–149, 151, 152, 154
Marine cyanobacteria
  direct nested PCR and, 197–202
  exogenous substrates and, 219–224
  molecular adaptation to stress by, 34–36
Marine genomes, 31–37
Marine macroalgal suspension cultures, 403–413
Marine photosynthetic bacteria
  light energy conversion by, 167–171
  screening of with *nifH*, 36

Marine photosynthetic microorganisms, 32–34
Mass cultivation
  of cyanobacteria, 467–473
  of microalgae, 425–428
  in RITE program, 503–505
Medaka, 36–37
Medium optimization, 483, 485–486
Membrane bioreactors, 387–388
Membrane photobioreactors, 384, 385–386
2-Mercaptoethanol, 496
Metabolic engineering
  of *Escherichia coli*, 73–78, 87–96
  of plant cell cultures, 483, 486–488
  toolkit for, 87–96
Metabolic fluxes, 93–94
Metabolite production
  in perfusion air-lift bioreactors, 475–481
  in plant cell cultures, 475–481
Metal-free hydrogenase, 244
Methane, 11, 23, 24
Methanobacterium, 55
Methanococcus, 55
Methanosarcina barkeri, 55
Methanothermus, 55
Methyl jasmonate (MJ), 484–485, 489
Methyl viologen, 175, 193, 260
Methyl viologen-dithionite system, 385
Metronidazole (MNZ), 228
Microalgae
  antenna size minimization and, 41–51
  CO_2 fixation in, 265–271, 311–317
  cultivation of, 320–321
  mass cultivation of, 425–428
  recycling test for, 323, 327
  starch accumulation in, 311–317, 319, 320, 321–323, 324, 326, 327
Microalgal biomass, 311–317
Microbiological Information System (MICRO-IS), 453–454, 455
Microcystis aeruginosa, 173–178
Milk industry waste water, 151–155
Ministry of International Trade and Industry (MITI), 501, 515
Molecular biology of hydrogenase, 65–71
Mutants
  Chlamydomonas reinhardtii, 227–233
  Escherichia coli, 73–74, 77–78
  hox gene, 194
  Rhodobacter sphaeroides RV, 123–130
  uptake-hydrogenase, 167–171

NAD, 279
NADH, 273, 275, 278–279
NAD(P), 189–194, 476
NAD(P)H, 476, 479
NADP+ reductase, 254, 261
*nar*G gene, 73, 77–78
National Institute of Bioscience and Human Technology (NIBH), 456, 501, 502

National Institutes of Health, 3
National Renewable Energy Laboratory (NREL), 22, 232, 416, 417, 420, 449
*ndh*D1 gene, 193
Near-tubular horizontal reactor (NHTR) photobioreactors, 391–401
Netherlands, 450
New Energy and Industrial Technology Development Organization (NEDO), 67, 501
Nickel-iron hydrogenase, 65, 66, 67, 69, 105, 106, 243
  characteristics of, 244
  Scenedesmus obliquuus, 246–247
  Synechocystis 6803, 189, 190
Nickel-iron-selenium hydrogenase, 244
nifD gene, 54, 203
NifF protein, 100, 101–103
nifH gene, 205
  direct nested PCR and, 197–202
  screening of marine photosynthetic bacteria with, 36
NifJ protein, 103
Nitrate, 175
Nitrate reductase, 73–74, 77–78
Nitric oxide, 265, 268–269
Nitrogen, 135, 157–160, 483
Nitrogenase, 27
  biophotolysis and, 25
  electron transport and, 99–103
  exogenous substrates and, 222–224
  in helical tubular photobioreactors, 432, 438
  hydrogenase vs., 448–449
  light energy conversion and, 14, 15, 16
  nifH gene of, 36
  plant cell cultures and, 489
  purple bacteria and, 286
  in RITE program, 502
Nitrogen fixation, 182, 203, 204, 282
Nitrogen-fixing cyanobacteria, 21, 27
  biophotolysis and, 26
  direct nested PCR and, 197–202
Nitrogen-fixing photosynthetic bacteria, 19
p-Nitrophenyl phosphate, 476
Nitrosoguanidine, 229, 230, 231
Non-nitrogen-fixing cyanobacteria, 27
  hydrogenase-mediated hydrogen metabolism in, 173–178
Nostoc, 432
Nostoc MiamiBG 039501, 198, 202
Nostoc MiamiBG 104401, 198, 202
Nostoc muscorum, 58–60, 210, 281, 288
Nostoc PCC 73102, 53–62
  biotechnological potential of, 62
  transconjugable plasmids for insertion mutagenesis of, 209–216
Nuclear energy, 296
nuo gene, 191

Oak Ridge National Laboratory, 22, 420, 449
Open pond systems, 449

Operon technologies, 91–92
Organic acid substrates, 13, 23–24
    in floating-type photobioreactors, 370, 371, 373
    in plane-type photosynthetic bioreactors, 362, 363–365
    Spirulina platensis NIES-46 and, 307–308
    starch decomposition and, 321–323, 324
    waste water treatment and, 152
Orthophosphate, 476
Oryzias latipes, 36–37
Oscillatoria NKBG 091600, 32, 34
Oxygen
    CO_2 exchange with in photobioreactors, 441–446
    photosynthesis in green algae and, 235–240
Oxygen-evolving complex (OEC), 254
Oxygenic photosynthesis, 244–245
Oxygen sensitivity, 227, 228, 233, 236–237, 449
Oxygen tolerance, 254
    in *Chlamydomonas reinhardtii* mutants, 227–233
    in hydrogenase-based algae systems, 420–421

P3, 83, 85
    characterization of in photohydrogen production, 123–130
    in RITE program, 503
Palmitoleic acid, 32
Parallel plate photobioreactors, 325, 326
Perfusion air-lift bioreactors (PALB), 475–481
Perfusion cultures, 478–479, 481
pH, 136
    cyanobacterial mass cultivation and, 469
    Enterobacter aerogenes E.82005 and, 273, 278, 279
    Halobacterium halobium and, 300, 301
    isolation of photosynthetic bacteria and, 164
    in photobioreactors, 147, 373, 445, 461
    polyhydroxybutyrate accumulation and, 158, 159–160
    Scenedesmus obliquus and, 254, 259
    Spirulina platensis NIES-46 and, 308
Phaeophyceae, 245
Phormidium NKBG 041105, 32
Phosphate-starvation response, 92–93
Photobioreactors, 27–28, 426, 428, 509–510
    A-type, 346–347, 348–351
    batch operation in: *see* Batch operation
    B-type, 346, 347, 348, 349, 350, 351
    closed, 393–394
    construction of 20-L prototype, 335
    continuous operation in, 143, 145, 148–149
    continuous-recycle operation in, 404, 406, 412
    correlation of growth rates and light parameters, 334
    C-type, 346, 347, 348, 351
    determination of optimum size, 334–335
    development of efficient large-scale, 329–342
    externally illuminated, 334
    floating-type, 369–373, 504, 505
    glasstube, 384–385
    helical tubular (bench-top), 431–439
    incorporating cyanobacteria, 384–387

Photobioreactors (*cont.*)
    incorporating photosynthetic bacteria, 387
    induced and diffused, 353–358, 504–505
    internal gas exchange, 441–446
    internal irradiation, 505
    internally illuminated, 331, 332, 334
    light distribution inside, 332–333
    light penetration and wavelength effect on, 345–352
    light supply index and, 331–333
    membrane, 384, 385–386
    near-tubular horizontal reactor, 391–401
    parallel plate, 325, 326
    Rhodobacter sphaeroides O.U.001 and, 143–149, 151, 153, 154
    in RITE program, 503, 504–505
    scale-up parameters in, 338–339
    semi-batch operation in, 154
    semi-continuous operation in, 404, 406, 412
    spiral PVC tubular, 384, 386–387
    sulfolipid production in, 459–466
    thin-layer plane, 504
    tubular: *see* Tubular photobioreactors
    tubular recycle, 403–413
    waste water treatment and, 151, 153, 154
Photodynamic therapy (PDT), 140
Photoelectrochemical hydrogen production, 295–303
Photofermentation, 21, 392
Photohydrogen evolution, 245, 273–274
Photohydrogen production, 117–122; *see also* Photosynthesis; Photosynthetic *entries*
    bacterial photosynthetic pigment ratio and, 81–85
    CO_2 fixation and, 265–271, 311–317
    from fermentates of algal biomass, 267, 270–271
    by light-harvesting mutants, 123–130
    by *Scenedesmus obliquus*, 243–249
    sunlight light cycle simulation and, 375–380
    by *Synechococcus* 043511, 219–224
    Synechocystis 6803 hydrogenase and, 189–194
    waste water utilization and, 151–155
Photohydrogen uptake, 117–122
Photosynthesis, 19
    anoxygenic, 244
    antenna size minimization and, 41–51
    in green algae production of hydrogen and oxygen, 235–240
    light-saturation curve of, 44, 45, 47, 237–239
    microalgae starch accumulation in, 319, 320, 321–323, 324, 326, 327
    oxygenic, 244–245
    quantum yield of, 46, 47, 48
    solar energy and, 9–10
Photosynthetic bacteria, 12–13, 20, 21, 24, 27, 319, 320, 323, 326
    alteration of pigment ratio in, 81–85
    5-aminolevulinic acid and, 133–141
    CO conversion to hydrogen and, 22–23
    CO_2 fixation and, 266, 267
    energy conversion and, 11
    in floating-type photobioreactors, 370

Photosynthetic bacteria (*cont.*)
  of Hawaii, 163–166
  light energy conversion and, 14, 16
  light penetration and wavelength effect on, 345–352
  marine: *see* Marine photosynthetic bacteria
  nitrogen-fixing, 19
  in photobioreactors, 345–352, 387
  production and uptake of photohydrogen in, 117–122
  in RITE program, 503–504
  Spirulina platensis NIES-46 and, 305–309
  sunlight light cycle simulation and, 375–380
Photosynthetic microorganisms, 319–327
  in RITE program, 502–503
  useful products from marine, 32–34
Photosynthetic pigment ratio, 81–85
Photosynthetic reaction center (RC): *see* Reaction center
Photosystem I (PSI), 15–16
  antenna size in, 42–43, 44, 45, 47, 240
  Chlamydomonas reinhardtii and, 239–240
  Microcystis aeruginosa and, 173, 176
Photosystem II (PSII), 15–16, 20, 26
  antenna size in, 42–43, 44, 45, 46, 47, 48, 50
  Chlamydomonas reinhardtii and, 239
  Microcystis aeruginosa and, 173, 176
  Nostoc 73102 and, 54
  Scenedesmus obliquus and, 254–255
Photosystems, *see also* Photosystem I; Photosystem II
  interaction with reversible hydrogenase, 192–193
  minimizing antenna size of, 41–51
Pigment-protein analysis, 125–126, 128–130
Plane-type photosynthetic bioreactors, 359–367
Plant cell cultures
  bioproduct optimization in, 483–490
  metabolite production in, 475–481
Plant growth-promoting factors, 140–141
Plant growth regulators, 32
Plastocyanin, 261
Plastoquinone, 173, 176, 192
Pollution, 425, 449–450, 511–512
Polyhydroxyalkanoate (PHA), 16, 502
Polyhydroxybutyrate (PHB), 157–160, 503
Polymerase chain reaction (PCR)
  direct nested, 197–202
  of *hup* genes, 181, 183, 184–185
  of *Scenedesmus obliquus* hydrogenase, 247, 248
  single cell, 200, 201–202
Polyphosphatase, 92
Polyphosphate, 93
Polyphosphate kinase, 92
Pond systems, 415–416, 421, 423–424, 449
Portugal, 5
ppk gene, 92, 93
ppx gene, 92, 93
Propionate, 363–365, 369, 370, 371, 373
Protoporphyrin IX (PPIX), 140
puf genes, 83, 126, 128

Purple bacteria, 516–517
  in ArBAS project, 281–293
  carbon/nitrogen ratio and, 288–290
  directed evolution of, 282–286
  polyhydroxybutyrate accumulation by, 157–160
Purple membrane fragments, 295, 296, 297, 298–299
Pyrococcus furiosus, 65, 67–68, 502
Pyruvate
  Microcystis aeruginosa and, 176
  polyhydroxybutyrate accumulation and, 159, 160
  Synechococcus 043511 and, 219–224
Pyruvate-ferredoxin oxidoreductase, 176
Pyruvate oxidoreductase (POR), 99

Quantum yield of photosynthesis, 46, 47, 48

Reaction center (RC), 82, 123, 124, 125, 126, 128–130
Recycling test, 323, 327
Renilla reniformis luciferase-modified GFP fusion protein, 493–499
repA gene, 35–36
Research and development
  in Italy, 502
  in Japan, 503
  in the United States, 1–3
Research Institute of Innovative Technology for the Earth (RITE), 376, 449, 501–505
Reversible hydrogenase, 25, 55, 192–193
RG protein, 494–495, 497–499
Rhizobium leguminosarum, 70, 71, 182, 183, 187
Rhodobacter capsulatus, 56, 69, 70, 71, 182, 183, 185, 187, 204, 205, 206
  in ArBAS project, 281–293
  electron transport and, 99–103
Rhodobacter CB, 387, 388
Rhodobacter rubrum, 69
Rhodobacter sphaeroides, 133–141
Rhodobacter sphaeroides DSM 9483, 291
Rhodobacter sphaeroides O.U.001
  in photobioreactors, 143–149, 151, 153, 154
  waste water utilization by, 151–155
Rhodobacter sphaeroides RV, 65, 67, 144, 163, 164, 165–166
  alteration of photosynthetic pigment ratio in, 81–85
  CO_2 fixation and, 117, 118, 121–122, 311, 316–317
  in induced and diffused photobioreactors, 353, 354
  light-harvesting mutant of, 123–130
  in photobioreactors, 345–352
  in plane-type photosynthetic bioreactors, 359, 360, 365–366
  polyhydroxybutyrate accumulation by, 157–160
  production and uptake of photobiohydrogen in, 117–122
  in RITE program, 502
  Spirulina platensis fermentation products and, 305, 306
  sunlight light cycle simulation and, 375–380
  uptake hydrogenase of, 69–71
Rhodobium marinum, 265, 267, 270–271

Index

Rhodophyceae, 245
Rhodopseudomonas capsulata, 121
Rhodopseudomonas palustris R-1, 369, 370, 371
Rhodopseudomonas RV, 134, 135
Rhodopseudomonas TN3, 135
Rhodospirillaceae, 163–166
Rhodovulum sulfidophilum, 36, 169
Rhodovulum sulfidophilum NKBG190471R, 34
Rhodovulum sulfidophilum NKPB160471R, 167–171
Rhodovulum sulfidophilum WIS, 319
RITE program: *see* Research Institute of Innovative Technology for the Earth
RKC Codes, 453–454
Roundtable (discussion), 511–518

Saccharomyces cerevisiae, 205
Salinity stress responsive gene, 35–36
Salmonella typhimurium, 67
Salt tolerance, increasing, 141
Scenedesmus obliquus, 243–249
  electron transport and, 248, 253–262
Scenedesmus sp., 306
Screening
  for oxygen tolerant green algae, 232–233
  of photosynthetic microorganisms, 502–503
Seaweeds: *see* Marine macroalgal suspension cultures
Secondary metabolites, 483, 484, 485–486, 488
Selective pressure
  on hydrogen production, 230–231
  on hydrogen uptake, 231–232
Semi-batch operation, 154
Semi-continuous operation, 404, 406, 412
Sheanella putrefaciens SCRC-2738, 34
Shift reaction bioreactors, 387–388, 389
Shikonin, 486
Single cell polymerase chain reaction (PCR), 200, 201–202
Single-crossover experiments, 214–216
Single-stage biophotolysis, 392
Small Business Innovation Research (SBIR) grants, 3
Sodium bicarbonate, 41–42, 44–47, 49, 50, 347
Sodium D,L-lactate, 347, 354, 360, 361, 362, 363, 365
Sodium glutamate, 133
  in bioreactors and photobioreactors, 143, 144–145, 147, 148–149, 347, 354, 360, 361
  waste water treatment and, 152
Sodium succinate, 360
Solar energy, 7–8, 9–10, 330, 508
Spiral PVC tubular bubble-train bioreactors, 388
Spiral PVC tubular photobioreactors, 384, 386–387
Spirulina, 32, 384, 426
Spirulina maxima, 112
Spirulina platensis, 175, 334, 338; *see also Arthrospira platensis*
Spirulina platensis NIES-46, 305–309
Standards workshop, 507–510

Starch
  in CO_2-fixing microalgal biomass, 311–317
  photosynthetic accumulation in microalgae, 319, 320, 321–323, 324, 326, 327
Stopped-flow kinetic analysis, 100
Stress adaptation, 34–36
Structured query language (SQL), 453
Substrates
  effect of exogenous on *Synechococcus* 043511, 219–224
  for mass cultivation, 503–504
  organic acid: *see* Organic acid substrates
Sucrose, 478, 480
Sugar, 24
Sulfhydrogenase, 67–68
Sulfolipids, 459–466
Sulfoquinovosyl diacylglycerol (SQDG), 459–466
Sunlight light cycle simulation, 375–380
Synechococcus MIAMI BG 043511, 219–224
Synechococcus NKBG 042902, 32, 34, 35, 36
Synechococcus PCC 6301, 210
Synechococcus PCC 7942, 111–114
Synechocystis, 454–455
Synechocystis PCC 6803, 55, 61, 210
  hydrogenase in, 189–194
  Scenedesmus obliquus and, 243
  sulfolipid production from, 459–466

Taxol, 483, 484–485, 488
Taxus, 483, 484–485
Technoeconomic analysis, 415–424
Technology Reinvestment Program (TRP), 2
Temperature
  cyanobacterial mass cultivation and, 468–470
  direct nested PCR and, 200
  in photobioreactors, 369, 373, 461
Thermal tolerance studies, 468–470
Thin-layer plane photobioreactors, 504
Thiobacillus ferooxidans, 187
Thiocapsa roseopersicina, 55
α-Tocopherol, 336, 341
Tokyo University of Agriculture and Technology, 501
Transconjugable plasmids, 209–216
Transformed cells, 493–499
Tricarboxylic acid (TCA) cycle, 89, 94, 277, 278–279
Tridecenic acid, 32
Trifolum repense, 139
Tubular integral gas exchange photobioreactors: *see* Near-tubular horizontal reactor photobioreactors
Tubular photobioreactors
  internal gas exchange, 441–446
  in RITE program, 504
Tubular recycle photobioreactors, 403–413
Turkey, 5
Two-stage biophotolysis, 392

Ultraviolet A stress responsive gene, 34–35
Undecylenic acid, 32

United Nations (UN), 450, 512, 513, 514
United States, 1–3, 5, 21, 22, 450, 514
University of Hawaii, 22, 442, 501
University of Miami, 22
Uptake hydrogenase, 69–71, 502
  gene cluster from *Anabaena* 7120, 181–187
  light energy conversion by mutant of, 167–171
  Nostoc 73102, 55–58, 209–216
  Rhodobacter sphaeroides RV, 69–71

Vinblastine, 486
Vincristine, 486
Vindoline, 486
Volatile fatty acids (VFA), 133, 136, 137–138, 141, 314

Waste water
  floating-type photobioreactors and, 369–373
  microalgal treatment of, 426, 427–428
  plane-type photosynthetic bioreactors and, 359
  Rhodobacter sphaeroides O.U.001 utilization of, 151–155
Water splitting, 20, 26, 254, 281, 282, 287–288, 305, 517
World Wide Web, 451, 452, 453, 455

XisA, 203, 205
XisC, 203, 205

Yeast extract, 10, 360, 373

Z-scheme, 239–240